PRECIOUS STONES

A POPULAR ACCOUNT OF THEIR CHARACTERS,
OCCURRENCE AND APPLICATIONS
WITH AN INTRODUCTION TO THEIR DETERMINATION,
FOR MINERALOGISTS, LAPIDARIES, JEWELLERS, ETC.
WITH AN APPENDIX ON PEARLS AND CORAL

BY

Dr. MAX BAUER

Translated from the German with Additions by
L. J. SPENCER, M.A. (Cantab.), F.G.S.

With a New Foreword and Addenda by
DR. EDWARD OLSEN
Curator of Mineralogy
at the Field Museum of Natural History
Chicago

IN TWO VOLUMES

Volume II

New York
DOVER PUBLICATIONS, INC.
London
CHARLES GRIFFIN & COMPANY LTD

International Standard Book Number: 0-486-21911-9
Library of Congress Catalog Card Number: 68-19167

Manufactured in the United States of America
Dover Publications, Inc.
180 Varick Street
New York, N. Y. 10014

CONTENTS

VOLUME I

PAGE   PAGE

VOLUME II

THIRD PART

DETERMINATION AND DISTINGUISHING OF PRECIOUS STONES

APPENDIX

EXPLANATION OF PLATES

Plates I, XII-XVI, XVIII, and XX, all in color, are in volume II following page 376.

LIST OF TEXT FIGURES
VOLUME I

VOLUME II

CORUNDUM.

Some of the most beautiful and valuable of precious stones, including the red ruby and the blue sapphire, belong to the mineral species corundum. All such stones are alike in the possession of those physical characters which essentially define a mineral species. Their appearance, however, owing to the great variety of colour displayed, may be very diverse, and thus the species furnishes a large number of gems, each with a distinct and characteristic colour, but having the same chemical composition and crystalline form.

Chemically considered, corundum is pure alumina, the oxide of the now much used metal aluminium. The chemical formula Al_2O_3, by which this oxide is represented, corresponds to 53·2 per cent. of metal and 46·8 per cent. of oxygen. Chemical analyses of naturally occurring corundum, however, always show the presence of some impurity, the amount of which is smaller the clearer and more transparent is the material used in the analysis. Foreign impurities are sometimes present in large amount, up to 10 per cent. or even more, and when this is the case the stone is rendered cloudy and loses its æsthetic value. The analysis of natural corundum has demonstrated the presence of iron oxide, silica, and occasionally traces of chromium oxide. The chemical composition of a beautiful transparent red corundum, the so-called " oriental ruby," and of a blue corundum or " oriental sapphire " of equally fine quality, is given below :

	Ruby.	Sapphire.
Alumina (Al_2O_3) . . .	97·32	97·51
Iron oxide (Fe_2O_3) . . .	1·09	1·89
Silica (SiO_2)	1·21	0·80
	99·62	100·20

It is on the presence of such foreign substances as iron oxide and in part probably on chromium oxide that the variety of colour found in this species depends.

Corundum occurs not infrequently in well-developed crystals belonging to the rhombohedral division of the hexagonal system. A series of the more frequently occurring forms, which are of two different habits, is shown in Fig. 53, a—i. In some a hexagonal prism is more or less largely developed and terminated at the two ends by basal planes perpendicular to the prism planes, faces of the primitive rhombohedron occupying alternate corners. Most of the crystal-faces are smooth ; the basal planes, however, bear regular, triangular striations, as in the crystal shown in Plate I., Fig. 5. Crystals with these three forms are shown in Figs. 53, a, b, c ; in a and b the prism predominates, the only difference between the two being the greater size of the rhombohedron faces in b ; in c the prism faces are narrow, while the rhombohedron faces and the basal planes are all about equal in size. Fig. 53 d shows a combination of the same forms with a double hexagonal pyramid in addition, the twelve faces of which replace the edges between the prism faces and the basal planes. This hexagonal bipyramid is present in all the remaining figures, e to i. In Fig. 53 e it occurs alone, and in the remaining forms it is predominant; in f it is in combination with the basal planes, and in g with the basal plane and a rhombohedron, the faces of which replace alternate corners above and below. In h there are three such hexagonal bipyramids with different inclinations, each successive pyramid being less steeply inclined the nearer it is to the end of the crystal; in combination with these three pyramids are the basal planes and a narrow hexagonal prism. The frequent repetition or oscillation between pyramids of this kind gives rise to horizontal striations on the faces, such as is shown in *f*. In Fig. 53, *i*, a combination of two hexagonal bipyramids with the hexagonal prism and an acute rhombohedron is shown. As mentioned above, crystals of corundum occur in one of two habits, that is with either the prism or the hexagonal pyramids predominating. The former is the more characteristic for red corundum or ruby, and the latter for blue corundum or sapphire. A crystal of ruby is shown in Plate I., Fig. 5, and a sapphire crystal in Fig. 7 of the same plate, both stones being represented in their natural colours.

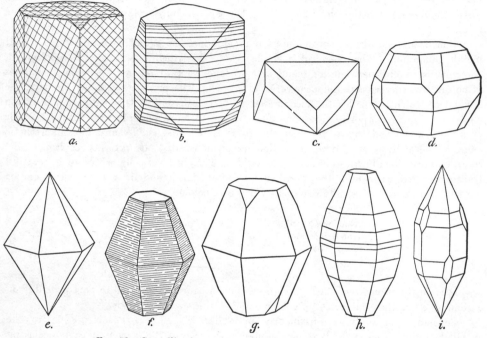

FIG. 53.   Crystalline forms of corundum (*a—d*, ruby; *e—i*, sapphire).

Twin intergrowths are often met with in cloudy corundum, but less frequently in transparent material suitable for use as gems. These are of two kinds: in the one, large numbers of thin plane lamellæ, parallel to a face of the primitive rhombohedron, occupy twin positions in the crystal, as indicated in Fig. 53, *a*; in the other, illustrated by *b* of the same figure, the lamellæ are parallel to the basal planes. Owing to this cause the faces of crystals often bear fine striations, which differ in direction in the two kinds of twin intergrowth. In the first, the striations are present on the basal plane in three sets of parallel lines intersecting at 120° or 60°, and on the prism and rhombohedron faces in two sets intersecting at an oblique angle. In the other kind of twin intergrowth, the basal planes are smooth, while the faces of the prism and rhombohedron are horizontally striated, the striations on the prism faces being perpendicular to the prism edges.

This twin lamination of corundum crystals makes the mineral appear to have two cleavages, one parallel to one or more faces of the primitive rhombohedron and the other parallel to the basal plane. Such a crystal when dropped or struck with a chisel will separate along plane surfaces in these directions. This splitting up, which is known as a platy separation, is not a true cleavage but is due to want of cohesion between the individual lamellæ. If the division were due to cleavage the lamellæ after separation could be split up again in the same direction, which is not the case. Moreover, the property of splitting along surfaces parallel to the faces of the rhombohedron and to the basal plane is possessed only by those crystals which are penetrated by twin-lamellæ parallel to these faces. In crystals in which twin-lamellæ are absent all trace of platy separation is also absent, the fracture being conchoidal as is the case with many other non-cleavable minerals.

Corundum is brittle, and among other physical characters it possesses that of hardness to a high degree. After diamond, it is the hardest of all minerals, having the number 9 assigned to it in Mohs' scale of hardness. It is, however, much more nearly approached in hardness by certain other minerals than it itself approaches diamond in this respect. Its hardness renders this mineral very valuable as a grinding and polishing material, the cloudy and opaque varieties, especially the compact black emery, being much used for this purpose, while rather superior qualities are utilised for the pivot supports of watches and various delicate instruments. There are degrees of hardness among the different varieties of corundum, the blue sapphire standing first in this respect.

The specific gravity of corundum is very high ; that of pure material is very near to 4, values varying between 3·94 and 4·08 having from time to time been determined on pure material. Greater variations on either side of the mean value of 4·0 are due either to errors of observation or to the determinations having been made on impure material. No difference has been observed in the specific gravity of differently coloured varieties. Corundum is thus one of the densest of precious stones, and this character renders it easily distinguishable from other minerals which may resemble it in general appearance. It sinks rapidly, not only in methylene iodide but also in the heavier liquid (sp. gr. = 3·6) obtained by dissolving iodoform and iodine in methylene iodide.

It is not attacked by acids either in the cold or when warmed, and is completely infusible before the blowpipe. Many specimens of corundum when heated in a dark room display a beautiful phosphorescence. When rubbed with cloth or leather the mineral acquires a charge of positive electricity, which it retains for a considerable time.

Corundum varies greatly in appearance. Most frequently it is cloudy and opaque, and only a small proportion is clear and transparent and valuable from an æsthetic point of view. The common cloudy and opaque varieties of corundum will not be here considered in detail, but our attention will be directed to the transparent varieties known as noble or precious corundum.

The lustre of precious corundum is very fine, and is displayed to great advantage by the polished facets of a cut stone. It approaches that of diamond in brilliancy, but differs from it in character, being not adamantine but vitreous, like that of many other precious stones and glass. So brilliant and perfect a lustre and so marked a fire is displayed by no other precious stone with the exception of diamond, and perhaps also of colourless hyacinth (zircon), so that by these characters alone the stone may be distinguished from other gems of the same colour. Even when a specimen of corundum is of a poor colour it will still be cut as a gem, since the fine lustre of the mineral will redeem the stone from insignificance. Moreover, owing to the great hardness of this stone, the lustre is retained even after rough usage and hard wear.

Corundum is optically uniaxial and doubly refracting. Its refraction, though strong, is considerably less than that of diamond ; its double refraction is small, the refractive indices for the ordinary and extraordinary rays differing but slightly. In a crystal from Ceylon they have been determined for yellow sodium light to be : $\omega = 1\cdot7690$; $\epsilon = 1\cdot7598$. The dispersion produced by corundum is also small, the refractive indices for different colours of the spectrum differing but slightly ; hence the mineral shows no marked flashes of prismatic colours as does the diamond.

The high index of refraction, small dispersion, and considerable hardness of corundum render the colourless or pale-coloured varieties very suitable for the preparation of microscopic lenses. Pritchard who, as we have already seen, made use of diamond for this purpose, in 1827 constructed lenses of very pale blue sapphire ; they have never, however, come into general use. Transparent corundum finds its most extensive application in jewellery, the fine appearance of the stone being due to the combination of transparency and high lustre with fine colour. The mineral corundum includes some of the most beautifully coloured precious stones known, and we will now consider the different colour-varieties of the mineral.

Absolutely pure crystallised alumina has no colour and is perfectly water-clear. It is then known as leuco-sapphire ; it is, however, but rarely found in this condition, more usually showing a more or less pronounced colour. In many cases the colouring matter is irregularly distributed, sometimes being aggregated in patches surrounded by colourless portions of material ; not infrequently also different portions of one and the same crystal are differently coloured. As has been already stated, the various colours shown by corundum are due to the intermixture with the ground substance of various foreign substances, the nature of which in particular cases has never been determined. The colours vary in tone from light and pale to dark and intense ; stones showing pale tones of colour have been described as " feminine," and those coloured more deeply as " masculine."

The phenomenon of dichroism is always observable in deeply coloured specimens of corundum, it is not seen in very pale coloured varieties, and is the more noticeable the deeper the colour of the specimen.

Several varieties of precious corundum are recognised ; these are distinguished from each other by their colour, and from the jeweller's point of view differ much in importance and value. The two varieties which occur with the greatest frequency are red corundum, or ruby, and blue corundum, or sapphire, the latter being more abundant than the former. All other coloured varieties are in comparison almost rare. They are known by the same names as certain other stones of the same colour, being distinguished from these by the prefix " oriental," which is meant to signify the possession of specially noble qualities, such as great hardness and fine lustre. The same prefix is often applied to ruby and sapphire themselves with the object of distinguishing them from other precious stones of the same colour. Moreover, in the case of sapphire, it serves to distinguish stones of the true sapphire-blue from other colour-varieties of the species, to which, with the exception of ruby, the term sapphire is sometimes extended. It should be remembered that the only character which the different colour-varieties of corundum have in common with the stones after which they are named is their colour ; thus, by the term " oriental emerald," must be understood transparent corundum of a green colour.

The different colour-varieties of corundum are tabulated below :

Variety.	Colour.
Ruby (" oriental ruby ")	red.
Sapphire (" oriental sapphire ")	blue.
Leuco-sapphire	colourless.
" Oriental aquamarine "	light bluish-green.

Variety.	Colour.
"Oriental emerald"	green.
"Oriental chrysolite"	yellowish-green.
"Oriental topaz"	yellow.
"Oriental hyacinth"	aurora-red.
"Oriental amethyst"	violet.

The different colours of these varieties of corundum usually appear just as beautiful when viewed by artificial light as by daylight, which is not always the case with the stones, after which they are named. The colour of certain varieties of corundum can be in some cases destroyed, and in others changed by the application of heat; this point, however, will be considered later on.

The different colour-varieties of corundum occur as irregular grains and as well-developed crystals embedded in the mother-rock, which, as a rule, is an old crystalline rock, such as granite or gneiss. The gem-varieties are found with especial frequency as secondary contact minerals, which have been developed in limestone by contact with a molten igneous rock. By the weathering and denudation of such rocks, the embedded crystals are set free, and are subsequently found with other water-worn débris in the beds of streams and rivers. It is in these derived deposits that the most beautiful specimens of the varieties mentioned above are found in all countries in which the original mother-rocks occur.

Having briefly considered the characters common to all corundum, we pass now to the consideration of the varieties suitable for use as gems, and begin with the most costly of all stones, namely, the ruby.

RUBY.

CHARACTERS.—Of all the colour-varieties of precious corundum the red, or ruby ("oriental ruby"), is the most highly prized. It is probably identical with the anthrax of Theophrastus, and is one of the stones referred to in mediæval times as carbuncle. It has all the general characters of corundum, and is only distinguished from other varieties by its red colour. A natural crystal of ruby is shown in Plate I., Fig. 5, and a faceted stone in Fig. 6 of the same plate.

The tone of colour differs in different specimens, being sometimes deep and intense ("masculine" ruby), sometimes pale and light ("feminine" ruby). The lighter shades vary from pale rose-red to reddish-white, some specimens being so faintly tinged with red as to appear almost colourless. The darker colours are either pure red, carmine-red, or blood-red; the red of the majority of rubies, however, has a more or less distinct tinge of blue or violet, this being specially noticeable in transmitted light. The shade of colour which is most admired is the deep, pure carmine-red, or carmine-red with a slight bluish tinge. This colour has been compared by the Burmese to that of the blood of a freshly-killed pigeon, hence the references to such stones as being of "pigeon's-blood" red. The various shades of red of the ruby are remarkable in that they lose none of their beauty in artificial light, a statement which cannot be made respecting any other precious stone of the same colour.

The colouring of rubies is not always perfectly uniform, colourless layers being sometimes interposed between portions coloured red. In such cases, the stone will often become uniformly coloured throughout after heating. Provided the stone is gradually heated it may be raised to the highest temperatures with no fear of fracture. The interesting changes in colour exhibited by certain gems when gradually heated and then allowed to cool have been already described. During cooling the ruby becomes first white and then green, finally regaining its original red colour, so that in this stone the colouring-matter is neither permanently changed nor destroyed by exposure to high temperatures. It is otherwise, however, with the sapphire, for this gem at a high temperature loses its beautiful blue and takes on a dull grey colour.

The red colouring-matter of the ruby is therefore certainly not organic in nature, as seems to be the case with those gems which lose their colour on heating. It is more likely to be some compound of chromium, an element whose presence has been detected in the analysis of some rubies. That the colouring of the ruby is due to chromium is also suggested by the fact that the colour of the so-called " ruby " glass is obtained by adding a small amount of chromium oxide to the other constituents of the glass. The same substance is also used by M. Frémy for the production of the red colour of his artificially prepared rubies. Some of the crystals produced by this investigator were partly red and partly blue, resembling in this respect certain natural rubies which occur rarely in Burma.

The dichroism of deeply coloured rubies is very noticeable; with the exception of stones of very pale colour, a difference in the colour of every ruby can be observed when viewed in different directions. On looking through a dark-coloured crystal of ruby, such as is illustrated in Fig. 53, a—d, in a direction perpendicular to the basal planes, it will appear of an intense red colour, either pure red or with a slight tinge of violet. If, however, the light received by the eye has passed through the crystal in any direction perpendicular to a prism face or edge, the stone will appear much lighter in colour. On allowing the light which has passed through the crystal in this direction to enter the dichroscope, the two images, in that position of the instrument in which the greatest difference in colour is shown, will be one light, and the other dark red usually tinged with violet. In all other directions in which the light may travel, with one exception, the two images will be more or less differently coloured. This exceptional direction is perpendicular to the basal planes and coincides with the direction of the optic axis. Along this direction the crystal is singly refracting, and the two images seen in the dichroscope are of the same deep red colour as the crystal appears when viewed in this direction without the intervention of the dichroscope. The dichroism of the ruby affords a means whereby it may be distinguished with certainty from other red stones, such as spinel and the different varieties of garnet, which crystallise in the cubic system, and thus being singly refracting can show no dichroism.

The fact that the colour of the ruby varies with the direction in which it is viewed, makes it necessary that the form of the cut gem should have a certain definite relation to that of the crystal in order to obtain the finest colour-effect. The plane of the largest facet of the cut stone, namely, the table, must coincide as closely as possible in direction with the basal planes of the crystal in order to obtain the greatest depth in colour of which the stone is capable. The greater the angle at which the table is inclined to the basal plane of the crystal the poorer will be the colour-effect produced, and when the table is perpendicular to the basal plane, and therefore parallel to the prism faces of the crystal, the minimum colour-effect is the result.

Some rubies show on the basal plane, or still more plainly on a cut and polished curved surface approximating to the basal plane in direction, a six-rayed star of glimmering reflected light. Such stones are known as *star-rubies*, or asteriated rubies, sometimes also as ruby cat's-eye. The appearance is similar to that seen in the star-sapphire, but, as a rule, less marked; it will be therefore considered in greater detail under sapphire.

VALUE.—A clear, transparent, and faultless ruby of a uniform deep red colour is at the present time the most valuable precious stone known. Except in ancient times, it is probable that the ruby has always held a foremost place in the estimation of connoisseurs. This, however, is not true of stones of a pale red colour, which are always less highly prized, on account both of their light shade of colour and of the fact that they occur more abundantly and in larger size than do stones of a true pigeon's-blood red.

The value of the finest ruby, therefore, far exceeds that of a diamond of corresponding size and quality. One-carat diamonds of the first water are of far more frequent occurrence than rubies of the same size and quality, while large rubies are still rarer than large diamonds. A fine deeply-coloured ruby of 3 carats is a great rarity, whereas it is by no means unusual to come across fine diamonds of this size. Again, while 10-carat diamonds are of moderately frequent occurrence, rubies of the same weight scarcely ever occur, while only very few specimens of still larger stones are known. It is therefore to be expected that larger rubies should command exceptionally high prices; indeed, the prices of stones of ordinary sizes may be arrived at very closely by the application of Tavernier's rule. The relation between the ratio of the weight and the value of a large and a small stone is very different when, on the one hand, the two stones are diamonds, and when, on the other, they are rubies. Thus, while a 1-carat ruby is worth twice as much as a 1-carat diamond, a 3-carat ruby of the first quality is worth ten times as much as a diamond of the same description, that is to say, that while a 3-carat brilliant of the first water would be valued at about £150, a ruby of the same description would be worth about £1500. The value of a 5-carat diamond of the first quality would be about £300, while that of a similar ruby would be £3000. These values, of course, apply to stones in the cut condition, the weight of which uncut would be about doubled. For rubies of still larger size there is no fixed market price; almost fabulous sums have been paid for very fine stones of large size required for some special purpose. A fine ruby of $9\frac{6}{16}$ carats has been recently valued by Mr. G. F. Kunz, the American gem expert, at 33,000 dollars (£6776). Again, £10,000 is stated by Mr E. W. Streeter, the London jeweller, to have been paid for a cut ruby of $32\frac{5}{16}$ carats, and double this amount for another weighing $38\frac{9}{16}$ carats; both of these stones were faultless specimens of magnificent colour.

It is recorded by Benvenuto Cellini in the middle of the sixteenth century that a carat ruby was eight times the value of a carat diamond, the price of the former being 800 golden scudi (£160) and that of the latter 100 scudi (£20). The ratio at the present time is only about two to one, the market price of a fine 1-carat ruby being about £25, and that of a brilliant of the same weight about £15, only in very exceptional cases £20 or £25 being paid. The particular shade of colour shown by a ruby exercises an enormous influence on its value, thus a carat stone of a pale rose colour is worth at the most but £1, which contrasts strangely with the value of a stone of equal size, but of a deep red colour.

The value of any particular ruby does not reach the high figures mentioned above unless it is an absolutely faultless specimen. The faults most commonly met with are lack of clearness; existence of cloudy portions (so-called " clouds "), specially frequent in light coloured stones; milk-like, semi-transparent patches (" chalcedony patches "); small internal cracks and fissures (" feathers "); unequal distribution of colour, and so on.

Just as some few diamonds, on account of their singular beauty, large size, or unique colour, have become famous and well known all the world over, so certain rubies on account of their exceptional size have acquired more or less fame and renown. Tavernier states that he saw two rubies, in the possession of the King of Bijapur, in India, which weighed $50\frac{3}{4}$ and $17\frac{1}{2}$ carats, and which he valued at 600,000 and 74,550 francs respectively. Other large rubies have been occasionally met with in India and specially in Burma. The King of Ava was reported to be in the possession of a ruby mounted as an ear-pendant which is the size of a small hen's egg. A few specimens of similar exceptional size are known in Europe. Kaiser Rudolph II. of Germany possessed a ruby of flawless beauty and of the size of a hen's egg, which was valued by the gem expert Boetius de Boot at 60,000 ducats (about £28,000). It is related that in 1777 Gustavus III. of Sweden presented to Catharine II. of Russia a beautiful ruby of the size of a pigeon's egg; the present whereabouts of this stone is, however, unknown. The largest of the fine rubies set in the French crown, according to the inventory made in 1791, weighed 7 carats, and was then valued at 8000 francs (£320). Another weighed $25\frac{11}{16}$ carats, but on account of its pale colour was valued at no more than 25,000 francs. Other rubies of large size will be mentioned under the description of localities. The largest ruby known is said to be from Tibet; it weighs 2000 carats, but is not perfectly transparent. The largest ruby yet found in Burma is also a little cloudy; its weight is given by Streeter at 1184 carats. A tabular crystal of ruby of a rich red colour, and in part perfectly transparent, was presented to the British Museum by Professor John Ruskin; it has a weight of $162\frac{2}{3}$ carats.

RUBY AS A GEM.—The facets of a cut ruby are ground on a rotating iron disc precisely as in the diamond. The use of diamond-powder as a grinding material is now very general in Europe since it considerably expedites the process. The operation of grinding is followed by that of polishing, which is effected on a copper disc charged with tripolite moistened with water.

The forms of cutting adopted for the ruby are those generally used for the diamond. The brilliant form (Plate I., Fig. 6) is frequently chosen, since this displays the beauties of the stone to the best possible advantage. In order to increase the transparency of the ruby, however, the brilliant is cut thinner and flatter than is allowable in the case of the diamond. Owing to the strong refraction of the ruby, the rays of light which enter the stone by its front facets are totally reflected by the back facets and pass out by the front of the stone, the fine red colour of the ruby having been imparted to them during their passage through it. It is this colouring of the rays of light, together with the brilliant lustre of the stone, which gives the ruby its effectiveness. Owing to the small dispersion of corundum, the magnificent play of prismatic colours characteristic of the diamond is almost absent in the ruby. This being so, the step-cut or trap-cut form of cutting (Plate III., Figs. 2—4) is just as effective as the brilliant for the ruby, or indeed for any coloured stone which shows no play of prismatic colours. The mixed-cut, of which the upper portion consists of brilliant facets and the lower those of the step-cut, is also an effective form (Plate III., Fig. 5). Table-stones, point-stones, and similar forms are scarcely ever cut now; the few examples met with are the work of former times. Flat and thin rubies are usually cut as roses (rosettes), since this form involves little loss of material and, at the same time, produces a good effect. Very small stones are irregularly faceted; they are used to form a contrasting border round some larger precious stone.

In Burma, the chief home of the ruby, the stones are cut en cabochon, that is to say with a rounded surface, before they come on the market. When this form of cutting does not display the beauties of a stone to the best advantage it is re-cut in Europe. It is obviously to the purchaser's advantage to buy a ruby cut en cabochon rather than an uncut stone, since in the former case it will be possible to detect any faults in the interior. With the exception of the asterias or star-rubies, this gem is seldom in Europe cut en cabochon; in the exceptional case mentioned, the rounded form of cutting is obviously the most suitable for displaying the six-rayed chatoyant star for which the stone is peculiar.

Clear and transparent stones of a full deep colour are usually mounted in open settings (à jour); those of poorer quality are often backed by a foil of gold or copper or red glass, which materially improves their appearance.   In Burma it is customary, instead of setting such a stone on a foil, to hollow out the underside and fill it in with gold.

Besides being faceted and cut *en cabochon*, rubies are sometimes engraved with inscriptions or figures, this being most frequently done in the East.   Such antique gems of ruby engraved with the head of Jupiter Serapis and a figure of Minerva are known.

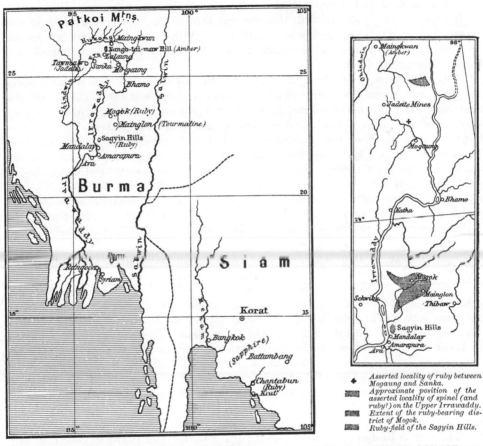

FIG. 54.   Occurrence of ruby and sapphire in Burma
and Siam.   (Scale, 1 : 15,000,000.)

FIG. 55.   Ruby-fields of Burma.
(Scale, 1 : 10,088,500.)

OCCURRENCE.—While the poorer qualities of ruby are widely distributed, clear, transparent material suitable for cutting is found in but few countries, of which Burma, Siam, and Ceylon are alone of commercial importance at the present time.

Now, just as in former times, **Upper Burma** furnishes us not only with the finest but also with the largest supply of rubies.   The distribution of precious stones (ruby, red tourmaline, jadeite, and amber) in this country is shown on the map in Fig. 54, the ruby localities being given in special detail in Fig. 55.   The ruby mines of Upper Burma were worked at least as early as the fifteenth century and have ever since supplied the greater part of the material used in jewellery, including the finest stones known.   The majority of the rubies which are now put on the market come from Burma. It is probable, however, that part of this supply is the gradually accumulated stock of former times, and that the yield of the mines is now smaller than formerly.

The Burmese ruby mines were mentioned long ago by Tavernier. According to his account, which, however, was not based on personal observation but cn second-hand information, they were situated in the "Capelan Mountains," in Pegu, twelve days journey in a north-east direction from the town of Syriam, now a small village close to Rangoon. The yield at that time (second half of the seventeenth century) was apparently not very great, and was estimated at 100,000 écus (£22,500) per annum by Tavernier, who adds that he found the importation of rubies from Europe into India a lucrative business.

Tavernier's error in describing the locality of the ruby mines has been repeated again and again, and is even now current in the text-books of the present day. There is not the least doubt that the mines referred to are those which are still being worked in Upper Burma, and which are very much further removed from Syriam than Tavernier stated them to be. The distance from here to Mandalay is at least thirty-six days journey, and from Mandalay to the principal ruby district of Mogok is another eight days journey, the less important district of the Sagyin Hills lying, however, a little nearer. Until recently the exact location of these mines was a secret jealously guarded by the Burmese, who thus rendered them practically inaccessible to Europeans. Since the annexation of the country in 1886 by Britain, more detailed information has been obtainable, and a part of the workings has been taken over by Europeans. The district was officially visited and reported upon in 1888 by Mr. C. Barrington Brown. The rocks and minerals collected there were examined by Professor J. W. Judd, the result of their joint examination being published in 1896 in the *Philosophical Transactions* of the Royal Society of London.

The district of Mogok is the most important "ruby tract," or "stone tract," and embraces an area of forty-five square miles, or, if some abandoned mines are included, sixty-six square miles. The ruby-bearing area is, in all probability, much greater than this, extending to the south and east into the independent Shan States, and has been estimated by Lockhart, who for two years was resident engineer to the Burma Ruby Mining Company, at 400 square miles. This opinion is supported by the recent discovery of an old ruby mine in the river gravels of the Nampai valley, near Namseka village, in the Mainglon State. The district, which is mountainous, and scored by deep valleys, lies to the east of the Irrawaddy, from which it is separated by a plain thirty miles in width, in which a few unimportant ruby mines are worked by the natives.

This district has formed a part of the kingdom of Burma since 1637 ; its chief town and centre of the trade in precious stones is Mogok, latitude 22° 55′ N., longitude 96° 30′ E. of Greenwich, thirty-four miles in a straight line (but fifty-eight by road) from the river, and ninety miles north-north-east of Mandalay. A little below Mandalay is Ava, formerly known as Ratanapura (= city of gems), the old capital of Burma, round which the trade in precious stones of the whole country centres. Mogok stands at an elevation of 4100 feet above sea-level, while the highest point of the district has an elevation of 7775 feet. In spite of this the country is covered with thick forests, and is unhealthy both for Europeans and natives. The principal mines are situated in the valleys in which stand the towns of Kathay and Kyatpyen (= Kapyun). The mountains surrounding the latter town have been conclusively proved by Prinsep to be identical with the "Capelan Mountains" of Tavernier.

The mother-rock of the ruby and of the minerals, such as spinel, with which it is associated, is a white, dolomitic, granular limestone or marble, which forms whole mountain ranges in this district, and which, according to the investigations of Dr. F. Noetling, of the Indian Geological Survey, is of Upper Carboniferous age. These rocks were originally compact limestones of the ordinary kind, which have been altered by contact with intrusive masses of molten igneous rock ; this caused the calcium carbonate to re-crystallise out as pure calcite, while the impurities contained in the original limestone crystallised out separately as ruby and its associated minerals. The alteration of rocks by contact with a mass of molten igneous material is known to geologists as contact- or thermo-metamorphism ; the results of the process are frequently to be observed in all parts of the world, but, although corundum is often to be found in such altered rocks, fine ruby of gem-quality is only rarely met with. Such were the conclusions as to the geology of the district and the mode of origin of the ruby arrived at by Professor Max Bauer, from information and specimens supplied to him by Dr. F. Noetling, and published in a scientific journal in 1896, and in the German edition of the present work. The point of view adopted by Mr. C. Barrington Brown and Professor J. W. Judd, explained in the paper published shortly before, and which has already been mentioned, must not, however, be passed over without notice.

These authors describe the white crystalline limestone, which alone contains the ruby and spinel, as occurring in thick bands interfoliated with gneisses. These gneisses are usually of intermediate chemical composition, but sometimes of more acid, and at other times of more basic character ; the crystalline limestones are more intimately associated with the basic gneisses (pyroxene-gneisses and pyroxene-granulites, with pyroxenites and amphibolites). These contain crystals of calcite, and as the proportion of calcite present increases, they merge gradually in the limestones. It is concluded, on these grounds, that the limestones have been derived by the alteration of the lime-felspar in these basic rocks. This felspar (anorthite), being a silicate of calcium and aluminium, would, on alteration, give rise to calcium carbonate and hydrated aluminium silicates, the former being deposited as calcite, and the latter as silica (opal), and various aluminium hydroxides (diaspore, gibbsite, bauxite, &c.). Under other conditions of temperature and pressure these may have been afterwards converted into crystallised anhydrous alumina, that is ruby.

In the masses of crystallised limestone occurring *in situ* precious stones are only sparingly present, being found in much greater abundance in the clayey and sandy weathered products of the mother-rock, which lie on the sides of the hills, fill up the bottom of the valleys, and are often overlain by similar detrital material containing no precious stones. This secondary gem-bearing bed consists of brown or yellow, more or less firm, clayey, and at times sandy, material, known to the Burmese as "byon," which may be regarded as the residue after the solution of the limestones by weathering processes. It contains beside ruby, sapphire, and other colour-varieties of corundum, spinel (Tavernier's "mother of ruby"), tourmaline, large fragments of quartz, grains of variously coloured felspars, nodules of weathered iron-pyrites, and other minerals of more or less value, together with fragments of the rocks which occur *in situ* in the neighbourhood. Sometimes in the river alluvium, instead of clayey and sandy material, there are pure gem-sands consisting mainly of minute sparkling grains of ruby.

The gem-bearing layer lies on a soft decomposed rock of characteristic appearance. When the natives reach this level in their excavations they know that the "byon" extends no further down, and that work at that spot must be abandoned. The "byon" lies about 15 to 20 feet below the surface of the floor of the valleys, and is from 4 to 5 feet in thickness, though occasionally it may thin off to a few inches. On the sides of the hills the bed of "byon" may be 15 to 20 feet thick, and sometimes as much as 50 feet.

A peculiar feature connected with " byon " is its occurrence in caves in the limestones. These can be traced for miles underground, now as wide and high chambers or vaults, now as small crevices and narrow cracks. These caves are either wholly or in part filled with the " byon," which is usually covered over with a thick deposit of calcareous tufa in stalactites and stalagmites of the most fantastic shapes.

Previous to 1886 the deposits were mined exclusively by the natives, who adopt methods differing according to the conditions. In alluvial deposits in the valleys the " byon " layer is reached by excavating pits ("twinlones "), 2 to 9 feet square. When the excavation is made in loose, crumbling material the sides of the pit are supported by bamboo. From the bottom of the pits are driven horizontal galleries from one pit to another, so that as much as possible of the gem-bearing earth may be excavated. The light, earthy part of this material is removed by washing, and the remaining sand then searched for precious stones. The actual work of excavating can only be undertaken in the dry season of the year.

The ruby-bearing layer on the sides of the hills is reached by means of open cuttings or trenches (" hmyaudwins "). Their excavation is usually effected by means of running water, which is led in bamboo pipes often over considerable distances to the spot where it is required. The flow of water thus obtained washes away the superimposed débris and all the lighter part of the gem-bearing layer, leaving the heavy precious stones behind. This kind of work is naturally carried on in the rainy season, since large volumes of water are required.

Finally, we come to the " loodwins," or workings in the gem-bearing material, filling the limestone caves. This material is excavated and washed for precious stones in the ordinary way. The workers in cuttings on the hillside occasionally come upon limestone caves ; a specially large one was found and worked about the year 1870 on the Pingudaung Mountain near Kyatpyen.

The mines situated in the alluvial deposits of the river valleys are the most important, and the greater part of the yield is derived from the mines in the valleys in which stand the towns of Kyatpyen, Kathay, and Mogok, the latter valley being especially rich. The cave deposits are probably rich enough to pay for the introduction of European mining methods ; the primitive efforts of the natives are attended by great danger and loss of life, and by very meagre returns.

In former times intending miners were obliged to procure a licence before undertaking any work ; they were also required to pay a tax and to hand over to the king all stones exceeding 1000 rupees in value. Whether the finder of such a stone received anything in return for it depended entirely upon the caprice of his sovereign. It was natural that attempts should be made to evade this obligation ; many large and valuable stones were broken up into pieces small enough to be legitimately retained by the finder, while others found their way into the hands of illicit dealers.

Persons desirous of trading in rubies were required to obtain the permission of the Government of Burma, and were subject to a special tax. The monthly yield of stones was about 50,000 to 100,000 rupees worth, and before being put on the market they had to be taken to the Ruby Hall in Mandalay. Side by side with the legitimate trade there flourished a trade in smuggled stones, the finest of which found their way through Lower Burma into India, being in many cases sold in Calcutta. In the time of the last king of Upper Burma (deposed in 1886) it is said that the illegitimate trade in rubies in Lower Burma amounted to between two and three lacs of rupees per annum.

Thousands of small rubies with rounded surfaces, rudely fashioned, are set in a great variety of articles belonging to the Burmese regalia. These were taken from the palace of

King Theebaw at Mandalay at the time of the British conquest of Upper Burma, and are now preserved in the Indian section of the Victoria and Albert Museum at South Kensington.

In 1886, when Burma became part of the British Empire, the monopoly of the ruby mines by the natives came to an end. In the neighbourhood of Mogok work on a large scale was carried on first by an Anglo-Italian and then by an English company. In return for the concession of mining rights the Indian Government demanded from the company a yearly payment of four lacs of rupees (about £25,000), and from each native miner at first twenty, and afterwards thirty, rupees. It would appear from the fact that attempts have been recently made to get the payment reduced, that the operations of the company have not been altogether successful. Not content with working the alluvial deposits of the valleys in which rubies have been searched for for centuries, the company has also tapped the ruby-bearing deposits on the Pingudaung Mountain, near Mogok, and on the Kyuktung Mountain.

The " byon " yields not only ruby but also other colour-varieties of precious corundum, namely, sapphire, " oriental topaz," &c., common corundum, and frequently precious spinel. Beside being the most valuable and beautiful of the different varieties of corundum the ruby is here also the most abundant, about 500 rubies being found to one sapphire, and other colour-varieties of precious corundum are still rarer. The unequal proportions in which ruby and sapphire exist in these deposits is partly counterbalanced by the much greater size of the crystals of sapphire. The majority of the rubies found here do not exceed ⅛ carat in weight, and large stones, when found, are often full of all kinds of faults. Flawless stones of 6 to 9 carats are rare, and very few reaching a weight of 30 carats have ever been found.

In the year 1887 a stone of 49 carats was found, and in 1890 one of 661 carats. The discovery in earlier times of two stones of 172 and 400 carats has been reported. The two most beautiful rubies found here were sold in Europe by the King of Burma in 1875. Both were of a magnificent colour, and before they were re-cut weighed 37 and 47 carats. After being re-cut in Europe they sold for £10,000 and £20,000 respectively. Cloudy corundum unsuitable for cutting as gems because of its lack of transparency, occurs in much larger pieces, some of which have weighed over 1000 carats. A stone of this description weighing 1184 carats is figured by Streeter. This, together with other stones of exceptional size, was found since the English occupation of the country. On the whole, the proportion of large rubies found recently in the mines of Burma seems to have increased, but in almost every case these large stones are unfit for cutting as gems.

The colour of the rubies of Upper Burma is, as a rule, some shade of deep red. In that country the shade most admired is pigeon's-blood red. Stones of this colour, transparent and free from faults, command high prices even at the mines. Rubies of a poorer tone of colour are also found, but not as frequently as in Ceylon, where they predominate over the stones of deeper shades.

When found embedded in its original matrix, namely, the white crystalline marble, the ruby has always a regular and well-developed crystalline form, diagrams of which are shown in Fig. 53, a to d. When found in the gem-earths this symmetrical form is not always missing, though, as a rule, rubies from this deposit are irregular in outline, while those picked out of the alluvial sands of the river valleys are usually much rounded. It is the custom among the natives not to sell a ruby in its natural form, but to give it some artificial shape or another, usually quite irregular and not calculated to enhance the natural beauties of the stone. Such specimens have to be re-cut in Europe. The two large rubies, for example, which came to Europe in 1875 were roughly cut en cabochon. By re-cutting the weight of the one was reduced from 37 to $32\frac{5}{16}$ carats and of the other from 47 to $38\frac{9}{16}$ carats. The native lapidaries are for the most part settled at Amarapura near Mandalay.

Almost all the Burmese rubies which now come into the market are found in the district around Mogok. Valuable stones, however, are reported to have been found in the river gravels of the Nampai valley near the village of Namseka, which lies fifteen miles south-west of Mainglon (Fig. 54), in latitude 22° 46′ N. and longitude 96° 44′ E. Assuming this occurrence to be a fact, Dr. F. Noetling explains it by supposing that the gravels in this outlying district have been washed down from the ruby-bearing area by the Mogok stream when in flood. At the time of Dr. Noetling's visit a large excavation had been made in these gravels; in them he found spinel, tourmaline, &c., but his search for rubies was unsuccessful.

A second ruby district in Upper Burma, less important, however, than Mogok, exists in the neighbourhood of Sagyin, twenty-one miles north of Mandalay. Here a range of low hills built of crystalline limestone rises up out of the alluvial plain of the Irrawaddy, itself two miles distant. This white marble is for the most part overlain by red clay; it is much creviced and penetrated by caves, and indeed in all respects closely resembles the marble of Mogok. The ruby has here two modes of occurrence. In places in which the marble contains few or no embedded rubies, its crevices are often filled up by fragments of the same rock, rich in rubies and firmly cemented together. In other cases the crevices and caves in the limestone are filled, just as at Mogok, with brown clayey material produced by the weathering of the limestone. In this are found the precious stones together with other minerals, namely, ruby, sapphire, red and black spinel, amethyst, brown chondrodite, very pale blue apatite in small grains and crystals, reddish-brown mica, &c. The precious stones are separated from this weathered material by washing. A more systematic working of the deposits would be likely to yield better results. The rubies found here are sometimes said to be paler and inferior in quality to those from the Mogok district, but this statement is disputed by many observers.

It is asserted that rubies have been found, associated with spinel and as usual embedded in granular limestone, at a place further north, near the village of Nanyetseik, between Mogaung and the jadeite mines of Sanka; also at still another locality on the Upper Irrawaddy. The position of these two localities may be seen from the map (Fig. 55); the occurrence of ruby here is, however, decidedly doubtful. According to native reports, rubies and spinels occur in the limestone of two hills lying a little to the north of the Sagyin Hills. It may be mentioned finally that during the construction of the railway from Rangoon to Mandalay, abandoned ruby mines were met with near the town of Kyoukse, about thirty miles south of Mandalay.

The occurrence of rubies in **Siam** has been long known, but it is only recently that the deposits have been carefully investigated and systematically worked. Prospecting for precious stones in this country was for long rendered impossible or, at least, extremely difficult by the exercise of royal and official privileges. Now, however, an English company, known as " The Sapphires and Rubies of Siam, Limited," has obtained a concession of mining rights in this country. Details of the work are given by Mr. E. W. Streeter in his book, *Precious Stones and Gems*.

Though some of the Siamese rubies equal those of Burma in beauty, the majority are very dark in colour and generally inferior. The mines are situated in the provinces of Chantabun and Krat; a few rubies are found also in the sapphire mines of Battambang, south-east of Bangkok (Fig. 54).

The mines of Chantabun are about twenty hours journey by steamer from Bangkok. They lie not far from the coast of the Gulf of Siam (Fig. 56), and near Chantabun, the capital of the province of the same name. The lofty mountains of this region consist of greyish granite, and the lowlands of limestone. The latter may possibly be, as in Burma, the mother-rock of the ruby, but this point has not as yet been determined. At present the precious stone is only known to occur in sands, which have hitherto been worked according to the most primitive methods by the natives. The workers are for the most part Burmese, and their usual method of working the deposits is to excavate a small shaft which never exceeds 24 feet in depth. Precious stones were at one time very abundant in this region; according to a missionary report of the year 1859, it was possible in half an hour to collect a handful of rubies from the "Hill of Gems," an eminence standing to the east of the town of Chantabun. This particular accumulation is now dispersed, but the town of Chantabun remains the centre of the trade in precious stones of this region.

More detailed information respecting the mines in the province of Krat has been given by Demetri and others. Krat, the capital of the province, lies not far from Chantabun, in a south-south-easterly direction, and is on the sea-coast. The mines of this region are scattered over a wide area, and are divided into two groups thirty miles apart. Those of one group are known as the mines of Bo Nawang, and of the other as the mines of Bo Channa. At the time of Demetri's visit the men employed in the two groups of mines numbered about 1250.

The mines of Bo Nawang, situated in the neighbourhood of the village of Nawang, near the eastern margin of the map (Fig. 56), cover an area of about two square miles. They are small pits, 2 to 4 feet deep, sunk in a coarse yellow or brown sand, which extends over a wide stretch of country and overlies a bed of clay. The rubies are found at the base of the sand in a layer of material 6 to 10 inches thick. As elsewhere in Siam they are accompanied by sapphires, and, though small, are said to be superior to rubies from other Siamese localities. The mines have only been systematically worked since the year 1875.

The mines of Bo Channa lie about thirty miles to the north-east of the other group, and are scattered over an area about a mile square. The ruby-bearing sand is 6 to 24 inches thick, and a few of the mines reach a depth of 24 feet. The natives are of opinion that the stones have been washed down by the river from the Kao Sam Nam, and many fine stones are reported to have been found in the rivers rising in this mountain. The mines have been worked since 1885, always under unfavourable conditions due to the unhealthiness of the climate.

Between the provinces of Chantabun and Krat lies the ruby district of the sub-province of Muang Klung (or shortly Klung), shown in the map (Fig. 56). It is situated to the north-east of the town of Chantabun, and is reached after traversing twelve miles of rough road. The centre of the district, which extends for a distance of seven miles, is the small Burmese village of Ban Yat. The valleys of this district are from 600 to 800 feet above sea-level, while the hills dividing them have an elevation of 500 feet more. The gem mines are situated in the valleys and on the sides of the hills. All the valleys are traversed by small mountain streams, affluents of the river Ven, upon the banks of which narrow patches of alluvium are laid down. It is these alluvial deposits which are worked for gems. No alluvium is laid down in the upper part of the valleys, since here the streams are too rapid; it is only in the lower and wider parts that the streams are sufficiently slow to lay down a deposit. These small patches of alluvium are worked only in the dry season; in the wet season the miners confine their attention to the deposits on the sides of the hills, which lie above the present high-water level of the streams. These deposits overlie a trap-rock of the nature of basalt, which is the principal, if not the sole, constituent of the hill ranges. The gem-gravels are made up of fragments of this rock, and the separating layer of tenacious grey, brown, or yellow clay is also, in all probability, a decomposition product of

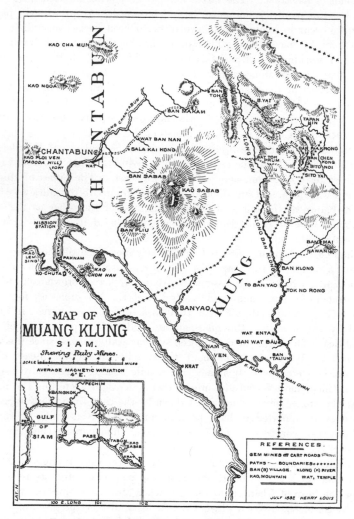

FIG. 56.   Ruby and sapphire mines of Muang Klung in Siam.

the same rock.   On these grounds it is concluded that the basaltic trap-rock is here the mother-rock of the ruby; this conclusion, however, requires the support of further evidence.

The gem-bearing layer varies in thickness from 10 inches to 5 feet, and is overlain by a sandy and clayey deposit from $2\frac{1}{2}$ to 12 feet thick, containing no precious stones.   In the clayey gravel are found ruby and sapphire, as well as common corundum; quartz, in good transparent crystals, and crystals of zircon and ilmenite are abundant, while topaz is very rare.   Of these minerals, the first two only are commercially valuable and sought after by the miners.   Rubies are much more frequently met with than are sapphires, the occurrence being in the proportion of about two to one. Good specimens of both are rare, the rubies being pale in colour and lacking in lustre, while the sapphires are opaque and dull.

The Burmese method of working the deposits is very simple. Small parties of three or four men working together sink a pit, usually about 4 feet in diameter, through the surface of the gem-bearing gravel. This they remove in baskets, leaving undisturbed all boulders too heavy to lift. The mines give employment to about 200 men, whose work consists of excavating the gem-bearing gravel, washing away, in the usual manner, the lighter earthy portions, and picking out from the residue any gems it may contain. Stones to the weight of about 500,000 carats are produced annually ; their aggregate value is, however, no more than from £2000 to £3000, so much of this weight being of inferior quality.

In the gem-sands of the island of **Ceylon** (Fig. 59) a few rubies, together with a far larger number of sapphires and other gems are found. Many stones preserve distinctly the outlines of their original crystalline form, which agrees completely with that of the rubies of Burma (Fig. 53, a—d) ; others occur as rounded grains. The gem-bearing sands lie on the hill-sides above the present high-water level of the streams, and also on the floors of the river valleys. The neighbourhood of Ratnapura and Rakwana and the district about the foot of Adam's Peak are specially rich. Though it is said that fine rubies of good colour, equal to or better than those of Burma, are sometimes found in Ceylon, yet, as a general rule, Cingalese rubies are pale in colour and not of very great value. The occurrence of sapphire in Ceylon is of far more importance ; it will be treated in detail in its appropriate place. According to Tennant, the mother-rock of the ruby in Ceylon, as in Burma, is a crystalline dolomite limestone or marble, which occurs *in situ* near Bullatotte and Budulla. The mother-rock of the sapphire is probably different, as we shall see later ; this is thought to be gneiss.

The mainland of **India**, though so rich in common corundum, is very poor in the precious variety. A few stones suitable for cutting have been found with common corundum in Mysore and in the Salem district of Madras ; also in the alluvium of the Cauvery river, which flows into the Bay of Bengal some distance south of Pondicherry. The occurrence of the precious stone in the sands and gravels of this river bears a striking similarity to its occurrence in the river alluvia of Ceylon. The ruby has been stated to occur in the gravels of other Indian rivers ; in many of these cases, however, it is probable that the supposed ruby is in reality garnet, a stone which is widely distributed in India. Many of the rubies preserved in the treasuries of Indian princes have probably been brought from Burma or from Badakshan, a locality for ruby which has yet to be mentioned.

In **Afghanistan** permission to work the ruby mines near Jagdalak, thirty-two miles east of Kabul, has been obtainable from the Amir since the 'seventies. The rubies found here lie in a micaceous crystalline limestone ; many show a distinct crystalline form, which is identical with that of rubies from Burma. These stones were originally described as being spinel, but specimens which have come to Europe have been proved to be indubitable rubies. The occurrence of ruby in this locality is strikingly similar to its occurrence in Burma ; as in Burma and Ceylon, so probably in Afghanistan also, spinel occurs associated with the ruby.

A ruby of 10½ carats was brought to Europe by a traveller from Gandamak, a place about twenty miles from Jagdalak, and in latitude about 34⅓° N. and longitude 70° E. Nothing further as to this occurrence is known, and it is possible that both place-names refer to the same occurrence.

The ruby mines of Badakshan were famous in olden times, and they supplied some of the vast store of treasure amassed by the Great Mogul.   They are situated

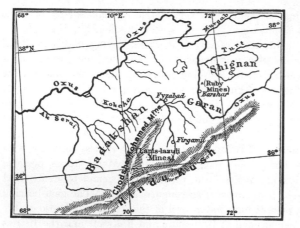

FIG. 57.   Ruby mines in Badakshan on the Upper Oxus.
(Scale, 1 : 6,000,000.)

(Fig. 57) in Shignan, on the bend of the Oxus river, which is directed to the south-west, in latitude about 37° N. and longitude 71½° E. They lie between the upper course of the Oxus and its right tributary the Turt, near Gharan, a place the name of which is said to signify " mine," sixteen miles below the town of Barshar, in the lower, not the higher, mountain ranges. This locality is by no means a familiar one, and reports as to the mode of occurrence of the ruby here are very conflicting. According to one they are found in a white earth; according to another in a red sandstone; while yet a third states them to be found in a magnesian limestone.   From analogy with the Burmese occurrence, the last-named mode of occurrence seems the most probable.

Rubies are said to have been found formerly in these mines in large numbers, and associated with the variety of spinel known as " balas-ruby."   Marco Polo, who visited the mines in the thirteenth century, states that the output from them was strictly limited by the ruler of the country in order to keep up the value.   Part of the output was paid away as tribute to the Mongol Emperor, another part to other rulers, while the remainder was put into the market.   The yield appears to have fallen off in later times, till in the end work was altogether discontinued.   It is stated that the mines were reopened in the year 1866 ; whether they are being worked at the present day or not is unknown, but, according to a recent report, they are practically exhausted, and give employment to only about thirty persons, the few stones that are found being the property of the Amir of Afghanistan.   A stone the size of a pigeon's egg is said to have been found here in 1873.

It is possible that the rubies and spinels which have recently come into the market through Tashkent, and which, according to the merchants, were mined in the Tian-Shan Mountains, are in reality from these same mines.   There is no reliable information as to the existence of ruby mines in the Tian-Shan Mountains or in Tibet, so that the 2000-carat ruby recently received by Streeter, and said to be from Tibet, may also have been found in these mines on the Oxus.

Compared with the importance of the occurrence of ruby in Asia, that in all other parts of the world is insignificant.

In **Australia** rubies of small size have been found in the gold-sands sometimes associated with the diamond, never, however, in large numbers.   Such occurrences have been noted in the sands of the Cudgegong river (Fig. 43), in a few of its tributaries near Mudgee, and at a few other places in New South Wales.   In Victoria the gold-sands of Beechworth and Pakenham have yielded rubies.   In all Australian ruby localities, however, this stone is very much less common than the sapphire.   Moreover, red garnets have been mistaken for rubies time after time.   Thus, many years ago, an abundant occurrence of rubies in the Macdonnell Ranges of the Northern Territory of South Australia was reported. No less than twenty-four companies were very shortly formed for the working of the deposits. A more thorough examination of the stones, however, showed them to be red garnets, of fine quality, but compared with the true ruby almost valueless. These same garnets are now sometimes sold as " Adelaide rubies."

The **United States** of North America yield a few rubies, occurring rarely associated with common corundum, which is very abundant in this country. In the Lucas mine on Corundum Hill in Macon County, North Carolina, small amounts of transparent red corundum, sometimes suitable for cutting, have been found. The boss of serpentine which forms the main mass of Corundum Hill is traversed by large veins of common corundum. Rubies of much better quality have been found more recently in Cowee Creek, also in Macon County, about five miles from Franklin, and have been described (1899) by Professor J. W. Judd and Mr. W. E. Hidden. These authors point out that there are three distinct modes of occurrence of corundum in North Carolina and the adjoining States :

(1) In the ordinary crystalline schists and gneisses of the district, as long prismatic crystals usually of a purplish tint and not of gem quality.

(2) In the olivine-rocks, and the serpentine derived from them, which are intrusive in the crystalline schists, as crystals often of large size and showing great variety of colour, but seldom or never clear and translucent.

(3) In certain garnet-bearing basic rocks at Cowee Creek, as small tabular and short prismatic crystals, which frequently exhibit the transparency and colour of the " oriental ruby."

These garnet-bearing basic rocks include eclogite, amphibolite, and other similar rocks of igneous origin. They have been much weathered, however, and are represented solely by a soft decomposition product known to American petrologists as " saprolite." In this material the rubies occur in " nests " and " bands," and also in what appear to have once been cavities in the original rock. These cavities, when the corundum is pale coloured, appear to have been filled with felspathic material ; but when the corundum is of a ruby-red colour, the surrounding space is filled up with chloritic material. The associated minerals are garnet, in great abundance, sillimanite, kyanite, staurolite (often very clear and gem-like in character), cordierite, zircon, monazite, and others, together with minute quantities of gold.

The corundum found here varies in colour from ruby-red through different shades of pink to white. Many of the red crystals exhibit the beautiful so-called pigeon's-blood tint, and are in no way inferior to the finest Burmese rubies. Enclosures of various kinds are, however, frequent ; these may be extremely minute (" silk " of jewellers), giving rise to a cloudiness (" sheen ") in the faceted gems, or they may be larger reniform masses of clear red rutile or black ilmenite. Some crystals of ruby have been found to enclose crystals of the newly discovered variety of garnet known as rhodolite, to be described in its appropriate place. Enclosures of this kind, however, in no way impair the transparency and beauty of the ruby. Some few specimens of ruby have been found perfectly free from enclosures and large enough to give a cut gem of very fair size.

Although the Cowee Creek rubies are very like Burmese stones, yet their mode of occurrence is totally different, for in the former locality the white crystalline limestone of Burma is absent, as are also the fine red spinels so characteristic an associate of the rubies of Burma.

In the State of Montana a few rubies have been found in association with sapphire ; they are usually of a pale rose-red colour like those of Ceylon, stones of a fine deep colour being only occasionally met with. It is hoped that a more systematic working of these deposits, especially at Ruby Bar, will result in more frequent discoveries of stones of good colour. In America, as in Australia, garnets have been frequently mistaken for the more costly ruby, and have been collected and sold as such.

In Europe red corundum, suitable for cutting as gems, is practically absent, and the same is the case in the continent of Africa, the so-called " Cape rubies," occurring in South Africa in association with diamond, being not ruby but garnet.

ARTIFICIAL PRODUCTION.—Ruby is the only valuable precious stone which hitherto has been produced by artificial means in crystals of fair size showing all the characters of the

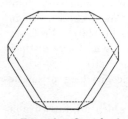

FIG. 58.   Crystal of artificially prepared ruby. Magnified. (After Fremy.)

natural minerals. The honour of this achievement belongs to the French chemist Fremy, whose efforts in this direction have, after many trials, at last been crowned with success. His object was attained by fusing together in an earthen crucible at a high temperature (1500° C.), a mixture of perfectly pure alumina (Al_2O_3), potassium carbonate, barium (or calcium) fluoride, and a small amount of potassium chromate, the whole mass being kept in a molten state for a week. The series of reactions which take place under these conditions probably begins with the formation of aluminium fluoride. This compound, as a result of contact with the moisture of the atmosphere and furnace gases—a contact rendered possible by the porous nature of the crucible—yields aluminium oxide (alumina). This, by taking up chromic oxide from the potassium chromate, assumes a red colour and crystallises out as ruby. When isolated, after cooling, from the fused mass, in which the crystals are embedded, they are found to differ in nowise from naturally occurring crystals of ruby.

The artificial crystals so formed have always the form shown in Fig. 58, which represents a rhombohedron in combination with extensively developed basal planes, faces which bound natural crystals of ruby also, as shown in Fig. 53, a—d. The thin tabular crystals produced by this method are always of small size, never exceeding ⅓ carat in weight. Their size is increased when larger amounts of material are allowed to interact in the crucible. The colour of the artificial product varies from pale to dark red, according to the conditions of the experiment. The most beautiful and characteristic ruby-red colour was produced by the addition of 3 to 4 per cent. of potassium chromate. The colour-results seem, however, somewhat difficult to control, for the crystals often more or less incline to a violet colour, sometimes, indeed, being quite blue, while crystals coloured red at one end and blue at the other have been occasionally produced. From these observations Fremy concludes that the colour of naturally occurring sapphires, as well as of rubies, may be due to chromium. More than 3 to 4 per cent. of the chromium salt is taken up only with difficulty, and the crystals receive a violet tint, differing very markedly from the colour of naturally occurring rubies.

Artificially formed rubies which, of course, differ in no way from naturally occurring stones, except in their mode of formation, have been mounted as gems, both in a cut and in an uncut condition. Having the same hardness as natural corundum, they have also been utilised as the pivot-supports of watches.

The cost of production of artificially made rubies is so high that they are no cheaper than stones formed by nature; moreover, their small size strictly limits their general application. Before his death Fremy expressed a hope that crystals of much greater size would result from experiments conducted in a crucible of 50 litres capacity. However this may be, there is no immediate prospect of the artificial ⟨ the natural product.

Other investigators, experimenting in the same direction, have been successf producing crystals of corundum, notably J. Morozewicz. His fused silicates yielded cry. of spinel as well as of corundum. The corundum crystals were tabular in habit and react a diameter of 1·5 millimetres. The various colours—red, blue, yellow, and greenish-yell⟨ —of these crystals must have been due to the presence of iron, for in the experiments of th investigator chromium was not an ingredient of the fused mass.

The peculiar, fine carmine-red rubies of considerable size and unknown origin, which appeared in the market in 1885 at Geneva, may be mentioned here. They have the hardness and specific gravity of the natural mineral, but are less brilliant, and in all probability are artificial products. The colour as seen in the spectroscope is more like that of the artificial crystals prepared by Frémy than of natural crystals, and, moreover, certain appearances under the microscope point in the same direction. The origin of these stones is mysterious, and, if artificial, nothing as to their mode of preparation is known. According to one report they have been formed by fusing together several small rubies; this, however, is scarcely credible, since at the extremely high temperature of the melting-point of corundum the ruby assumes a dull grey colour. Some authorities again have supposed them to be formed by a method analogous to Frémy's, while yet others have supposed each stone to consist of several small rubies held together in a matrix of glass of the same colour and refractive index. The success which has been attained so far in the artificial production of rubies is encouraging, and affords grounds for the hope that still greater achievements will be possible to future investigators.

DISTINCTION FROM OTHER RED STONES.—It is only natural that attempts should be made to substitute for the costly ruby some less valuable stone of similar colour. The two stones most frequently passed off as, and mistaken for, rubies are spinel and garnet. The so-called rubies of cheap jewellery are in reality either the variety of spinel, known as " ruby-spinel," or red tourmaline (rubellite), while topaz may be substituted for pale red ruby. Red quartz will scarcely pass as a substitute for ruby, but red glass (paste) is frequently so passed off.

The crystalline form of the above-mentioned substitutes will, if exhibited, serve to distinguish each from the ruby. In either the rough or cut condition, spinel, garnet, and glass may be readily distinguished from ruby by their single refraction and the absence in them of dichroism. Red tourmaline and quartz have a much lower specific gravity than ruby; the latter sinks heavily in methylene iodide, while the two former float easily. Rose-red topaz can only be substituted for pale rose-red ruby; since there is little difference between them in value it is not so important from a pecuniary point of view to be able to distinguish the one from the other. The specific gravity, however, affords a distinguishing characteristic, since topaz (sp. gr. = 3·5) floats in the heaviest liquid, while ruby (sp. gr. = 4·0) sinks. One can scarcely fail to distinguish the ruby, from any stone which may be substituted for it, by its great hardness. After the diamond corundum is the hardest of all known minerals, and will scratch any of the stones mentioned above with ease.

The word ruby is often used in the designation of stones belonging to mineral species other than corundum. Rose-quartz, for instance, is known as " Bohemian ruby," rose-red topaz as " Brazilian ruby," red garnet as " Cape ruby," and also as " Adelaide ruby "; " Siberian ruby " is the name given to red tourmaline (rubellite), and " false ruby " to red fluor-spar; while certain varieties of spinel are referred to as " ruby-spinel " and " balas-ruby."

In the manufacture of the so-called ruby-glass various pigments have been used for the purpose of reproducing the colour of the ruby. Manganese salts give a fairly close imitation, but the colour which results from their use is too strongly violet. The best results are obtained with gold salts, purple of Cassius, &c., which are fused with the glass or strass. The use of gold salts necessitates the greatest care, otherwise the glass will be cloudy. A glass coloured with gold salts after first being cooled is yellowish-green, the fine red colour only appearing after the glass has been annealed, an operation which is known as "tinting." By the use of gold salts, glass of the finest ruby-red colour can be obtained, and by varying the percentage of gold in the strass different shades of colour are produced. It is an interesting fact that fine ruby-glass has been found in ancient Celtic graves.

SAPPHIRE.

CHARACTERS.—Sapphire ("oriental sapphire") is the name given to blue corundum. Sapphire differs from ruby most essentially in colour, it is in addition, however, slightly harder, being the hardest of all the varieties of corundum; and, moreover, is stated to have a slightly higher specific gravity, the specific gravity of ruby being given as 3·99 to 4·06, and that of sapphire as 4·08. The form of a crystal of sapphire agrees completely in its general symmetry with that of a crystal of ruby; the two crystals differ somewhat in habit however. The prism and rhombohedron, usually well developed in the ruby, are subordinate in the sapphire, and here the hexagonal bipyramid predominates, as is shown in Fig. 53, e–i, and for a natural crystal in Plate I., Fig. 7.

While the ruby is usually coloured uniformly throughout its substance, the distribution of colour in the sapphire is often very irregular. A single stone may show an alternation of colourless, pale yellowish, and blue portions, or in the colourless ground-mass of a stone there may be patches of a blue colour. Such stones, compared with a sapphire of a uniform blue colour, are almost worthless as gems.

It would be quite possible to collect a series of sapphires showing small gradations in colour from deep blue to yellowish or colourless stones. These latter are known as *white sapphire* (leuco-sapphire); they are only rarely perfectly colourless and transparent, usually showing a bluish or yellowish tinge. Sapphires of a definite yellow colour are described as "oriental topaz."

The blue colour of the sapphire disappears on heating, hence it is possible to transform a patchy or pale-coloured stone into a leuco-sapphire, and at the same time greatly enhance its value.

The distribution of the blue patches in the colourless or yellowish ground-mass of a sapphire is usually quite irregular, only occasionally is there any definite arrangement to be observed. In such cases, the crystal may be blue at one end and colourless at the other, or the middle portion may be colourless and the two ends blue, or colourless and blue bands may alternate. Moreover, different portions of the same specimen may exhibit different shades of blue, such as pure blue and the greenish-blue peculiar to the sapphires of Siam, or even colours which are altogether different. Thus, crystals of sapphire are known which are blue at one end and red at the other, and others of which the two ends are blue and the middle part yellow. A crystal answering to this latter description and weighing 19⅛ carats is exhibited in the mineralogical collection in the museum of the Jardin des Plantes in Paris.

The peculiar distribution of colour in such sapphires has been sometimes ingeniously utilised; for example, the figure of Confucius, preserved in the museum at Gotha, is carved out of a parti-coloured sapphire in such a way that the head of the figure is colourless, the legs yellow, and the body pale blue.

Every shade of blue, from the palest to the darkest, is represented amongst sapphires. The very dark shade of blue, closely approaching black, is described as inky; very pale " feminine" sapphires are sometimes described as "water-sapphires," while stones of the darkest shade are variously known as " indigo-sapphire," " lynx-sapphire " or " cat-sapphire." So long as the depth of its colour does not interfere with the transparency of a sapphire, the darker it is in colour the more highly is it prized. The colour of sapphires shows as great a variety of tone as of shade. Thus, we have stones of indigo-blue, Berlin-blue, smalt-blue, cornflower-blue, greyish-blue, and greenish-blue, the last being specially common. The most admired tone of colour for a sapphire is an intense cornflower-blue. A really fine sapphire will combine with this colour a beautiful velvety lustre; the latter character, though occasionally seen, is by no means common. A fine blue crystal of sapphire is shown in Plate I., Fig. 7, and a faceted stone in Fig. 8 of the same plate.

The blue of the sapphire is always more or less tinged with green, this being very noticeable when one looks through the stone in certain directions. Like the ruby the sapphire is distinctly dichroic, the phenomenon being very marked in dark coloured stones but scarcely noticeable in stones of a light shade of colour. Looked at in the direction of the optic axis, that is to say, along the line joining the apices of the hexagonal bipyramid or perpendicular to the terminal basal planes, a crystal of sapphire appears of a pure blue colour, more or less intense or inclined to violet, according to the particular character of the stone. In a direction perpendicular to this the stone appears paler, and its blue colour is distinctly tinged with green; observed in intermediate directions the sapphire will appear of intermediate tints. The dichroism of the sapphires of Siam, which have lately come into the market in large numbers, as well as of those from Le Puy, in Auvergne, and from some other localities, is especially well marked.

If the light passing through a crystal of sapphire in the direction of its optic axis be received in a dichroscope, the two images formed by the instrument will be identical in colour—either pure blue or blue tinged with violet—and the colour will remain unchanged when the stone or the instrument is rotated. If examined in the direction perpendicular to this, the two images will, as a rule, be coloured differently. In the position in which the greatest difference in colour exists, the one will be of a pure dark blue and the other usually of a paler greenish-blue, but sometimes of a yellowish-green.

It follows from these facts that the pure blue colour of a sapphire crystal is best displayed in the cut stone when the table of the latter is perpendicular to the optic axis, and parallel to the basal planes, of the crystal. It will be remembered that it is advantageous for the same reason to cut crystals of ruby also in this manner.

The appearance of sapphire in artificial illumination varies in different specimens. In some no difference in colour can be detected, in others the colour becomes darker, or it may change to reddish, purple, or violet. The latter change in colour is rare, and stones which show it are valuable on this account.

While the colour of the ruby remains unaltered after the stone has been exposed to a strong red-heat, that of the sapphire under similar conditions disappears, although in other respects the stone remains unchanged. When exposed to very high temperatures, however, the sapphire, like the ruby, becomes grey and cloudy. The decolorisation of the sapphire does not take place with equal facility in all stones. Indian sapphires lose their colour most easily; and there are stones the colour of which it is impossible to completely destroy. From the fact that sapphires can be decolorised by heat, it has been argued that their colour is due to some organic compound. Some authorities have referred it to the presence of a small amount of iron, which has been detected in analysis, while others, relying on Frémy's experiments in the artificial production of rubies, consider it to be due to small quantities of some compound of chromium.

Asterias (Star-sapphires). There is often to be seen on the basal planes of sapphire crystals a six-rayed star of chatoyant light. This appearance, which is known as asterism, and is often very beautiful, is best displayed by cutting the crystal in which it exists *en cabochon*, the centre of the curved surface lying in the axis of the crystal. The rays of the star spread out to the margin of the stone, and the movements of the latter are followed to a certain extent by the star, the centre of which is always directed towards the light. A perfect six-rayed star is, however, seen less frequently than a patch of chatoyant light, which may be more or less rounded or elongated in outline, in the latter case being regarded as a single ray of the star. This patch or star of milky, shimmering, and opalescent light is sometimes tinged with a red or blue colour. The rays of the star may be narrow and sharply defined, showing up against the dark surface of the stone like silver threads, or they may be broad and ill-defined, merging imperceptibly into the darker portions of the stone. A sapphire in which the star is sharply defined, is much prized, and is known variously as a *star-sapphire*, an asteria or star-stone, an asteriated sapphire, or as a sapphire-star-stone. A stone which shows only an irregular patch of opalescent light is known as *sapphire-cat's-eye*, "*oriental girasol*," or opalescent sapphire. To rubies showing the same appearance corresponding names are applied, namely, *star-ruby*, asteriated ruby, ruby-star-stone, ruby-cat's-eye, and opalescent ruby. The phenomenon of asterism is not confined entirely to red and blue corundum, but is also occasionally seen in yellow corundum or " oriental topaz." A stone of this kind is, like ruby-cat's-eye, included in the term " oriental girasol " when it shows an elongated or round patch of opalescent light ; it is then called a "*topaz-cat's-eye*," and when it shows a regular star it is called a "*star-topaz*."

No exceptional value is attached to asteriated stones ; a fine star-sapphire is about equal in value to an equally fine stone of the ordinary kind, though large star-rubies fetch rather higher prices than ordinary stones ; small ones can be obtained for comparatively little. None of the asteriated varieties of corundum are confined to any special locality, being found wherever precious corundum is found.

The phenomenon of asterism has been variously explained. Some consider it to be due to the reflection of light from the surface of the twin-lamellæ, which are present in such crystals in large numbers, and are arranged so that their planes are parallel to the faces of the primitive rhombohedron (Fig. 53 *a*). The existence of these lamellæ is indicated by the striations on the basal planes. These striations are grouped in three sets inclined to each other at angles of 60°, and were considered by Babinet to be the cause of the star of opalescent light. Another and more probable explanation is that it is due to the reflection of light from the surface of a multitude of microscopically small tubular cavities or rifts enclosed in the crystal, and grouped into three sets, each of which is parallel to a face of the hexagonal prism ; corundum crystals also enclose minute tabular crystals, consisting of alteration products of the corundum, and arranged, like the rifts, in three sets inclined at 60° to each other. The six rays of the star are produced by the total reflection of light from the surfaces of these tubular cavities, and perhaps also from the surfaces of the tabular crystals. The phenomenon is only to be observed in stones in which there are large numbers of these enclosures, such stones being usually cloudy and having a metallic sheen. Star-stones are never in fact quite clear and transparent throughout their whole mass ;

frequently also they are built of alternate blue and colourless layers. The same phenomenon may be observed in many opaque specimens of common corundum, especially in the brown adamantine-spar, which is sometimes cut so as to show the asterism. Enclosures of the kind described above are more frequent in blue sapphire than in corundum of any other colour, and, as a consequence, star-sapphires are commoner than other varieties of asteriated corundum.

In Europe star-sapphires only, as a rule, are cut *en cabochon;* in India this form of cutting is much more frequently employed, not only for star-stones but also for others. Such stones, however, on their arrival in Europe are always re-cut with facets. Both the form and methods of cutting employed for the sapphire are identical with those used for the ruby, as is also the mounting of the stone. The colour of the sapphire is frequently intensified by placing a piece of blue silver-foil beneath the stone in its setting.

VALUE.—Sapphires of large size and fine quality are far more common than rubies of the same description, hence the latter always command higher prices than the former. Thus a flawless carat sapphire of perfect transparency, velvety lustre, and of a uniform deep cornflower-blue colour, will seldom fetch more than £10, while £25 will be easily obtained for a ruby of corresponding size and quality. Sapphires of this description, weighing between 2 and 3 carats, are about equal in value to diamonds of good quality and of the same weight. Faulty stones, the colour of which is pale or of irregular distribution, do not fetch more than a few shillings per carat. Since large sapphires are far more common than large rubies, there is a much smaller disproportion between the prices of large and of small sapphires than between those of large and small rubies. In the case of sapphires, indeed, the prices are almost proportional to the weight, a stone of double the weight being not much more than double the value, and so on. The flaws seen most commonly in sapphire are in general the same as in ruby, namely, clouds, milky and semi-transparent patches, white glassy streaks, alternation of differently coloured layers, areas showing silky lustre, &c.

Some few sapphires of exceptional beauty and size have acquired wide renown. The most magnificent of these is one of 951 carats seen in 1827 in the treasury of the King of Ava, as described by the English Ambassador at the Court of that monarch. It is reported to have been found in Burma, and to be not absolutely flawless. In the collection of the Jardin des Plantes in Paris is a rough stone of $132\frac{1}{16}$ carats; this is the "wooden-spoon-seller's" stone, and is said to have been found in Bengal by a man who followed that particular trade. It is known also as the "Rospoli" sapphire, after the family in whose possession it formerly was, and is one of the most magnificent of blue sapphires, free from all patches and faults. In the same collection is preserved another fine sapphire, 2 inches long and 1½ inches deep. A beautiful sapphire, weighing over 100 carats, is the property of the Duke of Devonshire; the lower portion of this stone is step-cut while the upper is cut as a brilliant. Among other noted sapphires may be mentioned a dark, inky, faultless stone weighing 252 carats, which was exhibited in London in 1862, and a fine blue stone, with a yellow patch on one side, which weighed 225 carats, and was exhibited in Paris in 1867.

OCCURRENCE.—The mode of occurrence of sapphire is practically the same as that of ruby. It is found in sands and in solid rock, frequently together with ruby, in the manner already described. There is probably no single locality where one stone is found without the other; they are invariably associated together, here one and there the other predominating, and with them are usually found other varieties of precious and common corundum. Ruby predominates at the localities specially described above for this gem.

Sapphire is the more abundant of the two in Siam (the two, however, coming from different mines), in Ceylon, at Zanskar in Kashmir, in the gold and diamond sands of Australia, especially of New South Wales, and in Montana in the United States. Other localities, such, for example, as the European, are unimportant. By far the largest number of sapphires which come into the market are from Siam, the production of other countries being in comparison with this quite insignificant.

Not only the largest number of sapphires, but also the finest quality of stones, come from **Siam** (see Map, Figs. 54 and 56). The most important of the long-known mines of this country, the systematic working of which has recently been undertaken by Europeans, are those of Battambang, in which a few rubies are found with the sapphires. A certain number of good stones are found in the ruby mines of Chantabun and Krat, mentioned above. It is estimated that the mines of Bo Pie Rin in Battambang alone yield five-eighths of the total sapphire production of the world. Many of the stones found here surpass those from all other localities in their intense blue colour and velvety lustre. Many, however, of the so-called inky stones, are so deep in colour that in reflected light they appear almost black. It is a remarkable fact that the larger stones exceeding one carat in weight are almost invariably of finer colour and quality than smaller stones. Although the occurrence of sapphire in Siam was known at least as early as the beginning of the nineteenth century, the mines have been regularly worked only since about the year 1875. It is possible, however, that stones from these mines came into the market through Burma and were sold as Burmese stones. The mines of Siam have, therefore, grown into importance with great rapidity. According to Streeter, to whom the present account is due, the sale of Siamese sapphires by a single firm of London gem merchants amounted, in 1889, to £75,000.

The sapphire in this locality is found in a slightly sandy clay, usually about 2 feet below the surface of the ground. The most important mines are situated in the sides and floor of the Phelin valley. Each is a rough pit almost 4 feet square and 5 to 12 feet deep. As usual in occurrences of this type, the clay is washed away from the excavated mass and the stones picked out of the sandy residue.

So far as is known at present, the sapphire-bearing deposit extends over an area of about 100 square miles. The centre of the trade both for rubies and sapphires is the town of Chantabun, on the Gulf of Siam, in latitude about $12\frac{1}{2}°$ N. In the neighbourhood of this town, besides the ruby mines already mentioned, there are deposits in which sapphire is the predominating gem, and these appear to have been known and worked longer than those of Battambang. The sapphire has not yet been observed to occur in Siam in deposits of any type other than gem-sands, so that little is known of the minerals associated with it in the mother-rock.

As to the occurrence of sapphire in **Burma**, there is little to be added to what has been already said respecting the occurrence of ruby in this country. Sapphires are found at the same localities and under the same conditions, but where one sapphire is found there will be 500 rubies. While, however, rubies of good quality and exceeding 10 carats in weight are of extremely rare occurrence, large sapphires are found with considerable frequency. The discovery of sapphires weighing 1988, 951, 820, and 253 carats respectively, has been reported. Stones weighing 6 to 9 carats, though common, are often faulty. The largest faultless stone yet found in Burma weighs $79\frac{1}{2}$ carats; all others show considerable faults. The colour of Burmese sapphires is usually so dark that they appear almost black, they are seldom comparable in quality with those from Siam, and do not command a high price.

The sapphire occurs in **Ceylon** associated with many other precious stones. The yield of gems of this island is not large, the total value of the annual production being said to be no more than £10,000. The locality is, however, remarkable for its variety of gem-stones, namely, sapphire, ruby, topaz, amethyst, cat's-eye and other varieties of quartz, garnet (almandine and cinnamon-stone), zircon (hyacinth), chrysoberyl in its different varieties, spinel, tourmaline, moon-stone, and others which are rarer and of less importance. In association with the precious stones, there are found fragments of common corundum, magnetite, felspar, calcite, &c. Of the above-mentioned precious stones the sapphire is by far the most frequent.

The precious stones and the minerals with which they are associated were originally, for the most part, constituents of certain granite and gneissic rocks, by the weathering and disintegration of which they have been set free. While the sapphire and garnet were originally embedded in gneiss, other precious stones, such as the ruby and spinel, have been derived from the crystalline limestones (marbles) which are associated with the gneiss. The gems occur in their mother-rocks only sparingly, and are never obtained directly from them, but from the sands, gravels, and clays formed by weathering. These secondary deposits, in which the gems weathered out of the solid rock have been accumulating for long periods of time, are found in the beds of the streams of the present day, and on the sides of the hills above the present high-water level.

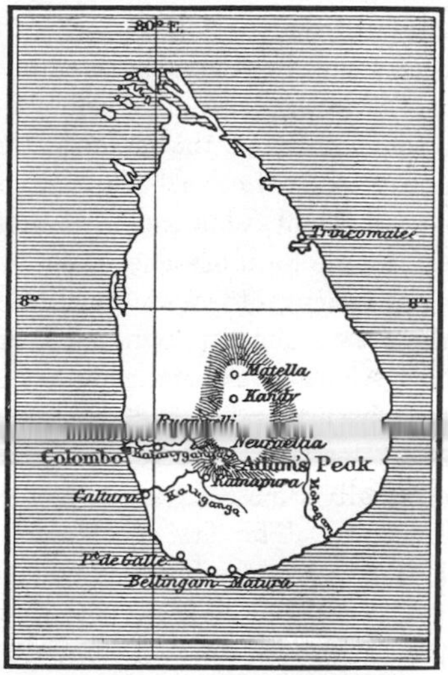

The richest locality for gems is in the south of the island, on the southern slopes of the mountains in the Saffragam district. On this account the principal town of the district has received the name of Ratnapura (or Anarhadnapura), which signifies the "City of Rubies." The occurrence of gems is, however, by no means confined to this one locality, stones being found in the western

FIG. 59. Occurrence of sapphire in Ceylon.

plain between Adam's Peak and the sea, near Neuraellia, Kandy, Matella, and Ruanwelli, and in the river-bed of the Kalany Ganga near Sittawake, six miles east of Colombo. Also near Matura, on the south coast of the island, and in the rivers on the east in the neighbourhood of the Mohagam river. The localities specially rich in sapphires are the Saffragam district and the neighbourhood of Matura, where a considerable number of stones of large size and fine quality are found.

The gem mines near Ratnapura were visited and described by Ferdinand Hochstetter, during the voyage of the Austrian frigate *Novara*. They are situated on the Kalu Sella, a small tributary of the Kalu Ganga, partly in the bed and partly on the right bank of the river. The mines, which reach a depth of 30 feet, were not being worked at the time of Hochstetter's visit, and were filled with water. The uppermost layer is a thick yellow clay with nodules of limonite resembling our boulder clay in appearance. Below lies unctuous black clay and clayey sand; then bituminous clay enclosing abundant plant remains, the teeth and bones of elephants, &c.; then sand, and finally a bed of pebbles with red, yellow, or sometimes blue clay. This constitutes the gem layer, and is known as the stone-gravel or "malave." The gems are found mainly between the large pebbles; they are specially abundant when the layer contains a greenish, talcose, partly decomposed mica.

In the Kalu Ganga, between Ratnapura and Caltura, most of the gems are washed from the sands above small rapids in the river.

The gem mines of Ukkette Demy, near Ratnapura, were visited in 1889 by J. Walther, of Jena, who was kind enough to furnish the following unpublished details : The mines lie in a valley basin about 3 kilometres wide, in which several side streams deposit the débris weathered from the surrounding ancient crystalline rocks, such as gneisses, &c. The strata in which the mines are sunk include an upper layer of 80 centimetres of mud, then 50 centimetres of white sand, with a few bands of black vegetable matter; beneath this a metre of dark yellow clay, and then the gem layer consisting of a tough clay, which may be white, yellow, red, or green, and encloses much-decomposed boulders of the surrounding crystalline rocks. The gem-bearing clay, which rests on a bed of gravel 3 metres thick, is richest when white and poorest when green. The precious stones have doubtless been derived from the gneisses, &c., of the neighbourhood, since grains of sapphire have frequently been found in decomposed boulders of these rocks contained in the deposit.

The sapphires of Ceylon are not of very good quality; though a few stones of a rich colour are found, the majority are too pale to be of any great value. Star-sapphires are not of unusual occurrence, and yellow (" oriental topaz") and white (leuco-sapphire) stones are abundant, while parti-coloured sapphires are not infrequently met with. The original crystalline form of some stones is distinctly recognisable, although the edges are usually not quite sharp ; others are much worn and rounded. Large stones are rarely found here, and the ruby is far less abundant than the sapphire.

Another important locality at which, since 1881 or 1882, sapphires in large numbers have been found, is the Zanskar range of **Kashmir**, in the north-west Himalayas. The exact locality of these finds was for a long time a secret, which was jealously guarded, especially from Europeans, first by the original discoverers and then by the Government of Kashmir. The first geologist who succeeded in visiting the locality was Mr. T. H. D. LaTouche, of the Indian Geological Survey.

According to his report these deposits are situated in a small upland valley in the upper part of the district of Pádar, about thirteen days' journey south-east of Srinagar, the capital of Kashmir, a few miles to the east of the village of Machél, and a little west-north-west of the village of Soomjam. Soomjam is higher than any other village on the south-western slopes of the lofty Zanskar range. It is about half a day's journey down from the Umasi Pass, and has an altitude of 11,000 feet. It lies in latitude 33° 25′ 30″ N., and longitude 76° 28′ 10″ E., on the Bhutna river, a tributary of the Chináb.

The valley in which the sapphires are found is 1000 yards long and 400 yards wide at its lower end ; it has an elevation of 13,200 feet above sea-level, and its floor rises towards the north-west, the average angle of slope being about 20°. The first find is said to have been made in the sapphire-bearing rock which forms a precipice at the head of the valley. This rock, which was laid bare by a landslip, is at an altitude of 14,800 feet, and lies very near the limit of perpetual snow. A large number of gems were at first won from the solid rock ; very soon, however, it was discovered that they existed in equal abundance in the loose detrital material weathered from these rocks and deposited on the floor of the valley.

The rocks of the district, mainly mica-schists and garnetiferous gneiss with interfoliated crystalline limestone, are penetrated by veins of granite, and it is in these veins that the sapphire, associated with an abundance of dark-brown tourmaline, is found. The material formed by the weathering of the granite is laid down in the valley as a white bed of little thickness, and is described as being overlain by a reddish-brown earth. The gems can be picked out by hand from this deposit " like potatoes," though they are, of course, also won by washing. The dark-brown tourmaline, mentioned above as being present in the granite veins, is also found in these secondary deposits.

The fine blue colour of the sapphires of this locality first attracted the attention of the inhabitants, who, not knowing the value of the stones often used them for striking fire. They were so abundant at first that large numbers were collected by the natives and sold to the gem merchants of Simla and Delhi, who, supposing them to be blue quartz or amethyst, purchased them very cheaply. When their true nature became known many expeditions were sent out to the Zanskar range with the object of collecting as many of these valuable stones as possible. The prices, of course, rose, and very quickly reached the figure at which sapphire is usually sold, namely, about £20 per ounce. Later on the stones fell again in value owing to the large number which were put on the market. Soon the Maharajah of Kashmir, in whose dominions the deposit is situated, began to interest himself in the matter. Those persons who had already found stones were allowed to retain them, but any further search could only be made by duly licensed individuals, who had to pay for the privilege. This arrangement still holds good.

The sapphires found in the Zanskar range are frequently in well-developed crystals, of the forms shown in Fig. 53, e to i. Numerous dark-brown or green tourmalines of small size are often observed enclosed in, or growing on the surface of, the crystals of sapphire. The crystals are sometimes very large, specimens suitable for cutting having been found measuring 5 inches in length and 3 inches in thickness, while a few are said to have attained a length of a foot. Irregular grains and fragments of the gem are frequently met with, but many of these are probably due to the fracture of crystals during their extraction from the mother-rock. The stones found in the loose weathered material on the floor of the valley are more or less rounded, showing that they have been transported some distance by running water. Some are of considerable size, weighing 100 or even 300 carats.

The crystals of sapphire are often bluish-white or bluish-grey, but specimens of a finer and richer colour are also frequently found. Single crystals often show a difference of colour in different portions; thus the centre of a crystal may be of a fine blue colour, and the two ends colourless. The majority of the stones found here possess, wholly or in part, a milky cloudiness; silkiness of lustre is also a common fault. Only transparent and finely-coloured stones are valuable as gems. Large cloudy crystals often have a small portion clear and transparent, which is carefully cut away by the lapidary and transformed into a gem. The yellow, brown, and red varieties of corundum are rare at this locality.

These mines are not the only places in this remote region where sapphires worth cutting have been found. At some distance away, but still in the same neighbourhood, are several places at which sapphire occurs under exactly similar conditions as far as is known. Thus, stones, which were not at first recognised as sapphires, were brought down from the Sacha Pass to the gem-market at Delhi, and others have been found in the gneiss and mica-schist of the upper Raini valley, below the Hamta Pass in Kulu, Punjab, as well as at other places.

All varieties of precious corundum—ruby, sapphire, " oriental topaz," " oriental emerald," &c.—are found in the **United States** of North America, being specially abundant in two particular regions. The first of these regions includes the western portions of North Carolina and of South Carolina and extends into Georgia and Alabama. Almost all the precious corundum found in this region comes from Macon County in North Carolina, where the crystals, which are usually well developed, are enclosed in an olivine-rock (dunite). The occurrence of corundum in rocks other than dunite in North Carolina, and specially at Cowee Creek in Macon County, has already been dealt with under ruby. In these localities the pure mineral often forms the nucleus of large masses of common corundum. In the Culsagee mine on Corundum Hill, near Franklin in Macon County, a crystal weighing 311 pounds was once found. This, however, was not of gem quality and was coloured partly red and partly blue. At the same mine rubies, sapphire, "oriental topaz," and a few "oriental emeralds," &c., suitable for cutting as gems, have been found. Fine star-stones occur here also, as well as in Delaware County, Pennsylvania.

The other region which is specially rich in precious corundum is situated in the west. Sapphire and other colour-varieties of corundum have been known since 1865 to occur in the neighbourhood of Helena on the upper reaches of the Missouri river, in the State of Montana, being first discovered during the process of gold-washing. Again the true nature of the stones was not at first recognised, and they were sold at much below their actual value. Since 1891 these deposits have been systematically worked for gold, and at the same time large numbers of the precious stones have been collected. They are found in masses of glacial débris known as "bars," which are laid down on the sides of the valleys parallel to the river-courses and at a height of 300 feet above the present high-water level of the upper Missouri. These glacial sands and gravels containing gold overlie black shales, probably of Lower Silurian age, which are associated with limestones, quartzite, and rocks of igneous origin. It is in the lowest layer of these sands and gravels, with a thickness of only a few inches, that the sapphire is principally found. The sapphires are most abundant at Eldorado Bar, Spokane Bar, French Bar, and Ruby Bar, and these deposits are still being worked. Spokane Bar near Stubb's Ferry, twelve miles to the east of Helena, is approximately the central point of this district, which extends along the Missouri for at least fifteen miles and embraces an area of certainly no less than eleven and a half square miles.

The sapphires frequently occur as well-developed crystals, having the form of a short hexagonal prism with basal planes, an unusual type for this gem. Irregular grains are also found which, like the crystals, are more or less rounded. Neither crystals nor grains attain to any considerable size, measuring at the most from $\frac{1}{4}$ to $\frac{1}{2}$ inch in diameter and rarely exceeding 9 carats in weight. Though small in size the stones are abundant in number, as evidenced by the fact that an acre of the deposit at Eldorado Bar yielded no less than 2000 ounces of sapphire. Many of these stones, however, would be unsuitable for cutting, since the predominant tints of the sapphires of this locality are all pale.

The colours, though almost always pale in shade, show great variety of tint, red, violet, yellow, blue, green, bluish-green, and all possible intermediate colours being met with. Bluish-green and green corundum is specially abundant, while the pure blue and the red varieties are absent. Occasionally a stone with a red nucleus and a border of another colour is met with. Some green and blue stones appear red by artificial light. Almost all the colour-varieties of corundum from this region, which are suitable for cutting, have a peculiar metallic sheen, which is very characteristic and is not seen in stones from any other locality. They are remarkable also for the brilliancy of their lustre, and, according to the statements of lapidaries, are specially hard.

Corundum is associated in these glacial sands with many other minerals, among which are crystals of white topaz not exceeding $\frac{1}{4}$ inch in length, fine ruby-red garnets the size of a pea (which have often been mistaken for true rubies), kyanite, cassiterite in small, rounded grains (stream-tin), iron-pyrites altered to limonite, chalcedony, and small rounded fragments of calcite.

As already mentioned, the rocks occurring *in situ* in the district and underlying the gemmiferous sands are penetrated by dykes of igneous material. In one of these dykes, consisting of mica-augite-andesite, crystals of sapphire, garnet (pyrope), and sanidine have been found ; and it has been argued from this that in every case the sapphires originated in similar situations and have been set free by the weathering of the igneous rock. This origin for the sapphire is not universally accepted, although parallel cases may be found in the occurrence of fine blue sapphire in the volcanic rocks of other regions, such, for example, as the basalts of Unkel on the Rhine, Niedermendig on the Laacher See, Calvarienberg near Fulda, and Expailly near Le Puy-en-Velay in France, &c.

More recently sapphires have been found at Yogo Gulch in Fergus County, also in the State of Montana, and seventy-five to one hundred miles east of the Missouri bars. According to G. F. Kunz and others they occur here in a yellow earthy material, which also may owe its origin to the weathering of an igneous rock. The blue stones vary in shade from light to dark, some being of the true sapphire or cornflower-blue, while there are others which incline to an amethyst or almost ruby shade of red. The crystals are rhombohedral in habit, and in this respect differ from the sapphires found near Helena.

The amount of corundum of a quality suitable for cutting which comes into the market from **Australia** is not altogether insignificant. The mineral is found in gold-sands with diamond and in stanniferous and other similar sands and gravels in Victoria, South Australia, Queensland, and especially in New South Wales. In the last named State, sapphires are found in the north-east corner in the New England district, especially in the neighbourhood of Bingera and Inverell, and indeed at all the localities which have been already mentioned for diamond (Map, Fig. 43). Sapphire occurs here under exactly the same conditions as does the diamond, and it is even more widely distributed. An occurrence of the stone in Tasmania has also been recently reported.

Australian sapphires, as a rule, are too dark to be of much value as gems ; they vary from perfect transparency and absence of colour through various shades of blue and grey to almost absolute opacity and dark blue colour. Crystals showing a fine sapphire-blue colour are met with occasionally, and fine star-sapphires are not uncommon. A few rubies are found, but corundum of a fine green colour, that is " oriental emerald," is more abundant, every hundred stones always including two or three specimens of " oriental emerald." The original crystalline form of the stones, a hexagonal bipyramid (Fig. 53, *e*, &c.,) is frequently well preserved, but more often they are in the form of irregular grains or rolled pebbles, like the other constituents of the sands. From a commercial point of view the Australian output of sapphires is unimportant.

In **Europe** a well-known and often mentioned locality for sapphires is the Iserwiese, the district in which is the source of the Iser river, which drains the Iser mountains in northern Bohemia. Sapphire, together with ceylonite, zircon, garnet, and iserine, is found here in loose, alluvial material derived from the weathering of granite. The sapphires sometimes occur as small hexagonal prisms, but more often as water-worn grains of various shades of blue and with various degrees of transparency. While pale blue stones are usually cloudy and opaque the darker ones are, as a rule, transparent. Single sapphires of the finest quality are said to have been found here, all, however, of small size ; stones over 4 carats in weight are extremely rare. The deposit, never very extensive, has been systematically worked for many years and is now practically exhausted.

Single stones suitable for cutting have also been found in the garnetiferous sands of Meronitz in Bohemia, in the auriferous sands of Ohlapian in Transylvania, of the Urals, of Madagascar, of Borneo, and of some other regions. It is, however, unnecessary to give a detailed account of the occurrence at these localities since in each case the stones are present in such small numbers.

COUNTERFEITING.—The blue stones which may be mistaken for, or passed off as, sapphire are cordierite ("water-sapphire"), kyanite (sapparé), blue tourmaline ("indicolite"), blue topaz, and blue spinel.   Amongst such stones may perhaps be included also haüynite, blue diamond, and aquamarine, which in some cases may resemble the sapphire.   All, however, without exception, differ from the sapphire in density, most of them being considerably lighter and floating in the heaviest liquid, while corundum sinks heavily; spinel and kyanite alone have a density near that of this liquid (sp. gr. = 3·6).   With the exception of diamond these stones, too, are all considerably softer than corundum, by which they can easily be scratched; many of them, indeed, may be scratched even by topaz.

Blue tourmaline, moreover, may be distinguished from sapphire by the difference in the tone of its colour, which is an indigo-blue.   Kyanite again is characterised by the existence of a system of fine rectangular cracks, which are absent in the sapphire, and are due to the presence of perfect cleavages and twinning.   The blue of kyanite, however, is very similar to that of sapphire, hence the name sapparé, but its transparency is less perfect. Cordierite is characterised by its very strong dichroism, far stronger than that of sapphire. The specific gravity of topaz is its most salient distinguishing feature.   Diamond, spinel, and haüynite are singly refracting, and show no dichroism; the same is true also for blue glass, but this substance may be recognised also by its softness.

The colour of sapphire is easily imitated in glass by adding to the strass a little cobalt oxide, one part of cobalt oxide to seventy or eighty parts of strass giving a very fine sapphire-blue colour.   The same effect may be produced under certain conditions by the use of iron.   Thus, chemical analysis has shown that the beautiful blue colour of an antique vase, ornamented with bas-relief in white, now preserved in the British Museum, is due not to cobalt but to iron.   The blue colour of the slag from iron furnaces is also due to the same metal.   It is not, however, customary at the present day to use iron for the purpose of colouring glass.

White sapphire, diamond, colourless spinel, zircon, topaz, rock-crystal, and phenakite, as well as colourless strass, may each be mistaken the one for the other by the uninitiated. Of these, sapphire, zircon, and spinel sink slowly in the heaviest liquid (sp. gr. = 3·6), while all the rest float.   Diamond only is capable of scratching leuco-sapphire, while this itself scratches all the others.   Glass, diamond, and spinel are singly refracting, and can thus be distinguished from the other stones mentioned.   Taking into account all these differences, it should not be a matter of great difficulty to distinguish a colourless sapphire from the colourless stones it somewhat resembles.

OTHER COLOUR-VARIETIES OF PRECIOUS CORUNDUM.

In addition to the true or oriental ruby and the oriental sapphire there are other varieties of transparent corundum, which are distinguished from these and from each other solely by their colours.   We have already seen that these varieties are known by the name of some precious stone which they resemble in colour with the qualifying prefix " oriental." Thus certain of these colour-varieties of corundum are referred to as " oriental aquamarine," " oriental emerald," " oriental chrysolite," " oriental topaz," " oriental hyacinth," and " oriental amethyst."   The precious stones from which these varieties take their names are sometimes given the prefix " occidental "; all are softer than corundum and are easily scratched by it.   With the exception of zircon (hyacinth), the specific gravity of which is greater than that of corundum, all the " occidental " are lighter than the " oriental " precious stones; while the latter sink heavily in the heaviest liquid (sp. gr. = 3·6) the former float in this, some indeed floating in pure methylene iodide. Very little familiarity with the appearance of "oriental" and "occidental" precious stones enables one to distinguish the former from the latter solely by the difference in lustre; and this difference has led to the term "oriental," conveying by its use an impression of great hardness and brilliant lustre in the stone to which the term is applied.

None of the colour-varieties of precious corundum now under consideration are abundant in nature. They occur as more or less isolated examples, together with ruby and sapphire, at the localities where these precious stones are found, namely, in Burma, Siam, Ceylon, Montana, North Carolina, &c. Together with the ruby and sapphire they are collected from the various deposits, and are cut and mounted in the same manner as are these stones, so that further comment on this subject is superfluous.

"**Oriental aquamarine**" is pale bluish-green or greenish-blue in colour, and resembles in this respect, and also in transparency, the variety of beryl known as aquamarine. "Oriental aquamarine" sometimes inclines most to green and other times to blue; specimens are also met with of a dark greenish-blue colour, a transition shade between the colour of the sapphire and that of the aquamarine. Such stones are remarkable for their specially strong dichroism.

"**Oriental emerald**" is corundum of a more or less intense green colour resembling that of the emerald, another colour-variety of beryl. While the "oriental emerald" always shows a tinge of yellow, and is thus inferior to the true emerald in purity and depth of colour, it surpasses the latter in transparency and lustre. So rare is this variety of corundum that its very existence has been doubted, and it has been suggested that the supposed specimens of "oriental emerald" are in reality true emerald or beryl. This idea, however, is negatived by the well-established occurrences of the stone not only in Burma, Siam, and Ceylon, but also in New South Wales, Montana, and at the Culsagee mine in Macon County, North Carolina, where a crystal measuring 100 by 50 by 35 millimetres was once found. On account of its great rarity the "oriental emerald" far surpasses in value the finest sapphires, but falls short of the value attached to the ruby. It is distinguished from the true emerald by its greater hardness and specific gravity, and by the fact that it is much more markedly dichroic; the two colours shown by the dichroscope are blue and green. This variety of corundum sometimes varies in colour according as it is viewed in reflected or transmitted light. Thus a stone from Chantabun, in Siam, appeared in reflected light of a deep bottle-green and in transmitted light of a bluish-violet colour.

"**Oriental chrysolite**" is of a pale yellowish-green colour; its tint is more yellow than that of the last variety of corundum considered, and corresponds very closely to that of chrysolite (olivine) or to pale coloured chrysoberyl. It is much commoner than "oriental emerald." Clear and transparent greenish-yellow chrysoberyl, with no chatoyant lustre, is sometimes referred to as "oriental chrysolite"; it is distinguished from chrysolite proper by its much greater hardness.

"**Oriental topaz**" ("topaz-sapphire," yellow sapphire) is of a pure yellow colour. The value of this stone depends upon the particular shade of its colour; specimens of a saffron-yellow tinged with red or of a pure citron-yellow are most highly prized. In the majority of stones the colour is a pale straw-yellow, or it may incline to green or brown; in the former case it approaches the colour of "oriental chrysolite." Precious corundum with a more or less pronounced yellow colour is fairly common; the finely coloured "oriental topaz," however, is much rarer, and being both rarer and more beautiful in colour than "oriental chrysolite" is more highly prized. The price of stones showing great depth and intensity of colour is as high as that of the finest sapphires. A not uncommon fault in these stones is the existence of a peculiar, avanturine-like, glittering appearance, probably due to the presence of small enclosures. The appearance which gives their names to star-ruby and star-sapphire is sometimes seen in " oriental topaz," which is then referred to as " asteriated topaz " or " topaz-cat's-eye." Tavernier states that he saw among the jewels of the Great Mogul an " oriental topaz " of 157¾ carats, which he valued at 271,600 francs (£10,777). Another stone of this kind weighing 29 carats was in the possession of the Parisian jeweller, Caire. It was remarkable for the Arabic inscriptions it bore, not engraved merely on the surface but penetrating the whole thickness of the stone, and was probably an Eastern amulet.

" **Oriental hyacinth** " (" vermeille orientale ") varies in colour from pale aurora-red to reddish-brown. The presence of a pronounced tinge of yellow or brown makes its colour very different from that of the ruby. This colour-variety of corundum is not an important one ; it sometimes shows the sheen already mentioned as being present in " oriental topaz." Its specific gravity of 4·0 distinguishes it from true hyacinth (zircon), the specific gravity of which is 4·6 to 4·7.

" **Oriental amethyst** " (violet ruby, " amethyst-sapphire " or purple sapphire) is violet in colour and is of more importance than the last-named variety of corundum. Its tint is often of a bright violet-blue, closely resembling the various shades of colour of the true amethyst (a variety of quartz). Sometimes, however, its colour inclines to rose-red or purple, and when this is the case the stone appears either like certain almandine-garnets or like certain spinels. This stone, therefore, may be of almost any shade of colour between the red of the ruby and the blue of the sapphire. It is distinguished from the true amethyst by its strong dichroism, which is apparent even to the naked eye. The light which reaches the eye along the axis of the crystal and out by one of the basal planes is of a warm violet colour, while that which travels through the crystal in a direction perpendicular to this is pale and almost colourless. This is a point which must be remembered when the stone is cut as a gem, the lapidary arranging that the table is parallel to the basal planes of the crystal, otherwise the stone will appear pale and insignificant.

The " violet ruby," which by daylight always appears more or less red, has a still more pronounced colour, and is even more beautiful by candle-light. Caire has described such a stone, which was blue like the sapphire by day and of a fine purple-red by artificial light. We may contrast with this the dull grey appearance of the true amethyst in candle-light. The Maltese Cross, shown in Plate III., Fig. 8, is the form of cutting best suited to the " oriental amethyst " ; it is cut, however, in all the forms employed for other coloured stones including ruby and sapphire. An " oriental amethyst " of a full deep colour is worth approximately as much as a good sapphire.

All the varieties of corundum hitherto considered are clear and transparent. Cloudy and opaque corundum, when it possesses some beautiful feature, such as a fine colour, may be cut as a gem. A case in point is that of **adamantine-spar**, a semi-transparent, hair-brown corundum, the basal planes of some crystals of which show, like star-stones, a beautiful bluish-white sheen. When such a stone is cut *en cabochon* in this direction it presents an appearance very similar to that of asteriated ruby. China is considered to be the principal locality for adamantine-spar ; it is also found at other places together with precious and common corundum.

SPINEL.

The precious stones which are most appropriately considered after corundum are those belonging to the spinel group of minerals.   Their colour often resembles that of the ruby, but in all other characters the two minerals are perfectly distinct, so that the names " ruby-spinel " and " balas-ruby," which are sometimes given to certain colour-varieties of spinel, are misleading and incorrect.   Spinels of a sapphire-blue colour are also known, but, like black spinels, they are of little importance.

In scientific mineralogy the spinel group includes a very large number of minerals of varied composition but of identical crystalline form and chemical constitution.   Of all these minerals of the spinel group, differing widely from each other in chemical composition, hardness, colour, transparency, &c., there is but one which is generally used as a gem, and this particular stone is therefore distinguished as *precious spinel*, or noble spinel.

Precious spinel is a compound of alumina (the sole constituent of ruby) with magnesia, its composition being represented by the formula $MgO.Al_2O_3$ or $MgAl_2O_4$.   This compound is in itself colourless, so that the various colour-varieties of spinel owe their tints to the presence of small quantities of foreign substances.   The colour both of the ruby and of the red spinel is thought to be due to the presence of chromic oxide (Cr_2O_3), but, according to Doelter, the presence of the small amount of iron, as shown in the following analysis, is sufficient to account for the colour of these stones.   The analysis referred to was made by Abich on a red spinel from Ceylon.

Alumina (Al_2O_3)	70·43 per cent.
Chromic oxide (Cr_2O_3)	1·12 ,,
Magnesia (MgO)	26·75 ,,
Ferrous oxide (FeO)	0·73 ,,

Spinel crystallises in the cubic system, and the form of the crystals can often easily be made out, even when they are very much rounded and water-worn.   This form is most

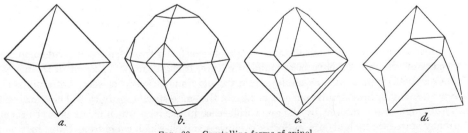

a.   b.   c.   d.

FIG. 60.   Crystalline forms of spinel.

frequently an octahedron (Fig. 60 *a*), uncombined with other forms and usually with faces developed on all sides.   Fig. 60 *c* illustrates the truncation of the edges of the octahedron by faces of the rhombic dodecahedron, the result of which is a not uncommon form.   A less frequent form, except in the case of black spinel, is shown in Fig. 60 *b*, in which each of the corners of the octahedron are replaced by four faces of an icositetrahedron.

Twin-crystals of spinel are common. In Fig. 60 d, the two individuals are united in such a manner that one face of the octahedron is common to both, and the two individuals are symmetrical about this face. The other octahedral faces form alternate salient and re-entrant angles at the plane of junction. The twinning together of two octahedra in this manner is of such frequent occurrence in spinel that such compound forms, which occur in several other minerals, including the diamond, are said to be twinned according to the spinel-law. The two individuals of a spinel-twin have often very little thickness in the direction perpendicular to the common octahedral face. The crystal, then, has the form of a triangular plate with alternate salient and re-entrant angles at the edges. One of the individuals of the twin may be twinned in the same way with a third octahedron, and this again with a fourth, and so on. This may give rise to very complicated groups.

Spinel has no cleavage, or, if any, a very imperfect one, and its fracture is irregular to conchoidal. The mineral is brittle and hard, being the fourth hardest precious stone known ; it immediately follows chrysoberyl, and is very little harder than topaz, the number of which, on Mohs' scale of hardness, is 8. The specific gravity is fairly high (sp. gr. = 3.60 to 3·63), not far removed from that of diamond. Spinel becomes positively electrified when rubbed, but acquires no pyroelectrical charge when heated or cooled. It is unattacked by acids, and infusible before the blowpipe.

The lustre of spinel, though of the common vitreous type, is very brilliant, especially on polished facets. It is suceptible of a very high degree of polish, not, however, equalling the ruby in this respect. It is stated to be possible for an expert to distinguish a ruby from a spinel by the difference in brilliancy of lustre alone. Some spinels are beautifully clear and transparent, while others are cloudy and opaque, the latter being, of course, valueless as gems. All varieties of spinel, since they crystallise in the cubic system, are singly refracting, that is to say, optically isotropic. Thus the polariscope offers a simple and ready means for distinguishing a spinel from a ruby. The refractive power of spinel is fairly high, being about the same as that of corundum. The refractive indices for different colours of the spectrum do not differ widely ; they were determined for a pale red spinel as follows :

$$n = 1\cdot71 \text{ for red light.}$$
$$n = 1\cdot72 \text{ for yellow light.}$$
$$n = 1\cdot73 \text{ for blue light.}$$

The dispersion, as in the ruby, is therefore small, and the cut gem produces no marked display of prismatic colours, such as one is accustomed to see in the diamond.

When absolutely pure the substance of spinel is perfectly free from colour, and colourless octohedra of the mineral have been found in nature, though rarely. Except in the matter of lustre, such crystals will resemble octahedra of diamond very closely, both minerals, moreover, being optically isotropic and of nearly the same specific gravity. They are very easily distinguished, however, by the great difference in hardness which exists between them. Dufrénoy stated the weight of a perfectly colourless spinel, which came from India in a cut condition, to be 12·641 grams (61½ carats).

Precious spinel is commonly, however, red in colour, of shades which incline to violet and blue on the one hand and yellow on the other. It is stated that in all red spinels, whatever be their shade of colour, a tinge of yellow reflected from the interior of the stone, especially if it be cut, can be detected, by which they can be distinguished from the ruby. Spinel being optically isotropic is not dichroic, so that, in contradistinction to the ruby, it appears of the same colour in whatever direction it is viewed. Specimens of spinel showing every possible gradation of colour, from deep red to white or colourless, are in existence; stones of the deepest tone of colour sometimes appear almost opaque. In contrast to the ruby the colour of spinel is very uniform, and spots are rarely seen, so that far fewer stones need to be thrown out from a parcel of spinels than would be the case with a similar parcel of rubies. Faults of other descriptions are also less frequent in spinel than in ruby. Like the ruby, the colour of the spinel remains unaltered by heat; at very high temperatures, however, the stone loses its colour, regaining it as it cools, but not passing through an intermediate condition, in which the colour is green, as does the ruby. Experiments in this direction made on valuable stones must be performed with care as the spinel is easily cracked.

The more deeply coloured a spinel is the more highly is it prized, provided, of course, that the depth of its colour does not interfere with the transparency of the stone. The colour of deep red stones is sometimes almost indistinguishable from that of the ruby. Such stones are known as "*ruby-spinels*" ("spinel-rubies") (Plate I., Fig. 10), and are not infrequently sold as rubies. Carmine-red, blood-red, and poppy-red are the shades of colour most admired in the "ruby-spinel." Very fine stones of cochineal-red or blood-red are known to jewellers as "gouttes de sang."

Spinels of a rose-red or light shade of colour inclined to blue or violet are referred to as "*balas-rubies*" (rubis balais) (Plate I., Fig. 9). They not infrequently combine with this character a peculiarly milky sheen which considerably detracts from their value. Stones the colour of which is more decidedly blue or violet resemble, although much paler, some almandines, and are known as "*almandine-spinels*." Violet spinels, which are not too pale in colour, often resemble both the true amethyst and the "oriental amethyst," and indeed have sometimes been put on the market under the latter name. There should, however, be no danger of mistaking the one for the other, since the spinel is optically isotropic and is not dichroic. To distinguish spinel from almandine and other red garnets is more difficult, since both agree in being optically isotropic and in the absence of dichroism. The colour of the garnet, however, is usually much deeper than that of spinel, while, in the case of almandine, the specific gravity (sp. gr. = 4), which is greater, and the hardness (H − $7\frac{1}{2}$), which is less than that of spinel, may be relied upon. Rose-coloured topaz may closely resemble "balas-ruby," but then topaz is doubly refracting and strongly dichroic. Generally speaking, it cannot be said that "ruby-spinel," "balas-ruby," and "almandine-spinel" are very sharply marked off one from another, since stones showing intermediate characters are usually to be found.

The variety of spinel of a more or less pronounced yellow shade of colour is known as *rubicelle*. This variety, which may be hyacinth-red, orange-yellow, or even straw-yellow, is not esteemed very highly. The rubicelle which accompanies topaz and other precious stones in Minas Novas, in Brazil, has received the name of "vinegar-spinel," on account of its yellowish-red colour. At the particular locality at which it is found it is also known as "hyacinth," also on account of its colour. The name "vermeille," though more commonly applied to certain garnets, is also bestowed upon rubicelle of a decided orange-red shade.

Practically the same forms of cutting are used for the spinel as for the ruby. Finely coloured, transparent stones are cut as brilliants, but the step-cut or a mixed-cut is frequently employed, especially for darker stones. Foils of burnished gold or copper are sometimes used for the purpose of improving the colour and lustre of cut stones.

Spinels of good quality and up to 1 carat in weight are of ordinary occurrence; stones, the weight of which lies between 1 and 4 carats, are not infrequent, but large stones exceeding 8 or 10 carats in weight are rare.   As we have already seen, spinels are far more free from faults than are rubies, so that it is not surprising to find that spinels of fine quality are much more abundant than rubies of the same description.

The value of a spinel varies with its colour and transparency, but is always less than that of the ruby.   A perfectly transparent " ruby-spinel," up to and not exceeding 4 carats in weight, is worth, at the most, about half as much as a ruby of equal weight ; thus a 1-carat " ruby-spinel " will fetch from £5 to £7 10s.   The value of larger stones is considerably greater.   " Balas-rubies " and other varieties of spinel of the best quality are about half the value of " ruby-spinels " ; when exhibiting the milky sheen, described above, they are worth still less.

Although small specimens of precious spinel are much more abundant than large stones, yet a certain number of the latter are in existence.   A " ruby-spinel " of $56\frac{1}{16}$ carats, valued at 50,000 francs (£2000), is mentioned in the inventory of the French crown jewels drawn up in 1791 ; also a " balas-ruby " of $20\frac{6}{16}$ carats, valued at 10,000 francs, and two smaller stones of $12\frac{6}{16}$ and 12 carats, valued respectively at 3000 and 800 francs.   Another famous spinel of large size, the " Black Prince's Ruby," is set in the English crown ; it is often, though erroneously, referred to as a true ruby.   Probably the two largest spinels known are those which were shown at the Exhibition of 1862 in London ; both were flawless stones of perfect colour and were cut *en cabochon*.   The one weighed 197 carats, and on re-cutting gave a gem of 81 carats ; the other weighed $102\frac{1}{4}$ carats before and $72\frac{1}{2}$ carats after re-cutting.

With respect to mode of occurrence spinel is closely related to corundum, since it is found under the same conditions and chiefly at the same localities as are ruby, sapphire, and the other varieties of corundum.   Like corundum, spinel is essentially a mineral of the primitive rocks, being found in gneisses and schists, and especially in crystalline limestones interfoliated with gneisses ; it also occurs in limestones formed as the result of contact-metamorphism.   The most important localities for spinel are briefly mentioned below, the reader being referred for descriptive details to the account of these same localities given under ruby and sapphire.

In Upper Burma the different varieties of precious spinel are, in association with the ruby, of common occurrence both in the white crystalline limestone or marble and in the gem-sands derived from this.   Three-fourths also of the precious stones which are brought for sale by the natives from the neighbouring Shan States are spinels.

Next in importance to Burma come the gem-gravels of Ceylon, especially of the interior of the island in the neighbourhood of Kandy.   Fine crystals are not common, but a beautiful, transparent blue spinel, which will be referred to again below, is peculiar to the locality.   Here also the mother-rock of the spinel is a granular limestone.

Spinels are abundant in the ruby mines of Badakshan.   As early as the thirteenth century, spinels, together probably with rubies, were collected by the famous Venetian traveller, Marco Polo, in the province of Balascia on the Upper Oxus.   This province is identical with the Badakshan of the present day, and the term " balas-ruby " is said to be derived from the place-name Balascia.   Spinels are found in association with the rubies, which are sold in Tashkent, and which are said to come from the Tian-Shan Mountains ; also with those found at Jagdalak in Afghanistan, and with the rubies and sapphires of Siam.

Spinels occur in the gem-gravels of Australia, usually as rounded grains.   Here again they are specially abundant in the New England district and the Cudgegong and Macquarie rivers of New South Wales; also in Victoria, for example in Owen's river, and in other States. Single stones have been found also in Tasmania.

The so-called " vinegar-spinel" has already been stated to occur in the gem-gravels of Minas Novas, in Brazil. In the United States of North America, a few spinels suitable for cutting, although rather dark in colour, have been found at Hamburgh in New Jersey. The sapphires of Montana are remarkable in that they are not accompanied by spinels. Other localities for spinel are of even less importance and need not detain us.

Blue Spinel.—Among the colours of spinel, blue is not very prominent; stones of this colour exist, however, and may be referred to as " sapphire-spinels " in correspondence with the term " ruby-spinel" applied to stones of a ruby-red colour. The colour of these stones is due to a small amount of ferrous oxide in combination with alumina, which is present in addition to magnesia. Blue spinel occurs, often in large crystals, at Åker in Södermanland, Sweden; at this locality, however, the mineral is often opaque and scarcely of gem-quality. Transparent stones occur as isolated specimens, together with red spinel, at the localities mentioned above, that is to say, in Burma and especially in Ceylon. In the latter island are found very beautiful, dark blue octahedra. The lustre of these stones when cut as gems, although inferior to that of sapphire, is still very brilliant. In beauty of appearance they do not fall far short of sapphires, and they command a good price.

Black Spinel (ceylonite or pleonaste).—In this variety of spinel a part of the alumina is replaced by ferric oxide and the greater part of the magnesia by ferrous oxide, so that instead of a magnesia-alumina spinel, we get an iron-alumina spinel, the chemical composition of which may be represented by the formula $(Mg,Fe)O.(Al,Fe)_2O_3$. Being only another variety of spinel, its crystalline form is the same as that already described for precious spinel; the combination shown in Fig. 60 b is, however, commoner in this variety than in others. Ceylonite is greenish-black in mass and dark green in thin layers; like all spinels it takes a good polish and may be used in mourning jewellery. It is found in loose grains, sometimes exceeding an inch in diameter, in the gem-gravels of Ceylon, especially near Kandy; also as small brilliant crystals in some of the ejected blocks of Monte Somma, the ancient portion of Vesuvius. At many other localities it occurs as a contact-mineral in limestones, which have been subjected to the action of masses of molten granite and other igneous rocks. Large octahedra of black spinel, measuring 3 to 4 inches along the axes of the crystals, are found at Amity in the State of New York; small crystals occur in great numbers in the Fassathal in the Tyrol, and at many other places; few, however, are suitable for cutting. Green spinels, some of which are transparent and have been used as gems, have been found in Mitchell County, North Carolina, and in a lead mine in New Mexico.

CHRYSOBERYL.

In contrast to the variety in colour presented by corundum, a species which has furnished the jeweller with an extensive range of precious stones of identical composition, but dissimilar appearance, the mineral we have now to consider, namely, chrysoberyl, is characterised by the absence of any great range of colour, and displays only a few tints of green and its closely-related colour, yellow. As the consideration of the mineral spinel was appropriately made immediately to follow that of corundum, on account of the similarity in the colours of the two minerals, so now the next place is given to chrysoberyl on account of its hardness approximating to that of corundum.

The hardness of chrysoberyl is exceeded only by that of corundum and, of course, of diamond. In Mohs' scale of hardness it is placed between corundum (9) and topaz (8), and has a hardness of $8\frac{1}{2}$ assigned to it, being, therefore, the third hardest mineral known.

Like the other extremely hard minerals, corundum and spinel, chrysoberyl is composed very largely of alumina, containing 80·2 per cent. of this oxide and 19·8 per cent. of beryllia, a compound of oxygen with the metal beryllium, in the same way as alumina is a compound of oxygen with the metal aluminium. This percentage chemical composition corresponds to the chemical formula $BeO.Al_2O_3$. Chrysoberyl is, however, never found in nature in this ideally pure condition; most of the material which has been analysed contains iron in small amounts, while in alexandrite, a variety of chrysoberyl found in the Urals, a little chromic oxide is also present. The following are typical analyses of ordinary chrysoberyl from Brazil and of alexandrite from the Urals:

	Chrysoberyl (Brazil).	Alexandrite (Urals).
Alumina (Al_2O_3)	78·10	78·92
Beryllia (BeO)	17·94	18·02
Ferric oxide (Fe_2O_3) . . .	4·88	3·48
Chromic oxide (Cr_2O_3) . . .	—	0·36
	100·92	100·78

Chrysoberyl crystallises in the rhombic system. Simple crystals are rare, more or less complicated twin-crystals being most frequently met with; the former have the form of short, rhombic prisms in combination with other forms (Fig. 61 *a*). Both simple and twinned crystals are tabular in the direction of a pair of parallel faces, which are striated as indicated in the figure. Two crystals are frequently twinned together in such a way that the resulting form is symmetrical about a plane perpendicular to the striated face, that is to say, each individual of the twin is a reflection of the other in this plane, which is called the twin-plane. In the compound crystal the striated faces of the two individuals are co-planar, and the two sets of striations meet at the twin junction at an angle of very nearly 60° (Fig. 61 *b*, and Plate XII., Fig. 10). It frequently happens that three crystals are twinned together according to this same law, and when they also interpenetrate a very complicated group results (Fig. 61 *c*, and Plate XII., Fig. 8). In such compound forms the boundaries of the individual crystals may be easily traced by the intersection of the three sets of striations at angles of approximately 60°.

Chrysoberyl has no distinct cleavage but a conchoidal fracture. It is brittle and very hard (H = 8½), and has a specific gravity of 3·68 to 3·78. It is unattacked by acids and infusible before the blowpipe. When rubbed it becomes positively electrified and retains its charge for several hours.

The lustre of chrysoberyl is vitreous or slightly inclined to greasy in character; the mineral takes, and on account of its great hardness retains unaltered, a very brilliant polish. With regard to transparency the mineral is very variable, some specimens being beautifully

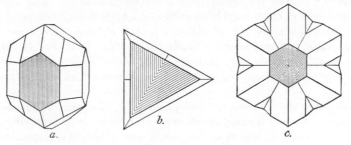

FIG. 61. Crystalline forms of chrysoberyl.

clear while others are cloudy and opaque. The transparency, even of the clearest specimens, is, however, usually only apparent when the stones have been cut and polished, since the mineral is found most frequently in nature as water-worn and apparently opaque pebbles.

It is easy by means of the polariscope to demonstrate the doubly refracting character of transparent fragments of chrysoberyl. The double refraction of this mineral is not, however, very strong, the two indices of refraction having been determined as 1·756 and 1·747 respectively. The strength of refraction possessed by chrysoberyl is thus also rather small, and about the same as that of corundum. The refractive indices of chrysoberyl for the various colours of the spectrum differ from each other but little, so that the dispersion produced is slight and no prominent play of prismatic colours is seen.

Chrysoberyl shows only a limited range of colour; in Brazil, which is the most important locality of the mineral, it varies from pale yellowish-green to golden yellow and brownish-yellow. The Uralian crystals are of an intense green colour, varying in shade from grass-green to emerald-green. There are thus two varieties of chrysoberyl to be recognised, the one of a pale yellowish-green colour, considered as chrysoberyl proper, and the darker emerald-green variety, which is distinguished by the name of alexandrite. The first named is the more abundant and typical variety, the second being comparatively rare and of less importance.

CHRYSOBERYL PROPER AND CYMOPHANE.

The common variety of chrysoberyl is typically pale in colour. Green strongly tinged with yellow (Plate XII., Figs. 10 and 11), more or less bright olive-green, asparagus-green, grass-green, and green inclining to grey or white, are all shades of colour of which examples may be found among chrysoberyls. The more usual green colour sometimes merges into golden-yellow, pale yellow, or brownish-yellow, occasionally even into brown or black. Chrysoberyl is only feebly dichroic, the difference in the two colours of a crystal is not very distinct even when it is examined with the dichroscope, and is barely perceptible to the naked eye. The colour of chrysoberyl is unaffected by heat. Its yellowish-green tints are very similar to those of chrysolite, and for this reason the mineral is often referred to by jewellers as chrysolite. After the greenish-yellow, brownish-yellow tints are most frequently seen in chrysoberyl.

The transparency of the mineral is very variable. Perfectly clear and transparent stones are free from the chatoyant sheen mentioned below and which is of such frequent occurrence in this precious stone. When of a yellowish-green colour it is sometimes referred to in the trade as " oriental chrysolite," as is also corundum of the same colour. Cloudy and opaque specimens of chrysoberyl often exhibit in certain directions a peculiar chatoyant or opalescent sheen similar to that of cat's-eye (quartz-cat's-eye), only usually much finer. This chatoyant variety of chrysoberyl is illustrated in Plate XII., Fig. 11 ; it is known to mineralogists as **cymophane,** and to jewellers as *chrysoberyl-cat's-eye*, *oriental cat's-eye*, *Ceylonese cat's-eye*, more briefly as opalescent or chatoyant chrysoberyl, or simply as cat's-eye.

The chatoyancy characteristic of cymophane and distinguishing it from ordinary chrysoberyl appears as a milky, white, bluish, or greenish-white, or more rarely golden-yellow, sheen which follows every movement of the stone, and is seen to best advantage when the stone is cut *en cabochon*. Extending across the curved surface of such stones is a silvery line or streak of light, especially obvious when the stone is viewed in a strong light. This streak of silvery light may be more or less sharply defined ; its boundaries are usually sharp and clear in small stones, while in large specimens they are often blurred and indistinct, merging gradually into the dark background. These latter are less highly prized than the former, while specimens which show only an irregular patch of light with no sharp borders are still less esteemed. The term cat's-eye should, strictly speaking, be limited to stones which show a sharply defined streak of light, others being referred to as cymophane.

The form taken by the chatoyant reflection depends partly upon the character of the stone, but partly also upon the form in which it is cut ; hence the effectiveness of any particular stone will be increased or diminished by judicious or injudicious cutting. Generally speaking, the greater the curvature of the cut surface the greater is the effect produced. With a slight curvature, the patch of opalescent light broadens out and becomes less well defined, while if the stone be cut with a perfectly plane surface nothing will be seen but a perfectly uniform sheen over the whole surface.

The appearance known as chatoyancy is strictly limited to cloudy chrysoberyl, never being seen in the transparent mineral, and, as a rule, the more cloudy the chrysoberyl the more marked is the chatoyancy. The cloudiness of this mineral is due to the existence in immense numbers of microscopically small cavities in the substance of the stone. Some idea of their abundance may be obtained from the fact that in an area of $\frac{1}{4}$ inch square Sir David Brewster estimated 30,000 of them to be present. Just as in star-sapphires, the optical effect is due to a certain definite arrangement of these cavities, and the stone must be cut with due regard to this arrangement, otherwise the chatoyancy will be enfeebled if not altogether lost.

Chrysoberyl is esteemed for the brilliancy of its lustre and the brightness of its colour. It shows no play of prismatic colours, but in place of this we get the chatoyant effect described above. Of transparent varieties, those which are bright in colour are most sought after, but chatoyant chrysoberyl, that is, oriental cat's-eye, is still more highly esteemed, its value, however, varying with the body-colour of the stone. In this, as in all other cases, the relative value of the different varieties of chrysoberyl is, of course, subject to the caprice of fashion, now the transparent varieties, then the chatoyant cat's-eye being most favoured.

The sudden popularity of chrysoberyl-cat's-eye caused by the use of this stone in the ring given by the Duke of Connaught to his *fiancée*, has been already mentioned. It resulted in a great rise in its price, the stone being now about equal in value to a " balas-ruby."

The finest and largest chrysoberyl-cat's-eye known was included in the famous Hope collection, and is figured by B. Hertz in his *Catalogue of the Collection of Pearls and Precious Stones formed by Henry Philip Hope*, published in 1839. This stone is nearly hemispherical in shape, the diameter in the direction of the splendid chatoyant band measuring 2 inches and slightly exceeding the other diameter. It is dark in colour and the band of light which crosses it is not absolutely perfect, being nicked slightly in the middle. Among other fine stones now known is a magnificent jewel, sold in America, measuring 23 millimetres (nearly 1 inch) in length and breadth and 17 millimetres in thickness, and weighing 80¾ carats. It is yellowish-brown in colour, and the band of light which crosses it is wonderfully sharp, narrow, and straight for so large a stone.

The transparent varieties of chrysoberyl, especially those of a fine golden-yellow colour, are cut usually as brilliants, the step-cut and mixed-cut, however, being sometimes employed. Unless the colour is very intense this gem is seldom mounted *à jour*, more usually a closed setting with a foil of burnished gold to deepen the colour of the stone is employed. The opaque varieties, or cat's-eye, are of course cut *en cabochon*. Such stones are usually cut with an oval outline so that the band of light coincides in direction with the major axis of the oval.

The most important locality for chrysoberyl is **Brazil,** the district of Minas Novas, in the north of the State of Minas Geraes, being specially rich. It is found associated with rock-crystal, amethyst, red quartz, green tourmaline, yellowish-red spinel (so-called " vinegar-spinel "), garnet, euclase, and especially with white and blue (but not yellow) topaz.

Chrysoberyl is one of the finest coloured precious stones found in Brazil. It is known to Brazilians and in the trade generally as " chrysolite." It exhibits considerable variety of colour within certain limits, greyish-white, pale ochre-yellow, citron-yellow, olive-green, grass-green, and pale green being shades of colour ordinarily seen, while pure wine-yellow, greyish-yellow, and colourless stones sometimes occur. The last-named variety closely approaches the diamond in brilliancy and transparency. Stones of perfect transparency are rare, the chatoyancy of cat's-eye being, as a rule, more or less prominent even in the transparent varieties. Chrysoberyl, especially the variety of a pure green colour, is very highly esteemed in Brazil, and consequently commands a high price, being more expensive in that country than in Europe.

The mineral occurs as pebbles, not, as a rule, larger than a bean. A block of supposed chrysoberyl weighing 16 pounds is reported to have been once found, but it is more likely to have been aquamarine (beryl). In spite of the fact that the pebbles are rounded and water-worn, the broad, striated face, which is shown in Fig. 61, and parallel to which the crystals are usually tabular, is often easily recognisable. They are found in an auriferous mud or clay derived from the weathering of granite and gneiss, rocks which are always to be found in the vicinity of deposits containing chrysoberyl. It would appear from this fact that chrysoberyl was originally formed in such rocks, but so far the mineral has been met with only in secondary deposits. Among the principal localities for the mineral may be mentioned the upper course of the Piauhy, and the neighbourhood about the source of the Calhão stream. These deposits, however, are less rich than in former times, and it is said that at the present day material suitable only for the pivot-supports of watches is found there. Chrysoberyl also occurs associated with diamond in small amount in Minas Geraes.

The occurrence of chrysoberyl in the gold-washings of the Sanarka river, in the country of the Orenburg Cossacks in the Southern **Urals** is very similar to that in Minas Novas, but is quite insignificant commercially.   The mineral here is, as a rule, of a fine sulphur-yellow colour, rarely greyish or greenish, and is associated with euclase, rose-red topaz, and other minerals.   It occurs almost invariably in very small pebbles, and is accompanied by a small amount of alexandrite, another variety of chrysoberyl.

Most of the chrysoberyl now cut for gems comes from **Ceylon,** and is of the chatoyant variety.   As these stones are frequently referred to in descriptions of the precious stones of Ceylon simply as cat's-eye, it is often impossible to decide whether chrysoberyl or the variety of quartz, also known as cat's-eye, is meant.   It seems that both the chatoyant and the transparent varieties of chrysoberyl are more abundantly found now than formerly, probably as a consequence of an increased demand and a more exhaustive search.   Among the finest specimens of chrysoberyl, deep golden-yellow, pale yellow, yellowish-green, greyish-green, dark green, greenish-brown, and other colours, may be seen, with or without the chatoyant effect.   Dark green stones show many of the characters, and specially the marked dichroism of alexandrite, a variety of chrysoberyl which will be described in detail below.   The largest chrysoberyl-cat's-eye hitherto known came from Ceylon, and until the year 1815 adorned the crown of the King of Kandy ; the other of the two large stones mentioned above was probably also found in this island.

The stones found in Ceylon vary in weight between 1 and 100 carats, and are found accompanying sapphire in the gem-gravels ; the principal localities are in the district of Saffragam and the neighbourhood of Matura in the south of the island.

Burma (Pegu) has also been given as a locality for chrysoberyl, but its occurrence here is not a well-established fact.   Although the mineral undoubtedly occurs both in India and rather more abundantly in the diamond-washings of Borneo, yet in neither country is the occurrence of any commercial value.

The tendency amongst jewellers to confuse chrysoberyl with chrysolite (olivine) has been already remarked.   This confusion is only possible, however, in the case of specimens of chrysoberyl from which chatoyancy, which is never seen in chrysolite, is absent.   The two minerals differ too in their physical characters ; thus chrysolite, with a hardness between $6\frac{1}{2}$ and 7, is much softer than chrysoberyl.   Again, the specific gravity of chrysolite varies from 3·34 to 3·37, while that of chrysoberyl lies between 3·65 and 3·75, so that although both minerals sink in methylene iodide, in the heaviest liquid (No. 1), the former will float and the latter sink.   Unless examined in some detail the optical characters give little help in distinguishing these two minerals.

Chrysoberyl-cat's-eye may also be mistaken for quartz-cat's-eye, and *vice versâ*, in spite of the fact that the former is usually more brilliant and finer in every way than the latter. Here, again, a difference in the hardness and specific gravity comes to our aid, the hardness of quartz being only 7 and its specific gravity 2·65.   Thus, while quartz-cat's-eye floats in methylene iodide chrysoberyl-cat's-eye quickly sinks.

ALEXANDRITE.

Alexandrite (Plate XII., Figs. 8 and 9) is the name given to that variety of chryso-beryl the colour of which varies between dark grass-green and emerald-green.   The colour is probably due to the presence of a small amount of chromic oxide.   Alexandrite differs from ordinary chrysoberyl also in the fact that it is very strongly dichroic, a crystal or cut stone appearing, when viewed in a direction perpendicular to the broad striated face, not green but a fine columbine-red inclined to violet. In ordinary diffused daylight, however, this colour is not perceptible, and the stone appears always of a green colour (Plate XII., Figs. 8 and 9, *a*). When suitably cut, then, the same stone which by daylight is green appears in artificial light of a red to violet colour; the alexandrite has therefore been described as an emerald by day and an amethyst by night. To accentuate this peculiar character the stone must be cut of a certain thickness, the difference in colour being much less marked in a stone cut with little depth. A crystal of alexandrite when viewed through the dichroscope in a direction perpendicular to the broad striated face gives an emerald-green and a yellow image; when the direction is parallel to this face one of the images is red.

Until comparatively recent times alexandrite was found only in Russia, in the emerald mines on the right bank of the Takovaya, a small stream eighty-five versts (about fifty-seven miles) east of Ekaterinburg in the **Urals,** a locality which will be considered in more detail under the description of emerald. The stone is found, together with emerald and many other minerals, embedded in mica-schist, close to the line of contact of this rock with granite. It occurs usually in star-shaped triplets, identical or very similar in form to that shown in Fig. 61 *c*, and Plate XII., Fig. 8, and consisting of three crystals twinned together. Simple or twin-crystals (Fig. 61, *a* and *b*,) are very rare. The triplets often measure as much as 9 centimetres across, and sometimes even more; they have a tendency to grow together in groups, one such group being found to contain twenty-two large crystals and many small ones of the same kind. The occurrence was accidentally discovered in 1830, on the very day on which the coming of age of the Czarevitch Alexander Nicolajevitch, afterwards Czar Alexander II., was celebrated, and the mineral received its name in honour of this personage. The, at first, exclusively Russian occurrence of the stone and the fact that it combines the national military colours, green and red, gives it a peculiar value in the eyes of Russians, by whom it is worn with great pride. Crystals of alexandrite are, as a rule, cloudy and full of fissures, and are therefore unfit for cutting as gem-stones; they may, however, contain pure and transparent portions free from cracks and markedly dichroic, and it is from such portions that gems are cut. It follows, then, that in Russia, at least, if not elsewhere, where the stone is used but little, much higher prices are demanded for alexandrite than for ordinary chrysoberyl, and the more so as the mines at the present day are almost completely exhausted.

For a long time alexandrite was known only at the locality mentioned, then it was found with pebbles of ordinary chrysoberyl and other precious stones in the auriferous sands of the Sanarka river in the southern Urals.

Still more recently alexandrites have been found in comparative abundance in the gem-gravels of **Ceylon.** These show the characteristic dichroism of the Uralian stones, while some display, in addition, the chatoyancy of cymophane, a feature never seen in specimens from the Urals. The Ceylonese alexandrites are, on the whole, finer than the Uralian, the columbine-red colour seen in artificial light being especially beautiful; the chatoyant stones, moreover, which may be called alexandrite-cat's-eye, are peculiar to this locality. The largest alexandrite yet found in Ceylon weighed 63⅜ carats, and those ordinarily found never weigh less than 4 carats. The large stone just mentioned was cut with double facets (Plate III., Fig. 6), and gave a gem measuring 33 by 32 millimetres at the girdle and with a thickness of 17 millimetres. Its colour by day is grass-green tinged with yellow, and by artificial light a fine raspberry-red. Another beautiful stone from Ceylon weighed 28²⁄₃ carats and measured 32 by 16 by 9 millimetres. Its colour by daylight was a fine sap-green with a trace of red, while in candle-light it appeared of a full columbine-red, scarcely distinguishable from a purplish-red Siamese spinel. Localities for alexandrite, other than Ceylon and the Urals, are not at present known.

BERYL.

The mineral species beryl includes, besides the emerald and the aquamarine, other precious stones of less importance, which are referred to generally by jewellers as beryl. The different varieties of beryl differ only in colour; their other characters are identical, just as ruby and sapphire are mere colour-varieties of the mineral species corundum. It will be convenient to consider the specific characters before passing to a more detailed description of each colour-variety.

The oxide alumina, which enters so largely into the composition of corundum, spinel, and chrysoberyl, is also present in beryl, but in smaller amount and in combination with silica and beryllia. The oxide beryllia, so called on account of its presence in beryl, is also, as we have seen, a constituent of chrysoberyl. Beryl is thus a silicate of the metals aluminium and beryllium, the chemical composition of which is expressed by the formula $3BeO.Al_2O_3.6SiO_2$, and the percentage composition by, silica $(SiO_2) = 66.84$, alumina $(Al_2O_3) = 19.05$, beryllia $(BeO) = 14.11$.

In several analyses of this mineral the presence, in small amounts, of water, iron, alkalies, chromic oxide, and other substances have been determined, while in some beryls, as, for example, the beautiful emerald from Muzo in Colombia, South America, traces of organic matter have been found. The result of the analysis of this stone made by Lewy, together with an analysis by Penfield of aquamarine from Adun-Chalon, in Siberia, is given below. Chromic oxide, which is absent from this specimen of aquamarine and exists as a trace in the emerald, is sometimes present to the extent of 3 per cent.

	Emerald (Colombia).	Aquamarine (Siberia).
Silica (SiO_2)	67·85	66·17 per cent.
Alumina (Al_2O_3)	17·95	20·39
Beryllia (BeO)	12·4	11·50
Chromic oxide (Cr_2O_3) . . .	trace	—
Ferrous oxide (FeO)	—	0·69
Magnesia (MgO)	0·9	—
Soda (Na_2O)	0·7	0·24
Lithia (Li_2O)	—	trace
Water (H_2O)	1·66	1·14
Organic matter	0·12	—

Beryl crystallises in the hexagonal system. The crystals (Fig. 62, *a* to *e*) are usually rather long, six-sided prisms with smooth faces, terminated in many cases, and nearly always in emerald, by a single plane at right angles to the faces of the prism (Fig. 62 *a*), this being known as the basal plane. Not infrequently the edges of the hexagonal prism are truncated by the faces of a second hexagonal prism, and these again by a twelve-sided prism; the resulting form, though in reality a prism, bounded by many small faces (Fig. 62 *d*), has the appearance of a longitudinally-striated cylinder. For this reason the prism faces of beryl are usually striated in the direction of their length, that is, parallel to their mutual intersections. Moreover, in many cases, the crystals are terminated not only by the basal plane but also by six-sided and twelve-sided pyramids in combination with the prism. Fig. *b* shows a hexagonal pyramid of the second order, and Fig. *c* a hexagonal pyramid of the first order, while in Fig. *e* there are two hexagonal pyramids of the first order, one of the second order, and a dihexagonal pyramid in combination with a hexagonal prism and the basal plane. These more complicated crystals are more characteristic of aquamarine.

The crystals are either attached to the matrix by one end, in which case they often form beautiful druses, or they are embedded in it, and are then developed regularly in all directions. In the former case, the free end alone bears regular crystal-faces, while in the latter both ends are developed; these terminal faces are, however, sometimes small and irregular.

The cleavage of beryl is not an important character, crystals of this mineral cleaving only indistinctly in certain directions. The mineral is brittle, and its fracture conchoidal.

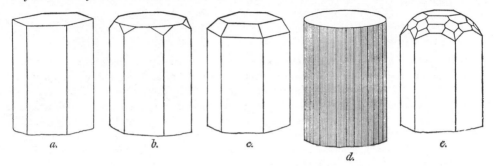

a.   b.   c.   d.   e.

FIG. 62.   Crystalline forms of beryl (emerald and aquamarine).

With respect to its hardness it stands a little above quartz, but below topaz, the degree of its hardness being represented on the scale by $H = 7\frac{1}{2}$; this is rather low for a precious stone, and beryl is, in fact, one of the softer of the more valuable gems. The different varieties of the mineral show small differences in hardness among themselves, the Colombian emerald being, for example, a little softer than the Siberian aquamarine. Although the hardness of this stone is not great, it is sufficient to render it susceptible of a fine polish, which, however, is not retained for as long as is the case with harder stones.

The specific gravity of beryl, like its hardness, is rather low, its mean value being 2·7. That of precious beryl varies between 2·67 and 2·75, being always slightly higher than the specific gravity of quartz (2·65). The specific gravity of certain specimens of emerald from Muzo has been found to be 2·67, while for Siberian aquamarine the values 2·68 to 2·75 are given. Beryl will thus always float in methylene iodide, and when pushed beneath the surface quickly rises again; in liquid No. 4 with the specific gravity of quartz it slowly sinks. This character affords a ready means by which it may be distinguished from certain stones of similar appearance.

With the exception of hydrofluoric acid, beryl is unattacked by acids. When a fragment is heated before the blowpipe it becomes white and cloudy, and fuses, only with difficulty, at the edges to a white, blebby glass.

Different beryls differ greatly in appearance, especially in respect to colour and transparency, some being cloudy and opaque, others beautifully clear and transparent; all possible gradations between these two extremes are known. The opaque variety, known as common beryl, usually occurs as crystals in coarse grained granite. Such crystals have been known to measure 6 feet in length, and to weigh $2\frac{1}{2}$ tons; they are useless, however, for gems, since, besides being opaque, they are usually of an unpleasing yellowish- or greenish-white colour. Only the transparent or semi-transparent precious beryl is used for cutting as gems ; this is usually of a beautiful colour, often green or blue, but sometimes yellow. All varieties of beryl have the common vitreous lustre.

In correspondence with its crystalline form, all hexagonal crystals being birefringent, beryl is doubly refracting, only, however, to a small extent, for the greater and lesser indices of refraction for the same colour differ but slightly. Its refraction is also small. In the case of emerald from Muzo the greater and lesser refractive indices have been determined to be 1·584 and 1·578, and in the case of Siberian aquamarine 1·582 and 1·576. The dispersion is also small, the refractive coefficients given by the same crystal for differently coloured rays of light differing but slightly. This may be seen by comparing the refractive indices of a crystal of beryl for red, yellow, and green light which are given below :

	Red.	Yellow.	Green.
Greater refractive index . . .	1·566	1·570	1·574
Lesser „ „ . . .	1·562	1·566	1·570

It follows, then, that scarcely any play of prismatic colours, such as is characteristic of the diamond, is seen in beryl ; its beauty depends mainly on its strong lustre, and on its fine body-colour. There is a certain amount of variety in this latter character, but very much less than in corundum. Green and bluish-green beryl is most common, yellow rather less so, pale red and water-clear stones rare. Different varieties of beryl are distinguished by their colour ; bright grass-green beryl being known as *emerald*. Other varieties, which are always of a pale colour, are referred to as *precious* or *noble beryl*. Of these light-coloured varieties, the pale blue, bluish-green, or yellowish-blue, is distinguished as *aquamarine*, the yellowish-green as " *aquamarine-chrysolite*," while the yellow variety is known to jewellers as *beryl*, and when of a pure golden-yellow as *golden beryl*. Of all these varieties the emerald is by far the most important as a precious stone, ranking, indeed, with the costliest of gems ; aquamarine is also much used, while the other varieties are of less importance.

All transparent beryls, whatever their colour if not too pale, are distinctly dichroic ; differences in colour can often be observed with the naked eye, and with the dichroscope are, as a rule, unmistakable. On account of this property, it is possible, therefore, to distinguish genuine beryl from coloured glass imitations, or from other gem-stones which it may resemble in appearance.

The characters by which the different varieties of beryl are distinguished must now be considered. These varieties, which differ from each other principally in colour, are by no means equally valuable as gems, the emerald being by far the most costly.

EMERALD.

Emerald is the name given to beryl of a pure and intense green colour ; the particular shade of colour seen in this variety of beryl is often alluded to in ordinary language as emerald-green, but emeralds may be also grass-green, green tinged with yellow, or celadon-green tinged with grey. Beryls of a bluish-green colour are not included in this variety. Many specimens of emerald are very pale in colour, varying in intensity down to greenish-white ; these are, however, not cut as gems ; only those of a beautiful and deep emerald-green to grass-green are highly prized. The particular shade of green characteristic of the emerald is shown in Plate XII., Figs. 1—3 ; it is almost unrivalled in depth and brilliancy, and is often compared to the fresh green of a meadow in spring. The finest stones possess a peculiar velvety lustre, like that shown by some dark blue sapphires.

According to F. Wöhler, the colour of emerald from Muzo in Colombia withstood subjection for an hour to a temperature at which copper readily melts. Although emerald from this locality has been shown by Lewy to contain a small amount of organic matter, its colour cannot, therefore, be due to this, but probably depends upon the presence of chromic oxide, 0·186 per cent. of this substance having been found in the specimen examined by Wöhler. That the deep green of the emerald can be produced by so small an amount of chromic oxide has been proved by fusing white glass with the same percentage of this metallic oxide, a glass having the intense green of the finest emeralds being produced. More recently it has been shown that the colouring-matter of Uralian and Egyptian emeralds also is very probably due to chromic oxide.

Emeralds of good, full colour are distinctly dichroic; the images shown by the dichroscope are respectively emerald- or yellowish-green and bluish-green.

The colour of the emerald is not always distributed uniformly through its substance. The differently coloured portions may occur irregularly or in layers; in the latter case the layers are, as a rule, parallel to the basal plane of the crystal, that is to say, perpendicular to the prism faces.

The transparency of the emerald is perfect only in rare cases. The majority of crystals are rendered cloudy and dull, not only by fissures and cracks, but also by the presence of microscopic enclosures, which in places are accumulated in large numbers. These enclosures may be fluid or solid, scales of mica being specially common. Cloudy and opaque crystals of emerald are usually dull in colour, approaching the characters of common beryl and being useless for cutting as gems. Perfectly clear and transparent stones are naturally the most valuable, but fissured and cloudy specimens, provided they possess the fine emerald-green colour, have a certain value.

Compared with other precious stones, the rarity of perfect specimens of emerald is unique. The most common faults are those which have been just mentioned, fissures being almost invariably present. Stones which are clouded by fissures are described as "mossy." Irregularities in the distribution of colour, and dull and cloudy patches, are also frequently to be seen.

The disparity between the value of a perfect and of an imperfect emerald is enormous. A faultless emerald is worth as much, or nearly as much, as a ruby, and certainly more than a diamond. A one-carat stone, perfect in colour and transparency, is worth at least £20, and large stones, on account of their rarity, have a value out of all proportion to their size. As a matter of fact, a perfect emerald weighing but a few carats is so rare that almost any price will be given for it by collectors. Fissured stones, which are cloudy but of good colour, are much cheaper; when the colour is pale they are worth no more than £5 or even £2 10s. per carat. The value of such stones is more or less proportional to their size, large stones of this description being by no means uncommon.

Flawless emeralds of large size are extremely rare, so that only small stones are available for cutting as gems. Emeralds of considerable size have been known, but their quality leaves much to be desired; moreover, in the case of large stones found and described in early times, it must be remembered that the name emerald was applied to other stones of a green colour. The ancient Peruvians are said to have numbered among their deities an emerald the size of an ostrich's egg. Again, there is reported to be in the treasury at Vienna an emerald which weighs 2205 carats, while Schrauf states that in the same place is preserved an ink-well cut out of a single stone, besides other large emeralds cut as table-stones. One of the largest and finest emeralds known belongs to the Duke of Devonshire. It is a natural crystal of the form characteristic of emerald, namely, a hexagonal prism with a basal plane. This stone (Fig. 63) measures 2 inches across the basal plane, and weighs $8\frac{9}{10}$ ounces or

1350 carats ; it is of the finest colour, clear and transparent, and almost faultless. This stone came from the emerald mines at Muzo in Colombia, where crystals of the length and thickness of a finger are by no means rare. Crystals of equal size are found in the Urals and are not specially rare ; one measuring 8 inches in length and 5 inches in diameter is preserved in the collection of the Imperial Institute of Mines in St. Petersburg, and still larger crystals have been reported. Probably the largest is in the possession of the Czar of Russia ; it is said to measure 25 centimetres (nearly 10 inches) in length and 12 centimetres in diameter. One or two very large stones, formerly thought to be emeralds, have on closer examination proved to be green glass ; such, for example, is one weighing $28\frac{3}{4}$ pounds in the Reichenau monastery above Chur, in the Rhine Valley, Switzerland.

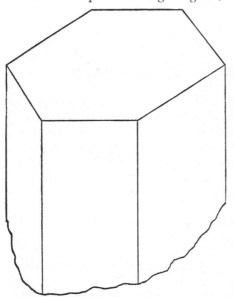

The form in which an emerald is cut depends upon the character of the rough stone. Perfectly faultless, transparent fragments, when not too dark, are cut as brilliants or as rosettes. Most frequently, however, the step-cut (Plate XII., Fig. 3) with brilliant facets on the upper portion is adopted. The emerald, though not infrequently cut as a simple table-stone, is probably never, at least in Europe, cut *en cabochon*. Cut gems, perfect in colour and transparency, are mounted *à jour*; paler stones are provided with a green foil placed beneath them, while fissured or otherwise faulty stones are mounted in a closed setting blackened inside.

Natural crystals of emerald are, as a rule, too large and too much flawed to be cut as single gems; they are, therefore, sawn into portions of suitable size and purity, great care being taken to avoid unnecessary loss of material. In

FIG. 63.   Crystal of emerald belonging to the Duke of Devonshire. (Actual size.)

many crystals, clear and transparent portions suitable for gems have to be cut out of the main mass of the crystal, and in this operation special care is required. Each portion so cut out of a crystal is faceted in the form best suited to its particular shape.

Compared with the precious stones hitherto considered, the emerald, in its mode of occurrence, is unique, for it is found exclusively in its primary situation, that is to say, in the rock in which it was formed. It is one of the minerals characteristic of crystalline schists, and in many places is found embedded in mica-schists and similar rocks. The famous occurrence at Muzo in Colombia is the only exception to this rule, the emerald being here embedded in calcite veins in limestone. This occurrence has called forth the perfectly groundless supposition that the emeralds here were originally formed in crystalline schists and were afterwards deposited in the calcite veins. The emerald practically never occurs in gem-gravels, in the way in which diamonds, rubies, &c., occur.

The earliest known emerald locality is doubtless that in **Upper Egypt,** not far from the coast of the Red Sea and south of Kosseir. Though the occurrence of emerald in Ethiopia was known to the ancients, the locality, in course of time, became completely forgotten, and ancient accounts of the occurrence were regarded as erroneous. It has been supposed that true emeralds were first introduced into Europe at the end of the sixteenth century from South America; they had been found previously, however, both with Egyptian mummies and also among the ruins of the two Roman cities, Herculaneum and Pompeii. These latter, which were discovered long before the end of the sixteenth century (1566), could not have been brought from South America, the most important locality at the present day, but probably came from Egypt or, as also mentioned in ancient writings, from Scythian lands, and thus perhaps from the Urals, where they are still found.

The ancient Egyptian mines were re-discovered in the second decade of the nineteenth century by Cailliaud, a member of the expedition organised by Mehemet Ali Pasha; they have been frequently visited since by European travellers. The workings were partly surface and partly underground, the timbering of the latter being frequently found in a well preserved state. The deposit was worked to a considerable extent, some of the mines being large enough to admit of 400 men working together at the same time. The facts which led these ancient miners to suspect the existence of emeralds in these deposits, and the date at which the workings were commenced, are alike unknown. The appliances and tools which have been found in the mines date back to the time of Sesostris (1650 B.C.). It is recorded in ancient inscriptions that, in the time of Alexander the Great, Greek miners were employed in these mines; and it is evident that they were worked during the reign of Cleopatra, for emeralds bearing an engraving of herself were used for presentation by this queen.

There is no subsequent record of the mines until their re-discovery by Cailliaud, who, with the permission of Mehemet Ali, re-opened them in 1819, the actual work being performed by Albanian miners. Perhaps on account of the poorness of quality of the stones the work was soon abandoned, and apparently with great suddenness, for a number of baskets filled with material ready to be drawn up to the surface have been discovered in the mine just as they were left by the Albanian miners.

These ancient mines are situated in a depression of the long range of mountains which borders the west coast of the Red Sea; in the same range are to be found gold and topaz mines. The emerald mines are in two groups, one being known as the Jebel (= Mount) Sikait (also called Sakketto), and the other, about ten miles to the north, as the Jebel Sabara (Zabara, Zubara, &c.), both being a little south of latitude 25° N. The most important and extensive of the two groups of mines is that of the Jebel Sikait; it is connected with the Red Sea fifteen miles to the east by the Wadi Chamal, and judging from the ruins of houses, temples, and other buildings which are still to be found there, must have been the site of a town of no inconsiderable size. Hundreds of shafts of various depths have been driven into the hill, which is 600 to 700 feet in height. These so-called Cleopatra's emerald mines have recently (1899) been again visited by Mr. D. A. MacAlister with the view of re-working the old mines. The ancient workings on the Jebel Sabara are similar, but less extensive.

The emeralds found here are of a fine, though not very deep, colour; at both places the mother-rock is a dark mica-schist interfoliated with talc-schist, and containing in the Jebel Sikait district augite and hornblende in addition. The mother-rock of the emeralds found in the Urals and in the Salzburg Alps, to be described later, is precisely similar in character.

It has occasionally happened that fine emeralds of a good colour, some cut and some rough, together with other precious stones, have been thrown up by the sea on the beach near Alexandria. These stones are apparently part of a sunken treasure, and probably came originally from the ancient mines in Upper Egypt, being similar both in quality and in the character of the minerals with which they are associated to stones known to have come from these mines.

Emeralds from **South America** were first introduced into Europe at the end of the sixteenth century, and from this period up to the year 1830 all the emeralds which came into the market were brought from this country.

At the time of the Spanish conquest of South America many large and beautiful emeralds were found in the possession of the Peruvians. The mines from which these stones had been derived were probably at the time of the invasion deserted and filled in by the natives, for the search made for them by the Spanish conquerors was altogether unavailing. They are supposed to have been situated in the Manta valley near Puerto Viejo, from whence is said to have come the emerald the size of an ostrich's egg, which was worshipped by the ancient Peruvians as a deity. However that may be, it is certain that at the present day no emeralds are found in Peru.

The number of stones which the Spaniards took from the natives and shipped to Europe must have been enormous. José d'Acosta relates that the ship by which he voyaged from Peru to Spain in 1587 carried two cases, each of which contained no less than a hundredweight of emeralds. This large importation of emeralds from Peru, together with the abundant yield from the mines soon afterwards discovered in Colombia, had the effect of very considerably lowering the price of these stones, which up to then had been so rare in Europe. The South American emeralds were far finer than any previously introduced into Europe whether from Egypt or elsewhere, and hence emeralds of good quality came to be distinguished as " Peruvian," or " Spanish," just as the finest specimens of other precious stones were given the prefix " oriental," whether they came from the Orient or not. Many of the emeralds now in use as gems are the same stones as those brought over to Europe by the Spaniards from Peru. In most cases, however, their shape has been altered from time to time in order to conform to the passing fashion of the day. It is said that the Spaniards were possessed of the idea that a genuine emerald would withstand a blow from a hammer, and that many Peruvian stones were in consequence reduced to splinters by being subjected to this test.

The Spaniards found the natives of Mexico also in possession of very beautiful emeralds, in many cases cut with great skill into peculiar and characteristic forms which are not seen elsewhere. Five stones cut into the shapes of fantastic flowers, fishes, and other natural objects were brought to Europe by Cortez. Since nothing is known as to the natural occurrence of emeralds in this country, it is inferred that the ancient Mexicans obtained the rough stones either from Peru or from the mines in Colombia.

Though the Spaniards were unsuccessful in searching for naturally occurring emeralds in Peru and Mexico, and could obtain the beautiful green stones only from the treasure stored up in the graves and temples of the ancient Peruvians and Mexicans, they were more fortunate in the country now known as **Colombia** or New Granada. Here the deposits from which the natives obtained their stones were easily found, and it is from this same source that the emeralds which now find their way into the markets of the world are for the most part derived.

Besides the Colombian deposits there is no other well-authenticated occurrence of emerald in South America, the existence of the supposed Peruvian deposits being by no means unquestionable. This being so, it has been suggested that the emeralds found by the Spaniards in the possession of the natives of Venezuela and Ecuador, and specially of Peru, were all derived from the Colombian deposits. The term " Peruvian emerald," except when used to describe the quality of a stone, is therefore misleading, South American emeralds being more strictly described as Colombian. Whether emerald mines ever existed in Peru and other parts of South America or not, it is certain that at present the Colombian are the only deposits known.

The Spaniards first learnt of the existence of Colombian emeralds on March 3, 1537. A gift of emeralds was offered to the Spanish conquerors by the Indians, who, at the same time, pointed out the source from which the stones were derived. This spot, known as Somondoco, a name still in use, lies nine leguas (about twenty-three miles) distant from Guatequé, close to the waterfall of Nagar, over which the Garagoa flows before joining the Guario, a tributary of the Upia, which in its turn feeds the Rio Meta. The place is situated on the eastern slopes of the Cordillera of Bogotá, in latitude about 5° N. and about half a degree east of Bogotá (formerly Santa Fé de Bogotá) the capital of Colombia. The wild and inaccessible nature of the region soon drove the Spaniards to abandon the workings in spite of the richness of the deposit. No exact records of this occurrence and of the situation of the old mines are in existence, and doubt has sometimes been thrown on the authenticity of the occurrence. It is probable, however, that the majority of emeralds mined in Colombia in former times came from this spot. The deposit at Somondoco is now (1901) being worked by an English company, but as yet only second quality stones have been found.

A short time after the discovery by Europeans of Somondoco as a locality for emeralds, another, about 100 miles distant, richer and of greater importance than any now known, was discovered. This locality is the only one in Colombia at which fine emeralds are now met with. The stones occasionally found in ancient graves or mountain lakes, which latter were the sites of votive offerings, are all of poor quality, while the naturally occurring stones are frequently of admirable colour and transparency.

The mines now under consideration are situated in the country of the wild Muzo Indians, who for a long while successfully resisted the Spanish attempts at conquest. They were partially subdued in 1555 by the Spanish under Luiz Lanchero, who, in the same year, founded the town of Santissima Trinidad de los Muzos, the present village of Muzo, in the Itoco Mountains. This latter name was at that time applied to the town itself as well as to the mountains.

In spite of the continued hostility of the Indians, the mining of emeralds was commenced in 1558, an old mine in the mountains, of which all trace is at present lost, being first worked. Later, the centre of the workings was situated about a legua (about two and a half miles) from Muzo, work being commenced here in the year 1594. Numerous other mines were opened in the same district in this year, but were afterwards abandoned for various reasons. Some have been reopened and are being worked at the present day.

The district is situated in the Tunka valley in the eastern Cordilleras of the Andes, which branch away near Popayan from the main chain and stretch along the right or east bank of the Rio Magdalena in its northward course. It is a wild, mountainous region, and its inhospitable character, combined with the hot, damp, and unhealthy climate, renders the search for emeralds anything but an easy task. Crystals of emerald, not, however, of very good quality, have been found at not a few places in this region, so that it is probable that many other emerald localities are still to be discovered.

During the period which has elapsed since the discovery of these deposits they have been worked with varying success in many different spots; at one time under Government direction, at another by private enterprise. At one time service in the mines was made compulsory for the neighbouring Indians, and this short-sighted policy resulted in so serious a depopulation of the country that mining operations were appreciably hindered by lack of workers. The deposit was at first worked in underground levels; later, open workings were adopted, partly in order to render possible a stricter supervision of the workers and to avoid the loss through thieving of a large proportion of the output. The yield of these mines was on the whole small and extremely variable, the labour of several months being sometimes unrewarded by a single find, while on the other hand emeralds to the weight of 100,000 carats would be found in one day. The places at which workings are commenced are chosen entirely at random, for there is nothing to indicate the probability of one spot being more favourable than another. No really reliable statements as to the total yield of the mines are obtainable, but as far as is known it is very variable; thus, for example, in the year 1849 it averaged 12,400 carats per month, and in the 'fifties 22,386 carats per annum.

The most important mine at the present day is situated one and a half leguas to the west of Muzo, in latitude 5° 39′ 50″ N., and longitude 74° 25′ W., of Greenwich; it is about 150 kilometres (ninety-four miles) NNW. of Bogotá, and lies 878 metres (2897 feet) above sea-level. It has been worked for a long period, but not uninterruptedly; it ceased, for example, in the middle of the eighteenth century in consequence of a serious fire, and was only recommenced in 1844. The mine has been at one time worked by the Colombian Government, at another leased to natives or to European companies. In the interest of the whole locality much secrecy is observed in the granting of such leases, so that there are many points on which it is impossible to get information. From 1849 to 1861 the mine was worked by an English company, who paid the Government for this privilege 14,200 dollars and 5 per cent. of the net profits. From 1864 to 1875 a French company, under the direction of Gustav Lehmann, paid the Government 14,700 dollars per annum for permission to work all the mines. The number of workers employed in the mines has varied at different times from 100 to 300. The stones were at first sent to London, but later were placed on the market at Paris.

There are several detailed descriptions of the most important of these mines, which agree among themselves very completely. From them we learn that it is situated on the left side of a small mountain valley called Minero, or at the present time Carare, which joins the valley of the Magdalena river towards the north-east. It is 60 metres above the bottom of the valley, and has the form of a funnel, the upper diameter of which measures 200 metres, and the lower 50 metres. On the side towards the mountain it reaches a depth of about 120 metres, but on the opposite side only 20 or 30 metres; its walls are very steeply inclined. The rock in which the mine is excavated is a dark bituminous limestone; this rests on red sandstone and clay-slates, and contains ammonites, which show it to be of Lower Cretaceous (Neocomian) age.

The emeralds are found in this rock in " horizontal veins," or, more correctly speaking, in single nests embedded in calcite, which is either dark and bituminous or water-clear like Iceland-spar. A crystal of emerald in such a matrix is represented in Plate XII., Fig. 1. Associated with it are very fine crystals of quartz, some water-clear and others green; also brilliant, well-developed crystals of iron-pyrites, having the form of pentagonal dodecahedra, green gypsum, rhombohedra of black dolomite, and finally crystals of parisite, a fluo-carbonate of cerium and other rare metals, named after Paris, by whom the mine was re-discovered, and who held a lease of it for many years.

The emeralds are usually of the finest dark green colour, but paler or almost colourless stones are also found, as well as crystals which are quite black, the latter being specially remarkable for their velvety lustre. Occasionally crystals are found which are green on the exterior but colourless inside. The crystals are classified for trade purposes according to transparency and depth of colour. Almost all have the simple form of a six-sided prism with basal plane (Figs. 62 *a*, and 63). They rarely exceed the size of a man's thumb, and are usually smaller. Frequently they are broken across in one or more places, the cracks being filled up with thin layers of calcite, so that as long as the crystal remains in its matrix it appears whole and unbroken, the fracture only becoming evident when the crystal is detached from the matrix. Together with the well-formed crystals are found rounded fragments of emerald, a fact which affords a certain support to the theory that the calcite veins are not the primary situation of the emeralds, but that they have been washed into these veins from gneissic or granitic rocks.

Some crystals have the peculiarity of falling to pieces, with no apparent cause, after being taken from the mine. It has been sought to avoid this by placing the emeralds when first unearthed in a closed box, thus protecting them for a few days from the action of light and allowing them to dry slowly. This device is not, however, as a rule, successful. Moreover, most of the emeralds which come from the mine clear, transparent, and free from fissures, in course of time lose their transparency and assume the usual turbidity of this stone owing to the development of fissures within them. In this connection we may recall the statement that emeralds, both from this and from other localities, only acquire their own particular hardness after they have been taken from the mine for some time. Fine emeralds suitable for use as gems are known in Colombia as "canutillos," and poorer stones as "morallion."

Since there is nothing to indicate in what part of the mine emeralds are likely to occur, the workers simply loosen blocks of rock from any part of the walls until a nest of emeralds, the presence of which is indicated by green quartz crystals, is met with. This is then carefully broken out and taken away. The material loosened from the walls of the mine is allowed to fall to the bottom, and when it has accumulated to a certain extent it is washed out into a canal constructed for this purpose, by the sudden fall of a head of water stored on the heights above the mine. The canal empties itself into the Minero, which carries the mine débris still further away. This is the method adopted at the present day, but in former times when the Spaniards were in possession the workings were all underground.

The only other locality of importance for emeralds beside Colombia is the **Ural Mountains.** There is only one mine from which emeralds are obtained, the same in which Uralian chrysoberyl, that is to say, alexandrite, is found. It is situated on the right bank of the Takovaya, a tributary stream of the Bolshoi Reft (that is to say, Great Reft) which flows into the Pyshma, eighty-five versts (about fifty-seven miles) east of Ekaterinburg.

The finest of the Uralian emeralds are quite equal in transparency and beauty of colour to South American stones. In this, as in other localities, perfect crystals are rare, the majority being fissured and opaque, and the colour, in many cases, being irregularly distributed or too pale. The crystals commonly have the form of a hexagonal prism, often terminated irregularly, but sometimes with a basal plane like Colombian crystals; other forms scarcely ever occur. In size Uralian emeralds often exceed those from other localities, especially South America. Some of exceptional size have already been mentioned; the largest have a length of 40 centimetres (15¾ inches), and a thickness of 25 centimetres, but they are not, as a rule, of good quality.

The mode of occurrence of the emerald in the Urals is similar to that in Egypt, but differs from that in Colombia. The stones are found embedded in a mica-schist, which is interfoliated with chlorite-schist; this is also the mode of occurrence at Habachthal in the Salzburg Alps. A crystal of emerald in a matrix of mica-schist from the latter locality is represented in Plate XII., Fig. 2. Scales of mica are found on the surface of, or inclosed in, emerald crystals from all these localities. Uralian crystals occur singly or in groups; they are often grown together in parallel position, but occasionally radial aggregates of columnar crystals are met with.

The emerald locality on the Takovaya was discovered accidentally in 1830, a peasant noticing a few small green stones among the roots of a tree torn up by the wind. These stones were picked up and taken to Ekaterinburg, where the gem-cutting works of the Czarina, Catharine II., had been established as far back as 1755, and where the many beautiful stones found in the Urals were and are still worked. After this the locality was carefully examined, and a mine opened in the mica-schist. With the emeralds found in this rock are associated alexandrite, phenakite, apatite, rutile, fluor-spar, and other minerals, besides another variety of beryl, the pale-coloured aquamarine. Emerald was at first comparatively abundant at this locality, the yield, however, has gradually fallen both in quality and quantity, and the mines are by no means in full work. They have recently (1900) been rented by a British company; the venture, owing mainly to theft of stones, has not, however, been successful.

This is the only spot in the Urals at which emeralds occur in large numbers. Only once has a finely coloured and transparent stone been found elsewhere, namely, in the gold-sands in the valley of the stream Shemeika, in the Ekaterinburg mining district. The emeralds reported by the ancients to come from Scythian lands may actually have been found in the Urals, but nothing is exactly known as to their origin.

The occurrence of emerald in the **Salzburg** Alps is similar to that at Takovaya; the crystals found at the former locality are, however, smaller, and their lustre less brilliant, so that from a trade point of view they are unimportant. The spot at which they are found lies above the Sedlalp (or Söllalp) on a steep wall of rock, the "Smaragd-Palfen," on the east slopes of the Legbach ravine, a side branch of the Habachthal. It is 7500 feet above sea-level and very inaccessible. The deposit is not rich enough to justify any extensive .workings; the stones have been mined by irregular methods and at the risk of the workers' lives for a long period, it is said since the time of the Romans.

Here also the form taken by the emerald crystals is that of a hexagonal prism, and on the surface of the crystals scales of mica and needles of black tourmaline are often to be seen. In colour the crystals are sometimes of a fine, dark emerald-green, but more often of a pale grass-green or greenish-white; moreover, the colour is frequently irregularly distributed. Perfectly transparent crystals are rare, the majority are turbid, semi-transparent, or translucent to opaque, only a small proportion being fit for use as gems. The white or pale-coloured crystals are usually larger and purer than the green ones. The crystals vary from a line to an inch in length and from ⅛ to 3 lines in thickness; stones exceeding these dimensions are exceptional. The mother-rock in which the crystals are embedded is a finely granular mica-schist, dark-brown or greenish in colour and resembling clay-slate; it is interfoliated with a green mica-schist, which is rich, in some places, in chlorite, and in others in hornblende. A crystal from this locality is shown in its matrix in Plate XII., Fig. 2. Iron-pyrites occurs in association with the emeralds. The finest and largest stones are said to be found in comparatively thin veins of mica with a thickness of about 1 to 3 inches.

Besides the locality described above, there are a few other places in the neighbourhood at which emeralds are found; none, however, are of any importance.

Among other European emerald localities we may briefly mention Eidsvold, on the southern end of Mjösen lake in Norway. The crystals, which are here embedded in granite, are nearly all turbid and pale in colour, and are therefore not, as a rule, cut as gems.

All emerald localities other than those which have been mentioned are unimportant, and it would seem that in some supposed localities the occurrence of emerald is doubtful. Thus, for example, it appears that there has been no well-authenticated occurrence of emerald in India or Burma, although both countries are often described as emerald localities. The green stones, the occurrence of which in Rajputana, in north-west India, is fairly authentic, may very possibly be chrysoberyl.   The emerald is a stone which is highly prized in India, but the emeralds now in the country were probably brought either from South America or perhaps from the Urals.   This is probably the history of the emeralds which are now exhibited in the collection of jewellery in the Indian Section of the Victoria and Albert Museum, South Kensington.   Among the Burmese regalia there exhibited is a fine slice, about 2 inches across, of a large hexagonal crystal of emerald.   At the present day a large number of emeralds are sent from London to India, and it is also stated that South American stones are sent direct to India to be cut after the manner customary in that country, after which they are placed on the market as stones of Indian origin.

Emerald is said to occur as pebbles in Algeria, namely, in the Harrach and Bouman rivers, and also *in situ* in the neighbourhood, but according to other statements the mineral in question is green tourmaline.

The emerald localities in **Australia** are not important.   Mount Remarkable in South Australia is one, and there are a few in New South Wales.   At one of these, nine miles north-east of the township of Emmaville in the County of Gough, N.S.W., mining operations, first for tin-stone and then for emerald, have been carried on, not apparently, however, with very successful results.   The emeralds are found here in a pegmatite vein, which is an offshoot from a mass of granite penetrating clay-slates, probably of Carboniferous age.   The associated minerals are topaz, fluor-spar, cassiterite, and mispickel.   The colour of these emeralds ranges from a pale shade of green to a moderately bright green; an emerald-green, the crystals never have any great depth of colour, and they resemble beryl almost more closely than typical emerald.   The largest crystal which had been found previous to 1891 measured 1¼ inches in length, and the largest faceted stone weighed 2⅛ carats.

A number of fine emeralds have been found in North America.   Small crystals have been met with at numerous places in the eastern parts of the **United States**.   In the State of North Carolina they occur in druses in gneiss at many places in Alexander County, and especially at Stony Point.   Here also are to be found other varieties of precious beryl, together with hiddenite, the so-called " lithia-emerald."   During the course of a few years stones to the value of 15,000 dollars (about £3000) were obtained at this spot by the Emerald and Hiddenite Mining Company ; at the present time, however, the mine appears to be exhausted.   Only a few of the emeralds found here were suitable for cutting as gems; the largest and finest stone found yielded a faceted gem weighing 6 carats.   Russell Gap Road in the same county is also mentioned as a place where emeralds have been found.   A few good crystals have been met with near Haddam in Connecticut, and near Topsham in Maine, but the occurrence of emerald as a whole in North America has only local significance.

In former times Brazil was considered to be a country in which fine emeralds abounded, and after the Portuguese conquest strenuous efforts were made to discover naturally occurring stones.   Not a single emerald, however, has been found in this country, and it seems probable that green tourmaline, which abounds in Brazil, was mistaken for emerald.

It is certain that in ancient times the name emerald was applied loosely to a large number of green stones, such, for example, as green jasper, chrysocolla, malachite, and others.   Even at the present day the name with a distinguishing prefix is applied to several green stones; thus "oriental emerald" is green corundum; "lithia-emerald" is hiddenite, a green mineral belonging to the pyroxene group and found with the true emerald in North Carolina; "emerald-copper" is dioptase, a beautiful green silicate of copper. The two latter are both used as precious stones.

The green minerals which are sometimes substituted for the emerald, and which may be mistaken for it, include green corundum, known as "oriental emerald," green garnet, known as demantoid, hiddenite, diopside, alexandrite, green tourmaline, and perhaps also chrysolite and dioptase. Each of these minerals has a higher specific gravity than the emerald; each sinks in liquid No. 3, and some even in the heaviest liquid, while the emerald floats in both. Moreover, the "oriental emerald" is much harder; the demantoid, the colour of which has usually a yellowish tinge though sometimes very similar to that of the emerald, is singly refracting. Hiddenite is very rare, and is considered to be more valuable than the emerald; it is used as a gem practically in America only. Diopside is much more of a bottle-green colour than is the emerald. Alexandrite is distinguished from emerald by its hardness and its remarkable dichroism. The colour of green tourmaline, though often not dissimilar to that of pale emeralds, is frequently distinctly bluish in character; this mineral is easily distinguished from emerald, however, by its specific gravity, which is 3·07, slightly greater than that of liquid No. 3, in which, therefore, it sinks. Chrysolite is yellowish-green, and can be distinguished from emerald by its colour and its faint dichroism. Finally dioptase is always of a very dark emerald-green colour; it is only semi-transparent and far softer than the emerald. A more detailed account of the characters by which the emerald may be distinguished from other green stones which resemble it more or less in appearance is given in Table 14, Part III. of this book.

A glass of a fine emerald-green colour may be obtained by fusing together 4608 parts of strass, 42 parts of pure copper oxide (CuO), and 2 parts of chromic oxide. It differs from emerald in being optically isotropic, in the absence of any trace of dichroism, and in being much softer. An imitation of emerald which contains from 7 to 8 per cent. of beryllia is sometimes put on the market at the present time. It has a fine emerald-green colour, but is not perfectly transparent; it encloses numerous small air-bubbles and is not dichroic; its specific gravity is 3·19. It is obvious that we have here a glass to which beryllia has been added in order to give it a chemical composition similar to that of true emerald.

PRECIOUS BERYL.

AQUAMARINE, "AQUAMARINE-CHRYSOLITE," AND GOLDEN BERYL.

There are several colour-varieties of transparent, precious beryl. The most typical colours are the light blue, greenish-blue, or bluish-green of aquamarine, the yellowish-green of "aquamarine-chrysolite," and the yellow of yellow beryl, or golden beryl as the finest specimens are called. Rose-red and colourless beryl is less common, and is not, as a rule, faceted. These varieties differ from the emerald in colour; they are also, as a rule, richer in faces, as shown in Fig. 62, b to e, and in Plate XII., Figs. 4 and 5, the emerald, as already noticed, seldom showing forms other than the simple combination of a hexagonal prism and basal plane (Fig. 62 a). In the following pages aquamarine will be fully dealt with and other varieties somewhat briefly, since they do not differ essentially from aquamarine in character or mode and place of occurrence.

Aquamarine is characterised by a pure sky-blue, bluish-green, or greenish-blue colour, very similar to the tint of sea-water; hence its name, and the old saying that this stone when placed in the sea becomes invisible. Aquamarine of a deep shade of colour is very rare; it is found, in small amount and of a fine deep, sapphire-blue, at Royalston in Massachusetts, U.S.A. A distinction is sometimes made between sky-blue beryl and beryl of a greenish-blue or bluish-green colour, the former being considered as aquamarine proper, (Plate XII., Fig. 7), and the latter referred to as Siberian aquamarine. All pale bluish and greenish beryls are, as a rule, however, included in the term aquamarine.

The colours shown by precious beryl, and especially by aquamarine, are fine and brilliant, and their beauty is still more noticeable in artificial light. The colour in every case is supposed to be due not to chromium, as in the emerald, but to iron, which is always present to the extent of $\frac{1}{2}$ to 2 per cent. Experiments have been made to test the stability of the colour of common greenish and yellow beryl from granite in the neighbourhood of Dublin. After an hour's exposure to a temperature of 357° the crystals were observed to be still translucent, but to have lost all their colour. When the crystals were fused a colourless, cloudy mass was obtained.

The dichroism of aquamarine of a sufficiently deep colour is appreciable; of the two images seen in the dichroscope, one is a pure but pale blue and the other very pale yellowish-green, almost colourless. The dichroism of specimens of a still deeper colour can be observed with the naked eye.

Aquamarine, and indeed all precious beryl, is, as a rule, very uniformly coloured, irregularities in the distribution or character of the colouring being rare. Flawless and perfectly transparent stones are also very much less rare than are emeralds of the same description; fissures and turbid or cloudy patches are sometimes to be seen, however. These latter are caused by the enclosure of numerous microscopically small cavities closely aggregated and either empty or containing a liquid. When these cloudy patches are present it is impossible to produce a good polish, although beryl, free from such faults, is susceptible of a very brilliant polish. In some crystals transparent and cloudy portions alternate, in which case the latter must be removed by cutting before the former can be utilised.

Precious beryl, including aquamarine, is, as a rule, cut in the brilliant form or in some modification of the step-cut (Plate XII., Figs. 6 and 7). On account of the paleness of its colour the cut stone must have a certain depth or it will show too faint a colour. The lustre and colour of beryls is often improved by the use of foils; thus aquamarine is placed upon a silver foil or is mounted in a closed setting with a black lining. Very beautiful stones with not too pale a colour are mounted in an open setting (à jour). A magnificent faceted aquamarine, weighing 179½ grams (875½ carats), is to be seen in the mineral collection of the British Museum.

Crystals of beryl often have the form of long and relatively thin prisms, and in correlation with this the girdle of the faceted stone has often an elongated outline. Large prisms of beryl are often cut in the East into dagger-handles and other articles of considerable size. In gems of such an elongated form the direction of greatest length coincides with the direction of the principal axis of the crystal, and the stone is set in such a position that its dichroism is most apparent. Aquamarine has been, and is, much used as a vehicle for the expression of the engraver's art; its comparative softness renders the work less arduous than is the case with many other precious stones. It is said that in ancient times beryl was the material of which the lenses for spectacles were constructed, and that from this originated the German word " Brille " (spectacles).

Transparent, finely coloured and flawless crystals of precious beryl, specially of aquamarine of considerable size, are by no means uncommon. Prisms of beryl of gem quality, and of the length and thickness of a man's thumb, are frequently met with, and the discovery of still larger specimens is not unusual.   In his work on precious stones, Barbot, the late Parisian jeweller, mentions a rough aquamarine of rare beauty of which the weight was about 10 kilograms (22 pounds), and for which 15,000 francs (£600) was asked.   Again, a beautiful grass-green beryl, weighing 15 pounds, was found, in 1811, in Minas Novas, Brazil, and similar finds are often reported.   It is not surprising, then, to find aquamarine and the precious beryls generally among the lowest priced gems, a carat stone of medium quality being obtainable for a few shillings.   Only those stones which are exceptionally beautiful in colour and perfect in every other respect command higher prices, and even these fall far short of the value of a fair emerald.   In this connection it may be mentioned that the value of beryl, unlike that of emerald, is proportional to the size of the stone.

Precious beryl, and especially aquamarine, is a mineral of somewhat wide distribution, occurring in gem-quality at many localities.   Like the emerald it is met with, for the most part, in its primary situation in druses in coarse-grained granite and similar rocks.   Its occurrence in secondary deposits, such as gem-gravels, is less usual, but not so rare as is the case with emerald.

Brazil is a country in which fine beryl is abundant.   The stones are often cut before they are exported, but as the form they are given leaves much to be desired they are usually re-cut when they reach Europe.   The mineral is found in great abundance, associated with chrysoberyl, white and blue topaz, &c., as pebbles in the sands of the Minas Novas district in the north-east corner of the State of Minas Geraes ; also, though sparingly, associated with diamond in the diamond-sands of the same State.   These localities, which have been already mentioned under chrysoberyl, will be described in greater detail when Brazilian topaz is under consideration.   Among the pebbles of aquamarine are sometimes some of considerable size ; one weighing 15 pounds, which was found in the year 1811 near the source of the Rio S. Matheus, in Minas Novas, has been already mentioned.   Another fine pebble weighing 4 pounds was found soon after at the same place ; but, as a rule, the pebbles are much smaller, their greatest diameter being no more than from 2 to 5 lines. The character of the rock in which the pebbles were originally formed is not certainly known ; it is probably, however, a coarse-grained granite, since aquamarine is often found in a similar situation, and other precious stones found in Minas Novas are known to have been formed in a rock of this description.

In the neighbourhood of Rio de Janeiro aquamarine occurs in coarse-grained granite veins penetrating gneiss.   At Vallongo in the year 1825 a fine crystal, weighing 4 pounds and valued at £600, was found.   Previous to this had been found at the same place a transparent, faultless stone, which measured 7 inches in length and 9 lines in thickness.

Beryl is abundant also in the Ural Mountains and elsewhere in Siberia.   At many places crystals of gem-quality, associated, as in Brazil, with topaz, are to be found, so that the importance of Siberia as a locality for beryl is comparable to that of Brazil.

In the **Urals** it is found at various places in the neighbourhood of Ekaterinburg in the Government of Perm, also on the Ilmen Lake near the Ilmen Mountains, as well as in gold-washings on the Sanarka river in the Southern Urals, the two latter localities being in the Government of Orenburg.

In the Ekaterinburg district it is found principally in the neighbourhood of the villages of Mursinka (Mursinsk) and Shaitanka (Shaitansk), occurring in drusy cavities in coarse-grained granite, which is penetrated by veins of fine-grained granite.

The finest beryl to be found in the Urals occurs in the neighbourhood of Mursinka. It is usually in transparent, well-developed hexagonal prisms, which may be wine-yellow, greenish-yellow, yellowish-green, bluish-green, or pale blue in colour, and which range in length from a few millimetres to three decimetres (1 foot). The crystals are, as a rule, single, but intergrowths are occasionally met with in which the crystals are arranged irregularly or in parallel position. A group of fine yellowish-green or asparagus-green crystals, perfectly transparent and grown together in parallel position, was found in 1828. The group, which measures 27 centimetres in length and 31·2 centimetres in circumference, is now in the collection of the Imperial Institute of Mines at St. Petersburg, and has been valued at 43,000 roubles (£6800). The cavities in the rock, to the walls of which the crystals are attached, are usually filled with brown clay, and the presence of this substance is considered to indicate that beryl is to be found not far away. Associated with the beryl are quartz, felspar, mica, and black tourmaline, also topaz and amethyst, of which more will be said later. There are numerous pits or mines from which these variously coloured stones are won. For the most part the stones are worked in the gem-cutting establishments of Ekaterinburg. Formerly all the mines clustered round the village of Mursinka, but later other mines were opened in the neighbourhood of the villages of Alabashka, Sisikova, Yushakova, Sarapulskaya, and others, the population of which consists almost exclusively of gem-seekers. The beryls of Shaitanka were known as far back as the year 1815; they are all colourless or of a pale rose shade, and therefore of less importance as gem-stones.

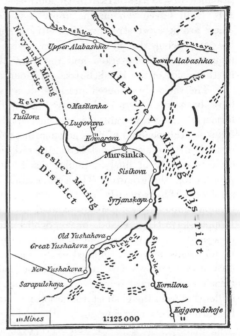

FIG. 63a.   Occurrence of beryl near Mursinka, Urals.
(Scale, 1 : 125,000.)

Magnificent specimens from this and other localities are to be seen and admired in all mineral collections. The position of the mines in the neighbourhood of Mursinka is shown in the accompanying map (Fig. 63a), of which more will be said when amethyst is dealt with.

The pale apple-green beryls which accompany the emerald at Takovaya are of less value. Those found on the Ilmen Lake also are only in part of gem-quality. They occur on the eastern shores of this lake, six versts (four miles) north-east of the smelting works of Miask in the Ilmen Mountains, and to the south of Ekaterinburg, in the Zlatoust mining district, latitude about 55° N. Crystals of beryl from this locality sometimes reach a length of 25 centimetres; they are bluish-green inclining to leek-green in colour, much fissured, and, as a rule, only translucent. They occur, together with topaz crystals, also fissured, and green felspar (amazon-stone), in pegmatite veins penetrating the rock known as miascite.

The occurrence of beryl in the gold-washings of the Sanarka river in the Southern Urals is also of small importance; here the mineral is found as pebbles associated with topaz, chrysoberyl, &c.

The beryl of the **Altai Mountains** is distinguished less for the beauty of its crystals than for their size, prisms with a length of 1 metre (39⅓ inches) and a thickness of 15 centimetres being met with. These crystals, which have the usual form, namely, a hexagonal prism terminated by a basal plane perpendicular to the prism planes (Figs. 62a, and 63), range from sky-blue to greenish-blue in colour, and occur in brown, much fissured quartz, the exact locality being a spot in the Tigirezh Mountains. The mineral is here at the best only translucent, and therefore rarely of use as a gem.

Of greater importance is the occurrence of beryl in the Nerchinsk district of the province of **Transbaikalia** in south-east Siberia, Nerchinsk itself being in longitude 116° E. of Greenwich on the upper course of the Shilka river, a tributary of the Amur. There are here two stretches of country in which beryl, and especially aquamarine, abounds, the one being the mountain range Adun-Chalon and its southern continuation, the mountains of Kuchuserken, and the other the neighbourhood of the Urulga river on the northern side of the Borshchovochnoi Mountains.

The variously coloured precious stones which occur at Adun-Chalon (Adun-Tschilon) have been known since the year 1723. The output of gems from these deposits was formerly very considerable; it reached its highest in the year 1796, when no less than 5 poods (180 pounds) of pure aquamarine, suitable for cutting as gems, was obtained. The crystals of beryl are found here attached to the walls of cavities in a topaz-rock, which consists mainly of finely granular quartz and small topaz crystals, and occurs as veins penetrating the granite. The aquamarine in these cavities is accompanied by topaz and smoky-quartz, frequently also by other minerals. The highest mountain of the Adun-Chalon range has two peaks, separated by a narrow valley. The western peak is known as Hoppevskaya Gora, that is to say, Schörl Mountain; it consists almost entirely of topaz-rock, and is scarred from foot to summit with the workings of gem-seekers. The mineral is by no means, however, confined to this mountain, numerous mines being scattered about an area of two square versts in the neighbourhood. These mines are nothing but open pits or trenches of the most primitive kind, without timbering, and never more than three fathoms in depth; from these, short tunnels are worked in the rock in all directions. Immediately beneath the turf covering the southern slopes of the Hoppevskaya Gora is a layer of loose material, containing much iron-ochre, derived from the weathering of the topaz-rock. In this layer fine specimens of aquamarine, and its customary associate topaz, are to be found. A hexagonal prism of transparent beryl, 31 centimetres (over 1 foot) in length and 5 centimetres in diameter, from Adun-Chalon, is preserved in the British Museum collection of minerals.

The beryls of Adun-Chalon differ from the smooth-faced prisms of the Urals and of the Borshchovochnoi Mountains (or Urulga river) in that the prism-faces are deeply striated (Fig. 62 d). The crystals are, as a rule, greenish-blue in colour, but sky-blue, yellowish-green, wine-yellow, and colourless specimens are met with; and every degree of transparency is represented. The crystals are often united in groups, which are frequently invested with a thin surface layer of iron-ochre, the substance with which the drusy cavities are, as a rule, filled.

The country between the rivers Shilka and Unda in the Borshchovochnoi Mountains abounds with fine beryl. A large amount of the mineral was obtained about the middle of the nineteenth century, for the most part from the granite mountains which border the Urulga river, a tributary of the Shilka on its right bank. Beryls from the neighbourhood of the Urulga are remarkable for their size, transparency, and beautiful colour. The majority are yellowish-green, the remainder being variously tinted or colourless. The crystals may reach a length of 10 centimetres and a thickness half as great; they are frequently developed with great regularity. Beryl from the Urulga river is in general very similar to that from Mursinka in the Urals.

In other parts of Asia precious beryl occurs but sparingly. Aquamarine has been found at some places in **India,** and various objects worked in this mineral have not infrequently been found in ancient tombs, temples, &c. Most of it appears to have been obtained in the Coimbatore district of the Madras Presidency, as at Paddur or Patialey, where, at the beginning of the nineteenth century, the mineral was obtained from cavities in a coarse-grained granite. When all the more easily obtained stones had been taken, work was abandoned. Later on, aquamarine was discovered at Kangayam in the same district; specimens from this locality were shown at the Vienna Exhibition of 1873, and others are preserved in the British Museum. Here was once found a stone of the most perfect transparency, which weighed 184 grains (900 carats) and sold for £500.

Pale blue crystals of fair size, sometimes measuring as much as $3\frac{1}{4}$ inches in length, are found at many places in the Punjab in granite veins penetrating gneiss. They are, however, almost invariably much fissured and unsuitable for gems. In the Jaipur State in Rajputana aquamarine is mined in the neighbourhood of Toda Rai Sing in the Ajmer district, in the Tonk Hills, and at various places lying within a radius of 38 miles from Rajmahal on the Banas river. Most of these crystals are quite small and therefore, in spite of their fine colour, of little value. They are found buried in marshy ground, and have probably been derived from the granite veins which penetrate the sedimentary rocks of Rajputana in large numbers. Small crystals of yellow beryl occur embedded in a thick vein in the Hazaribagh district in Bengal. Other reputed Indian localities require authenticating.

In Burma, pebbles of aquamarine are reported to have been found in the Irrawaddy. Whether this is so or not, it is certain that beryl is of only sparing occurrence in Burma; while in Ceylon, a locality so rich in other precious stones, it is practically non-existent.

Although in **Europe** many localities for common beryl are known, precious beryl of gem quality occurs but sparingly. In the Mourne Mountains, in County Down, Ireland, crystals of aquamarine of a beautiful and comparatively deep blue colour occur, together with topaz, in cavities in granite; these, however, are rarely perfectly transparent.

In the **United States** of North America numerous localities are known, from which fine stones of various colours, and of a quality suitable for cutting, have been obtained. The mineral is found, for example, with the emerald in Alexander County, North Carolina, while at Russell Gap Road, in the same county, more aquamarine of gem quality was found than anywhere else in the United States. Fine blue aquamarine is found also in Mitchell County, North Carolina, and green beryl at Stoneham, in Oxford County, Maine; a fine bluish-green fragment, found recently at the latter place, gave an almost faultless brilliant, weighing $133\frac{3}{4}$ carats, and measuring 35 millimetres in length and breadth and 20 millimetres in thickness. Golden-yellow beryl of good quality is found at Albany in Maine, in Coosa County in Alabama, and at a few other places. At Royalston in Massachusetts there occur, with other varieties of precious beryl, some of a fine blue colour comparable to the blue of the sapphire; it is by far the most beautiful blue beryl known, but, unfortunately, occurs only in quite small crystals. Beryl is also found in Colorado, namely, on Mount Antero, ten miles north of Salida, at a height of 12,000 to 14,000 feet above sea-level. The crystals, which range in colour from a pale to a dark shade of blue, are found, together with phenakite and other minerals, attached to the walls of drusy cavities in granite. They vary in length from 1 to 4 inches, and in thickness from $\frac{1}{10}$ inch to an inch; from the largest a faceted stone of about 5 carats can be cut. There are many other localities in America at which beryl is found, but none of any commercial importance.

A small amount of beryl occurs also in Australia; at several places in New South Wales for example. Here, again, the occurrence has no economic significance.

Precious beryl of a yellow colour, and also the yellowish-green "**aquamarine-chrysolite**," come principally from Brazil, although it is to be found in good quality at some of the localities already mentioned, for example, in Siberia associated with aquamarine. Beryl of a deep, pure yellow, such as is represented in Plate XII., Fig. 4, is known as **golden beryl**. It occurs at many beryl localities in North America, especially at Albany in Maine; it has been collected also in the vicinity of New York City, and in Litchfield County, Connecticut. It is always of sparing occurrence in the States, and, though highly prized there, does not in general command high prices, only exceptionally fine stones costing more than a few shillings per carat.

Certain of the several varieties of precious beryl are liable to be mistaken for other precious stones which they resemble in appearance; the exceptionally low specific gravity of beryl, however, prevents any serious confusion. Aquamarine resembles in colour "oriental aquamarine," euclase, some tourmalines, and blue topaz; its resemblance to blue topaz is so close that the latter is often known in the trade as aquamarine. Each of the four stones mentioned above, however, sinks in liquid No. 3 (sp. gr. = 3·0), while beryl floats. In the same way yellow beryl, that is to say, "aquamarine-chrysolite" and golden beryl, may be distinguished from other yellow and greenish-yellow stones of similar appearance, namely, from yellow topaz, "oriental topaz," "oriental chrysolite," chrysolite, and chrysoberyl, all of which sink in liquid No. 3. To distinguish between yellow beryl and yellow quartz (citrine) is less easy, for there is no great difference between the hardness and specific gravity of these two minerals. In liquid No. 4 (sp. gr. = 2·65) citrine remains suspended while beryl slowly sinks; moreover, a smooth surface of quartz will be untouched by citrine, but will be distinctly, though not deeply, scratched by beryl. The stronger dichroism of beryl may also serve sometimes to distinguish it from citrine.

A glass resembling aquamarine in colour may be obtained by fusing together 3456 parts of strass, 24 parts of glass of antimony, and 1½ parts of cobalt oxide. The single refraction, entire absence of dichroism, and low degree of hardness of this imitation, are the characters whereby it is distinguished from genuine aquamarine.

EUCLASE.

Euclase is one of the rarest of minerals and is only occasionally cut as a gem, when it commands fancy prices. It resembles beryl, and specially aquamarine, in many ways; its chemical composition, for example, differs from that of beryl only in the presence of a little water and in the proportions of the constituents, its chemical formula being $H_2O.2BeO.Al_2O_3.2SiO_2$.

The mineral crystallises in the monoclinic system. The crystals are prismatic in habit and the prism faces are deeply striated parallel to their mutual intersections; they are terminated at the two ends by obliquely placed faces, as shown in Fig. 64. The crystals have a perfect cleavage parallel to their one plane of symmetry; this cleavage plane truncates the two acute edges of the rhombic prism. Owing to the perfect cleavage, crystals of euclase are liable to become fissured and then broken, and when being cut, unless exceptional care is taken, they chip at the edges.

The mineral has a hardness (H $= 7\frac{1}{2}$) slightly exceeding that of beryl; it is somewhat heavy, having a specific gravity of 3·05 to 3·10; when rubbed it acquires a not inconsiderable charge of electricity. Its lustre is vitreous, but in the direction of cleavage is sometimes pearly; the mineral is susceptible of a high polish and is frequently perfectly clear and transparent. Its refraction, double refraction, and dispersion are all small. Its colour resembles in many respects that of precious beryl, being either green with a tinge of blue (Plate XIII., Fig. 5), or green with a yellowish tinge; it is almost invariably pale in shade, deeply coloured stones being rare and perfectly colourless ones quite unusual. Stones of a rather deep blue-green colour, as represented in the plate just quoted, are most admired; they resemble some aquamarines and blue topaz very closely, but may be easily distinguished from either of these by the difference in the specific gravity and by the existence of a distinct dichroism in euclase.

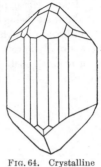

FIG. 64. Crystalline form of euclase.

Very few localities for euclase are known, and nowhere is it found in abundance. It was first met with in Brazil, in the neighbourhood of Ouro Preto (formerly known as Villa Rica), in Minas Geraes. It occurs here, associated with yellow topaz, in nests in the quartz veins by which the clay-slates, which accompany itacolumite, are traversed. It would appear, however, that topaz and euclase are never found actually side by side in the same nest or druse. The principal locality is Boa Vista, near Ouro Preto (Fig. 67). A euclase weighing over $1\frac{1}{2}$ pounds was reported by L. von Eschwege from this district, but the majority of the stones found there are much smaller, and, moreover, they are often broken into fragments along the cleavage plane.

Euclase is also found in the gold-washings of the Sanarka river in the Ural Mountains, situated in the Government of Orenburg. It occurs here in loose crystals, many of which, as in Brazil, are merely cleavage fragments; they may reach a length of $1\frac{1}{2}$ inches, but, as a rule, are much smaller. They vary in colour from grass-green to greenish-blue, and are associated with topaz, chrysoberyl, and other minerals.

Small crystals of euclase, of a pale yellowish colour, have been found in recent years in mica-schist in the Grossglockner district of the Austrian Alps; this occurrence, however, is solely of mineralogical interest.

PHENAKITE.

Like euclase, phenakite has but little importance as a gem. It contains beryllia but no alumina, being a silicate of beryllia with the chemical formula $2BeO.SiO_2$. It crystallises in the rhombohedral system, and the crystals, which are hemihedral with parallel faces, usually have the form of hexagonal prisms terminated by the faces of a rhombohedron or of a hexagonal pyramid, sometimes also by small faces of other forms, as shown in Fig. 65, a to c.

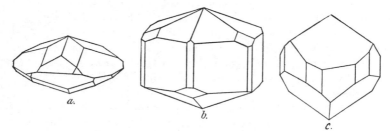

FIG. 65.   Crystalline forms of phenakite.

The cleavage of phenakite is very imperfect and its fracture is conchoidal. Its hardness ($H = 7\frac{1}{2} - 8$) is slightly greater than that of either beryl or euclase; while its specific gravity (sp. gr. = 2·95 to 3·0) is rather less than that of euclase, and the mineral just floats in liquid No. 3.

The lustre of a fractured surface of phenakite is brilliant and vitreous; that of the natural crystal faces is usually, however, much duller. It is susceptible of a very brilliant polish, and has then a lustre comparable to that of the sapphire. The mineral is frequently water-clear, but may be cloudy or only translucent. Usually it is colourless, but yellow, brown, and rose-red phenakite have been found. Except for its brilliant lustre, colourless, water-clear phenakite resembles rock-crystal in appearance; the refraction and double refraction of the two stones are very much the same, the former being a little greater and the latter a little less in phenakite. Water-clear phenakites, and indeed all stones of this description, are cut in the brilliant form, a form which displays the lustre and brilliancy of the stone to the best advantage. A phenakite brilliant has certain resemblances to the diamond, but never shows the brilliant play of prismatic colours characteristic of this gem. Two very fine faceted phenakites, weighing 43 and 34 carats respectively, are exhibited in the British Museum of Natural History.

Phenakite is a less rare mineral than is euclase, but the number of localities at which it is found is almost as limited as for euclase. The white phenakite, which occurs with emerald and alexandrite embedded in mica-schist at Takovaya in the Ekaterinburg district of the Urals was the first to be met with. At this locality crystals with a thickness of 10 centimetres and a weight of $1\frac{1}{2}$ pounds are found. Phenakite also occurs with topaz and green felspar (amazon-stone) at Miask, on Lake Ilmen, in the Urals, but this locality is of less importance. The Takovaya stones, those at least which are sufficiently transparent, are usually cut in Ekaterinburg, and are placed on the market at the fairs of Nizhniy Novgorod. Not many of these gems leave Russia, but some find their way to the Orient (Persia, India, &c.), through the dealers who frequent the Nizhniy Novgorod fairs.

Phenakite has been found comparatively recently in North America, chiefly in Colorado. One of the localities in this State is Topaz Butte, near Florissant, sixteen miles from Pike's Peak, where it occurs as flat rhombohedral crystals, and, as at Miask, associated with topaz and amazon-stone in veins penetrating granite. The other locality is Mount Antero, in Chaffee County, ten miles north of Salida, where it occurs as prismatic crystals, sometimes an inch in length, on quartz and beryl. These American phenakites are cut as gems, and are valued on account of their national origin. Other American localities, like the European, have no trade importance. In Europe small brown crystals, scarcely suitable for cutting, were formerly found in the iron mines of Framont, in the Vosges Mountains. The small crystals of phenakite, which have been found in recent years in mica-schist in the Canton Valais in Switzerland, are of mineralogical interest only.

TOPAZ.

Topaz is the most familiar of yellow stones, and for this reason its name is often applied to other minerals of the same colour. Thus, yellow corundum, as we have seen, is known as "oriental topaz," yellow quartz (citrine) is referred to variously as "occidental topaz," "Bohemian topaz," and "Spanish topaz," while yellow fluor-spar is sometimes known as "false topaz." The mineral species to which mineralogists apply the name topaz includes not only the stones known as precious topaz, or as Brazilian, Saxon, Siberian, or Tauridan topaz, but also blue, red, and colourless stones, which are known to dealers in precious stones by other names.

Topaz is a fluo-silicate of aluminium with the formula $(AlF)_2SiO_4$ and the percentage composition of, silica 33·3, alumina 56·5, and fluorine 17·6; from this it will be seen that alumina forms a large part of the mineral, as it does also of most of the precious stones hitherto considered. Besides the constituents already mentioned, other substances, such as ferrous oxide, lime, alkalies, water, &c., are sometimes present in small amount. Until recently the water, which is often present, was considered to be an impurity due in part to the alteration of the material by hydration. Penfield and Minor, however, have shown (1894) by a series of carefully conducted analyses combined with detailed determinations of the optical and other physical constants of the mineral, that water is one of its essential constituents and not a mere impurity. The amount of water present in the specimens which they analysed varied from 0·18 to 2·50 per cent. The specimen which contains only 0·18 per cent of water is almost pure fluor-topaz, and its composition is expressed by the formula already given, $(AlF)_2SiO_4$. Specimens containing more water may be regarded as hydro-fluor-topaz, in which the water is present as hydroxyl (OH), which replaces fluorine isomorphously, so that the formula becomes $[Al(F,OH)]_2SiO_4$. A mineral in which the whole of the fluorine is replaced by hydroxyl, and which would have the formula $(Al.OH)_2SiO_4$, has not yet been met with. The isomorphous replacement of fluorine by hydroxyl in this mineral is accompanied by small variations in its physical characters, such as specific gravity, refraction, double refraction, &c.; these variations are very slight and of purely scientific interest.

Topaz crystallises in the rhombic system ; all crystals have certain features in common, but show differences in habit.   A combination of two rhombic prisms forming elongated, eight-sided columns, often deeply striated parallel to their length, is almost invariably to be seen.   The terminal faces differ according to the locality from which the crystal comes.   As a rule, they are developed regularly at one end only of the crystal, the other end having been attached to the matrix in the drusy cavity in which the crystal grew.   A few forms taken by topaz crystals are shown in Fig. 66, a to d.

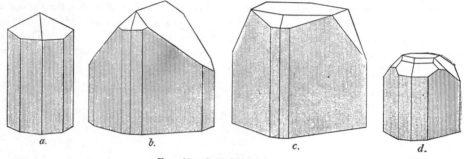

FIG. 66.   Crystalline forms of topaz.

Fig. 66 a, and Plate XIII., Figs. 2 and 4, show a simple form of crystal especially characteristic of topaz from Brazil and Asia Minor ; here the only terminal faces are those of a rhombic octahedron or pyramid.   In Fig. 66 b, the faces of this rhombic octahedron are small, and two large dome faces, give a roof-like termination to the crystal ; this habit is characteristic of topaz crystals from the Adun-Chalon Mountains, near Nerchinsk, in Siberia.   Crystals from Mursinka, in the Urals (Fig. 66 c, and Plate XIII., Fig. 1), have, in addition to these faces, a largely developed basal plane at right angles to the prism faces. The crystal shown in Fig. 66 d, is a combination of two prisms, three rhombic octahedra, a dome, and the basal plane ; this habit is characteristic of crystals from Schneckenstein in Saxony (Plate XIII., Fig. 3).   More complicated crystals, with a much larger number of faces, are to be found at other localities ; the examples cited are sufficient, however, to give a general idea of the crystalline forms of topaz.

Crystals of topaz as small as a pin's head have been found, but very large ones weighing several pounds are not at all unusual.   Thus, for example, a beautiful transparent crystal of topaz, weighing more than 25 pounds, was found in the neighbourhood of the Urulga river in Siberia.   A crystal of topaz, 2 feet in length and 137 pounds in weight, has been found quite recently (1901) in Sætersdalen, Norway, and is now exhibited in the British Museum.

Topaz differs from the majority of precious stones in the possession of a very perfect cleavage.   There is only one direction of cleavage, and this is parallel to the basal plane— that is to say, at right angles to the length of the striated prism.   In consequence of this cleavage, topaz crystals, when removed from the matrix, almost invariably break away with a smooth, shining, plane face ; the lower ends of the crystals in Fig. 66 are terminated by such cleavage planes.   A crystal of topaz, which is too long to be cut as a single stone, may be readily cleaved with a chisel into fragments of suitable size, and thus much laborious work avoided.   The drawback connected with this perfect cleavage is that it is the cause of a tendency in the stone to develop plane, even fissures.   The presence of such fissures, which is often indicated by brilliant iridescent colours, detracts considerably from the beauty and value of the stone.   To avoid the development of these fissures, the stone must not be allowed to fall or to be jarred in any way; and when undergoing the process of cutting great care is necessary, since the perfect cleavage renders the stone liable to chip at the edges and to become fissured or broken by the jarring of the grinding disc.

The hardness of topaz is represented by 8 on Mohs' scale. It scratches quartz with ease, but is itself readily scratched by corundum. Of the minerals hitherto considered, it is surpassed in hardness only by diamond, corundum, and chrysoberyl. On account of its hardness it takes a good polish and exhibits a brilliant lustre, which is also to be seen on the natural crystal-faces.

Topaz is a comparatively heavy mineral; its specific gravity, determined on different varieties, ranges from 3·50 to 3·57. Colourless topaz is often rather heavier than the coloured varieties; its specific gravity has been determined at from 3·53 to 3·56. This value is almost exactly the same as that of diamond, so that it is impossible by the density alone to distinguish between a diamond and a colourless topaz, which, when cut, have a certain resemblance. The specific gravity of the reddish-yellow topaz of Brazil (Plate XIII., Figs. 2 and 2a) and Asia Minor has been determined to be 3·50 to 3·55, while that of the greenish-blue from Nerchinsk is 3·53. Such small variations in the specific gravity are due to differences in the chemical composition depending upon the replacement of fluorine by hydroxyl. According to some determinations the specific gravity may vary between 3·4 and 3·6; this greater departure from the mean value, 3·5, is probably due either to impurity of material or to inaccuracy of determination.

When rubbed, topaz becomes strongly electrified and capable of attracting to itself any light bodies such as shreds of paper. Some topazes possess this property in a more marked degree than do others; thus, for example, those from Schneckenstein in Saxony acquire a charge of electricity when merely rubbed between the fingers, while in the case of certain Brazilian topazes a pressure of the fingers, exercised in the direction of length of the prismatic crystal, is sufficient; also, when heated and allowed to slowly cool, topaz becomes electrified and acquires a greater charge than would any other precious stone under similar circumstances, with the exception of tourmaline; on this account it is said to be pyroelectric. The charge may be retained for thirty hours or more after the stone has cooled down to its original temperature. This phenomenon, under certain circumstances, affords a means by which topaz may be distinguished from other stones which it may resemble in general appearance.

In the blowpipe flame topaz does not fuse, but becomes cloudy and opaque owing to the loss of water and of fluorine; coloured stones lose their colour. Acids, whether hot or cold, have no action on topaz.

More important than any other feature, however, are the optical characters of topaz, that is to say, its behaviour towards light. In this connection we may begin by distinguishing between cloudy and opaque topaz, the so-called "common" topaz, and that which is clear and transparent, "precious" topaz. The former, besides being opaque, is usually nondescript in colour, so that it is unsuitable for gems. A variety of common topaz, known as pyrophysalite, occurs as large crystals in granite near Fahlun in Sweden; another variety, known as pycnite, is found as columnar aggregates in the tin mines of the Erzgebirge between Bohemia and Saxony. Our attention must be devoted, however, not to these varieties of common topaz but to precious topaz, the transparency, colour, and lustre of which combine to make it a very beautiful gem.

Its lustre is of the ordinary glassy or vitreous type; on the cleavage face, however, it is pearly. The brilliant lustre of the natural crystal-faces has been already mentioned, and is specially noticeable on the prism faces. The lustre of a cut topaz which has been well polished is almost comparable with that of the diamond.

Topaz is not a highly refractive substance, its indices of refraction only slightly exceeding 1·6. Being a rhombic mineral it is doubly refracting, but here again the two indices of refraction differ from each other only slightly, so that the double refraction is not strong. The refractive indices for variously coloured light also differ but slightly, so that the dispersion is likewise small. Very little play of prismatic colours is shown, therefore, by a faceted topaz, which in other respects somewhat resembles the diamond. The greatest, mean, and least value of the refractive index of one and the same crystal for red and for violet light is given in the table below:

Red light	.	.	.	1·618	1·610	1·608
Violet light	.	.	.	1·635	1·627	1·625

These values will differ slightly for other crystals, especially when the latter differ in colour or in place of origin, owing to variations in chemical composition.

The range of colour exhibited by topaz is considerable. The purest variety is perfectly **colourless** and pellucid, and is of frequent occurrence. It is found as crystals at Miask, in the Urals, among other places, and in still greater abundance, in the form of rounded, water-worn pebbles, in the streams and rivers of Diamantina and specially of Minas Novas, in the State of Minas Geraes, Brazil. These pebbles, which are met with also in Australia, especially in New South Wales, are often perfectly colourless and transparent, and are then known to the Brazilians as "*pingos d'agoa*" (drops of water). Colourless topaz is sometimes known to the trade as "*goutte d'eau*," the French equivalent of the term.

The "Braganza," a supposed diamond belonging to the Portuguese crown jewels and weighing 1680 carats, is probably nothing more than one of these topaz pebbles of unusual transparency and beauty. These stones are often called "slave's diamonds" on account of their resemblance to the diamond. When cut they are not infrequently passed off as diamonds, and as the specific gravity is almost exactly the same as that of diamond, resort to some other means by which they may be distinguished is necessary. This is a matter of no great difficulty, for topaz, beside being much less hard than diamond, is also doubly refracting, while diamond, like other cubic minerals, is singly refracting. Topaz may, however, be distinguished from other colourless stones, such as rock-crystal, phenakite, and colourless sapphire, by the difference in specific gravity. Thus, in methylene iodide rock-crystal (sp. gr. = 2·65) and phenakite (sp. gr. = 2·98−3·0) both float, while topaz quickly sinks. In the heaviest liquid, on the other hand, colourless sapphire sinks while topaz floats. Moreover, topaz differs from the minerals just mentioned in its capacity for acquiring charges of electricity.

The colour of topaz, though usually pale, is sometimes deep and intense, and in this case the dichroism of the mineral, though scarcely apparent to the naked eye, can be easily observed in the dichroscope.

Topaz is very frequently **blue** in colour, either a pure blue or a blue tinged with green, but scarcely ever a pure green. A blue crystal of topaz is represented in Plate XIII., Fig. 1, and the gem cut from it in Fig. 1a of the same plate. Dark blue topaz is very unusual; the mineral is almost invariably pale in shade, sometimes so pale that it may be more correctly described as blue-white. Such stones are common among the "pingos d'agoa" of Brazil as well as amongst crystals from Mursinka, near Ekaterinburg, in the Urals; these latter are known as *Siberian* or *Tauridan topaz*. Stones of a darker shade of pale blue are referred to as "*Brazilian sapphire*," a term which is applied also to the blue tourmaline, which occurs in Brazil in association with white and blue topaz. Bluish-green and greenish-blue topaz is so very similar in appearance to aquamarine, that a careful examination is sometimes necessary in order to distinguish between them. The difference in specific gravity is here a valuable aid; in pure methylene iodide topaz sinks, while aquamarine floats. Topaz bearing this resemblance to aquamarine occurs at various localities, but specially in the district about Nerchinsk in Siberia, and when cut is always passed off as aquamarine. True aquamarine is more abundant and more widely distributed than is topaz, especially topaz of this particular colour, so that here we have the very unusual case of a rarer mineral substituted for one less rare. The dichroism of blue topaz is most apparent when its colour possesses a tinge of green, and the greatest contrast between the two images of the aperture of the dichroscope exists when one is practically colourless and the other almost pure green. In the case of aquamarine, the two images are coloured yellowish-white and clear sky-blue respectively. Topaz of a yellowish-green colour like chrysolite is rare; the typical colour of the mineral is the one we are now about to consider, namely, yellow.

Yellow topaz exists in a great variety of shades, ranging from the palest possible shade of pure yellow up to dark brownish yellow, usually tinged more or less with red. Yellow topaz is the only variety which is recognised by jewellers as topaz; it is by no means always of the same tint, and stones showing different shades of yellow differ in value and are distinguished by special names.

A fine saffron-yellow topaz, the so-called *Indian topaz*, occurs in Ceylon, not, however, in abundance, and as a great rarity in Brazil. Very beautiful topaz of a dark yellow colour, tinged with red or brown, occurs in great abundance at the Brazilian localities. A crystal of this description is represented in Plate XIII., Fig. 2, while Fig. 2a illustrates a faceted stone of a somewhat different tint. Topaz of gold-yellow, honey-yellow, wine-yellow, and other shades is also found in Brazil though in less abundance; the gold-yellow variety is distinguished as *Brazilian topaz*.

Saxon topaz is of a pale wine-yellow colour; a crystal of this variety is illustrated in Plate XIII., Fig. 3, and a faceted stone in Fig. 3a. It occurs at Schneckenstein, near Auerbach, in Saxon Voigtland. Occasionally it is tinged with green, and is then known as "*Saxon chrysolite.*"

The dichroism of dark yellow topaz is fairly well marked; the two images of the dichroscope aperture being coloured respectively light and dark yellow, or yellow and red. The paler the stone the less marked is the contrast in colour of the two images, and with quite pale yellow stones the difference is scarcely apparent at all.

The stones which are most likely to be mistaken for yellow topaz are yellow sapphire, the so-called "oriental topaz," and yellow quartz, the so-called citrine or "occidental topaz." The latter exhibits the same fine tints and is often substituted for topaz. The fraud may be easily detected by the difference in the specific gravity of the two minerals, just as rock-crystal is distinguished from water-clear topaz. The difference in specific gravity enables us also to distinguish "oriental topaz" from true topaz in the same way that colourless sapphire is distinguished from colourless topaz.

Topaz of a pronounced **red** colour occurs in nature but rarely. It is met with occasionally in Brazil associated with crystals of a yellow colour, and is usually of a light rose-red inclining to a lilac shade of colour, very similar to the colour of "balas-ruby." In spite of this resemblance the two gems need never be mistaken the one for the other, since the "balas-ruby" (spinel) is singly refracting and not dichroic, while topaz is dichroic and doubly refracting. This variety of topaz, a crystal and faceted stone of which is represented in Plate XIII., Figs. 4 and 4*a*, is known to jewellers as *rose-topaz.* When of a deep red colour it is sometimes referred to as "*Brazilian ruby.*"

Rose-topaz, which is so rare in nature, may be produced artificially by subjecting yellow topaz, especially the Brazilian, to a gradual rise of temperature, when it assumes the red colour of rose-topaz. Most of the rose-topaz sold by jewellers as "burnt" topaz is the yellow variety altered by heating. The rise and fall of temperature to which the stone is subjected must be very gradual, otherwise the crystal will be fissured. The darker the original colour of the stone the darker will be its colour after heating. There are various methods in use for the artificial production of the red colour of rose-topaz; in one, for example, the stone is packed in a crucible with powdered charcoal, sand, ashes, or any other powder, slowly heated and then slowly cooled; in another, it is enclosed in many wrappings of tinder; this material is then fired and the change in colour thereby effected. When the latter method is adopted too great a rise of temperature must be avoided, otherwise the stone will be completely decolorised, besides being rendered fissured and cloudy.

"Burnt" topaz is much more strongly dichroic than is naturally occurring rose-topaz, or, indeed, than topaz of any other colour; the two images seen in the dichroscope, when showing the greatest possible contrast in colour, are respectively dark cherry-red and honey-yellow. It has been supposed that naturally occurring rose-topaz has been derived from yellow-topaz by the action of heat, but L. von Eschwege, and others familiar with the mode of occurrence of the mineral in Brazil, have shown that crystals of the rare rose topaz occur together with the common yellow variety; this view, therefore, cannot be correct.

A change of colour is induced in some topazes simply by exposure to sunlight. Such a change has been observed in crystals from the Urulga river, Siberia, the original dark wine-yellow colour changing after a few months' exposure to dirty white. It may be mentioned here that the finest topaz crystals in the British Museum collection came from this locality, and for this reason are protected from the action of light. Some pale blue stones have been observed to become pale yellow after exposure to sunlight. These and similar instances indicate that the colour in such cases is due to an organic substance; those stones, on the other hand, which are unaffected by light, but become red on exposure to heat, probably owe their colour to a metallic oxide, since exposure to high temperatures would destroy any organic pigment.

All these varieties of precious topaz are made use of as gems; that is to say, all specimens which are sufficiently transparent, finely coloured, and free from faults. Inferior stones, the so-called "fallow topaz," are crushed and powdered, and in this form utilised as a hard, grinding material. The form of cutting best suited for coloured topaz is seen in the faceted stones represented in Plate XIII., Figs. 1*a*, 2*a*, 3*a*, 4*a*; these are step-cut, the table being somewhat small and the steps narrow and equidistant from each other. The brilliant form is sometimes adopted for coloured stones, but is more often seen in colourless topaz, "pingos d'agoa," &c. Yellow topazes are not infrequently table-cut. In the case of light-coloured stones, like the Saxon topazes, for example, an added brilliancy and depth of colour is given by the use of a burnished gold, or in some cases of a red, foil. Blue topaz is always backed with a pale blue shining foil; on a dark foil it presents a peculiar and not altogether attractive appearance. Only the finest and most transparent of stones, whatever be their colour, are mounted *à jour.*

Topaz varies in value according to its quality; large crystals are found quite as frequently as small, consequently the value of cut stones is proportionate to their size. Topaz of any colour is not at the present time a gem favoured by the votaries of fashion, and hence can be purchased at very moderate prices. This applies more particularly to the common yellow topaz; the red, dark brownish-yellow, colourless, and fine blue varieties command somewhat higher prices. The finest topaz is, at the present time, not worth more than 10s. per carat, while much less will be paid for inferior qualities. About thirty-five years ago topaz had quite three times its present value; thus, for a water-clear or rose-topaz weighing 1 carat about 30s. would be paid, for a burnt topaz about 18s., and for ordinary yellow 12s. The wholesale price of uncut yellow topaz is now from 1s. to 20s. per pound.

The faults, the presence of which reduces the value of a stone, are principally impure colour, fissures in the direction of cleavage ("feathers"), and turbidity. Cavities, either vacuous or filled with liquids of various kinds, are also frequently present.

The artificial production of topaz has not at present been achieved with certainty. Good imitations of topaz can be made by fusing strass with a certain amount of glass of antimony (antimony oxide) and with a trace of purple of Cassius (a compound containing gold), or with a little iron oxide. Purple of Cassius gives a darker, more reddish-yellow, and iron oxide a paler yellow. Such imitations may be distinguished from genuine stones by their single refraction, lower specific gravity, much lower degree of hardness, and by the entire absence of dichroism.

Topaz occurs commonly in the old crystalline silicate rocks, namely, in gneiss and crystalline schists as well as in granite. The crystals are attached to the walls of cavities or crevices in these rocks and are often accompanied by tin-stone (cassiterite), aquamarine (beryl), &c. The general conditions of the occurrence are such that topaz must be regarded as the product of fumarole action, the mineral having probably been formed by the interaction of vapours containing fluorine, which were liberated in crevices at the time of the intrusion of the igneous rock. By the weathering and breaking down of the mother-rock the topaz crystals are set free, carried away with the débris, and, as rounded pebbles, find a final resting-place in the alluvial deposits of rivers and streams. In recent years topaz crystals have been met with in the drusy cavities of later volcanic rocks, such as rhyolite; this is a much less common mode of occurrence and has no commercial significance.

Topaz of gem-quality occurs at several localities, many of which have already been briefly mentioned, but must now receive more detailed consideration.

The most important European locality is the Schneckenstein, near Gottesberg, in the neighbourhood of Auerbach, in Voigtland, Kingdom of **Saxony**. The Schneckenstein, which is situated four kilometres south-east of the railway station of Hammerbrück, is a steep wall of rock projecting from the surrounding mica-schists, and in appearance resembling an old ruin. It consists of comparatively small fragments of schists rich in tourmaline, cemented into a firm and hard mass by quartz and topaz; the whole rock-mass is known as a topaz-rock. Crystals of topaz, together with quartz, tourmaline, &c., are attached to the walls of cavities in this rock, and the cavities are often partly filled up with white or yellow kaolin. A portion of the wall of such a drusy cavity is represented in Plate XIII., Fig. 3. The free ends of the topaz crystals have a moderately complex termination, the terminal faces including a large basal plane (Fig. 66 d). The crystals vary in size, the smallest having a length and thickness of a few lines, while the largest measure 4 inches in length and 2 inches in thickness. The majority have a length and thickness of about ⅜ inch; larger crystals are rare.

In colour these topazes are mostly pale wine-yellow, rarely a dark wine-yellow, colourless or white; the darker the colour the more valuable is the stone. Crystals of a greenish tint, known as "Saxon chrysolite," are sometimes met with; those of a pure yellow are distinguished from them by the name Saxon topaz. This so-called " Schnecken topaz " was at one time much admired and sought after ; the specimens of decorative art, now to be seen in the " Green Vaults "·at Dresden, bear witness both to the exceeding beauty of some of these stones and to the favour in which they were at one time held.

During the eighteenth century, certainly as far back as the year 1737, Schneckenstein topaz was systematically mined and placed on the market. The stones were sorted into three groups ; the largest and purest were referred to as ring-stones (*Ringsteine*), the next quality as buckle or clasp-stones (*Schnallensteine*), and inferior stones as *Karmusirgut*. No mining has been carried on here for a long period, and the terms just mentioned have long since been forgotten in the locality.

Brazil, where, as we have seen, are to be found diamond, beryl, and chrysoberyl, is no less rich in topaz. All the colour-varieties of this mineral are found there in abundance, especially in the State of Minas Geraes, other Brazilian localities being of small importance compared with this.

Brazilian topaz is either blue, yellow, or colourless. The colourless and blue varieties always occur together as water-worn pebbles in secondary deposits ; yellow topaz, on the other hand, is met with only in its primary situation. Moreover, the localities for the colourless and blue and for the yellow varieties are widely separated.

White and blue topaz has already been mentioned as occurring in the district of Diamantina in association with diamond, and with beryl and chrysoberyl in the district of Minas Novas. This latter, which is known also as the district of Arrassuahy, is the most important locality for the blue and white varieties of topaz. It is situated in the north-east of the State of Minas Geraes, to the north-east of Diamantina, in the middle reaches of the diamond-bearing river Rio Jequetinhonha, known in its lower course as the Rio Belmonte, and to the south of this river. The gem-bearing deposits of this region extend over the plateau between the Rio Jequetinhonha and the Rio Arrassuahy towards the south and east as far as the Serra das Esmeraldas, a part of the Serra do Espinhaço. The precious stones which occur here are, or at least were at one time, of great commercial importance. The colourless topaz pebbles found here are known as "pingos d'agoa," or as "minas novas," after the district, while the blue stones are known in Brazil as " safiras " (that is to say, sapphires). In association with these two varieties of topaz are found garnet, chrysoberyl, aquamarine, rock-crystal, red quartz, amethyst, transparent spodumene, andalusite, and green tourmaline. The latter mineral, on account of its colour, was supposed to be emerald, hence the name Serra das Esmeraldas given to a mountain range in the neighbourhood. The principal localities are in the wooded and inaccessible wilds between the Rio Jequetinhonha and the three source-streams of the Rio S. Matheus, usually known as the Rio Americanas. Topaz pebbles are found loose in the débris of these streams, as well as in others which flow into the Jequetinhonha ; one of these, the Ribeirão Calhão, is well known and has been already mentioned as a locality for chrysoberyl. Workings for precious stones are reported to be in existence also on the upper Rio Piauhy.

Topaz is extremely abundant here, much more so than any of the other precious stones mentioned. It occurs as broken fragments, or more frequently as rounded, water-worn pebbles, the size of which varies between that of a pea and that of a chestnut, Larger fragments or pebbles the size of a man's fist or head, and weighing several pounds, have been met with, but are rare. The best quality of white topaz is said to be found in the Rio Utinga, but the " pingos d'agoa " are by no means confined to the bed of this river. The blue topaz is sometimes dark in shade, and sometimes pale or almost colourless ;

it is found as pebbles ranging in weight up to several ounces. Those of a dark shade are the most valuable.

The original situation of the blue and colourless topaz pebbles has not yet been ascertained. The commonest rocks in the neighbourhood are granite and gneiss, and it is probably from such that the stones have been derived. In many parts hereabouts there is a surface layer, sometimes as much as 14 feet thick, of weathered material, consisting mainly of quartz fragments, containing the same precious stones as are to be found in the river beds. This weathered material has been derived from the underlying granite and gneiss; it seems very probable, therefore, that the precious stones contained in it had a

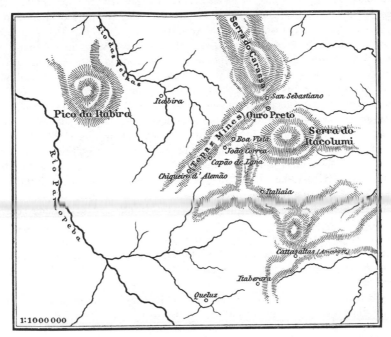

Fig. 67.   Occurrence of yellow topaz near Ouro Preto in Brazil.   (Scale, 1 : 1,000,000.)

similar origin, having been formed, perhaps, in the quartz veins by which the underlying rocks are penetrated. Small isolated grains of topaz and other precious minerals are to be found actually embedded as constituents in these rocks. The thick mantle of weathered material, together with the precious stones it contains, is gradually carried away by the action of running water, and during the transportation the stones assume the rounded form characteristic of the pebbles of a river bed. The mode of occurrence of topaz in these secondary deposits has a great general resemblance to that of diamond at Diamantina, though at present diamond has not been met with in Minas Novas.

The distribution and mode of occurrence of yellow topaz is quite different. Fig. 67 is a map of the district in which it occurs. The precious stone was here discovered about the year 1760 in the neighbourhood of Ouro Preto, then known as Villa Rica, in the south-west of Minas Geraes, of which State it is the capital. At that time yellow topaz was in great demand, and the search for it was prosecuted with great vigour. It is most abundant in a chain of hills extending for one and a half legoas (about six miles) in a south-west to north-east direction from Capão de Lana through João Correa and Boa Vista to Ouro Preto. The deposit has been traced as far south as Chiqueiro d'Alemão.

The topaz is here confined to a band, a few hundred yards wide, which extends with little interruption throughout the range of hills in the direction mentioned. This is the primary situation of the stone, that is to say, the place where it was formed. Water-worn pebbles of yellow topaz are rare, but have been met with in the " tapanhoacanga " of the neighbouring streams, a term which was explained in connection with the occurrence of diamond in Brazil. Although the topaz now lies unchanged at the place where it was originally formed, yet the rock in which it occurs has undergone considerable decomposition and alteration. These rocks consist of clay-slates associated with itacolumite. They are penetrated by quartz veins, and it is probable that the topaz, like the diamonds of Diamantina, was formed in cavities in these quartz veins. The clay-slates are altered in places by weathering into a soft clayey mass, and at isolated spots in this mass are found druses or nests, containing detached topaz crystals embedded in a clay or scaly kaolin, which ranges in colour from white to dark brown. These nests are probably isolated portions of the quartz veins by which the rock was penetrated and in which the topaz crystals were formed ; crystals are also found, however, in the clayey mass itself.

The minerals associated with topaz are practically the same as have been frequently observed to occur with diamond in the quartz veins which intersect the itacolumite and accompanying rocks. Besides quartz (rock-crystal and smoky-quartz) they include ilmenite, hæmatite, rutile, black tourmaline, and the rare euclase ; all occur in broken fragments embedded in the clay. It should, however, be noted that euclase is never found actually with topaz, but always in a druse by itself.

A different opinion as to the origin of yellow topaz in Brazil has been recently (1901) expressed by Orville A. Derby. He states that it occurs near Ouro Preto, usually in nodules in a clayey matrix which has resulted from the decomposition of a mica-schist. This latter, he considers, has been formed by the metamorphism of an igneous rock of the augite- or nepheline-syenite group, in the drusy cavities of which topaz had crystallised out.

The mode of occurrence of topaz at Saramenha, half an hour's journey from Ouro Preto, is somewhat different. Here the crystals are embedded in a deposit of brown iron-stone (limonite) intermixed with micaceous iron-ore (hæmatite), in which, after removal, they leave sharp, bright impressions. The mineral is here pale yellow in colour and abundant in quantity.

The topaz which occurs in the neighbourhood of Ouro Preto varies in shade from a pale yellow to a dark wine-yellow. As a general rule, the darker the kaolin in which the topaz crystals are embedded the darker are these crystals. A finely coloured crystal from this district is represented in Plate XIII., Fig. 2, and a faceted stone in Fig. 2a, of the same plate. The most beautiful of the stones found in this district are those having the rich colour of old Malaga wine. Red crystals (Plate XIII., Fig. 4,) also occur ; they are usually of a pale rose shade, but may be a dark ruby-red, when they are known as " Brazilian ruby," a term which is also applied to stones the red colour of which has been produced by artificial means. A faceted " Brazilian ruby " is represented in Plate XIII., Fig. 4a ; the stone is highly prized by connoisseurs.

The degree of transparency varies in different specimens, and the crystals are not all of gem-quality ; thus among a thousand stones there may perhaps be but one perfect example, all the rest being faulty in some way or another.

With regard to size, crystals with a length of 6, or even of 10 inches, and a thickness of 2 or 4 inches, have been described by L. von Eschwege. Such large stones, however, are almost always more or less faulty, and are rarely suitable for cutting. The majority are much smaller, about the length and thickness of a little finger. The form of the crystals is usually quite simple, like that shown in Fig. 66a.

In this locality excavations are made in the clayey mass in which the topaz occurs, and the larger nests, when met with, are carefully removed and opened. The crystals which lie loose in the clay are obtained by allowing a stream of water to play upon the loosened masses of clay in the mine. By this means the lighter material is washed away, and the heavier topaz crystals caught in the meshes of a net spread out for that purpose.

The annual output of topaz at one time amounted to as much as 18 hundredweight, but on an average was not more than 7 or 8 hundredweight, a large proportion being yielded by the estates of Capão de Lana and Boa Vista, which are specially rich in topaz. It is said that the mining of topaz at one time afforded employment to as many as fifty persons. The stones find their way into the market by way of Rio Janeiro, some being cut on the spot. The valley of Ouro Preto is studded with innumerable abandoned mines, mute witnesses of former activity in this district. As the demand for these stones gradually fell off the mines were one by one abandoned, and systematic work has now ceased for a long time. Many are of opinion that the locality is practically exhausted; others aver, however, that there are still rich treasures to be found.

The occurrence of topaz in **Mexico** is of little commercial importance. It is found at La Paz in the State of Guanaxuato, at San Luis Potosi, and at Durango, in stanniferous deposits. The crystals are pale in colour or colourless and water-clear.

The mineral is widely distributed in the **United States** of North America, but crystals of gem-quality are somewhat rare, the best material coming from the Western States. Transparent and water-clear, bluish, and greenish crystals are found, together with beryl and other minerals, in granite, at Harndon Hill, near Stoneham, in Maine; also at other places in the neighbourhood, and at North Chatham in New Hampshire. Topaz crystals, very similar to those from Saxony, are found in the granite of Trumbull, Connecticut, but they are usually cloudy and rarely of gem-quality.

In Colorado fine crystals of a pale blue colour, or colourless and water-clear, and occasionally of considerable size, are met with. They occur with phenakite and other minerals in drusy cavities in granite at various points in the Pike's Peak region, in El Paso County. Thus, for example, at Florissant, twelve miles north of Pike's Peak, they are found embedded in green felspar (amazon-stone); and in the neighbourhood of Devil's Head Mountain, about thirty miles from Pike's Peak, colourless, reddish, wine-yellow, and pale blue crystals, similar to those of Mursinka in the Urals, are found in the solid rock or loose on the ground. Another, not altogether unimportant, locality is Mount Antero, about ten miles north of Salida, in Chaffee County, Colorado. The Colorado localities have yielded the best specimens of North American topaz of gem-quality; two of the largest after cutting weighed respectively 125 and 193 carats. At several places topaz has been found also in younger volcanic rocks, namely, in rhyolites; for example, at Nathrop in Chaffee County, and on Chalk Mountain in Colorado.

Very fine colourless crystals are met with embedded in solid rock, or loose in its weathered product in the Thomas Range, forty miles north of Sevier Lake in Utah, and at the same distance north-west of the town of Deseret on the Sevier river. The topaz found at these localities in the State of Utah is perhaps the finest in the United States.

A few stones of gem-quality have been met with at all the localities mentioned, and at some others.   They are prized by the American as a production of his native country, but North America as a source of topaz has no commercial significance.

Crystals of **Russian** topaz are remarkable both for size and beauty, stones of fine quality and as much as 31 pounds in weight having been found.   They are often cut at Ekaterinburg, together with the variously coloured precious stones with which they are found ; they find their way into the markets, in the rough or the cut condition, by way of the fairs of Nizhniy Novgorod.

As in Minas Novas, so also at most of the Russian localities, topaz and beryl occur together ; the one mineral is never found without the other, except in the Altai Mountains where beryl occurs, but no topaz has at present been found.   The distribution of Russian topaz is practically the same as that of Russian beryl, which has been already dealt with. It will be unnecessary, therefore, to give here anything more than a few facts relating specially to the occurrence of topaz.

Topaz is specially abundant in the neighbourhood of the village of Alabashka, near Mursinka (Fig. 63a), in the Ekaterinburg district of the **Urals.**   It is found in druses in granite, together with smoky-quartz, beryl, large yellow crystals of felspar, small crystals of white albite arranged in spherical groups, and red plates of lepidolite ; these minerals all occur in well-developed crystals, and the combination of different colours renders the druse a very beautiful object.   The smallest crystals of topaz are about the size of a pin's head, while the largest are several centimetres in length.   They are usually bluish in colour, as represented in Plate XIII., Figs. 1 and 1a, sometimes light bluish-grey or greyish-white, rarely colourless.   As a rule, they occur singly in the druses, but sometimes grouped together in parallel position.   The usual crystalline form is the simple one represented in Fig. 66c, and in the coloured figure just cited.   With regard to transparency, some crystals are perfectly clear, while others are only translucent ; the transparent ones are cut at the works in Ekaterinburg and fetch a moderately high price.   The gem-mines near Mursinka will be again considered when we come to treat of amethyst.

Another Uralian locality for topaz is the neighbourhood of the smelting works of Miask, on the east side of Lake Ilmen.   Its mode of occurrence here is the same, namely, in drusy cavities in pegmatite ; these cavities are sometimes filled with a white clay, embedded in which are topaz crystals which have been detached from the walls of the cavity.   The pegmatite-veins are here found at four places, traversing a rock known as miascite. Associated with the topaz is green felspar (amazon-stone), in which it is frequently embedded, also phenakite, mica, and other minerals.   Two varieties of topaz occur here.   One is colourless and perfectly transparent, like the " pingos d'agoa " of Brazil, and occurs as symmetrically developed crystals rich in faces.   The other variety is of a dirty yellowish-white colour, translucent only at the edges, and so fissured and decomposed or, as it is described at the place, rotten, that the crystals, which are bounded by only few faces, may be easily crushed between the fingers.   Both these varieties occur in crystals of about the same size as those found at Alabashka.

Topaz is also found in the gold-washings, belonging to a merchant named Bakakin, in the valley of the Sanarka (a tributary of the Ui, which itself feeds the Tobol), as well as in a few tributary streams in the Southern Urals (Government Orenburg).   The crystals found here are so very similar to Brazilian topaz that their Uralian origin was at first doubted.   They usually retain their crystalline form, which is simple, like that shown in Fig. 66a.   Their colour is generally some shade of yellow ; some, however, are red and a few quite colourless. Many are beautifully transparent.   The largest crystals have a length of $2\frac{1}{2}$, and a thickness of $\frac{3}{4}$ centimetres.   The topaz in these river-sands is associated with a great variety of precious stones, some of which have been already mentioned. They include quartz (amethyst), corundum (ruby), chrysoberyl (alexandrite and cymophane), spinel, chalcedony (carnelian, agate, &c.), staurolite, kyanite, euclase, tourmaline, garnet, beryl, &c. In the case of rose-topaz its place of origin is known ; it occurs with green chromiferous tourmaline and green chromiferous mica (fuchsite) in quartz veins or nests in carboniferous limestone, which in this district forms a deposit extending over a wide area.

The topaz found in the Adun-Chalon Mountains, in the Nerchinsk district of **Transbaikalia**, is much fissured and far from being perfectly transparent. It forms with quartz the so-called topaz-rock, veins of which penetrate the granite. It has been already mentioned, in the description of beryl, that cavities in this topaz-rock are lined with crystals of beryl, smoky-quartz, and topaz, and that owing to the weathering of the rock these minerals lie loosely scattered in the surface soil.

In the mountain range, Kuchuserken, topaz was first met with at the beginning of the fifties of the nineteenth century. Although this range may be considered as a continuation of the Adun-Chalon Mountains, yet the topaz found here is more like that which occurs with beryl in granite near the Urulga river, in the Borshchovochnoi range.

The topaz of the last-named locality is distinguished by its exceptional beauty of colour and transparency and by the size of its crystals. In respect to the large size of the crystals it exceeds all other Russian topaz. Thus a perfectly transparent, dark honey-yellow crystal found here weighed 3 pounds, another fine transparent crystal of a pleasing dark wine-yellow colour weighed over 25 pounds, while a third measured 19 by 21 centimetres, and weighed 31 pounds ; this, however, which has been already mentioned, was only translucent and of a dirty yellow colour. Several very fine crystals from this district are preserved in the British Museum collection of minerals. In the majority of crystals the colour is something between the brown of smoky-quartz and the yellow of Brazilian topaz ; unfortunately this colour is speedily bleached on exposure to light. Sometimes it is dark honey-yellow ; stones showing light tints of this and other shades of yellow are also seen, as well as pale-blue, bluish-white, and perfectly colourless examples. The crystals occur singly or in groups, the individuals of a group having grown together in parallel position.

The Daurien district in the southern part of Transbaikalia is another locality for topaz, fine water-clear and well-developed crystals are found in the Shilka river, the upper course (or main supply stream) of the Amur.

Fine topazes also occur elsewhere in Asia. Those found in the neighbourhood of Mukla, or Mugla, in **Asia Minor**, resemble the yellow Brazilian topaz so closely both in form and colour that they can scarcely be distinguished from them. No details are known as to the locality or mode of occurrence. The stones vary in shade from a dark honey-yellow to pale wine-yellow ; sometimes they are rose-red, rarely blue. Their form is the same as that of the Brazilian crystals, Fig. 66a, and Plate XII., Figs. 2 and 4.

The occurrence of topaz in India is not well authenticated. In those instances in which it is supposed to have been found, it is probable that quartz or some other mineral has been mistaken for it. The occurrence of topaz in **Ceylon** is, on the other hand, well established. It occurs in abundance as colourless and pale or dark yellow pebbles in the gem-gravels, together with sapphire and other precious stones, which are all collected and sent to market together. The fine saffron-yellow variety of topaz, mentioned above, occurs as a great rarity in Ceylon ; it is distinguished as " Indian topaz." A large pebble, weighing 12 pounds 13 ounces, of perfectly colourless and transparent topaz, probably from Ceylon, is to be seen in the British Museum collection of minerals.

In recent times the mineral has been found in **Japan** as water-clear, pale-yellow, or greenish-blue crystals of moderate size. They occur in river gravels at various places, and have been derived from pegmatite veins intersecting granite and gneiss. Many of the crystals are well suited for cutting as gems, and Japanese topaz will probably become of importance commercially. Blue, green, and yellow topaz has been found also in Kamchatka.

In **Africa** topaz was found in former times in the same district of Egypt in which emerald occurs, namely, on Jebel Sabara, near the Red Sea. Numerous ancient topaz mines have been rediscovered here, but have been scarcely worked at all in modern times owing to the low price of topaz. Risk Allah is the only place in this region where topaz is mined at the present time. The mineral occurs also in German South-West Africa, sometimes turbid and cloudy, but mostly transparent and water-clear, though not of a quality suitable for cutting. It varies in colour from wine-yellow to brownish-yellow and is rarely distributed.

Finally, **Australia** as a topaz locality must be mentioned. The mineral is distributed widely in this continent, and occurs especially in gravels. Colourless, bluish, greenish, and yellow pebbles, the latter very like Brazilian stones, are found associated with cassiterite and diamond in the stanniferous gravels of the granitic region of New England in the north-east corner of New South Wales. The cassiterite (tin-stone) and topaz have both been derived from the granite. Topaz occurs, in a similar manner, in the rivers further south as an associate of diamond, the distribution of which is shown in the map, Fig. 43. The topaz pebbles found here are often beautiful and of considerable size, the largest weighing several ounces. They are colourless or blue, sometimes yellow. Yellow topaz pebbles have been found also in Owen's river in Victoria and at other places. At all these Australian localities topaz is sought for and cut as gems; this is probably scarcely the case with the topaz accompanying cassiterite in Tasmania, which is of poor quality.

ZIRCON.

Although zircon is of less importance and is less frequently cut as a gem than the precious stones hitherto considered, yet it has a certain vogue, the transparent yellowish-red variety, distinguished by the name hyacinth (jacinth), being most used.

Zircon is a compound of the oxides of silicon and zirconium; that is to say, a compound of silica and zirconia. It contains 23·77 per cent. of silica (SiO_2) and 76·23 per cent. of zirconia (ZrO_2), a composition corresponding to the chemical formula $ZrO_2.SiO_2$.

The forms taken by the crystals of this mineral belong to the tetragonal system and are usually very simple; four of the commonest forms are represented in Figs. 68a to d. The crystals are usually short, comparatively thick, and with faces symmetrically developed on all sides; they are bounded by square prisms and tetragonal octahedra of two orders. Hyacinth, which is practically the only variety used as a gem, scarcely ever occurs in any form other than that shown in Figs. 68b and c, and in Plate I., Figs. 11 and 12. We have here a square prism of the second order, with its edges sometimes truncated by narrow faces of a square prism of the first order (Fig. 68c), and terminated by a tetragonal octahedron, or pyramid, of the first order, so that at each end there is a four-faced pyramidal termination. The other forms shown in the figures are those assumed by common zircon.

Zircon has a very imperfect cleavage, scarcely observable in fact; its fracture is distinctly conchoidal. The mineral is harder than quartz but softer than topaz, its hardness being represented on the scale by $7\frac{1}{2}$; this is not very great, but is sufficient to

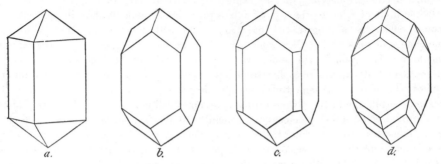

a.   b.   c.   d.

FIG. 68.   Crystalline forms of zircon.

admit of a brilliant polish, so that an artificially polished facet, like a natural crystal-face, shines with a brilliant, adamantine-like, vitreous lustre. The specific gravity is very high; it varies between 4·610 and 4·825, the mean value for the hyacinth variety being 4·681. Zircon is thus denser than any other precious stone, heavier, indeed, than any mineral not containing the heavy metals (lead, silver, copper, &c.) in large amount, and when placed in the heaviest liquid it quickly sinks.

Zircon may be cloudy and opaque, or clear and transparent. The opaque, common zircon is usually brown or grey in colour, sometimes between green and black, and is little used as a gem. A fire-red, cloudy zircon, called after its place of origin " Ceylonese zircon," is sometimes cut as a gem. Precious zircon is not always perfectly transparent; in such cases the stone, though markedly translucent, is still pleasing in appearance by virtue of its brilliancy and lustre, which is comparable to that of the diamond. Zircon is only rarely perfectly colourless and water-clear; crystals answering this description occur implanted on chlorite-schist at Wildkreuzjoch, in the Tyrol, and also in Ceylon. More frequently it is green, brownish-red or brown, and sometimes violet; but by far the commonest tint is a brownish shade of orange resulting from a mixture of red and yellow in about equal proportions. The latter colour inclines in some stones more to red, and in others to yellow, and some are dark and others lighter in shade. Zircon of this colour, which is known as *hyacinth*, and is practically the only variety commonly cut as a gem, is represented in Plate I., Figs. 11 and 12. Transparent green zircons, like the one illustrated by Fig. 13 of the same plate, are sometimes cut as gems, as are also the reddish-brown, brown, and violet varieties. It is only exceptionally, however, that these latter are sufficiently transparent for this purpose, and when they are, they are sometimes mistaken by dealers for tourmaline.

The reddish-yellow colour characteristic of hyacinth is known as hyacinth-red. The same colour is seen in cinnamon-stone or hessonite (Plate XIV., Figs. 7 and 8), a variety of garnet which occurs in association with hyacinth in Ceylon. So much alike are these two gems that hessonite is often sold by jewellers for hyacinth; indeed, it has been stated that practically the whole of the so-called hyacinth bought and sold in European markets is in reality hessonite, although this stone is far inferior both in brilliance and lustre to the true hyacinth-red zircon. The means by which the two stones may be distinguished will be given under hessonite ; it may be stated here, however, that the distinction is based on the fact that hessonite is singly refracting and hyacinth doubly refracting.

The very pale straw-yellow or colourless zircons from Ceylon are also cut as gems, and are called by jewellers "cerkonier" or *jargon* (jargoon). Though this variety of zircon occurs but rarely in nature, it can be produced artificially to any amount by heating hyacinth of the ordinary colour, this being very easily decolorised by heat. Thus, a stone on being brought near to the tip of a blowpipe flame suddenly loses its red colour, and becomes colourless or tinged with grey, sometimes with rose-red or straw-yellow. The same change of colour takes place when hyacinth is heated in a glass-tube. If the experiment is made in a dark room it will be observed that the stone, although at a temperature below red-heat, suddenly emits a phosphorescent light and becomes decolorised. Other important changes take place at the same time ; the specific gravity of the stone is increased by 0·1, or even more, while the lustre, always brilliant, becomes still more so. So brilliant indeed are these ignited, colourless or pale-coloured hyacinths, that when cut as rosettes it is impossible for any one other than an expert to distinguish them on mere inspection from diamonds, and there is no doubt that they are occasionally passed off as such. The stones found at Matura, in Ceylon, in the eighteenth century, were, in fact, regarded as diamonds of inferior quality, and were known as " Matura diamonds." The diamond, like hessonite, differs from hyacinth in that it is singly refracting ; its hardness, moreover, is, of course, enormously greater, and its specific gravity is less ; in the heaviest liquid (sp. gr. = 3·6) diamond floats and hyacinth rapidly sinks.

The phenomena attending the decolorisation of hyacinth have been somewhat closely investigated with a view to learning the nature of the colouring substance. It is found that crystals when heated in the presence of oxygen—for example, in the oxidising flame—do not lose their colour completely, but become paler. When, on the other hand, they are heated out of contact with oxygen—for example, in the reducing flame—they are completely decolorised ; and if these decolorised stones are again heated in the presence of oxygen they assume a pale red colour. Moreover, it has been observed that a very strongly ignited hyacinth becomes dark brown in colour. From these observations it is inferred that the colour of zircon is due to the presence of iron ; as a matter of fact, every specimen of zircon hitherto analysed has been found to contain small amounts, ranging up to 2 per cent., of iron oxide.

The colour and lustre of some hyacinths is liable to change even at ordinary temperatures if the stones are exposed to light, especially to the direct rays of the sun. In some cases the colour becomes pale, while in a few stones it changes to a brownish-red which gradually becomes more decidedly brown. At the same time the adamantine lustre becomes gradually more vitreous in character. Such altered stones, if kept in darkness, will recover their original colour and lustre to a large extent, but not altogether. Although these changes are not undergone by all hyacinths, it is advisable not to expose the gem to sunlight unnecessarily.

Zircon is more strongly refracting than any other precious stone with the exception of diamond. Being a tetragonal mineral it is doubly refracting ; the values for the greatest and least refractive indices of a crystal of a hyacinth from Ceylon were determined to be 1·97 and 1·92 respectively. The difference between the two, 0·05, is a measure of the double refraction, which is thus specially strong. On the other hand, the refractive indices for different colours do not differ much, hence the dispersion of zircon is small and its play of prismatic colours correspondingly insignificant, so that, although in brilliancy and lustre t may compare with the diamond, yet in the former respect no comparison is possible.

The dichroism of hyacinth, as of all varieties of zircon, is very feeble; it is almost impossible to detect any difference between the two images seen in the dichroscope. This instrument, which is frequently so useful in discriminating gems, is therefore of no assistance whatever in distinguishing the feebly dichroic hyacinth from hessonite, which has no dichroism at all. The dichroism of zircon of other colours, though stronger than that of hyacinth, is feebler than that of any other coloured doubly refracting precious stone.

A characteristic feature of many zircons, discovered by Prof. A. H. Church, is the presence of black absorption bands in the spectrum of white light which has traversed the stone. These absorption bands are attributed to the presence of small quantities of uranium compounds.

Some few characters of zircon still remain to be mentioned. It is infusible before the blowpipe, and is not attacked by acids, not even by hydrofluoric acid. When rubbed it becomes slightly electrified, but the charge is not sufficient to render this character of use for purposes of determination.

Zircon occurs in the older crystalline silicate-rocks, such as granite, gneiss, and other similar rocks of mountainous districts. Opaque zircon is a common constituent of such rocks, and precious zircon, including hyacinth, is found under similar conditions. The crystals are, as a rule, embedded in the rock, and are only rarely found attached to the walls of drusy cavities. Some few of these rocks contain such an amount of common zircon that the name of the mineral is used as a distinguishing prefix, as in the zircon-syenite of the neighbourhood of Fredriksvärn and Laurvik in the south of Norway; while some other localities, notably in North America, yield common zircon by the hundredweight. Isolated crystals of zircon of the hyacinth variety are sometimes found embedded in the younger volcanic rocks; for example, in the basalt of Expailly near Le-Puy-en-Velay (Department Haute-Loire) in France; in Germany in the so-called mill-stone lava of Niedermendig on the Laacher See, in the basalt of Unkel on the Rhine, and in some basalts in the Siebengebirge. A crystal of hyacinth, partly freed from the black basalt in which it is embedded, is shown in Plate I., Fig. 12. It is improbable that such zircon crystals were actually formed in the basaltic rock; their presence there may be explained by supposing that fragments of granite, or rock of similar nature containing these crystals, were caught up by the glowing, fluid basalt magma and all their constituents melted down except the resisting hyacinth, which remained unaltered, thus becoming an apparently normal constituent of the basalt.

Crystals of zircon are set free by the weathering of the mother-rock, carried away with the rock débris by running streams, and thus eventually become a constituent of sands and gravels. Owing to the unalterable nature of their substance, such crystals remain perfectly fresh and unweathered. The zircon used for cutting as gems is derived exclusively from such sands and gravels, never from the solid rock.

Various localities in the island of Ceylon send to the markets of the world almost the whole of the supply of hyacinth and other gem varieties of zircon. They are quite abundant, and are collected from the gem-gravels, together with spinel, sapphire, cat's-eye, and other Cingalese gems. The principal sources are the deposits in the neighbourhood of the town of Ratnapura in the Saffragam district, and those of Matura in the south of the island (Fig. 59); from the latter place come the colourless and pale-coloured, ignited hyacinths known as " Matura diamonds."

The crystalline form of the zircons found in these sands is still recognisable, although the crystals, like those of the other precious stones with which they are found, are much rounded. The crystals of hyacinth found in Ceylon are, as a rule, rather small, at most no larger than a lentil; specimens the size of a pea are rare, and still larger stones quite exceptional.   The dimensions of two exceptionally large stones have been given as $5\frac{1}{2}$ and 6 lines in length and $4\frac{1}{2}$ and 7 lines respectively in thickness.   Other colour varieties of precious zircon, which accompany hyacinth in Ceylon, occur in crystals of larger size, fine stones a centimetre in length being by no means unusual.   There is no doubt that the different varieties of zircon found in the gem-gravels of Ceylon have been derived from the same rocks in which the sapphires, for example, with which they are associated, also took their origin.   In a few rare cases crystals of hyacinth have been found actually in the matrix, which is a granite or gneissic rock, like the mother-rock of sapphire already described.

Compared with Ceylon, other localities for zircon are unimportant.   Many supposed occurrences in India are by no means well authenticated ; such as, for example, in the alluvial gravel at Ellore in the Madras Presidency, and in granite at Kedarnath on the Upper Ganges.   The occurrence of hyacinth with ruby in Upper Burma has been reported.

The occurrence of zircon in Europe is well established, but unimportant.   It is found with sapphire in sands, very similar to those of Ceylon, in the Iserwiese in the north of Bohemia ; the crystals from this place are, however, smaller in size and fewer in number than in the sands in Ceylon.   These zircons, sapphires, and other precious stones with which they are found have been derived from gneiss.   The hyacinths which occur in the streams near Expailly, in France, have been weathered out of basalt ; here again the mineral occurs under the same conditions as in Ceylon, but in small crystals few in number.

The occurrence of zircon in Australia, though of little importance in the trade, must not be forgotten.   It is found in auriferous, and also in diamantiferous, sands, more especially at various places in New South Wales ; fine specimens have been met with at Mudgee in this State.

Fine, richly coloured zircons are found in the gold-sands of North Carolina in the United States ; the crystals are unfortunately, however, too small for cutting.

In conclusion we must consider the manner in which zircons are used in jewellery. The coloured stones are most frequently cut as table-stones or thick-stones, or sometimes they are given step-cut or brilliant-cut forms, according to their transparency and depth of colour.   Fine stones of a pure colour require no special devices to improve their appearance, but inferior specimens are mounted on a gold foil or in a closed setting lined with black. Colourless stones, especially those which have been burnt, are usually cut as rosettes, a form which, on account of the brilliant lustre and absence of any play of prismatic colours, is more suitable than the brilliant, though the latter is sometimes employed.   These burnt hyacinths were at one time, on account of their peculiar dusky lustre, used in mourning jewellery in preference to diamonds.

Hyacinth is not worth much at the present time ; the demand for the gem has fallen off very considerably, and the genuine stone is rarely met with in the trade.   Quite small stones are, of course, cheap, on account of the supply being large ; it is only larger stones that are of any appreciable value.   A good, faceted 1 carat stone of fine colour is valued at from 50s. to 75s., while a number of small stones of similar quality and weighing 1 carat together would not fetch at the most more than 10s. or 12s.   Corresponding values also hold good for zircons of other colours.

The similarity in general appearance between hessonite, or cinnamon-stone, and hyacinth, and between diamond and burnt hyacinth, together with the methods by which they may be distinguished, have been already mentioned.   Another stone which may resemble hyacinth, more or less, is yellowish-red corundum, known on this account as " oriental hyacinth."   Its lustre is as brilliant and strong as that of true hyacinth, from which it is distinguished by its greater hardness ($H = 9$) and lower specific gravity (sp. gr. $= 4.0$).   The two stones differ also in another respect ; for " oriental hyacinth " is distinctly, though feebly, dichroic, while the dichroism of true hyacinth is so feeble as to be scarcely observable.

The name hyacinth is sometimes also applied to red crystals of ferruginous quartz of the variety known as " Eisenkiesel," which occurs, for example, embedded in gypsum at Santiago de Compostela in the north of Spain, and is known as " Compostela hyacinth." It is much inferior to true hyacinth both in lustre and transparency ; moreover, being quartz, its specific gravity is only 2·65 and it floats in methylene iodide.   Some of the colour-varieties of tourmaline are somewhat similar in appearance to certain zircons, less often, however, to hyacinth.   They may be readily distinguished, however, for tourmaline is strongly dichroic, and being much lighter than zircon floats in methylene iodide.   Rose-topaz is also sometimes substituted for hyacinth ; here again the former is of quite a different shade of colour and is strongly dichroic.

Imitations of hyacinth in coloured glass may be easily distinguished from genuine stones by the fact that they are singly refracting and much less hard.

THE GARNET GROUP.

Garnet is extensively used as a gem, and is to be found adorning some of the costliest as well as the simplest articles of jewellery.   If the ordinary display in a jeweller's window is observed, it will often be found that at least one half of the gems exposed for sale are garnets of various kinds.

Garnet differs from diamond, corundum, and many other precious stones hitherto considered in that its chemical composition is not fixed and unchangeable, but is subject to considerable variation.   The minerals grouped together as garnet have some physical characters in common ; they all have the same crystalline form and the same type of chemical constitution.   The several members of the group differ from each other, however, in the chemical elements of which they are composed ; the garnets, in fact, form what is called an isomorphous series, such as is often met with in the mineral kingdom.

The point of agreement in the chemical composition of all garnets is the association of three molecules of silica (SiO_2) with one molecule of a sesquioxide, represented generally by R_2O_3, and with three molecules of a monoxide, represented by MO.   The chemical formula which expresses the composition of the whole garnet group in general terms is therefore $3MO.R_2O_3.3SiO_2$.   The various members of the group differ from each other, however, in the nature of the monoxide and sesquioxide which take part in their constitution ; thus the monoxide may be lime (CaO), ferrous oxide (FeO), or magnesia (MgO), occasionally also manganous oxide (MnO), or chromous oxide (CrO), while the sesquioxide may be alumina (Al_2O_3), or ferric oxide (Fe_2O_3), or sometimes chromic oxide (Cr_2O_3).

These different monoxides and sesquioxides are associated together in the most varied proportions, always, however, conforming to the general formula given above.   All the varieties of garnet, which it is theoretically possible to construct by different combinations of the oxides mentioned above, are not actually known to exist in nature ; the small number, which are known, are distinguished according to the particular oxides present. Omitting those of less importance, these are :

1. Calcium-aluminium garnet   .   .   .   . $3CaO.Al_2O_3.3SiO_2.$
2. Iron-aluminium garnet .   .   .   .   . $3FeO.Al_2O_3.3SiO_2.$
3. Magnesium-aluminium garnet   .   .   . $3MgO.Al_2O_3.3SiO_2.$
4. Calcium-iron garnet   .   .   .   .   . $3CaO.Fe_2O_3.3SiO_2.$
5. Calcium-chromium garnet   .   .   .   . $3CaO.Cr_2O_3.3SiO_2.$

The results of chemical analyses show that each of these garnets may occur in nature in a moderately pure condition.   In most garnets, however, the part of the monoxide or of the sesquioxide is not played exclusively by calcium oxide, ferrous oxide, &c., or by aluminium oxide, ferric oxide, &c., but by mixtures of two or more oxides.   Thus, there are garnets which contain, beside silica and alumina, two monoxides—lime and ferrous oxide—of which now the first, and now the second, predominates.   Thus in one case the mineral approaches in character to a calcium-aluminium garnet, and in the other to an iron-aluminium garnet, and it may be considered to be a mixture in varying proportions of the two molecules $3CaO.Al_2O_3$ $3SiO_2$ and $3FeO.Al_2O_3.3SiO_2$.   The members of the garnet group are thus isomorphous mixtures of certain fundamental compounds, five of the most important of which have been given above ; a few others will also be mentioned later when each kind of garnet receives special consideration.   The garnet group is divided into a number of species known by particular names, the division being based upon differences in chemical composition.   The following table of the analyses of a few gem-varieties of garnet gives an idea of the diversity in their chemical composition, a composition which can, however, be reduced to one common type by applying the principle of isomorphous mixtures :

—	Hessonite, Ceylon.	Almandine.	Demantoid (emerald-green), Urals.	Pyrope, Bohemia.	Pyrope (hyacinth-red), Cape.	Pyrope (dark-red), "Cape ruby."
Silica (SiO_2) . . .	40·01	40·56	35·50	41·35	40·90	39·06
Alumina (Al_2O_3) . .	23·00	20·61	—	22·35	22·81	21·02
Chromic oxide (Cr_2O_3) .	—	—	0·70	4·45	1·48	—
Ferric oxide (Fe_2O_3) . .	—	5·00	31·51	—	—	2.69
Ferrous oxide (FeO) . .	3·31	32·70	—	9·94	13·34	18·70
Manganous oxide (MnO) .	0·59	1·47	—	2·59	0·38	0·58
Lime (CaO) . . .	30·57	—	32·90	5·29	4·70	5·02
Magnesia (MgO) . .	0·33	—	0·21	15·00	16·43	12·09
Total . .	97·81	100·34	100·82	100·97	100·04	99·16

The different varieties of garnet, though so diverse in chemical composition, nearly all occur in well-developed crystals of the same kind.   The crystals are met with embedded in rock, with faces fully developed on all sides, like the crystal represented in Plate XIV., Fig. 3, from which the surrounding matrix has been partly removed ; also attached to the walls of drusy cavities or crevices in rocks, such a druse being shown in Fig. 7 of the same plate.   The crystalline forms are those of the cubic system ; the most important are represented in Figs. 69a to d.   The rhombic dodecahedron (Fig. 69a) is so common and characteristic of garnet that it is sometimes known as the garnetohedron.   The twenty-four edges of this form are often more or less widely truncated by planes, which, as shown in

Fig. 29b, are usually delicately striated in the direction of their length. Sometimes the faces of the rhombic dodecahedron are more largely developed than are the truncating faces (Fig. 29b); at other times the reverse is the case, and the rhomb-shaped faces are small. These truncating faces belong to the icositetrahedron (Fig. 69c), a simple form, uncombined with others, frequently taken by garnet. The edges of the rhombic dodecahedron are in many cases not only truncated by the icositetrahedron, as in Fig. 69b, but the edges of

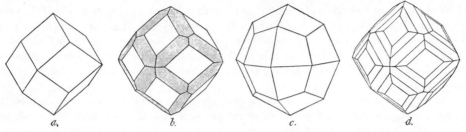

a.   b.   c.   d.

FIG. 69.   Crystalline forms of garnet.

intersection of these two forms are further truncated by delicately striated faces, the result being a form like Fig. 69d. This second series of faces are those of a hexakis-octahedron, a solid bounded by forty-eight faces, which is the greatest number possible on any single uncombined form. The simple hexakis-octahedron uncombined with other forms has not been observed in garnet. The mineral rarely takes a form other than those mentioned; those forms having a lesser number of faces, such as the octahedron and the cube, which, as a rule, are commonest in other minerals, are rarely seen in garnet.

The cleavage of garnet is more imperfect than in most other minerals; the fracture is sub-conchoidal to uneven. It is fairly hard, but this character varies in different varieties. All red garnets, which are the varieties chiefly used as gems, are harder than quartz but less hard than topaz, that is to say, for red garnets H = 7−8. The hardness of some green garnet is rather less; the demantoid, for example, which is sometimes used as a gem, has a hardness of $6\frac{1}{2}$ only, and is scratched by quartz; it is sufficiently hard, however, to scratch glass, so that it may be distinguished by this means from glass imitations. The pure emerald-green calcium-chromium garnet known to mineralogists as uvarovite is rarely cut as a gem, but is very nearly as hard as topaz. Garnet in the form of powder is a valuable grinding agent for precious stones and other hard substances, and is also used in the manufacture of the so-called emery-paper. It is the most widely distributed of any mineral with a hardness greater than that of quartz, and can be sold in large quantity at a low price. Good stones free from fissures serve for the construction of the pivot supports of watches, &c.

The specific gravity of garnet is another feature which varies to a large extent in different varieties. The variation is due to the diversity in chemical composition, the greater the proportion of heavy metal, such as iron, present in any one variety, the heavier will be that variety. The calcium-aluminium garnet, with a specific gravity of 3·4, is the lightest, while iron-aluminium garnet, with a specific gravity of 4·3, is the heaviest. The specific gravities of other varieties lie between these two extremes, and will be mentioned with the description of each. The specific gravity is a feature which enables us to distinguish garnet easily from other stones of similar appearance and from glass imitations.

The colour of garnet is not due to any intermixed pigment, but depends on the chemical composition of the mineral itself. This being the case, the colour, whatever it may be, is distributed throughout the substance of the stone with perfect uniformity and without patchiness.   When heated the mineral retains its colour, or if this be altered, it returns to the original shade when cool again.   Pure calcium-aluminium garnet is perfectly colourless ;   this white, so-called leuco-garnet is, however, never cut as a gem.   The commonest colour for garnet is red.   Every shade between the palest and one dark enough to be almost black is represented ; while the most diverse tones of red are met with, the most usual being tinged with brown, yellow, or violet.   Red garnet, the colour-variety which is almost exclusively used for gems, was formerly one of the stones to which the name carbuncle was applied.   It is probable that the term carbuncle formerly included all red stones, and not only the ruby, to which the application of the term is preferably limited at the present time.   The name of the mineral is said to have been derived from the colour of the gem-varieties, red garnets having been compared to the flowers and seeds of the pomegranate-tree.

Green garnet, the colour of which is comparable to a certain extent to that of the emerald, though usually more yellowish or brownish, is also cut as a gem, but much less frequently.   The finest emerald-green variety, the calcium-chromium garnet, or uvarovite, unfortunately occurs in such small and imperfectly transparent crystals that it is useless as a gem.

The colour of garnet is due to iron, and to a lesser extent to manganese and chromium. Iron is responsible for the reds, yellowish-greens, and the commonly occurring yellows and browns, while the colour of the emerald-green garnet, uvarovite, is due to chromium.

Black garnets also occur, their colour being also due to the presence of iron ; they are occasionally used in mourning jewellery.   Blue is a colour conspicuous by its complete absence in the garnet group.   The subject of colour will be further considered when the different varieties of garnet are dealt with.

As usual, in all minerals, different specimens of garnet show different degrees of transparency ; most crystals are turbid and opaque ; but amongst all varieties, numerous specimens of perfect clearness and transparency, sometimes combined with the deepest and darkest shade of colour, are to be found, and it is these alone which are cut as gems.

The natural faces of crystals vary in brilliancy, the lustre being sometimes very strong, at other times more feeble, owing to the roughness of the surface and to other causes.   The freshly fractured surface of a transparent stone is, however, always brilliant, and by grinding and polishing the lustre is still further increased, cut stones being very brilliant.   The lustre of garnet is the ordinary vitreous kind, so strongly inclined, however, to resinous lustre that some garnets closely resemble a piece of resin in external appearance.

Garnet, being a cubic mineral, is optically isotropic, that is to say, it is singly refracting ; anomalous double refraction is to be observed in a few rare cases, but scarcely ever in perfectly clear and transparent stones such as are cut as gems.   The mineral has a somewhat high index of refraction ; this, however, like the other physical characters, varies with the chemical composition.   The index of refraction for red light ranges from 1·74 to 1·79 in the different varieties.   The dispersion is almost always small ; indeed, the only variety which shows any very appreciable play of prismatic colours is the green calcium-iron garnet known as demantoid.   The value of the garnet as a precious stone depends upon its strong lustre and the depth and fulness of its colour.

In some cases when a candle-flame is observed through a stone, a four- or six-rayed star of light, similar to that shown by a star-sapphire, is seen.   This appearance, however, is rare and does not add to the beauty of the stone as a gem, nor to its value as a precious stone.

The fact that garnet is singly refracting, prevents its being mistaken for other stones which it may resemble in appearance, such, for example, as ruby or emerald, although this character does not distinguish it from a glass imitation. The absence of dichroism in a cubic mineral like garnet is, moreover, a further aid in its discrimination. The colour is the same in whatever direction light passes through the crystal, and the two images seen in the dichroscope are absolutely identical in shade. The absence of dichroism is most useful in distinguishing garnet from ruby. So similar in colour may the two stones be that even the practised eye of an expert fails to discriminate between them. As we have already seen, the garnet is frequently mistaken for ruby; for example, on account of the similarity in colour the fine red garnet associated with the diamond in South Africa is known as " Cape ruby."

As a rule, garnet fuses before the blowpipe with moderate ease; a few varieties fuse with more difficulty, and some are quite infusible. Crystallised garnet is unattacked by acids, except hydrofluoric acid ; after fusion, however, the material does not resist their action. The specific gravity of the mineral is also considerably less after fusion; thus the specific gravity of a yellowish-red calcium-aluminium garnet fell from 3·63 to 2·95. The varieties richest in iron are the most easily fusible; these, moreover, have a slight action on the magnetic needle, and the mass which remains after fusion is attracted by a magnet. All garnets when rubbed on cloth acquire a feeble charge of positive electricity.

The forms of cutting employed for garnet are those best suited to more or less darkly coloured stones. Most varieties are cut *en cabochon*, with a circular or oval outline, and with a very considerable curvature, so that the stone is high and hemispherical in shape. In the case of very dark specimens, the underside is hollowed out so as to render the stone thinner and more transparent, a device adopted for scarcely any other precious stones. These so-called garnet-shells (Plate IV., Fig. 17*b*) have been found in numbers in Roman ruins, so that the device was evidently known in ancient times. The table-cut and the step-cut are the most usual forms, while the mixed-cut is not infrequently seen. With dark-coloured stones each of these forms must be cut as thinly as possible. The table or large facet at the front of the stone, instead of being plane frequently has a convex curvature. The rosette or brilliant is used for some kinds of garnets and irregular, fanciful forms are also met with. A few cut garnets are represented in Plate XIV., Figs. 4, 6, 8, and 10, and in Plate XVIII., Fig. 7. Grains of garnet are often provided all round with small facets, arranged regularly and systematically or in no particular order; these are bored and strung together for bracelets and necklaces. Stones of the lighter shades of colour are mounted *à jour*, while the darker specimens are often mounted upon a burnished foil of silver or copper.

The different varieties of garnet differ very considerably in value. The worth of a stone depends for the most part on the beauty of its colour and the rarity of its occurrence. Particulars as to prices will be given later on under the description of different varieties.

The commonest fault in garnet is the presence of fissures along which the stone is apt to fracture. Garnets are often, however, almost ideally pure; they are transparent and faultless much more frequently than are many other precious stones.

Garnet is one of the most important of minerals, the common, imperfectly transparent varieties being widely distributed throughout the earth's crust. Transparent precious garnet is less abundant, but occurs under exactly the same conditions. Garnet is for the most part a mineral characteristic of the ancient crystalline-silicate rocks, especially crystalline schists, such as gneisses and mica-schist; it is found also in eclogite, serpentine, and other rocks.   As already stated, the crystals are either attached to the walls of drusy cavities in the rock or embedded in it.   Garnet is much less common in intrusive rocks, such as granite, and all varieties, with the exception of black garnet, or melanite, are of sparing occurrence in volcanic rocks.   Garnets are also found in some limestones at places where the rock has been baked by contact with a molten igneous rock, the garnets being a so-called contact product in the limestone.   The occurrence of precious garnet in its original situation is, however, of little importance from the gem-seeker's point of view.   But few garnets are extracted from the solid rock; for the most part they are picked up in the sands and gravels of rivers and streams, having been set free by the weathering of the mother-rock and carried away with the rest of the débris.

We have already seen that there are many varieties of garnet differing from each other in chemical composition. Garnets which agree in chemical composition are further classified according to their external characters and are distinguished by particular names.   Many of these varieties are solely of mineralogical interest, but a few are sufficiently transparent to be used for gems, and these we must now consider in some detail.   They include the light yellowish-red hessonite or cinnamon-stone, the dark violet-red almandine, the blood-red pyrope from Bohemia, the magnificent "Cape ruby," the fine green demantoid from the Urals, and, as rarities to the ordinary jeweller, the yellowish-red spessartite, the brownish-green grossularite, and the black melanite.

HESSONITE.

Hessonite, or *cinnamon-stone*, as may be seen from the table of analyses above, is essentially a calcium-aluminium garnet containing small quantities of ferrous and manganous oxide, or, in other words, mixed with smlla amounts of iron-aluminium garnet and manganese-aluminium garnet.   The rich, warm yellowish-red colour of the stone is due to these two constituents, calcium-aluminium garnet being of itself colourless.   The colour of this garnet is hyacinth-red, sometimes inclined to orange or to honey-yellow. The colour varies somewhat according to the distance at which the stone is held from the eye.   It appears distinctly red only when held at some distance away; close to the eye it often appears nearly pure yellow, the red being almost completely invisible.   Hessonite is also remarkable in that its colour by lamp-light is considerably more brilliant and fiery than by day.   The colour is well seen in Figs. 7 and 8 of Plate XIV.; the former is a representation of a druse of crystals having the form of Fig. 66d, which is not unusual in hessonite; and the latter of a faceted gem.

The appearance of hessonite has been compared to sugar-candy, but more often to the bark of the cinnamon-tree.   It is from the latter resemblance that it has derived its name, which is very appropriate, since most of the hessonite of gem-quality comes from the cinnamon island of Ceylon.   The colour of hessonite is, however, most strikingly similar to that of hyacinth.   So much alike are the two stones that they were only discovered to belong to distinct species at the end of the eighteenth century. Previous to this, hessonite had always been regarded as hyacinth (zircon), a mistake easily made, seeing that the stones are found together in the same gem-gravels in Ceylon. Even now the stones are never distinguished in the trade, and a large quantity of hessonite is sold as hyacinth. This is especially so in the case of large specimens, for hyacinth scarcely ever occurs in crystals of large size, while fragments of hessonite of good quality and considerable size are by no means rare.   When any difference is made by dealers in the application of the two terms, the darker specimens of hessonite are referred to as hyacinth and the lighter as cinnamon-stone. The substitution of hessonite for hyacinth is not done with the object of deception, but arises solely from confusion of names: although the lustre of hyacinth is far superior to that of hessonite, yet the colour of the latter is quite equal to that of hyacinth. A perfectly transparent, finely coloured, and faultless cinnamon-stone cannot, therefore, be any more inferior to hyacinth in value than it is in beauty of appearance.

Although hyacinth and hessonite are frequently mistaken the one for the other in the trade, yet their discrimination is really a simple matter. Hyacinth is much heavier than hessonite, for while the specific gravity of the latter lies between 3·6 and 3·7, that of hyacinth ranges from 4·6 to 4·7. Hessonite again is singly refracting, while hyacinth has strong double refraction. The lustre of the two stones differs also, that of hyacinth being stronger and more adamantine in character, while that of hessonite is strongly vitreous inclining to resinous, grains of hessonite being particularly resinous in appearance. Moreover, hyacinth is somewhat harder than hessonite; for the former $H = 7\frac{1}{2}$, while for the latter $H = 7\frac{1}{4}$, so that hessonite is only slightly harder than quartz. To distinguish between hessonite and yellowish-red spinel (rubicelle), on the other hand, is less easy, for the colour of the latter often closely resembles that of hessonite; moreover, the two stones both crystallise in the cubic system and are therefore singly refracting; they are of about equal hardness and very nearly the same density, the specific gravity of spinel being slightly less, namely, 3·60 to 3·63, and its hardness rather greater, namely, $H = 8$. Under these circumstances it is sometimes impossible to say whether a faceted stone or an irregular fragment is hessonite or spinel. When the crystalline form is observable it is easy, however, for spinel always crystallises in octahedra, while hessonite rarely takes this form, but more often that of Fig. 69d. The refractive indices of the two stones are also almost the same. The only certain means of distinguishing between stones similar or identical in so many respects is to make a chemical analysis, and this in the case of faceted stones is not always feasible. A glass imitation can, of course, be easily distinguished by its hardness and specific gravity.

A few other characters of cinnamon-stone remain to be mentioned. The index of refraction is slightly less than that of other garnets; for red light $n = 1·74$; anomalous double refraction is sometimes to be observed. Hessonite fuses to a greenish glass somewhat readily when heated before the blowpipe. Although it contains only a small amount of iron, the mineral has a slight action on the magnetic needle.

It has been mentioned already that hessonite occurs in **Ceylon**; this island (Fig. 59) is almost the only locality for hessonite of gem-quality, and it is probable that all the stones on the market come from thence. It is found in smaller or larger fragments, some lying loose, some still in the solid rock, which is gneiss containing actinolite, magnetite, and other minerals. Blocks of considerable size containing portions of gem-quality are found, amongst other localities, at Belligam, a few miles from the Point de Galle. It is the rounded pebbles found in the gem-gravels, especially in the district of Matura, which are most used for cutting as gems. These pebbles, the largest of which are several pounds in weight, are finer and purer and less fissured than are the angular fragments which have not been water-worn. These also, however, are traversed by cracks and fissures, the presence of which is the commonest, and almost the only fault usually present in cinnamon-stone.

The hessonite which occurs at other localities is less suitable for cutting, the specimens being small or imperfectly transparent. Still, there are a few localities in **Europe**, especially in the Alps, which yield, or have yielded, a small amount of hessonite of gem-quality. In early times fine " hyacinth-garnets from Dissentis " or " from St. Gotthard " were sometimes cut as gems. They occur with epidote in quartz in a small crevice in mica-schist in the Alp-Lolen in the Maigels-thal on the boundary between Cantons Uri and Graubündten (Grisons). The crystals vary in size, but none are much bigger than a pea ; each usually encloses a grain of quartz. At the present time they are scarcely sought for at all. Beautiful druses of hessonite occur in crevices in serpentine (Plate XIV., Fig. 7), on the Mussa-Alp, in the Ala valley in Piedmont. The crystals of hessonite are here associated with crystals of dark green chlorite and of pale green diopside : a number of the latter are represented among the garnets in the figure just quoted. These, as well as very similar specimens from Achmatovsk, in the Urals, are beautiful objects, to be seen in all mineral collections, and illustrate the natural occurrence of this stone in a very striking manner.

Hessonite is rarely cut *en cabochon* ; more usually it is cut with facets (Plate XIV., Fig. 8) in one or other of the forms mentioned above. The colour of this stone being pale it is unnecessary to hollow it out at the back or to make it specially thin. Cut hessonites are usually mounted upon a burnished foil, they are rarely set *à jour*.

SPESSARTITE.

The spessartite from Amelia Court House, Virginia, U.S.A., is very similar in colour to hessonite. It is a manganese-aluminium garnet, containing manganous oxide in place of lime. At the locality mentioned it is found in the mica mines in the granite as beautiful, clear crystals, which have yielded stones of very good quality weighing from 1 to 100 carats. Spessartite from other localities is scarcely suitable for cutting as a gem.

ALMANDINE.

Almandine is the deep red variety of garnet to which the name carbuncle used to be given. The name was applied to other red stones, but most frequently to garnet. It is generally believed that the word almandine is derived from Pliny's name for the stone, which was carbunculus alabandicus, since, according to his statement, it was found near the town of Alabanda, in Caria (Asia Minor), where also it was cut. At the present time transparent specimens of this kind of stone are commonly referred to simply as precious garnet.

Almandine, as shown in the analysis quoted above, is an iron-aluminium garnet containing, beside the predominating ferrous oxide and alumina, small quantities of ferric oxide, manganous oxide, &c.

The red of almandine is always dark in tint, but varies somewhat in different specimens. The colour is no doubt due to the large amount of iron which enters into the composition of the mineral. It is often distinctly tinged with violet, and is then described as columbine-red (Plate XIV., Figs. 3 and 4 ; Plate XVIII., Fig. 7). Shades of colour between brownish-red and reddish-brown are also not infrequently seen. The brownish-red stones are sometimes known to jewellers as vermeille garnet (vermilion garnet), or simply as "vermeille"; but the term is somewhat loosely applied, and often includes the Bohemian garnet or pyrope, which is deep red with a tinge of yellow. In artificial light almandine loses but little of its beauty ; the colour, however, is then more inclined to orange or hyacinth-red and approaches that of hessonite. When heated, almandine becomes black, but returns to the original red colour when cooled, though the fine appearance of the stone is somewhat impaired by this treatment.

A remarkable phenomenon connected with almandine and with only one other precious stone, namely zircon, is the presence of dark absorption bands in the spectrum of light which has passed through the stone. In the case of almandine the characteristic black bands occur in the green part of the spectrum, and can be seen with the aid of an ordinary pocket spectroscope. These absorption bands are perfectly characteristic of almandine and distinguish it from all other precious stones. The phenomenon affords an easy means of discriminating between almandine and red spinel, both of which are singly refracting and therefore devoid of dichroism.

Almandine in colour often approaches very closely to ruby, but may be readily distinguished from the latter by its single refraction and absence of dichroism. The two minerals differ also in specific gravity, that of ruby scarcely exceeding 4·0 while that of almandine varies between 4·1 and 4·3, a value which is higher than for any other garnet. Again, almandine is only slightly harder than quartz, its hardness being $7\frac{1}{4}$, so that it is scratched by topaz and still more easily by corundum or ruby.

Not only is almandine heavier than hessonite, but its optical refraction is greater also, the refractive index of the former for red light being $n = 1·77$. Before the blowpipe almandine fuses to a magnetic mass with moderate ease. Even before fusion almandine has a slight action on the magnetic needle like hessonite, the action of which is feebler, however, since it contains less iron.

Though not as brilliant as ruby, almandine, when cut and polished, has a fine lustre. It is cut usually in the forms adopted for other garnets. Fig. 4 of Plate XIV. represents a rosette of almandine. This stone is frequently also cut en cabochon (Plate XVIII., Fig. 7), more often than is hessonite, for example. Stones treated in this way are usually hollowed out, and are then known as garnet-shells; their colour is then not too deep and the lustre is advantageously displayed by the curved surface. Foils of burnished metal placed beneath the stone produce a fine effect.

The value of a stone depends on its size and purity, on the absence of faults, especially fissures, and in particular on its colour, which must show up brightly even if the stone be of considerable thickness. The nearer it approaches the fine, lustrous, velvety-purple of the ruby the greater will be the value, and the price of such stones may reach that of sapphires of medium quality. Stones of a brownish tinge, like the vermeille garnets, are little prized and are very cheap, as are also those of small size, impure colour, or with fissures or other faults.

Cloudy and opaque almandine of a quality not suitable for gems is the most widely distributed of all garnets. It is the common garnet of mineralogists, and is found in very well-developed crystals, sometimes many pounds in weight, in gneiss and mica-schist, sometimes also in granite and other rocks. This mode of occurrence is illustrated in Plate XIV., Fig. 3. Transparent precious garnet, such as is used for cutting as gems, occurs together with, and in the same manner as, the cloudy and opaque garnet, though the former is of more sparing occurrence than the latter. As is the case with so many other precious stones, almandine does not always remain in its original mother-rock, but is often set free by the weathering of the rock, and ultimately becomes a constituent of the sands and gravels of running streams, from which it is obtained in the form of rounded pebbles. Almandine of gem-quality is not only more abundant than hessonite, but is also more widely distributed. The more important localities are detailed below.

The locality standing first in importance, at which both hessonite and almandine occur, is **Ceylon**. Pebbles of almandine suitable for cutting are here, however, smaller and far less abundant than are those of hessonite. Near Trincomalee, on the eastern side of the island, almandine is said to occur embedded in a hornblende-schist. It is also found in gem-gravels in the southern and south-western parts of the island. On account of its similarity to ruby it is no doubt sometimes mistaken and substituted for that stone. It is consequently occasionally referred to as "Ceylonese ruby," a somewhat misleading term since the true ruby also occurs in Ceylon.

It is frequently stated that the most important of all localities for the finest almandine is **Syriam**, the ancient capital of the former kingdom of Pegu, which was conquered by the Burmese and now forms part of the British province of Lower Burma. Syriam is now only a small village near the important trade centre of Rangoon (see Map, Fig. 54). Both are situated on the alluvial deposits which form the delta of the Irrawaddy. According to Dr. F. Noetling, of the Indian Geological Survey, garnet does not occur here at all; it is probable, therefore, that almandine was brought from some other place to Syriam when the latter was the capital of the kingdom. But no precious almandine has as yet been observed to occur in any part of Burma. Nowhere in Pegu, that is to say, in the region of the lower Irrawaddy, are any precious stones met with, while in Upper Burma the only red stones found are ruby, spinel, and red tourmaline. It is said that the inhabitants of the neighbouring Shan States on the east bring almandine into Burma, where it is sold as ruby. However that may be, it is certain that the supposed occurrence of almandine at Syriam is by no means well authenticated.

In consequence of the supposed occurrence of almandine at Syriam (Syrian), this stone has sometimes been distinguished as "Syrian garnet" (not to be confused with Syria). In course of time, however, this term has come to signify a colour distinction, and is now applied to almandine, the colour of which inclines to violet and approaches that of the ruby or of the "oriental amethyst." "Syrian garnet," then, is a term which includes some of the finest and costliest stones, while "vermeille garnet" is the term given to the cheapest and least prized brownish varieties.

Garnet is of wide distribution in **India**. The precious almandine occurs there in such large amounts that it forms a not unimportant product of the country. It is collected and cut at many places, especially at Delhi and Jaipur. Whether all the garnets collected in India belong to the variety almandine is, however, doubtful, since chemical analyses have not been made; we shall therefore here treat of the occurrence of garnet generally, at such places in India where it is obtained in any amount. The stones suitable for cutting as gems appear to be all obtained by excavating and washing the weathered products of gneiss and similar rocks. There are workings of this kind at Kondapalli in the Godavari district (lat. 16° 38′ N., long. 80° 36′ E. See Map, Fig. 33). The garnets found here have been derived from a hornblende-gneiss, and have long been famous: those at present obtained are, however, of little value. At Bhadrachalam, on the Godavari river in the Central Provinces, as well as at Mahanadibett, in Orissa, garnets of the same kind are obtained. Stones of better quality than those of the last-named locality come from Gharibpeth, eight miles south of Paloncha in Haidarabad. They occur, with much kyanite, at a depth of 8 feet below the surface in weathered material, derived probably from granite or gneiss. The stones obtained are tested as to durability by a smart blow with a hammer; those which resist the blow are fit for cutting, and are sent for this purpose to Madras.

The garnet mines in Rajputana are a more important source of supply. The mines of Sarwar (lat. 26° 4′ N., long. 75° 4½′ E.), in the Kishengurh State, are often mentioned. The privilege of working in these mines is granted by the Rajah on a payment of one rupee per man per day. This source of revenue alone brings in 50,000 rupees per annum, so that there must be a daily average of 130 to 140 persons at work in these mines. According to

Tellery, the manager of the garnet works at Jaipur, which are mentioned below, stones from Sarwar are smaller than those from the garnet quarries of Kakoria, but in colour and lustre are surpassed by no others. Taste in Europe and America at the present day, however, inclines to stones of a more decided violet tint.

The garnet quarries of Kakoria have just been mentioned. Kakoria is situated in the State of Jaipur, and is probably identical with the place marked Kakor in the official Indian atlas, in lat. 26° 1′ N. and long. 75° 59′ E. of Greenwich. The quarries of Rajmahal (lat. 25° 23½′ N., long. 75° 21½′ E.) are also situated in this State, but their yield is less abundant. Garnets are obtained at Meja (lat. 25° 25′ N., long. 74° 37′ E.), in Udaipur, and also at several places in Meywar, but the yield is not as abundant as at Sarwar or at Kakoria, and the stones are of quite ordinary quality. Garnet of gem-quality is found at many other localities, none of sufficient importance, however, to warrant special mention. From the quality and size of the stones Tellery concludes that the Indian garnets described in ancient writings came from Rajputana.

Of **American** localities those in Brazil must be first mentioned. Almandine occurs here in rounded grains, which, though small, are finely coloured and transparent. They accompany the topaz found in the Minas Novas district of Minas Geraes. Stones of gem-quality are also said to occur in Uruguay. Numerous localities are known in the United States of North America, at some of which transparent stones suitable for cutting are found, though never in abundance. Among such may be mentioned the purple-red pebbles found in the Columbia river in the States of Washington and Oregon, some of which are of good quality and considerable size, with a weight between half a carat and half an ounce· The occurrence of almandine in Greenland is more important; the stones, which are of a fine colour and very transparent, although much fissured, occur, as a rule, embedded in chlorite- or mica-schist.

In **Australia** almandine, together probably with other varieties of garnet, is widely distributed. It is very abundant in the rivers of the Northern Territory of South Australia, the larger stones being of a bright cherry-red or yellowish-red colour, and the smaller light red inclined to violet. These stones, which were at first supposed to be rubies, were found in large numbers in the gravels of the Maude, Florence, and Hale rivers. They fetched a high price, and no less than twenty-four ruby companies, working some hundreds of claims, were floated. When it became known that the supposed rubies were in reality garnet, the companies instantly collapsed and work was suspended. At the present time very few garnets are collected in Australia for cutting as gems. The mistake made in the identification of these stones has led to their being sometimes known as " Adelaide rubies."

Garnet of gem-quality has recently been met with in German East **Africa**, where it is found in the Namaputa stream, a tributary of the Rovuma. It has here been weathered out of hornblende-gneiss, in which it is irregularly distributed as rounded enclosures up to the size of a man's fist. Most of the stones are clear and transparent and of a columbine-red colour with a tinge of brownish-red. Many fine gems, said to surpass the Indian in quality, have been obtained, but whether the deposit could be worked on a large scale is as yet doubtful. A chemical analysis of this garnet proves it to be almandine, with much of the ferrous oxide replaced by magnesia; the specific gravity is 3·875.

Almandine suitable for cutting is also found in **Europe**, but not in large amount nor of specially good quality. A certain number of these stones are collected every year in the Alps, which is the most important European locality. Specially remarkable are the rhombic dodecahedral crystals, measuring as much as an inch across, which occur in the dark mica-schist and the chlorite-schist of the upper Zillerthal in the Tyrol, the exact locality being on the Rossrucken opposite the Berlin Hut in the Zemmgrund. They are first quarried and then freed from the mother-rock by grinding against each other in a rotating barrel. The·majority are sent to Bohemia, where, as we shall see presently, an important garnet-cutting industry has been developed. To this country are sent garnets from all parts of the world, and these, together with the stones which actually occur in the country, are cut and used in the fashioning of various articles of jewellery. Most of the garnets found in Bohemia, however, belong not to the variety known as almandine but to that known as pyrope. Almandine of gem-quality is nevertheless found in Bohemia, especially in the alluvial ground in the neighbourhood of Kuttenberg and Kollin, such stones being known as " Kollin garnets." The occurrence of almandine here, as elsewhere in Europe, is unimportant. Other European localities for almandine which may be mentioned are Mittelwald, in the Rohoznabach of Hungary, where crystals of considerable size are sometimes found, and Alicante in Spain.

PYROPE.

Pyrope, or *Bohemian garnet*, is distinguished by its deep, rich, blood-red colour, which has always an unmistakable tinge of yellow (Plate XIV., Figs. 5 and 6), sometimes even verging upon hyacinth-red. Violet tints are never present in pyrope, and a garnet which shows a tinge of this colour is almost certain to be almandine : in the case of other tints, however, it is not possible to judge by colour alone. On account of the yellowish tint of its colour, pyrope is included in the term vermeille garnet, while some practical jewellers limit the application of the term to pyrope. It is highly probable that this stone was one of those to which the name carbuncle was formerly given. Pyrope is very similar in colour to some rubies, but, like almandine, it can be distinguished from this gem by the fact that it is refracting and devoid of dichroism, as also by the fact that its specific gravity is less than that of ruby, being only 3·7 to 3·8. The specific gravity of pyrope affords a sure means whereby it may be distinguished from almandine when this cannot be done by colour alone, since almandine is considerably denser, having a specific gravity between 4·1 and 4·3.

Pyrope is essentially a magnesium-aluminium garnet, but of more complex composition than the garnets hitherto considered. Besides magnesia it contains a not inconsiderable amount of lime, ferrous oxide, manganous oxide, and chromous oxide, the latter of which figures in analyses as chromic oxide. The magnesium-aluminium garnet is thus mixed with calcium-, iron-, manganese-, and chromium-aluminium garnet. The colour, which is almost invariably deep and rich, depends on the presence of small quantities of iron and manganese, perhaps also of chromium.

In contrast to almost all other garnets, pyrope scarcely ever occurs in distinct crystals. The few which have as yet been found have the form of a cube with curved faces, a form which is unique as far as other garnets are concerned. Pyrope occurs usually in irregular grains, with a dull, rough surface, although that of a fresh conchoidal fracture is bright and shining. The hardness (H = 7¼) is slightly greater than that of quartz. The refraction is greater than that of any other red garnet, the index of refraction for red light being 1·79 ; the substance is perfectly isotropic and shows no anomalous double refraction. Pyrope differs also from other red garnets in its behaviour before the blowpipe, since it fuses with great difficulty ; it is possible to fuse only the thinnest splinters to a black, magnetic glass.

Pyrope is usually perfectly clear and transparent so far as its dark colour allows. That from Bohemia is all without exception, of ideal purity. This absolute purity and freedom from enclosures of every individual stone is unparalleled among other precious stones. Moreover, the transparency is rarely impaired by fissures, as is the case with other garnets. When heated, pyrope behaves like almandine and becomes black and opaque; on cooling again it recovers to the full its original transparency and colour, which is not the case with almandine.

As to its mode of occurrence, pyrope is always found in olivine-rocks or in serpentine, an alteration product of the former. It occurs embedded in these rocks as irregular grains, among many other places at Petschau, in Bohemia, and Zöblitz, in Saxony, from which locality come the specimens represented in Plate XIV., Fig. 5. In collecting pyrope places are sought where the serpentine is completely weathered to a loose earthy material. The grains of garnet, being little affected by the weathering process, lie scattered loosely through this material, from which they can be separated with little trouble. Such are the conditions under which pyrope occurs in the north of **Bohemia**, where it is specially abundant, and from whence it is almost exclusively obtained. This variety of garnet, at the present time so much admired, occurs nowhere else in just the same manner, and is hence described as Bohemian garnet. The occurrence has given rise to an important industry in northern Bohemia, where now are cut not only garnets found in Bohemia itself, nor only pyrope, but garnet of all kinds from all parts of the world, from the Zillerthal, from India, Ceylon, Asia Minor, Australia, the United States, Greenland, &c., and in addition all other precious stones with the single exception of diamond.

The garnet-cutting works of Bohemia are very old-established and have seen many vicissitudes. After a period of decay, the industry received a fresh impetus through the establishment of baths at Carlsbad, Teplitz, and other places. Thousands of persons from all parts of the world were attracted there to benefit by the waters, and many carried away with them as souvenirs of their visit pretty articles of jewellery set with Bohemian garnets. The increased demand so created led to their becoming an important article of export. The importance of this particular industry may be judged from the fact that at the present time in Bohemia there are 3000 men engaged in garnet-cutting, some hundreds of garnet-drillers, about 500 goldsmiths and silversmiths, and some 3500 working jewellers. The collecting of garnets employs some 350 or 400 persons, so that, including the many persons whose work is indirectly connected with the industry, there must be between 9000 and 10,000 persons gaining their livelihood by labour connected with the working of this precious stone.

There are a few cutting works at Prague, but many more in the neighbourhood between Reichenberg and Gitschin, of which those at Rovensko, Semil, Sobotka, and Lomnitz may be mentioned. The centre of the industry, however, is at Turnau on the Iser, and here a Government school, at which the working of precious stones is taught, has been established. There are a few cutting-works on the German side of the border, at Warmbrunn, in Silesia, for example, and other places.

In the district where cutting is carried on, garnets, though known to occur, are not abundant; they have been found, for example, at Neu-Paka, a little to the east of Gitschin, where the few crystals of pyrope hitherto found in Bohemia were met with. The material for the cutting-works is obtained almost exclusively, however, from the neighbourhood of Teplitz, Aussig, and Bilin, in the Bohemian Mittelgebirge, some distance to the west. The garnet-bearing deposit covers an area of over seventy square kilometres and the mineral is abundant over an area of about one-tenth of this. The principal localities are, among others, Stiefelberg, near Meronitz, also the neighbourhood of Chodolitz, Dlaschkowitz, Podsedlitz, Chrastian, Tremschitz, Starrey, Schöppenthal, Leskai, Triblitz, Jetschan,

Semtsch, Solan, and Schelkowitz; at all these places garnets are now systematically collected. In the year 1890 there were in this district 142 owners of garnet-fields, and stones to the value of 80,000 gulden (£8000) were obtained by the labour of 362 persons The trade in garnets was kept in the hands of about seventeen merchants.

The stones are found in a clayey or sandy gravel belonging to the glacial drift and resting on beds of Cretaceous age. In this gravel grains of garnet completely freed from their mother-rock occur in large numbers. Some, however, are found embedded in a brown semi-opal occurring in masses, the largest of which is about the size of a man's head. These masses of opal are to be regarded as the remains of the serpentine in which the pyrope was originally embedded, most of which, as serpentine, is now completely destroyed. The garnets embedded in the masses of opal are not utilised, only those lying free in the gravelly material being collected and cut.

The deposit of gravel in which the pyrope occurs differs somewhat in different places. At Chrastian, beneath a layer of soil, one metre in thickness, is a layer, two metres thick, of garnetiferous gravel with a light grey loamy base; below this is another layer of garnetiferous gravel 4 metres thick bound together with a yellowish-brown clayey material. The whole rests upon a bed of fuller's earth belonging to the Senonian division of the Upper Cretaceous formation. At Meronitz the garnet-bearing layer is a peculiar clayey calcareous conglomerate.

Garnets are sometimes washed out of the loose gravel by rain-storms, and, owing to this, good specimens are occasionally found lying loose on the surface of the ground. For the most part, however, the stones are obtained in excavations. The surface-soil is removed and the garnet-bearing layer penetrated by pits of greater or less depth, which are refilled when the valuable material has been excavated. Only at specially rich spots are large excavations undertaken, where also underground mining operations are carried on to a small extent. The garnet-bearing earth is washed in suitable vessels to remove the lighter clayey particles, after which the stones are picked out by hand and sorted according to size by means of sieves. The stones are classified according to the number required to make up a loth (= $16\frac{2}{3}$ grams, or rather less than $\frac{1}{2}$ oz. avoirdupois), and are referred to as "sixteens," "thirty-twos," "hundreds," and so on, according as 16, 32, or 100 stones make up a loth. Most of the stones are very small, 500 or more in the aggregate weighing a loth. Those of which 400 make up the weight of a loth are very numerous and of little value. Stones the size of rice grains are worth more, as also are those the size of a pea, which, however, are not found every day, while several years may elapse between the finding of pyropes of the size of a hazel-nut. It is estimated that in every 100 kilograms (220 lbs.) of garnets there are only two to three "thirties," and in 2000 kilograms only one "sixteen."

It will be seen from this that a moderate number of stones of fair size are met with, but that really large stones are rare. In the *Gemmarum et Lapidum Historia* of Boetius de Boot, published in 1609, is mentioned a pyrope, the size of a pigeon's egg, in the possession of Kaiser Rudolph II., which was valued at 45,000 thalers (£6750). A very fine stone, the size of a hen's egg, is now preserved in the Imperial Treasury at Vienna. In the "Green Vaults" at Dresden is one measuring 35 millimetres in length, 18 millimetres in breadth, and 27 millimetres in thickness, that is, about the size of a pigeon's egg; it weighs $468\frac{1}{4}$ carats, and is set in an Order of the Golden Fleece.

Since all Bohemian pyropes are of the same quality and purity, their value varies only with their weight. Small stones are very cheap, but the price of larger ones is by no means inconsiderable; no rough stone found in recent years has realised more than 500 florins

(£25). Boetius de Boot states that large stones were worth as much as a ruby of equal size; this is not now the case, in spite of the estimation in which pyrope is held, and of the fact that its colour is equal in beauty to that of some rubies.

The very smallest grains of pyrope are not cut as gems, but are utilised in a variety of ways, for example, as counterpoises for delicate balances, for the preparation of grinding powder, and even as ornamental gravels for garden walks, which gives a good idea of the abundance in which they are found. There is nothing attractive about the appearance of the stones in the rough; the process of cutting, however, brings out the brilliancy of their colour, which is displayed by smaller stones just as advantageously as by larger. In spite of this, only stones exceeding a certain minimum size are cut as gems. Almost all the usual forms of cutting are utilised for pyrope. As in the case of almandine, it is frequently cut with a curved surface, *en cabochon*, when it is usually hollowed out at the back, but may or may not be provided with small marginal facets. The light reflected from the curved surface blazes with a wonderful fiery red colour. Still more frequently pyrope is cut in a faceted form, either as a table-stone or a low step-cut, often with a curved table. Brilliants and rosettes are also to be seen as well as fanciful forms in which the facets have no recognised arrangement. Pyrope is more frequently used for the manufacture of beads than are other garnets; only the smaller stones are used for this purpose, each has a hole drilled through it and is faceted regularly all over. The cut stones are either mounted upon a burnished copper or silver foil in a closed setting blackened inside, or *en pavé*. In the latter case the stones are fixed by means of small claws or pins over close-set perforations in a metal plate, the whole forming a kind of garnet mosaic.

There are only a few other localities where pyrope of gem-quality is found. Other than Bohemia the only European locality is Elie in Fifeshire; the so-called " Elie rubies " are of purely local interest, however.

The occurrence of pyrope in the western part of the **United States** is more important. The mineral is specially abundant in Arizona, New Mexico, and southern Colorado, and, as frequently happens, was at first mistaken for ruby, a mistake out of which arose the local trade names " Arizona ruby " and " Colorado ruby." In New Mexico pyrope occurs as angular or rounded grains in sands at Santa Fé, but most abundantly in the Reservation of the Navajo Indians, together with olivine and chrome-diopside. Pyrope is here collected by the Indians from the sands of ant-hills and scorpion-hills, as well as from the mother-rock. In Arizona pyrope occurs loose in sands, and, in the north-east of the State, embedded in the mother-rock. Here also it is collected by the Indians and occasionally by soldiers stationed there. The angular or rounded fragments measure $\frac{1}{8}$ to $\frac{1}{4}$ inch across; larger grains, ranging up to $\frac{1}{2}$ inch in diameter, are rare. The quality of these stones is good, about half being fit for cutting; of these about one quarter are of ordinary quality, exceptionally fine stones, especially those exceeding 3 carats in weight, being rare. Many enclose a network of fine needles, probably of rutile.

American pyropes on an average are smaller than the so-called " Cape rubies," to be described presently. They have an equally fine appearance by daylight, but in artificial light the American stones are superior to the African, the latter appearing somewhat dull. The so-called " Arizona rubies " and " Colorado rubies " are rather extensively used, more so than is the case with the pyrope found, for example, in the gold washings of the Counties of Burke, MacDowell, and Alexander in North Carolina, and from other districts in the United States.

The occurrence of pyrope in Mexico is of no more importance; it is known to occur in the State of Sonora and in that of Chihuahua, especially on the Jaco Lake, where its mode of occurrence is the same as in Arizona, and where it is collected by the Comanche Indians.

Of all varieties of garnet one of the finest is the dark, blood-red pyrope, which occurs in association with diamond in **South Africa.** This also was at first supposed to be ruby, and was collected and sold as such for some time, hence the term " *Cape ruby.*" Several different kinds of garnet are found in association with diamond at the Cape. Some are of a deep wine-red colour, some of a hyacinth-red, almost the colour of hessonite, while others, fewer in number, are brownish-yellow and deep blood-red.

The last named is the much prized " Cape ruby " and is the only one cut as a gem. It is a magnesium-aluminium garnet containing some manganese oxide and ferrous oxide, and differing but slightly in chemical composition from the Bohemian pyrope, as may be seen from the analysis quoted above. The " Cape ruby " must therefore be classed with pyrope and not with almandine, as is sometimes incorrectly done. Not only on account of its chemical composition, but also on account of its specific gravity, which is 3·86 (that of Bohemian pyrope being 3·7 to 3·8, and that of almandine 4·1 to 4·3) and of its colour, should this classification be adopted. The colour approaches indeed much more closely to that of pyrope, being an almost pure carmine-red more or less tinged with yellow and not very deep in shade (Plate XIV., Fig. 6), thus differing distinctly from the columbine-red of a good almandine. The hardness is $7\frac{1}{4}$, the same as that of Bohemian pyrope and of almandine. The " Cape ruby " fetches larger prices than any other garnet, stones of moderate size being worth £10 to £12 10s. per carat.

It has already been stated that the " Cape ruby " occurs in association with diamond in South Africa. It is found in the form of irregular angular grains with an uneven surface in the diamond-bearing rock known as the " blue ground " and the " yellow-ground." The mother-rock is thus an olivine-rock, or, more commonly, the weathered equivalent of this, namely, serpentine, just as is the case with the pyrope of Bohemia, North America, and all other localities. The " Cape ruby " is far less abundant than is the paler red pyrope by which it is accompanied. The grains are larger on the whole than are the grains of pyrope found in Bohemia and America ; they never exceed a certain maximum size, however, which is much less than that of the largest diamonds found at the same place. The residue of heavy minerals obtained by washing the diamantiferous material contains, besides diamonds, red garnets and green grains of an augitic mineral ; and from this residue the diamonds and " Cape rubies " are picked out. In the diamond-bearing rock of the " dry diggings " the " Cape ruby " is somewhat of a rarity, being far more abundant, though still uncommon, in the " river-diggings," that is to say, in the sands and gravels of the Vaal river, where it sometimes occurs in pebbles so smooth and rounded that they appear to have been polished. Here, as in the " dry-diggings," the stone is collected as a secondary product.

There still remains to be described a new variety of red garnet, recently (1898) described by Messrs. W. E. Hidden and J. H. Pratt, for which the name **RHODOLITE** is proposed. In many respects it is intermediate between almandine and pyrope, but more closely related to the latter, though differing from both in colour. Its occurrence in association with ruby at Cowee Creek and Mason's Branch in Macon County, North Carolina, has been mentioned already in the description of corundum. It is found as water-worn pebbles in the gravels of these streams, and also, together with ruby, in a decomposed, basic igneous rock, known as " saprolite," and, in the form of small crystals, enclosed in crystals of ruby. The colour is pale rose-red inclining to purple like that of certain roses and rhododendrons, hence the name rhodolite. It lacks the depth and intensity of colour which makes garnets, as a rule, such dark-looking stones especially by artificial light. The peculiarly beautiful rose tint of rhodolite combined with its transparency and brilliancy renders it an even more striking object by candlelight than by daylight. The lustre of rhodolite is comparable with that of demantoid, a green garnet from the Urals; this, together with its freedom from internal flaws and inclusions, makes it when cut a very striking and beautiful gem. The chemical composition of this new variety of garnet is shown by the following analysis:

SiO_2.	Al_2O_3.	Fe_2O_3.	FeO.	MgO.	CaO.
41·59	23·13	1·90	15·55	17·23	0·92 = 100·32

The chemical formula which represents this composition is:

$$2(3MgO.Al_2O_3.3SiO_2) + 3FeO.Al_2O_3.3SiO_2;$$

in other words, rhodolite is a combination of two pyrope molecules with one almandine molecule. The specific gravity, 3·837, is more in agreement with that of pyrope than with that of almandine; on the other hand, in spite of the preponderance of the pyrope molecules, an examination by Professor Church of the absorption spectrum of rhodolite shows the existence of the bands which are characteristic of almandine.

Mr. G. F. Kunz reports, on the authority of Mr. W. E. Hidden, that several crystals of rhodolite were found, during the summer of 1901, embedded in a decomposed saprolitic rock; these crystals are of considerable size, one weighing 3½ pounds and yielding 300 carats of fine red material, free from flaws and suitable for cutting. The yield of rhodolite in that year was about 200,000 carats, valued at about £4000.

DEMANTOID.

Demantoid is a beautiful green precious stone belonging to the group of calcium-iron garnets, as is shown by the analysis quoted above. The stone ranges in colour from a fine emerald-green to a brownish- or yellowish-green, and is sometimes indeed almost colourless. Demantoids of two shades of colour are represented in Plate XIV.; Fig. 9 illustrates the mineral in its rough condition, and Fig. 10 shows three cut stones. The colour most commonly seen is a light yellowish-green. The emerald-green variety, as shown by the above analysis, contains a small amount of chromium, and the beauty of its colour is no doubt due to the presence of this element. The paler green and yellowish-green stones contain no chromium; their colour, therefore, must be due to the iron which is present.

The lustre, the brilliancy of which is heightened by polishing, is strongly vitreous inclining to greasy, while the transparency and purity of the mineral is usually perfect. The index of refraction and the dispersion are both high, and by artificial illumination a faceted stone often shows a fine play of prismatic colours.

Demantoid is softer than any other garnet, its hardness being only 6½, which is less even than quartz. The specific gravity ranges from 3·83 to 3·85. It fuses before the blowpipe to a black magnetic glass, but only when in the thinnest of splinters. Demantoid differs from all other garnets in being easily and completely decomposed, even in its natural condition before being fused, by acids.

The mineral has hitherto been found only in the Ural Mountains. It was discovered in the 'sixties in the form of greenish-white or almost colourless pebbles in the gold-washings of Nizhni-Tagilsk. It was met with subsequently in the Sissersk (Syssertsk) district on the western slopes of the Urals in the stream Bobrovka, which flows into the Chussavaya, at a spot about ten versts south-west of the village of Poldnevaya, and twenty versts to the south of the smelting works of Polevskoi, first as pebbles in the gold-washings and afterwards *in situ* in the underlying mother-rock. It is distinguished as Bobrovka garnet from the locality at which it is found.

The demantoid occurs here, with dolomite, a little clayey material, and magnetite, in veins of chrysolite, which penetrate a peculiar grey to greenish-grey serpentine rock ; it is found sometimes also in the serpentine itself. The garnet is embedded in this fibrous chrysolite and coated with a layer of the same ; it occurs either as isolated irregular grains or more commonly in nodules, the surface of which is irregularly grooved and furrowed (Plate XIV., Fig. 9). These nodules measure from $\frac{1}{4}$ to 2 inches across and are greasy and cloudy in appearance. Each is built up of a large number of irregular grains of demantoid packed closely together, but separated from each other by a coating of serpentine. Each grain has a brilliant lustre and a perfectly conchoidal fracture. As a rule, each nodule is divided by deep, prominent grooves into a small number of portions, and the grains which build up these several portions are separated by finer grooves. Distinct crystals are rarely found ; the rhombic dodecahedron and icositetrahedron, and also combinations of these two forms, have nevertheless been observed. The rounded outline of the grains sometimes appears to be due to the combination of numerous imperfectly developed crystal-faces.

Demantoid is frequently cut and worn as a precious stone at the place of its origin, that is to say, in the Urals and elsewhere in Russia ; outside that country it is little used. It is cut *en cabochon* (Plate XIV., Fig. 10), and frequently also in various faceted forms. On account of its yellowish-green colour, the shade which is most frequently seen, demantoid was at first thought to be chrysolite, and is even now known by this name in the Urals. It may be distinguished from chrysolite, however, by its single refraction and high specific gravity, demantoid sinking in the heaviest liquid (sp. gr. = 3·6) while chrysolite (olivine) floats.

The chromiferous emerald-green variety of demantoid is very similar in appearance to the emerald, and is therefore sometimes called " Uralian emerald "; this term, however, is somewhat misleading, since true emeralds are also found in the Urals. The characters mentioned above as serving to distinguish demantoid from chrysolite also serve to distinguish it from emerald. The manner in which demantoid occurs precludes its extensive use as a precious stone, the grains which build up the nodules described above being always very small. Were it not for this fact the lustre, colour, and play of prismatic colours of demantoid would no doubt render it one of the most highly prized of precious stones, its lack of hardness not being sufficient to seriously affect its application in this direction.

OTHER GEM-VARIETIES OF GARNET.

There are still to be mentioned a few other varieties of garnet which are used for ornamental purposes. The brownish-green calcium-aluminium garnet known as *grossularite*, fine crystals of which occur in the Vilui river, Siberia, is sometimes cut under the name of " gooseberry-stone." A beautiful rose-pink, though rarely perfectly clear and transparent, calcium-aluminium garnet occurs in large, well-developed rhombic dodecahedral crystals in a finely granular limestone at Xalostoc, in the State of Morelos in Mexico, and is sometimes employed as a gem.

The black calcium-iron garnet, *melanite*, is used to a limited extent in mourning jewellery ; it differs from all other garnets in occurring exclusively in volcanic rocks, such, for example, as those of the Kaiserstuhl, near Freiburg in Breisgau, and at Frascati, in the Albanian Hills, not far from Rome.

Another calcium-iron garnet, known as *topazolite* on account of its similarity in transparency and colour to yellow Brazilian topaz, occurs in well-developed crystals in the Ala valley in Piedmont.

TOURMALINE.

The name tourmaline, like garnet, is given to a group of isomorphous substances, the chemical composition of which is constant and definite in its general type but variable with regard to the elements which enter into it. These substances agree very closely in their crystalline form, but, owing to differences in chemical composition, differ somewhat in other physical characters, and it is these differences which serve to distinguish one tourmaline from another. To several varieties, distinguished from each other by differences in specific gravity, colour, transparency, and so on, mineralogists have given special names; and of these varieties, those which are sufficiently transparent and pleasing in colour find an extensive application as gems. To the ordinary jeweller the name tourmaline and the several variety names recognised by mineralogists are alike practically unknown. The gem-varieties of tourmaline are distinguished by jewellers solely by their colour, and are referred to by the names of better-known gems to which a qualifying prefix is added.

The physical characters of tourmaline depend more or less directly on the chemical composition; it is therefore advisable to consider this first. In order to give an idea of the chemical composition of different tourmalines a table of analyses of differently coloured specimens is given below. These analyses refer to: I., colourless tourmaline from De Kalb, St. Lawrence Co., New York; II., pale green tourmaline from Haddam Neck on the Connecticut River, U.S.A.; III., red tourmaline from Shaitanka in the Urals; IV., brown tourmaline from Dobrowa, near Unterdrauburg on the Drau in Carinthia; V., dark blue tourmaline from Goshen, Massachusetts, U.S.A.; VI., black tourmaline from Unity, New Hampshire, U.S.A.

	I. Colourless. De Kalb.	II. Pale green. Haddam Neck.	Red. Shaitanka.	IV. Brown. Dobrowa.	V. Dark blue Goshen.	VI. Black. Unity.
Silica (SiO_2) . . .	36·72	36·96	38·26	38·09	36·22	36·29
Titanium dioxide (TiO_2) .	0·05	0·03	—	—	—	—
Boron trioxide (B_2O) .	10·81	11·00	9·29	11·15	10·65	9·04
Alumina (Al_2O_3) . .	29·68	39·56	43·97	32·90	33·35	30·44
Ferrous oxide (FeO) . .	0·22	2·14	—	0·66	11·95	13·23
Manganous oxide (MnO) .	—	2·00	1·53	—	1·25	—
Magnesia (MgO) . .	14·92	0·15	1·62	11·79	0·63	6·32
Lime (CaO) . . .	3·49	1·28	0·62	1·25	—	1·02
Soda (Na_2O) . . .	1·26	2·10	1·53	2·37	1·75	} 1·94
Potash (K_2O) . . .	0·05	—	0·21	0·47	0·40	
Lithia (Li_2O) . . .	—	1·64	0·48	—	0·84	—
Water (H_2O) . . .	2·98	3·10	2·49	2·05	2·21	7 2
Fluorine (F) . . .	0·93	1·13	0·70	0·64	0·82	—
Total .	101·11	101·09	100·70	101·37	100·07	100·00
Specific gravity . . .	3·049	3·089	3·082	3·035	3·203	3·192

It will be seen from these analyses that tourmaline is a silicate of very complicated composition; in fact, no other precious stone is so complex in character; in this respect, therefore, there is a marked contrast between tourmaline and diamond, the composition of the latter being the simplest possible. No tourmaline has all the fourteen elements indicated, and those which do occur are always present in variable amounts; this tends to make the constitution still more complex. Besides silica, there is always present boron, aluminium, magnesium, sodium, potassium, and water, the last of which is expelled only at a red-heat. Fluorine is very rarely absent, but lithium and manganese enter into the composition of only a few tourmalines; iron, as ferrous oxide, is an important, but variable, constituent, which may be absent, present in small amount or in considerable amount, as much as 20 per cent. Tourmalines containing a large amount of iron are always dark in colour and imperfectly transparent, frequently quite black and opaque, and are therefore unfit for use as gems.

The number of the constituents of tourmaline, together with the difficulty of accurately determining the proportions in which they are present, and the fact that these proportions vary in every specimen that is analysed, make it impossible to arrive at a chemical formula which satisfies every condition and is unexceptionable in every way. Every formula hitherto proposed is based to a certain extent on supposition, and no single one has been generally accepted as final. The first two analyses quoted above have been recently made (1899) by Professor S. L. Penfield and Mr. H. W. Foote, with the object of establishing a formula for the mineral. Every care and precaution was adopted, and their results have given rise to much discussion. The general formula they propose is $H_9Al_3(B.OH)_2Si_4O_{19}$, the nine atoms of hydrogen being supposed to be replaceable by variable amounts of aluminium, alkalies, magnesium, and iron. The special formulæ constructed to suit particular cases are necessarily very complex. According to another view, all varieties of tourmaline are, like the garnets, isomorphous mixtures of a small number of fundamental molecules of perfectly fixed and definite composition, the differences in the chemical composition of the different varieties of the mineral being due to the relatively varying amounts in which these fundamental molecules are present, as are also the differences in physical characters. Even if this theory be correct, the determination of the exact constitution of the fundamental molecules is still one of the problems of mineralogical chemistry.

In contrast to the chemical composition, the crystalline forms of all varieties of tourmaline are in close agreement. The crystals belong to the hexagonal system with hemimorphic-rhombohedral symmetry. A prism of greater or less length is nearly always developed, and is terminated by rhombohedra, scalenohedra, or by the basal plane, singly or in combination. Corresponding faces of these forms are inclined to each other and to the prism faces at angles which vary in different crystals irrespective of the chemical composition, only, however, to a small extent, at most only about a degree of arc. Such close crystallographic agreement in substances of different chemical composition is explained on the principle of isomorphism, the different varieties of tourmaline being regarded as an isomorphous series of minerals.

There is a peculiar and characteristic feature connected with crystals of tourmaline which may be seen in the accompanying figures (Figs. 70a to e), the arrangement of the faces at one end of the prismatic crystals differs from that of the faces at the other end; such crystals are said to be hemimorphic. The hemimorphism of tourmaline is rarely very distinct, for the crystals are, as a rule, attached at one end to the matrix and terminal faces developed only at the free end. Doubly terminated crystals of tourmaline may be readily distinguished by this hemimorphic development from any other mineral which they may resemble in appearance. In Fig. 70, this hemimorphic development is shown most conspicuously by the crystals lettered *b*, *c*, and *d*, there being many more faces present on the upper than on the lower ends. Further, the hemimorphic development is shown not only by the number and arrangement of the terminal faces, but also by that of the prism faces. For example, in Fig. 70*a*, the prism has only three faces instead of six, the number usually present on crystals belonging to the hexagonal system ; again, in Figs. *b* to *e*, instead of a prism of twelve faces there are only nine, this being a combination of a hexagonal with a trigonal prism. This feature is so characteristic of crystals of tour-maline that they may be recognised with certainty by it alone, even when no terminal faces are present.

The faces of the prism are usually more or less deeply striated in the direction of their length ; that is to say, parallel to the principal trigonal axis of the crystal, as is represented in the figures. This striation is specially prominent when more than nine prism faces are

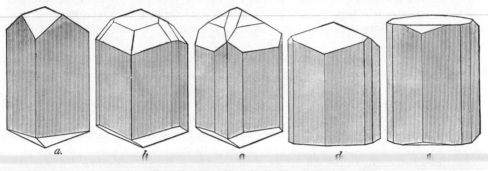

FIG. 70.   Crystalline forms of tourmaline.

developed, the crystals then having the appearance of cylinders, in which, however, a triangular arrangement can still be detected (Plate XV., Figs. 8 and 9). The terminal faces are usually smooth and not striated; some, however, are rough and dull. The form of crystals of tourmaline varies somewhat according to the locality at which they occur, but there is no essential difference between them. Of the diagrams of Fig. 70, *a* represents a brown crystal from Ceylon, *b* and *c* two green crystals from Brazil, *d* a red crystal from Shaitanka in the Urals, and *e* a rose-red crystal from the island of Elba. These are all forms taken by precious, transparent varieties of tourmaline suitable for cutting as gems ; such crystals are usually small in size, rarely exceeding the length and thickness of a little finger.

Tourmaline possesses no distinct cleavage. The fracture is uneven to imperfectly conchoidal ; and, the mineral being very brittle, most crystals are penetrated by numerous irregular cracks and fissures which tends to make them useless as gems. Tourmaline is just sufficiently hard to scratch quartz, but is itself easily scratched by topaz; its hardness is thus between 7 and $7\frac{1}{2}$, and may, as a rule, be taken as $7\frac{1}{4}$.

The specific gravity of tourmaline is not constant for all varieties, but ranges from slightly over 3·0 to 3·2. All tourmalines sink in liquid No. 3 (sp. gr. = 3·0), though some only slowly, and all float in pure methylene iodide. The density of the mineral varies with its chemical composition, increasing with the amount of iron present, as may be seen from the table of analyses with the corresponding values of the specific gravity, quoted above. Only in a few exceptional cases is the specific gravity slightly less than 3·0 or slightly over 3·2. It has already been pointed out that the greater the amount of iron present in any given tourmaline the darker will be its colour and the more imperfect its transparency.

These darker and more dense tourmalines are not cut, the paler and lighter varieties only being suitable for gems ; the specific gravity of the latter lies between 3·0 and a value slightly in excess of 3·1.

The common vitreous lustre of tourmaline is much heightened by polishing ; indeed, all tourmalines are susceptible of a good polish. In respect of brilliancy of lustre tourmaline surpasses beryl, especially by artificial light ; the latter mineral, however, is superior to tourmaline in the richness of its colours.

Tourmaline varies in transparency and colour to a considerable extent. Most of it is black, or, at least, very deeply coloured, and quite opaque in mass, but in sufficiently thin splinters it is transparent. Tourmaline which is transparent in mass is comparatively rare, and is distinguished as " precious tourmaline " in contrast to the darkly coloured and opaque " common tourmaline." Precious tourmaline is much lighter in colour, and is the only variety suitable for cutting as a gem ; perfectly clear and faultless specimens are rare and command a moderately high price. Tourmaline is very variable in colour, more so than most other minerals used as gems ; the colour variations will be dealt with in some detail below.

In striking contrast to black tourmaline, the so-called " schörl," is the colourless variety, which, however, even when transparent, is not perfectly water-clear, being usually faintly tinged with red or green. Colourless tourmaline is known to mineralogists as achroite ; it is rarely cut as a precious stone and is not of frequent occurrence. Red tourmaline is of greater importance : this variety is of a pale rose-red shade, lighter in some specimens, darker in others. It may even be of a fine ruby-red colour, and is then known as rubellite or, on account of its occurrence in Siberia, as siberite. A violet tint is sometimes seen owing to the admixture of blue with the typical red colour. The darker red tourmaline only is important as a gem, rose-red specimens not being used at all. Precious tourmaline of various tints of green, ranging in shade from pale to dark, is of much more frequent occurrence. A pure emerald-green is rarely seen, bluish-green and especially yellowish-green being much more common. Blue tourmaline, at least in transparent specimens, is rare ; it is known to mineralogists as indicolite, and is usually somewhat deeply coloured, being sometimes a pure indigo-blue, at other times showing a pronounced tinge of green. Brown tourmaline, to which the name dravite is applied, is also widely distributed ; it is either pure brown, greenish-brown, or reddish-brown, and ranges in shade from a colour of considerable depth to somewhat pale brownish-yellow and straw-yellow.

Crystals of tourmaline may be of one uniform colour throughout or differently coloured in different portions. Thus, for example, the terminal portions of prisms of colourless tourmaline from the island of Elba are frequently black ; such crystals are known as " negro-heads." The change from colourless to black material is abrupt, but there is no absolutely sharp boundary ; these conditions are sometimes reversed, a black crystal having white ends. Prismatic crystals coloured rose-red at one end and green at the other, as represented in Plate XV., Fig. 5, are not rare ; the transition from one colour to another in the middle of the crystal is gradual. Also of interest are the crystals from Chesterfield, in Massachusetts, and from other localities, in which there is a red central portion enclosed by a green shell, the transition between the differently coloured portions being sharp, as represented in Plate XV., Figs. 8 and 9.

The colour of tourmaline is not due to the mechanical intermixture of pigment, but is a property of the substance of the mineral itself, and even in the darkest crystals is distributed with perfect regularity. It may be supposed that the fundamental molecules mentioned above have each their own characteristic colour, and that the different colours of the isomorphous series of minerals, known collectively as tourmaline, are due to the association of these molecules in different relative proportions.

When a tourmaline contains, beside its constant constituents, small amounts of manganese and lithium with little or no iron, it is usually colourless, rose-red, darker red, or light green according, apparently, to the proportions in which iron and manganese are present. The darker red shade probably appears when a relatively considerable amount of manganese is present, while with a rather larger amount of ferrous oxide the colour is dark green, as is so often the case in silicates. Some green tourmalines, however contain chromic oxide, the substance to which the emerald owes its magnificent green colour.

In brown tourmaline there is practically no iron, manganese, or lithium, but a much larger amount of magnesium than is present in other varieties. The presence of a large amount of ferrous oxide is probably the cause of a blue colour, but apparently no transparent blue tourmaline has as yet been chemically analysed. The tourmaline richest in iron is, as has been stated, black and opaque in mass, but in thin sections is transparent and brown, green, or blue, in colour. The colours of tourmaline withstand the action of heat very markedly, in many cases remaining essentially unaltered after exposure to a red-heat. For example, dark-green stones after this treatment are still green, but of a pale or greyish shade, a change which renders them unsuitable as gems.

The dichroism of tourmaline is a very prominent feature; it is apparent in quite pale-coloured stones, but is more marked in the darker varieties. Tourmaline is, in fact, more strongly dichroic than any other precious stone, with the exception perhaps of cordierite (dichroite), but this is seldom cut as a gem-stone. In all cases the colour of the light which travels through a crystal of tourmaline in a direction parallel to the prism edges—that is to say, along the principal or optic axis—is darker than that which travels in a direction perpendicular to this. Again, a moderately thick slice of a dark-coloured crystal cut perpendicular to the optic axis may be quite opaque, while a plate of the same thickness, but cut parallel to the optic axis, may be transparent. The colours visible in these two directions are also usually different, as has been explained in the general account of dichroism given in the first part of this book. Very frequently it is possible to observe this difference in colour with the naked eye, unaided by the dichroscope. The maximum difference in colour of the two images of the dichroscope aperture will be seen when the crystal is viewed through in a direction perpendicular to the prism edges; one image will, as a rule, be darker than the other, and the particular colours shown will depend on the colour of the crystal. Thus with brown crystals the two images will be dark brown and pale brown to yellow; with red crystals they are darker and lighter red, and so forth. If the crystal is deeply coloured, one of the images will be almost or quite black and the other a light shade of the colour shown by the crystal.

In consequence of the strong dichroism of tourmaline, it is necessary that a crystal with a considerable depth of colour should be cut in a certain direction, namely, with the plane of the table or large central facet parallel to the prism edges. By this means the light which reaches the eye of the observer from the stone has travelled through it mainly in a direction perpendicular to the principal or optic axis of the crystal, thereby ensuring for the stone the best possible appearance. If, on the other hand, the stone is cut so that the table is perpendicular to the prism faces, that is to say, parallel to the basal plane, then the light will travel through the stone mainly in a direction approximating to that of the principal axis of the crystal, and the gem will have a dull, cloudy, and unpleasing appearance. Some of the Brazilian tourmalines, when cut in the manner first described, display a fine green colour, but in the other appear quite dark and only imperfectly transparent.   The same phenomenon may be observed in crystals from Paris, in Maine, U.S.A., which give gems of a fine dark-green or of an unpleasing yellowish-green according to the direction in which they are cut.   Only in the cases where the crystal is pale in colour is it advantageous to cut it so that the table facet is perpendicular to the prism edges, the resulting gem-stone being thus rendered darker in colour than it would otherwise appear.   With the requisite knowledge it is thus open to the lapidary to very materially improve the appearance of a stone by cutting it in a judicious manner.

The optical refraction of tourmaline is somewhat feeble, but its birefringence is fairly strong.   There is a slight difference in the refraction of stones of different colours ; that is to say, this character varies with the chemical composition.   An increase in the amount of iron present in the mineral is accompanied not only by an increase in the depth of its colour, but also by an increase in the refractive index.   The following are values for the greatest and least refractive indices which have been determined for tourmaline of different colours :

	Greatest refractive index.	Least refractive index.
Red tourmaline	1·6277	1·6111
Colourless tourmaline . . .	1·6366	1·6193
Green tourmaline	1·6408	1·6203
Blue tourmaline	1·6530	1 ·6343

A very characteristic feature of tourmaline is the readiness with which it becomes electrified.   When rubbed it acquires in a very short time a comparatively large charge of electricity, which it retains for some time ; and when subjected to changes in temperature it becomes electrified still more readily.   The behaviour of tourmaline in this respect is closely connected with the hemimorphic development of its crystals.   When a crystal is heated, one end becomes positively and the other negatively electrified, but on cooling its polarity is reversed.   A faceted stone will, of course, behave in the same way, the portions corresponding to the poles of the crystal being positively or negatively electrified.   The largest electric charges are acquired by light-coloured crystals of transparent, precious tourmaline, which are free from fissures.   Under favourable conditions the charge may be so strong that shreds of paper and similar light objects are energetically attracted.   This pyroelectrical property of tourmaline was observed in Holland at the beginning of the eighteenth century, when the mineral first became known.   Because it was observed to have the power, when cooling, of attracting ashes to itself, the name *aschtrekker*, meaning in Dutch ash-drawer, was bestowed upon it.   No other precious stone resembling tourmaline in the smallest degree, with the exception, perhaps, of topaz, is as strongly pyroelectric, so that this feature serves to distinguish red tourmaline, for example, from ruby and other precious stones.   The electrical properties of the mineral may be demonstrated in a convenient manner by dusting a mixture of red-lead and sulphur through a muslin sieve on to a cooling-stone, when the sulphur will be attracted to the positively electrified portion of the stone and the red-lead to the negatively electrified portion.

Another of the characters of tourmaline is that it is unattacked by acids.   The colourless, pale green, and red varieties are infusible before the blowpipe, while the darker varieties melt or run together and form a white to dark brown slag.

As regards the mode of occurrence, tourmaline is confined almost entirely to older crystalline rocks such as granite and gneiss.   Other modes of occurrence, for example, in granular dolomite at Campo-longo and in the Binnenthal, in Switzerland, are rare, and, as far as precious tourmaline is concerned, unimportant. The precious varieties are found almost exclusively in the rocks first mentioned, mainly in granite, and especially in its more coarsely crystallised variety known as pegmatite. The light-coloured crystals are attached to the walls of drusy cavities in such rocks, while in other druses of the same rock-mass there may be crystals of different colours, including black. Typical examples of this mode of occurrence may be seen in the granite of San Piero, in the island of Elba, of Penig, in Saxony, and specially at Paris, in Maine, U.S.A. Other localities, in particular those at which stones suitable for cutting are found, will be mentioned later. Darker crystals, brown, blue, or black in colour, are often found embedded in the rock itself, black crystals being met with very frequently. Crystals of tourmaline are often weathered out of their granitic mother-rock and are then to be found in sands and gravels. It is from such deposits that the finest gems are derived, these being collected along with other precious stones at various localities, especially in Brazil and Ceylon.

All varieties of tourmaline, so long as they are transparent and finely coloured, are suitable for cutting as gems, and, as a matter of fact, material of the most varied character is applied to this purpose. The colourless achroite is little used, and pale-coloured stones generally are not much prized. Stones of a full red, green, blue, or brown-colour are most admired, red stones being the most valuable, green the most abundant, and blue and brown the least important.

Tourmaline is rarely cut as a brilliant, the table-cut and a low step-cut being more generally adopted. The colour of cut stones is sometimes improved by the use of a suitable foil. Other than the frequently occurring fissures, already mentioned, tourmaline exhibits but few faults, enclosures of foreign substances, for example, being quite exceptional.

Each of the colour-varieties of precious tourmaline will be now treated of in some detail.

COLOURLESS TOURMALINE or **ACHROITE** occurs as crystals, which are perfectly, or almost, water-clear, but, as a rule, these are too small to give gems of much value. The acicular crystals are found, together with tourmaline of other colours, in the island of Elba, also as fine crystals associated with green tourmaline in the dolomite of Campo-longo, in Switzerland, and at some other localities, but everywhere as a rarity. Crystals of colourless tourmaline in some quantity and of moderate size have been found perhaps only in the neighbourhood of Richville, near De Kalb, St. Lawrence County, in the State of New York, where they occur, as in Elba, attached to the walls of drusy cavities in granite.

Achroite is readily distinguished from all other colourless and transparent stones by its specific gravity, which is 3·022, so that it just sinks in liquid No. 3 (sp. gr. = 3·0) and floats in liquid No. 2, which is pure methylene iodide (sp. gr. = 3·3). Phenakite, colourless beryl, and rock-crystal float in the former liquid, while diamond, colourless topaz, spinel, sapphire, and zircon all sink in the latter. In beauty of appearance, however, achroite has nothing to distinguish it from many other colourless and transparent stones, except in some cases its fine lustre. In common with all other varieties of tourmaline it may be distinguished from glass imitations by its double refraction and its hardness, as well as under certain conditions by its electrical properties.

RED TOURMALINE or **RUBELLITE** (siberite) may be of various shades of colour, from pale rose to dark carmine-red, sometimes tinged with violet. The colour may be so like that of certain rubies that it is difficult, even for an expert, to discriminate between these stones on mere inspection. The same similarity in colour may also exist between this variety of tourmaline and certain specimens of balas-ruby (spinel) and rose topaz. From all these stones tourmaline is distinguished by its specific gravity, which in the present variety is 3·08; it therefore floats in pure methylene iodide, while the stones mentioned above all sink.   The dichroism of red tourmaline is not very pronounced; the two images seen in the dichroscope vary in colour between pale rose and dark red, the former having sometimes a tinge of yellow and the latter usually a tinge of violet.   Other precious stones of a red shade show different pairs of colours in the dichroscope, so that tourmaline may be distinguished from them by means of this instrument.

The principal locality for this beautiful red stone is in the Ekaterinburg district of the **Ural Mountains**, namely, in the immediate and further vicinity of the village of Mursinka.   (Map, Fig. 63a.)   It is mined here with amethyst, topaz, beryl, and other variously coloured stones, and sent to be cut to the works at Ekaterinburg.   It is on account of its occurrence on the east side of the Urals that this variety of tourmaline has received the mineralogical name of siberite, while on account of its resemblance to the ruby it is known to jewellers as "Siberian ruby." The village of Shaitanka, thirty miles south of Mursinka and forty-five miles north of Ekaterinburg, deserves special mention as a locality for red tourmaline. The mineral occurs here with albite, quartz, green mica, and red lithia-mica (lepidolite), in druses in a very coarse-grained granite, and is usually implanted upon the albite and the lithia-mica; the crystals are also found lying in a yellow clay which is probably a disintegration product of the granite. They are prismatic in habit and deeply striated in the direction of their length. In colour they vary in shade between a paler or darker cherry- or carmine-red and violet-blue; tourmaline crystals of a pale olive-green or of some shade of colour between pale liver-brown and dark brown or black are also found at the same place. In nearly every case there is a slight difference in colour between the two ends of a crystal. At Sarapulskaya, seven and a half miles from Mursinka, groups of dark cherry-red crystals, usually of small size and radiating from a common centre or arranged in parallel, are found associated with tourmaline of other colours in a black earth, which occurs, mixed with granite débris, at the foot of a hill of granite.

There is also a sparing occurrence of red tourmaline in the district of Nerchinsk, in Transbaikalia.

The "Siberian ruby" is specially prized in Russia on account of its national origin, and is frequently worn as a gem; the nearer its colour approaches the red of the ruby the more valuable does the stone become.   These "Siberian rubies" find less favour outside Russia, at any rate in Europe.   A finely coloured specimen of the stone is represented in Plate XV., Fig. 6.

The places mentioned above are the principal sources from whence red tourmaline is derived; it occurs, however, at some other localities, but, as far as material of gem-quality is concerned, only in small amount.   It is found with ruby, sapphire, spinel, zircon, &c., in the gem-gravels of Ceylon and especially of Burma.

Its distribution in **Burma** differs from that of the ruby, it being found near Mainglon, twenty miles south-east of the town of Mogok, the centre of the ruby district. (See maps, Figs. 54 and 55.)   Tourmalines, both red and black, are found as water-worn pebbles in the sands of the Nampai valley, near Namseka village.   The Chinese work the deposits by excavating numerous small and shallow pits, and in the rainy season the stone is obtained by washing.   The whole of the material obtained in these ways is sent to China, where it is used in the making of buttons for the adornment of mandarins' caps, and is probably held there in as high esteem as the ruby itself.   Lower down the valley are numerous mines which are now quite abandoned.   The Nampai stream drains a district of gneissose rocks, and it is, therefore, not improbable that the rubellite found in its sands and gravels is derived from granite-veins penetrating the gneiss. Two very fine specimens of crystallised rubellite from Burma are exhibited in the Mineralogical Gallery of the British Museum, one of these, remarkable for its size and shape, being seven inches high and six inches across, was given by the King of Ava to Colonel Symes when on an embassy to that country in 1795; the other, not so large, but of a fine deep colour, was presented to the Museum in 1869 by Mr. C. S. J. L. Guthrie.

There is an important occurrence of beautiful rose-red tourmaline in the State of Maine in the **United States** of North America. The tourmaline of Maine is probably the finest in the world; rose-red, blue, green, and other varieties are obtained in great abundance. Amongst other important localities may be mentioned Mount Mica, about a mile east of Paris, where, as at Shaitanka in the Urals, the crystals occur with red lithia-mica attached to the walls of drusy cavities in a coarse-grained granite. Since the discovery of this locality in 1820, tourmaline of various colours has been constantly derived from this source, the value of the total production up to the year 1890 being estimated at 50,000 dollars. In the United States these stones are much prized on account of their national origin, just as in the case of Russian tourmaline. After Mount Mica, the most important locality in Maine is Mount Apatite at Auburn, discovered in 1882. Since that time, some 1500 crystals, colourless and of various shades of colour such as rose, lilac, pale blue, dark blue, green, and yellow, have been found. These when cut gave gems ranging in weight from 6 to 8 carats, but never more. Some of the tourmalines found here are peculiar in appearing of a darker shade of colour when cut than when in the rough condition. The red crystals from Paris are sometimes surrounded by an external layer of green tourmaline; the best examples of parti-coloured crystals, however, are the large prisms from Chesterfield in the State of Massachusetts (Plate XV., Figs. 8 and 9), and certain Brazilian stones. Sometimes in cutting such parti-coloured crystals, the central red portion only is utilised, the colour of which in American stones is generally very similar to that of the ruby. At other times, however, the crystal is cut so as to exhibit its parti-coloured character, and the contrast between the differently coloured portions of the gem has a peculiar effect.

Beautiful crystals of red and green tourmaline have been recently found at Mesa Grande in San Diego County, California. They occur in coarse-grained granite and are often well terminated. Many are of considerable size and transparency, and of great beauty. Rubellite is the commonest variety, but the characteristic zonal arrangement of colour, both concentric and in horizontal bands, is to be seen in many specimens.

A few fine red tourmalines are associated with the green tourmaline so abundant in Brazil, but they are of little importance.

The red tourmaline of the island of Elba and of Penig in Saxony is generally too pale in colour, and usually also occurs in crystals which are too small for cutting as gems.

GREEN TOURMALINE is the most widely distributed of the precious varieties of this mineral, and consequently is lower in price. It is rarely emerald-green, but when this is the case its colour lacks none of the depth of that of the true emerald. More commonly it is yellowish-green, but grass green, greenish-yellow, and, indeed, all possible shades, light and dark, are to be met with (Plate XV., Figs. 7, 8, and 9). The dichroism of this variety is very pronounced; the two images seen in the dichroscope are yellowish-green and bluish-green, which are usually very deep, to almost black, in shade; yellow, brown, and violet are also sometimes to be seen.

The principal locality for green tourmaline is **Brazil**, where it occurs, together with other colour-varieties, as prismatic crystals, the largest of which measure $1\frac{1}{2}$ inches in length and ½ inch in thickness. It is specially abundant in the small Ribeirão da Tolha, ten leguas from Chapada, and also in the district of Minas Novas in the State of Minas Geraes, where it occurs with white and blue topaz. It is sometimes of a fine emerald-green colour, and was indeed formerly thought to be emerald; and until its identity with the far less costly tourmaline was established, it was sought for with great eagerness. On account of this similarity the stone is known to jewellers as " Brazilian emerald"; it is emblematic of the priesthood in Brazil, and is much worn by priests as a ring-stone.

Another important source is the gem-gravels of **Ceylon**, where tourmaline of a yellowish-green colour, similar to that of chrysolite (peridot) occurs in abundance. Some of these stones are yellowish-green only on one side, the other being white with a milky opalescence. From the locality at which it occurs and its resemblance to chrysolite, this stone is known to jewellers as " Ceylonese chrysolite," and also as " Ceylonese peridot." In Ceylon, as in Brazil, it is accompanied by tourmaline of other colours, but the Ceylonese stones are less deeply coloured than the Brazilian. The name tourmaline originated in Ceylon, where turamali is the name used by Cingalese jewellers for hyacinth (zircon).

The occurrence of tourmaline of a fine green colour in the **United States** has been already mentioned. At Paris in Maine, and Chesterfield in Massachusetts, it is found not only as the external shell of red crystals, but also as crystals which are green throughout, and which, like the red stones, are often used as gems. The finest faceted stone cut from the green tourmaline of Paris measures 25 millimetres across and is 18 millimetres thick.

It is possible for green tourmaline to be mistaken for other precious stones of the same colour; for example, for emerald, chrysolite, hiddenite, and demantoid. From all these green tourmaline is distinguished by its specific gravity of 3·107, and by its strong dichroism, a feature which is completely absent in demantoid and insignificant in the other stones.

BLUE TOURMALINE or **INDICOLITE** (indigolite) is rare. It may be light or dark in shade, and of a pure indigo-blue, a smalt-blue, or a blue more or less markedly tinged with green. A faceted blue tourmaline is represented in Plate XV., Fig. 11. The colour is sometimes not to be distinguished from that of sapphire, while other specimens may closely resemble aquamarine in this respect. Indicolite may be distinguished from either of these stones by its specific gravity, which is 3·16, and by its dichroism, this feature being specially prominent in blue tourmaline. This variety occurs with green tourmaline, though in less abundance, in Brazil, and because of this circumstance is known to jewellers as " Brazilian sapphire." A few crystals suitable for cutting are found at Paris and other places in Maine, at Goshen in Massachusetts, and elsewhere in North America; also at Mursinka in the Urals. Fine specimens of indicolite occur with green tourmaline, lepidolite, and quartz in the granite south of Pahira, near Hazaribagh in Bengal. The largest crystals found here measure an inch in length; the central portion of the crystals is sometimes indigo-blue and the outer layers green. Blue tourmaline in association with the yellow and brown varieties occurs also with the sapphire of the Zanskar range in Kashmir. Everywhere, however, good transparent tourmaline of a fine blue colour is rare and much more valuable than the green variety.

BROWN TOURMALINE (dravite) still remains to be mentioned. This variety is sometimes clear and transparent enough to be cut as a gem and is then a very pretty stone. The colour varies from dark brown of different shades, through light brown to yellowish- or reddish-brown. Beautiful transparent stones, both brown and yellow, accompany the green tourmaline of Ceylon; from this island comes the brown faceted stone represented in Plate XV., Fig. 10. Among the brown crystals found embedded in mica-schist at Dobrowa, near Unterdrauburg on the Drau in Carinthia, are some which are clear enough for cutting;

the majority, however, are cloudy and unsuitable for this purpose. There is a similar occurrence at Crawford, in the State of New York, and some of the brown tourmaline found in the limestone of Gouverneur and Newcomb, in New York, is sufficiently clear and free from fissures to give good cut stones; but, as a rule, North American tourmaline of brown colour is rarely cut for gems.

OPAL.

The widely distributed mineral opal, like the still more frequently occurring quartz, consists mainly of oxide of silicon, that is to say, of silica, but differs from quartz in being not crystalline but amorphous. Besides silica, opal always contains water, the amount varying in different specimens. Various impurities are frequently present, and when in any amount render the stone cloudy and often of a deep colour, so that it is unfit for use as a gem. The variety almost exclusively used for this purpose is that known as the " precious " or " noble " opal; it is conspicuous amongst all others for the magnificent play of colours produced by the refraction and reflection of light in its colourless substance. This is the variety to which attention in what follows will be mainly directed; a few of the varieties, which show no play of colour and which are grouped together under the term " common opal," will receive brief mention later on.

The chemical composition of various kinds of opal used as gems may be seen from the table of analyses given below. The variableness in the amount of water present, and the diversity in the substances present as impurities, should be noticed:

____	Precious opal. Hungary.	Fire-opal. Mexico.	Milk-opal. Silesia.	Cacholong. Faroe.	Menilite. Paris.	Hydro-phane. Saxony.
Silica (SiO_2) . . .	90·0	92·0	98·75	95·32	85·50	93·13
Alumina (Al_2O_3) . .	—	—	0·10	0·20	1·00	1·62
Iron oxide (Fe_2O_3) . .	—	0·25	—	—	0·50	—
Lime (CaO) . . .	—	—	—	0·06	0·50	—
Magnesia (MgO) . .	—	—	—	0·40	—	—
Soda (Na_2O) . . .	—	—	—	} 0·13	—	—
Potash (K_2O) . . .	—	—	—		—	—
Water (H_2O) . . .	10·0	7·75	0·10	3·47	11·00	5·25
Organic matter . . .	—	—	—	—	0·33	—
Total . .	100·0	100·00	98·95	99·58	98·83	100·00

Since opal is an amorphous substance, it never possesses regular plane-faced boundaries, but occurs usually in rounded nodules, as botryoidal encrustations, as stalactites, and in other forms. There is, of course, a complete absence of cleavage; the fracture is conchoidal, often typically so. The mineral is moderately brittle, sometimes, indeed, very brittle, when it is easily fractured and broken. It is not very hard; $H = 5\frac{1}{2} - 6\frac{1}{2}$; that is to say, it is softer than quartz, so that though opal will scratch glass it is itself scratched by quartz. Because of its brittleness and low degree of hardness it is advisable to protect a gem of opal from a blow or fall of any kind, and from contact with harder substances, such as dust.

The specific gravity also is less than that of quartz; it lies between 1·9 and 2·3, the exact value depending upon the chemical composition, that is to say, upon the amount of water and impurities present.

The lustre is usually of the common vitreous type, though in some opals it may be greasy, resinous, or waxy. The lustre of natural specimens is only moderately strong; it is increased by cutting and polishing, but even then is in no way remarkable. One variety only, hyalite or glassy opal, is perfectly transparent; this is clear and colourless like glass, but is rarely used as a gem. Both the common and the precious varieties of opal are, as a rule, cloudy, being at the best only translucent or semi-transparent. In a pure condition the mineral is perfectly colourless, the tint of the coloured varieties being due to the presence of impurities. In colour these varieties are usually brown, yellow, or red, of various shades; green opal is rare, black is known, and the cloudy varieties of milk-opal are white. The optical refraction of the mineral is low, the index of refraction for precious opal having been determined to be 1·44. Being amorphous, opal is, of course, singly refracting.

When heated the mineral is easily fractured, so that it is desirable that cut stones should be protected from sudden changes of temperature. The constituent water is expelled below a red-heat, when the specimen, if not so already, becomes cloudy and opaque. Opal fuses in the oxyhydrogen flame, but is infusible in the ordinary blowpipe flame. It is attacked by only one acid, namely, hydrofluoric, but differs from quartz in being soluble in caustic alkalies.

Opal is found almost exclusively in the cavities and crevices of basaltic, trachytic, and other volcanic rocks; it is occasionally met with in serpentine, but never in rocks which contain no silica. All the different varieties of opal may occur in association with each other and with other minerals composed of silica, such as chalcedony and quartz, with which opal often forms a more or less intimate mixture. Both opal and these other minerals are in all cases alteration products of the rocks in which they occur. The silica dissolved out by water circulating through the rocks, which is sometimes, as in volcanic regions, very hot, is redeposited in the cavities and crevices of the rock when the solution cools or evaporates. The silica thus deposited is at first gelatinous, but on drying it takes on the characters of opal. The rounded form of the masses in which the mineral occurs is a natural consequence of its mode of formation, and is similar to that of the stalactites which originate by the deposition of calcium carbonate from water. Moreover, in some rocks silica has been found in the soft and wet, gelatinous condition in which it was deposited, and on exposure to the atmosphere has been observed to harden and dry up and eventually to become indistinguishable from opal.

Hitherto we have considered only the characters common to all opal; those varieties which are used for ornamental purposes will now be considered individually in more or less detail.

PRECIOUS OPAL.

The most important and valuable variety is the precious or noble opal, also known as the oriental opal and as the celestial opal. The features on which depend the value and beauty of other precious stones are in the opal insignificant or absent. Thus it is not transparent, has no pronounced body-colour, and compared with other stones has no very strong lustre and only a low degree of hardness. Its beauty depends solely upon the magnificent play of delicate colours seen on its surface, a feature which is to be found in no other stone, so that in this respect opal is unique.

Precious opal, as a general rule, is translucent or at most semi-transparent, a greater degree of transparency being exceptional. When seen by transmitted light the stone appears reddish-yellow in colour, but in reflected light it is colourless with a milky cloudiness, or milk-white with a faint shade of blue or pearl-grey. A pronounced body-colour, such as yellow, red, blue, green, or black, is very rarely seen. Of these, yellow, ranging from wine-yellow to sulphur-yellow, and red, especially a yellowish-red shade, are less infrequent than others. Rose-red opal, of which a magnificent example is preserved in the "Green Vaults" of Dresden, is very rare. Black opal is also rare and is sometimes of very great beauty, the play of colours showing up with striking effect against the dark background of the stone.

The play of colours characteristic of all precious opal is only shown when the stone is seen by reflected light; in transmitted light it is completely absent. The display of colour may be visible over the whole surface of the stone, or may be limited to isolated spots which merge imperceptibly into the surrounding uncoloured portions. Again, the whole surface may show a play of one uniform colour, yellow and green being in such cases much admired. In other stones there may be areas over each of which there is a play of a single colour, the play of colour over any one area differing from that over any other, and the different areas merging gradually into each other. Moreover, in some opals minute spangles of various colours are distributed in large numbers over the surface of the stone, giving a variegated kaleidoscopic effect, which has been compared to the iridescence of the neck-plumage of some pigeons or of a peacock's feather. The opals of which the general colour effect is more uniform resemble mother-of-pearl more closely. The colour of precious opal of good quality is always, however, more fiery than that of any of the objects to which it has been compared. An attempt at the reproduction of the colour effect of a few precious opals is given in Figs. 6 to 9 of Plate XVI.

In attempting a word-picture of the appearance of precious opal one cannot do better than quote Pliny, who described this stone as combining in itself the fiery red of the ruby, the magnificent green of the emerald, the golden yellow of the topaz, the deep blue of the sapphire, and the rich violet of the amethyst. All these colours may, as a matter of fact, be detected in one and the same stone; it often happens, however, that a few only are present, while some stones display but one. The play of colours is indeed very variable and never identical in any two stones, especially those from different localities. Thus, for example, in the Hungarian opal the colour is distributed irregularly in small patches and spangles giving a variegated effect (Plate XVI., Figs. 8 and 9), while in the Australian opal large areas of the surface display one uniform colour (Plate XVI., Figs. 6 and 7).

A number of varieties of precious opal are recognised, the distinction between them being based on differences displayed in the play of colours. Those stones in which close-set, angular patches of colour form a minute variegated mosaic, as it were, are known as *harlequin-opal*, a term which is also applied sometimes to stones in which the ordinary play of colours has a yellowish-red instead of a white background. In a *flame-opal* the colours are distributed more or less regularly in bands and streaks, while the whole surface of a *gold-opal* glows with golden light. The name *girasol* is sometimes applied to an almost transparent opal, over the surface of which there travels, as the stone is moved about, a wave of blue light; the same term is, however, also applied to other precious stones. *Opal-onyx* is built up of alternate layers of precious and of common opal. Other varieties also are distinguished, some of which will be mentioned under the various localities at which they are found.

The value of an opal depends in the first place upon the brilliancy and beauty of the colours it displays, and in the second place upon the uniformity with which the play of colours is distributed over the surface of the stone.   The stones which show a brilliant play of colours over the whole surface are most valuable, while in inferior stones the colours are dull and there are portions of the surface which show no play of colour.

When precious opal is exposed to the action of heat it loses its constituent water; the play of colours then disappears and the stone becomes cloudy.   Some stones indeed slowly lose water at ordinary temperatures, thus suffering a gradual diminution in beauty and value.   It is stated that in such cases the play of colours may be restored by immersing the stones in oil, but that, as the oil gradually decomposes, the stone becomes poorer and poorer in appearance until at last the play of colours completely disappears, and the stone itself assumes a dirty brown tinge.   Hungarian precious opal is least subject to such changes, and for this reason is worth more than that from other localities.

The substance of the precious opal is in itself colourless, and the brilliant play of colours so characteristic of this gem is due purely to changes effected in the incident rays of light during their passage through the substance of the opal, probably in a way which does not differ essentially from that whereby the iridescence of certain specimens of quartz and of other minerals arises.   The drying up and solidification of the gelatinous silica would be attended by the development of a network of cracks and fissures; these cracks, by their action on the rays of light at their surfaces, give rise to a display of the rainbow colours characteristic of thin plates.   Microscopical examination has demonstrated that these cracks are sometimes lined with a film of opal, the refractive index of which differs from that of the main mass.   It has therefore been conjectured that the play of colours of the opal may be due in part to this or to some other circumstance, seeing that it is so much more brilliant and magnificent in this stone than in any other iridescent mineral.   It is indeed possible that the phenomenon in opal has not as yet been completely explained by any of the various theories which from time to time have been promulgated.   There can be absolute certainty only on one point, namely, that the colours of the precious opal are effects of the interference of light and not of an admixture of pigment, since they are pale or completely absent when the stone is viewed by transmitted light.

On account of its much fissured condition, precious opal, although perhaps slightly harder, requires more care in handling than does common opal.   Sudden and extreme changes in temperature must in particular be guarded against, in order to avoid the fragmentation of the stone.   For the same reason, special care is necessary during the processes of cutting and grinding.

Precious opal is in almost all cases cut in a rounded form, for, not only does the existence of facets if anything detract from the colour effect of the stone, but owing to its softness the edges between the facets would very soon lose their sharpness.   Opal is, therefore, but rarely cut in faceted forms, though the table-cut and the step-cut are some-times to be met with.   According to the form of the rough stone, a cut opal may be circular or oval in outline and more or less convex, so that it may resemble the half of a pea, a bean, or an almond.   It is ever the aim of the lapidary to perform the operation of cutting with as little waste of the valuable material as possible, while at the same time he must contrive so that the play of colours is displayed to the best advantage.   This not infrequently requires great skill and much thought and consideration; the removal of the matrix and of those portions of the stone which show no play of colours with the least possible waste of precious material, and in such a manner as to ensure an uninterrupted

PLATE I. 1. DIAMOND, crystal in matrix (Brazil). 2. The same (South Africa). 3. DIAMOND (Bort). 4. DIAMOND (Carbonado). 5. RUBY, crystal. 6. RUBY, cut. 7. SAPPHIRE, crystal. 8. SAPPHIRE, cut. 9. SPINEL (Balas-ruby), crystal. 10. SPINEL (Ruby-spinel), crystal 11 ZIRCON (Hyacinth), crystal. 12. The same, in basalt. 13. ZIRCON, cut.

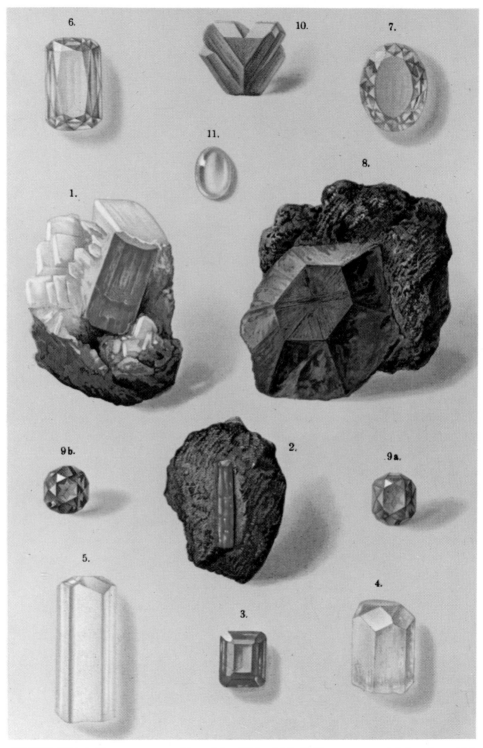

PLATE XII. 1. EMERALD, crystal in calcite (Colombia). 2. EMERALD, crystal in mica-schist (Salzburg). 3. EMERALD, cut. 4. BERYL (golden beryl), crystal. 5. AQUAMARINE, crystal (Siberia). 6, 7. AQUAMARINE, cut. 8. CHRYSOBERYL (Alexandrite), crystal (Urals). 9a. The same, cut-stone by day-light. 9b. The same, cut-stone by candle-light. 10 CHRYSOBERYL, crystal (Brazil). 11. CHRYSOBERYL (Cymophane), cut.

PLATE XIII. 1. Topaz, blue crystal (Urals). 1a. The same, cut. 2. Topaz, dark yellow crystal (Brazil). 2a. The same, cut. 3. Topaz, pale yellow crystal (Saxony). 3a. The same, cut. 4. Topaz, rose-red crystal (Brazil). 4a. The same, cut. 5. Euclase, crystal (Brazil).

PLATE XIV. 1. Epidote, crystals (Salzburg). 2. Epidote, cut. 3. Almandine, crystal in mica-schist. 4. Almandine, cut [see also Pl. XVIII, Fig. 7]. 5. Pyrope (Bohemian Garnet) in matrix. 6. Pyrope ("Cape-ruby"), cut. 7. Hessonite, crystals with diopside (Piedmont). 8. Hessonite, cut (Ceylon). 9. Demantoid, rough (Urals). 10. The same, cut. 11. Olivine (Chrysolite), crystal. 12. The same, cut.

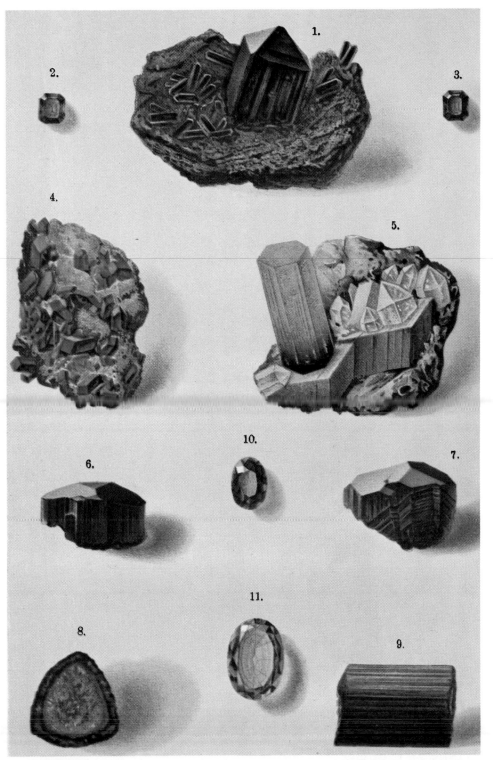

PLATE XV. 1. IDOCRASE, crystals (Piedmont). 2. The same, cut. 3. IDOCRASE, cut (Vesuvius).
4. DIOPTASE, crystals (Siberia). 5. TOURMALINE, rose-red and green crystal (Elbe). 6. TOUR-
MALINE, red crystal (Siberia). 7. TOURMALINE, green crystal (Brazil). 8, 9. TOURMALINE, red
and green crystal (Massachusetts). 10. TOURMALINE, brown cut-stone (Ceylon). 11. TOURMALINE,
blue cut-stone (Brazil).

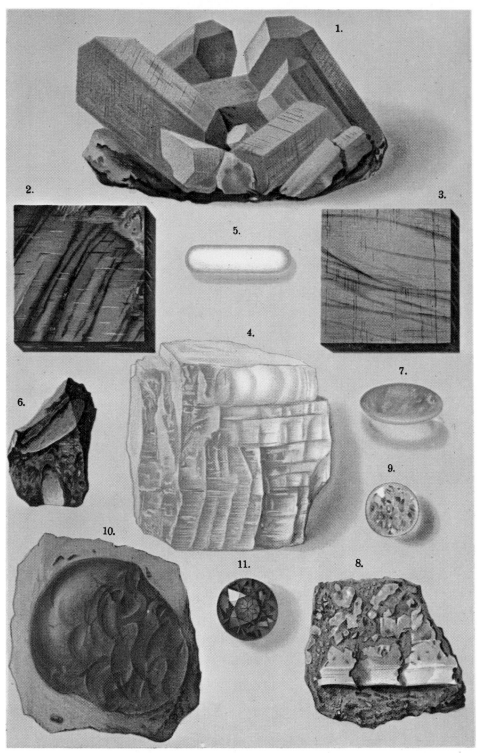

PLATE XVI. 1. AMAZON-STONE, crystals. 2. LABRADORITE, polished. 3. LABRADORESCENT
FELSPAR, polished. 4. MOON-STONE, rough. 5. The same, cut. 6. PRECIOUS OPAL, rough (Australia). 7. The same, cut. 8. PRECIOUS OPAL, rough (Hungary). 9. The same, cut. 10. FIRE OPAL,
rough. 11. The same, cut.

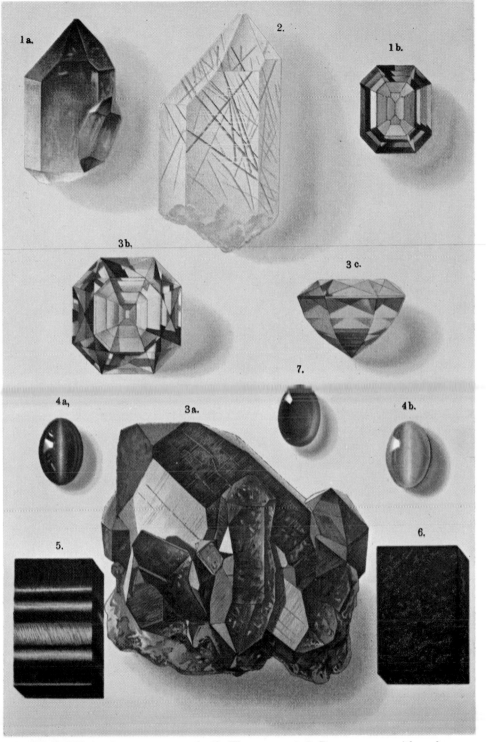

PLATE XVIII. 1a. AMETHYST, crystals. 1b. The same, cut. 2. ROCK-CRYSTAL with enclosures ("Needle stone"). 3a. SMOKY-QUARTZ, crystals. 3b, c. The same, cut. 4a, b. CAT'S-EYE, green and brown cut-stones. 5. TIGER-EYE, polished. 6. HELIOTROPE, polished. 7. ALMANDINE, cut [compare Pl. XIV, Fig. 4].

PLATE XX. 1. Lapis-lazuli, polished. 2. Turquoise, blue cut-stone. 3. Turquoise, green, in matrix. 4a. Malachite, rough. 4b. The same, polished. 5a, b. Onyx, cut. 6. Carnelian (intaglio). 7. Carnelian-onyx (cameo). 8. Chrysoprase, cut. 9. Amber, polished in part.

play of colours over the whole surface of the cut stone, often presents a problem of considerable intricacy.

By the employment of various devices the play of colours of precious opal can be increased to a certain extent; thus, for example, a cut stone, which is not too thick and opaque, may be placed upon a variegated foil, a piece of a peacock's feather, or a bright, polished plate of mother-of-pearl. The stone is rarely mounted à jour, but is best set in a black case and surrounded by a border of small diamonds or of coloured transparent stones of some sort. In the same way, the effect of a large diamond, ruby, or sapphire is greatly increased when set with a border of small opals.

Opal is a precious stone for which there is a considerable demand, and it consequently commands high prices. The value, as we have seen, depends in part upon the character of the play of colours, those in which red and green are predominating colours having been specially favoured, though this may be but a fleeting preference. A perfect stone should be neither too transparent nor too opaque, since in both cases the play of colours will be less brilliant; neither should it be cut too thick or too thin. Until recently the price of the larger and better specimens of opal was governed in the trade by their special qualities and size, and not according to their weight. Now, however, the weight of a stone expressed in carats is one of the factors which determines its price. Fine Hungarian stones are almost equal in value to brilliants of the same weight. Large opals, especially those with a considerable thickness, on account of the relative rarity of their occurrence, are dear, and the price of still larger stones is more than proportionate to the increase in size. A carat stone showing a brilliant play of colours is worth at least 50s.; stones which are inferior in this respect, as is the case with many of the so-called "Mexican" opals from Central America, are worth very much less. In the middle ages precious opal of fine quality was probably valued still more highly than it is now, and the stone was held in high esteem by the Romans.

No successful imitation of precious opal in glass has hitherto been achieved; any imitation can be instantly distinguished from the genuine stone by the appearance of the counterfeit play of colours. More successful attempts are sometimes made by mounting common opal upon a variegated foil in a blackened case, a fairly good imitation of the play of colours of precious opal being thereby produced. Black opals with a brilliant play of colours sometimes appear on the market; it is probable, however, that such stones have been treated in some way unknown.

The occurrence and mode of origin of precious opal differ in no wise from those of common opal. Everywhere the various kinds of common and precious opal occur in association with each other and with other minerals consisting of silica, such as quartz and chalcedony. It occurs as small patches in larger masses of common opal, and there is a gradual passage from one variety to the other; in the winning of opal those portions showing a play of colours are sought for and extracted from the main mass of valueless material.

By far the most important opal locality, and the one which yields the most valuable material, is the neighbourhood of the village of Czerwenitza (Hungarian Vörösvágás) in the Tokaj-Eperies Mountains, near Kaschan and Eperies, and in the Saros Comitat in northern Hungary. The mines are located on the Simonka mountain (Dubnik Hill), and especially on the Libanka mountain. The opals found here in former times were sent to Constantinople, and from thence found their way to the cities of the west, especially Amsterdam. This circumstance gave rise to the belief current till the end of the eighteenth century that the opal was found in the East, and accounted for the stone being known as "oriental opal,"

The expression is still in use even now, especially in the case of exceptionally fine stones, although it has long been known with certainty that the supposed oriental opal localities in Egypt, Arabia, Cyprus, Ceylon, &c., do not in fact exist. As early, indeed, as the end of the seventeenth century Tavernier, the French traveller and dealer in precious stones, stated his belief that precious opal was to be found only in Hungary, but this statement appears to have been overlooked.

The Hungarian opal mines are situated fifteen miles to the south-east of Eperies. At the foot of Simonka, the highest mountain of a wild, forest region, is the small settlement of Dubnik, which owes its existence to the occurrence of opal in that region. Here are found the many beautiful Hungarian stones so much prized as gems all the world over. There is little doubt but that the Romans obtained their opals from this identical spot, while records of the working of the deposit in the fourteenth century are actually in existence.

The mother-rock of the opal is here a brownish or greyish volcanic rock, technically described as a mica-hornblende-andesite. The portions of the mother-rock which contain opal are much weathered and bleached, the felspar being altered to kaolin and in part to opal. The precious opal occurs in nests in certain bands of the andesite, which are separated off from the barren rock by sharp lines of division or by open crevices. In these nests precious opal is accompanied by hyalite, milk-opal, and other varieties of common opal, into which it gradually passes. The precious opal occurs in much smaller proportion than the common opal, so that from a large mass of material after the removal of the common opal, only a small amount of opal of gem-quality will be left. Here, as elsewhere, the opal frequently occurs in rounded masses, indicating that it has been deposited from solution. The water, by the agency of which the opal was formed, was probably supplied by hot springs; although now dry in the immediate neighbourhood where the opals are found, such hot springs are met with at no great distance away. Moreover, it is said that specimens of opal, showing no play of colour, are found saturated with moisture underground in the mines, and that on exposure to the air they dry and the play of colours gradually appears.

In former times the mining of opal was carried on entirely by private enterprise. The deposits of this locality were worked principally by the inhabitants of the village of Czerwenitza, which is situated about an hour's journey to the south. They obtained the gem out of surface workings, traces of which remain at the present day in the form of heaps of débris. A certain proportion of stones have been set free by the weathering of the andesitic rock, and such are said to have been brought to the surface from time to time in the operation of ploughing or by the agency of rain-storms. The exclusive right of mining opal in this region was first claimed by the Government in 1788, and in place of surface workings, the method exclusively adopted till then, systematic underground mining operations were commenced, but in a very short time were entirely abandoned. For several decades after this all work ceased, when the present system of leasing the mines to private individuals began to be adopted. It was reported in 1877 by Professor Gerhard vom Rath that the rent per annum was 15,000 florins, while the working expenses for the same period amounted to 60,000 florins, so that to leave a margin of profit the yield could not have been very small. By improved methods of mining it was still further increased, so that the mines became very lucrative.

At the present time the workings are confined to the Libanka mountain, situated about half a mile west of Dubnik. On the east side of the mountain, opal mines and old débris-heaps extend for four and a half miles from north to south. The galleries, which lie at four or five different levels one above the other, have a total length of four and a half miles, and the hill is penetrated by an adit. The workings are excavated by preference in a conglomerate of andesite blocks, possessing great hardness and solidity. The miners, of which in 1877 there were 150, loosen the opal-bearing rock with great caution, and carefully free the precious material contained in the detached fragments from the mother-rock. On the same spot there are also cutting works, which find employment for six men; the stones are worked with emery on a leaden disc. There is a tradition that in the year 1400 no less than 300 persons were employed here in the mining of opal; even if this number be not exaggerated the yield would probably be less than it is at present with fewer workers but better tools and methods.

The yield is very variable, and depends to a great extent upon a fortuitous combination of circumstances. Not infrequently precious opal may be searched for in vain for a distance of 10 or 12 yards. Large specimens are now very rare, and several years may elapse between the finding of specimens of the size of a hazel-nut. Some of the large specimens met with in former times are exhibited in the collection of minerals preserved in the Imperial Natural History Museum at Vienna. In this collection is to be seen the largest known specimen of Hungarian opal; it is uncut, but quite free from the mother-rock, and exhibits a most beautiful play of colours. It is wedge-shaped in form and about the size of a man's fist, being $4\frac{3}{4}$ inches long, $2\frac{1}{2}$ inches thick, and $\frac{1}{2}$ to 3 inches high; it weighs 34 loths or nearly 600 grams (about 3000 carats). An Amsterdam dealer in precious stones is said to have offered for this fine opal half a million florins (£25,000); it has been valued at 700,000 florins, but Partsch in his guide to this collection (1855) values it at 70,000 florins only, and this has been copied into other works. The specimen was found in the seventies of the eighteenth century near Czerwenitza. A smaller stone, also remarkable for its purity and magnificent colour, of the form and size of a hen's egg, is perhaps a portion of the above-mentioned specimen; it is preserved in the Imperial Treasury at Vienna.

Another extensive find has more recently been made, namely, at the end of the eighties of the nineteenth century; in this case the opal occurred as a large mass, not as small nests as it usually does, in the andesitic mother-rock. The mass measured 15 metres in length and 20 centimetres in thickness; it consisted for the most part of milk-opal, but in two places was intersected by precious opal of fine quality and bordered here and there with the so-called oculus, that is to say, opal with a less brilliant play of colours. This particular occurrence was also remarkable in that the play of colours instead of being disposed in very small patches and spangles, as is characteristic of Hungarian opal, occupied large patches of the surface as in Australian opals. The difference in the appearance of opal from these two localities will be understood on comparing Figs. 6 and 7 with Figs. 8 and 9 of Plate XVI.

Besides pure opal, *mother-of-opal* is also mined in Hungary. This term is applied to a rock containing specks of precious opal of greater or less size, but always too small to be worth isolating from the rock in which they are embedded. The dark mother-rock flecked with the bright prismatic colours of the opal forms a material which can be applied very effectively to many decorative purposes, and when the flecks of opal are numerous and close-set, mother-of-opal may even be used in jewellery. The effectiveness of the stone is sometimes further enhanced by soaking the always more or less porous mass in oil, and afterwards exposing it to a gentle heat. The matrix is much darkened by this treatment and consequently forms a better background for the flecks of precious opal, which are unaffected by the process. It is possible that the black opal mentioned above is produced by similar treatment; nevertheless, there is no doubt that black opal does occur in nature, though very rarely.

Precious opal just as fine as that of Czerwenitza is found under the same conditions at other places in the north of Hungary. It occurs, for example, in a quartz-trachyte at Nagy-Mihály, east of Kaschau on the Laborcza in the Ujhely Comitat. The mineral, however, is not abundant and the occurrence has no commercial importance. The same is also true of other localities in Europe, such as, for example, in the neighbourhood of Frankfurt-on-the-Main in Germany, and at Neudeck in Bohemia; also in the basalts of the north of Ireland and the Faroe Islands, where precious opal has, on rare occasions, been found with the abundantly occurring common variety.

There are a few localities outside Europe at which precious opal has been found in considerable amount, but inferior in quality to that found in Hungary, which, in fact, at present is the finest known. These localities, all of which are situated in America and Australia, must now be enumerated.

The Central American State of **Honduras** may be first mentioned. The occurrence of opal here is already of some commercial importance, and the resources of the district are apparently not fully exploited. The stones resemble Hungarian opals in some respects, but are usually more transparent and less fiery. Their one undesirable feature is the tendency of the colours gradually to fade when exposed to the air. This, however, is not the case with all stones, and some have been found comparable, both in beauty and in the permanence of their play of colours, to the finest Hungarian opals. The precious opal found in Honduras occurs for the most part in the department of Gracias, in the west of the State, under the same conditions as in Hungary, namely, in a weathered volcanic rock. Here also all the varieties of opal occur in association together, in some districts in masses of gigantic size. The dark-coloured trachyte of the central districts, especially of the department named above, is penetrated by veins and bands of different varieties of opal, some being of large size and extent. Precious opal is found embedded in these veins and bands at numerous places, and here mines are worked. These mines are usually very inaccessible and far distant from lines of communication, and on this account the deposit is less extensively worked than is the Hungarian.

The best known mines are in the neighbourhood of the town of Gracias (Gracias á Dios), others are situated near Intibukat and a few others of importance at Erandique. Here also the mineral resources are undeveloped, and a largely increased yield would probably follow the employment of systematic methods of working. The different varieties of opal occur here in small irregular veins in the trachyte; these, which are almost vertical, stretch from north-east to south-west, and often divide, or string out, only to reunite further on. The precious opal occurs in isolated plates in the common opal, sometimes interlaminated with this so as to produce an onyx-like stone of peculiar but pretty appearance; at other times it is met with in larger masses. Mining operations are carried on principally in a hill of red trachyte three miles long and 250 feet high. For a distance of half-a-mile along this hill precious opal has been found at every point where search has been made. Mines have also been sunk in the neighbourhood of Erandique, but the deposits have nowhere been systematically worked.

At the places hitherto mentioned, opal is actually known to occur and has indeed been mined there. There are other places, however, in this country where opal has not been actually found, but where all the conditions point to its probable existence. There are some such promising localities between Intibukat and Las Pedras, also in the neighbourhood of Le Pasale and Yukusapa, and on the slopes of the large mountain of Santa Rosa. The opal mines in the valley between Tamba and the Pass of Guayoca also give promise of an abundant yield, large masses of opal of all kinds having been found. Among these different varieties is a pearl-grey opal with a play of red light; this has no market value, but its occurrence is taken as an indication of the existence of better stones.

It is evident that there are occurrences of opal in Honduras which are unknown to white men from the fact that fine stones are constantly brought into the towns for sale by Indians. These localities may extend over the border of Honduras into Guatemala, since precious opal from this State is to be seen in various collections. Neither the exact localities nor details connected with the mining of opal in this country are known. It is said that a belt of opal-bearing trachytic rock extends from Honduras northward through Guatemala, perhaps as far as Mexico, where opal is of frequent occurrence and is mined.

The occurrences in **Mexico** are of some importance, especially in the mines of Esperanza, where precious opal is so common that specks of it are often to be seen in the stones of buildings. These mines are situated ten leagues north-east of San Juan del Rio in the State of Queretaro and extend over an area measuring thirty leagues in length and twenty leagues in breadth. The occurrence was accidentally discovered in 1835 by an agricultural labourer, but systematic operations have been in progress only since 1870. The precious stone is found here as elsewhere in a trachytic rock of reddish-grey colour and porphyritic structure, of which Ceja de Leon and Peineta among other hills are built. In these hills are sunk many mines remarkable for the amount and variety of the material they yield. Thus a single block of rock from the Simpatica mine usually contains precious opal, harlequin-opal, lechosos-opal (a variety presently to be mentioned), milk-opal, and fire-opal. One of the largest mines is the Jurado; it is an excavation in the trachyte 150 feet deep, 100 feet wide, and some hundred feet long. Many other smaller workings are to be seen, though at present but few are in active operation.

The opal-bearing rock is sent to Queretaro, twenty-five leagues off, where almost the whole of the product of the mines is cut, but in so rough a fashion that the precious material is shown to very little advantage. The work is done by twenty native lapidaries in three cutting works. Very little uncut material is exported. In the mines themselves about a hundred Indians are employed; 50,000 cut stones are sold yearly, and with improved methods the production could be easily doubled. Large numbers are sent to the United States, where, in some districts, they are sold to travellers at the railway stations as home products. Others are exported to Europe, especially to Germany, where they are used in cheap jewellery. Central American and Mexican opals are very variable in price, but are always worth far less than Hungarian stones. The cheapest stones are worth no more than a few cents apiece, while the best will fetch a hundred dollars; parcels containing a hundred or more pieces of opal are often sold at less than ten cents apiece. Higher prices are paid for exceptionally fine stones, but they do not equal the sums paid for Hungarian opals.

Mexican precious opal often occurs in common opal in layers which are too thin for cutting. Cavities in the mother-rock are filled completely, or sometimes to only one-half or two-thirds of their extent, with these masses of opal. The different varieties are disposed in horizontal layers, and the uppermost is often a botryoidal layer of glassy hyalite. Such a mode of occurrence clearly indicates that the opal has been deposited from water containing silica in solution.

Precious opal occurs here in a considerable number of varieties: all are remarkable for the intensity of the colours reflected from their surfaces, and in this respect are comparable to Hungarian stones. Mexican opals frequently show extensive patches of one uniform colour, while over the whole surface of some stones there is a play of only a single colour, red, green, or yellow, which either remains unchanged as the stone is moved about or passes into other colours.

The Mexican harlequin-opal is often remarkable for the variegated colours of its surface markings. The variety known in that country as *lechosos-opal* is a beautiful fire-red opal with a magnificent emerald-green play of colour showing flashes of carmine-red and dark violet-blue. Some stones show a reflection of emerald-green light combined with a very fine dark ultramarine-blue. An opal with a magnificent rose-red play of colour was exhibited in Paris in 1887. These varieties vary in quality according to the place from which they come, the stones from certain localities having quite a distinctive character.

It would appear from what has been said as to the beauty of Mexican and Central American opal and the abundance in which it exists, that the introduction of improved methods of mining and of cutting the rough material would result in such an extension of the opal-industry as to seriously reduce the trade in Hungarian opal. This, however, is not the case, for all opal from this region has a tendency gradually to become either opaque or transparent, the play of colours in either case being more or less completely lost. This feature, it is true, can be restored by soaking the stone in oil, but the restoration is not permanent. Moreover, some of these stones have also a tendency to crack and fall into pieces in course of time with no apparent reason. When these considerations are taken into account, it is not surprising that Hungarian stones are preferred and that they fetch higher prices; also it is obviously advisable to exercise great care in the purchase of freshly broken Mexican opals.

In the **United States** there is an unimportant occurrence of opal in the neighbour-hood of the John Davis river, in Crook County, Oregon. The stones found here are greyish-white and reflect red, green, and yellow light; they are very similar to some Mexican stones, but do not appear to exist in great abundance. In the American continent it would seem that the abundance of precious opal decreases from south to north, Honduras being richest in this stone, Mexico considerably less rich, and the United States very poor. It was stated above that a great deal of the opal sold in the United States as a home product is in reality Mexican in origin.

Australia, especially New South Wales and Queensland, is an important source of precious opal. The stones, which are cloudy or milk-white, reflect light of the finest blue, green, and red colours ; these colours are respectively disposed over larger areas than is the case with Hungarian stones (compare Figs. 6–9, Plate XVI.). While in some opals different areas of the surface glow with different colours, merging into each other, however, at their margins, in other stones the whole area reflects light of one uniform colour. The difference between Hungarian and Australian opals has just been pointed out ; jewellers sometimes distinguish between them by referring to the latter as *opaline*. There occur also in Australia many stones with the same distribution of colour as in Hungarian opals and equal in beauty to the latter, but with a more decided tinge of yellow in the ground-colour. Many fine gems have been cut from Australian opal which, at the present time, is more abundant in the market than is the Hungarian mineral.

In New South Wales the finest opal is found on Rocky Bridge Creek, Abercrombie river, County Georgina, in a fine-grained, bluish-grey amygdaloidal basalt or trachyte, which has a thickness of 30 feet and is so altered that it may be scratched with the finger-nail. The precious opal, which occurs only in small amount, is deposited in amygdaloidal cavities or in crevices in the rock, and, as elsewhere, is accompanied by common opal and by hyalite. The precious opal, which forms in every case only a small part of the opal masses, is milk-white and reflects principally green, red, and rose-coloured light.

The most important occurrence in New South Wales of opal of fine quality, and the one exclusively mined, is at White Cliffs, on the farm Moomba in County Yungnulgra, about sixty-five miles north-north-west of Wilcannia. The following account of this occurrence is taken from a recent (1901) description by Mr. E. F. Pittman, Government Geologist of New South Wales:

The discovery of opals at White Cliffs was accidentally made by a hunter in 1889. Since that time mining operations have been carried on continuously, though sometimes under great difficulties, as in time of drought the locality is very badly provided with water; opal-mining has, however, now become a settled industry, and a thriving township has been established at White Cliffs. The area within which the mineral has been found in the district is about fifteen miles long and about two miles wide. Prospecting for precious opal is a decidedly speculative business because, as a rule, there are no indications whatever on the surface of the occurrence of the mineral below. It is only in very rare instances that an outcrop of the precious stone can be seen, and the usual procedure is to dig a trench or pit in such a position as fancy may dictate, and trust to luck. Fortunately sinking is easy, as the rock is of a soft nature, and in a fair number of instances the opal has been met with at a very short distance from the surface, though a large majority of attempts are unsuccessful. For several years the belief existed among the miners that it was useless to prospect for precious opal at a greater depth than 12 feet from the surface, but of late the incorrectness of this view has been proved, and the stones have been discovered at a depth of nearly 50 feet.

The precious opal occurs in a white siliceous rock, varying from a sandstone to a fine conglomerate in character, of Upper Cretaceous age. It is sometimes met with in thin flat veins between the bedding-planes of the rock; at other times it forms irregular-shaped nodules, or deposits occupying joints; occasionally fragments of wood are found converted into common opal, while where cracks have occurred in the wood they are filled by precious opal. Fossil bivalve shells and belemnites, entirely converted into precious opal, are not uncommon, and a fair number of opalised bones of saurians have also been found. Although these opalised fossils are of no intrinsic value as gems, they have acquired very high prices through the competition of dealers in curiosities, and are now extremely difficult to obtain. Another curiosity which is not uncommon is a pseudomorph of opal after groups of gypsum crystals; bunches of these, several inches in length, are sometimes found composed of precious opal, though the quality is usually poor. The really valuable opal, however, which is cut and polished as a gem, is found in the irregular nodules and seams in the joints and fissures of the soft siliceous rock. When the miner finds the first indication of such a deposit, he proceeds with great care to excavate the soft rock from all round it, and occasionally masses worth several thousand pounds have been found in this way. The output of the White Cliffs deposit in 1900 was valued at about £80,000.

The deposits of opal in Queensland are far richer than those in New South Wales. The occurrence here differs from that of other regions and is similar to that at White Cliffs described above. The opal occurs in thin strings and veins, and also in larger irregular nodules in a highly ferruginous sandstone or siliceous iron-stone known as the Desert Sandstone. The strings of opal are sometimes so thin that it is impossible to cut stones *en cabochon* out of this material; instead, it is cut in the form of flat plates. The body-colour of the opal found here is milk-white and the light reflected from the stone is dark blue, green, or red. As is the case at other localities, there is a considerable amount of common opal, which has to be cut away from the precious variety. It is difficult to obtain detailed accounts of the occurrence of opal at the different localities in Queensland. In recent years Queensland opal has come mainly from Bulla Creek, where it generally occurs as the nucleus of large nodules of iron-stone. Good specimens are also found on the Barcoo river.   The first find is said to have been made on Cooper's Creek.   Other localities are the northern part of Mount Tyre near Mount Marlow Station, Opal Range, Winton, Mayne river, Canaway Range, Bulgroo, Micavilla, and Listowel Downs.   The mineral is thus widely distributed, and, though inferior to Hungarian opal, is extensively used.

FIRE-OPAL.

Next in importance to precious opal comes fire-opal, so named from its fire-red colour. It is likewise known as sun-opal, also on account of its colour; while the term girasol is sometimes applied to this, as well as to other precious stones.

In colour fire-opal ranges from an almost colourless or light brownish-yellow shade to deep brownish-red.   The finest tints are like those of some topazes or of hyacinth.   The stone is much paler in thin splinters than in thicker pieces; also it is paler by transmitted than by reflected light.   Not infrequently several shades of colour are displayed by one and the same specimen, the different tints passing gradually into each other or into perfectly colourless opal.   The colour depends, no doubt, upon the presence of a small amount of iron oxide, the presence of which in fire-opal is shown by the analysis quoted above.

The stone is translucent to almost perfectly transparent.   The fracture is always markedly conchoidal; the lustre of a fractured surface is always high and can be enhanced by polishing.   When cut *en cabochon* or with facets, fire-opal, if not too light in colour, gives a pretty gem.   Fig. 10 of Plate XVI. represents a fire-opal in the rough condition and Fig. 11 a cut stone.

Many specimens of fire-opal exhibit a play of colours similar to that of precious opal, when the two can be distinguished only by the body-colour of the fire-opal.   The play of colours of the fire-opal has a pronounced yellow or red for background, but between this and yellow or red precious opal there are all possible gradations.   The light reflected by those fire-opals which show a play of colours is often less varied in colour than in the precious opal.   The most usual tints are red and green, which in the paler-coloured stones are often fine carmine-red and deep emerald-green; a combination of yellow and blue is also met with, but is much rarer.   As a general rule the colours reflected by fire-opal are less brilliant, and the play of colours taken as a whole is less striking than in the precious opal.

Fire-opal while one of the most beautiful of all varieties of opal is, at the same time, the least durable.   Thus it may be influenced by contact with water, by sudden changes of temperature, or by the action of light or atmospheric conditions.   Some changes in the stone are ascribed to the weather, and a fire-opal is stated to be more brilliant in summer than in winter.   If this is really the case, the difference is probably due to the brighter light of the warmer season of the year.   While some stones are more durable, others are easily fractured and often lose their lustre and colour for no apparent reason, even when they have been protected from external influences.   Very transparent stones exhibiting a play of colours are specially liable to become affected in these ways, and for this reason are unsuitable for use as gems.

The disadvantages of fire-opal as a gem are, however, of little consequence, since the stone, in spite of its fine appearance, is rarely met with in the trade.   The price is not as low as might be expected, probably because durable stones of any size are somewhat rare. For a fire-opal measuring $4\frac{1}{2}$ lines in length and $3\frac{1}{2}$ in width, 1200 francs is said to have been paid.

The fire-opal is mainly a product of Mexico, and was first brought to Europe by Alexander von Humboldt at the beginning of the nineteenth century.   It is found in any amount here, only in a porphyritic trachyte at Villa Seca, near Zimapan, in the State of Hidalgo, a little to the east of Queretaro and north of the city of Mexico, in latitude 20° 44½′ N. and longitude 81° 41¾′ W. of Greenwich. Together with common opal it fills the cracks and crevices of the mother-rock, and occurs also as isolated masses of larger size, the colour of which varies in the way described above. Many of these opal masses are invested with a layer of snow-white, greyish, or brownish porous material of greater or less thickness, due to the weathering of the stone (Plate XVI., Fig. 10). Besides the locality near Zimapan, the fire-opal is found near Tolima in Mexico, in Honduras, at a few places in North America, in the Faroe Islands, and elsewhere, always, however, together with other kinds of opal and under essentially the same conditions as at Zimapan. All these occurrences compared with that of Zimapan are of little importance, and to the trader in precious stones of none whatever, so that they require no further consideration.

OTHER VARIETIES OF OPAL (COMMON OPAL, SEMI-OPAL, &c.).

The other varieties of opal are not comparable in beauty of appearance either with precious opal or with fire-opal. They are used occasionally in cheap jewellery, but find a more extensive application in the manufacture of fancy goods, such, for example, as the knobs of umbrellas and sticks, for snuff-boxes, seals, knife-handles, &c. ; they will therefore receive here only a brief consideration.

One of the varieties of opal which show no play of colours is sometimes perfectly transparent, and when this is the case is either perfectly colourless and water-clear, or tinged slightly with red or brown. It occurs as a secondary formation in the crevices of basalt and other rocks containing silica, in the form of thinner or thicker crusts with a botryoidal surface. From its glassy aspect it is known as *hyalite*, glass-opal, or Müller's glass ; it is the purest and clearest variety of opal, but is rarely cut as a gem. Opal intermediate in character between hyalite and common or semi-opal also occurs ; it is neither as clear nor as colourless as hyalite, having a faint bluish or yellowish tinge and a slight milky cloudiness.

Opal in its purest condition is water-clear ; the presence of impurities of various kinds causes it to lose its transparency, colourlessness, and some of its lustre. Different specimens of opal may thus exhibit great diversity in appearance, while preserving unaltered the characters typical of the species. It is on such differences in transparency, colour, and lustre that the distinction between the varieties recognised by mineralogists is based. There is no sharp separation between these varieties, and, specimens intermediate in character are always to be met with. *Common opal* is translucent, and, as a rule, only slightly coloured. *Semi-opal* is less translucent, and ranges from colourless to deeply coloured. *Opal-jasper* or *jasper-opal* is very slightly translucent, and by reason of the large amount of impurities, especially of ferruginous material, which is present, is deeply coloured—reddish-brown, yellow, and different shades between green and black being met with. The usual vitreous lustre of opals is sometimes replaced by a greasy lustre which may incline to the waxy, the pitchy, or the resinous type. Yellow opal with a waxy lustre is known as *wax-opal*, brown opal with a pitchy lustre as *pitch-opal*, and opal with a resinous lustre as *resin-opal*. Wood, when silicified, furnishes another variety of opal known as *wood-opal*, and there are others which need not now be enumerated.

These different kinds of opal occur for the most part in the manner described above, and are associated together in large masses. The different kinds often occur in layers, or are otherwise regularly arranged with respect to each other and to other siliceous minerals, especially quartz and chalcedony, with which they may be associated.   Common opal is so widely distributed that it is scarcely possible to mention every locality.   The mineral occurs very abundantly in the basalts of Iceland, the Faroe Islands, the north of Ireland, the neighbourhood of Steinheim, near Hanau, and many other places.   It is found in trachyte at a few places in the Siebengebirge on the Rhine, in the volcanic region of the north of Hungary and Transylvania, in Honduras, and throughout the whole of Central America and Mexico to the United States, and in many other places.   It is found in serpentine in the neighbourhood of Frankenstein, Silesia, in great abundance.   Other modes of occurrence will be incidentally mentioned below.

Opal of all the kinds mentioned above, when cut and polished, usually acquires a good lustre, and, as it is often pleasing in colour, the rounded gem-stones into which it is cut are by no means unattractive in appearance.   As, however, the rough material occurs in nature in such abundance, and as, moreover, the finished product is so soft and brittle and lacking in durability, these stones are always low in price and are used only in the cheapest varieties of jewellery.   In connection with the abundance of the mineral, it may be mentioned that in former times a uniformly coloured, pale grey, translucent opal of very pleasing appearance was obtained in large quantity at Steinheim, near Hanau, and was cut at Oberstein ; when this deposit was exhausted, a supply of similar material was at once forthcoming from the Siebengebirge on the Rhine, so that there was no opportunity for a rise in the price of this variety.

The following different kinds of common or semi-opal are sometimes used for decorative and ornamental purposes :

Milk-opal is a cloudy but highly translucent opal of a milk-white, bluish- or greenish-white colour.   It occurs in large amount in decomposed serpentine at Kosemütz in the neighbourhood of Frankenstein in Lower Silesia, and at other localities.   Milk-opal sometimes exhibits black arborescent markings, or dendrites so-called, similar to those in certain varieties of chalcedony (compare Fig. 89).   Opal of this kind is known as *moss-opal;* it is cut so as to bring the markings as near the surface as possible.   Specially fine specimens measuring 3 or 4 inches across are found in Trego County, Kansas.

Opal-agate shows a banded structure, the bands being alternately light and dark in colour, or opal and agate may be banded together in the same way.   The arrangement of the layers is the same as in onyx, so that, like this stone, opal-agate may be used for making cameos, &c.   It is found at the Giant's Causeway, County Antrim, sometimes at Steinheim, also in the Siebengebirge, and, of specially fine quality, at Guayoca in Honduras.

Prase-opal is a highly translucent opal, the beautiful apple-green colour of which is due to the presence of a small amount of nickel.   It occurs at Kosemütz, near Frankenstein, Silesia.

Rose-opal is a semi-opal of a beautiful rose-red colour, probably due to the presence of organic matter.   It occurs interbedded with fresh-water limestone at Quincy, near Mehun (dep. Cher), in France.   An opal of the same character, which is cut at Oberstein, is said to come from Mokün in Upper Egypt.   A variegated rose-red, yellow, and green opal of the greatest beauty is found in large masses in the State of Jalisco in Mexico.

Wax-opal is yellow in colour and is characterised by its wax-like lustre.   It is specially abundant in trachytic tuffs in the neighbourhood of Tokaj and Telkibanya in Hungary, hence its name Telkibanya-stone.   At the same place occurs also the **pitch-opal**, a dark-brown opal with veins of a lighter colour, and with a brilliant, pitch-like lustre.

Wood-opal of a paler or darker colour arises from the opalisation of fossil wood. The structure of the wood down to the minutest detail is often to be seen in the polished surface of the opal, giving it a curious appearance. This variety occurs in large amount at the Hungarian locality just named, in Tasmania, in the Siebengebirge, and at many other places.

Menilite occurs as greyish-brown, rounded nodules in clayey shale at Menilmontant and St. Ouen, near Paris. When polished it acquires a brilliant lustre, and those stones which exhibit alternate bands of grey and brown are decidedly pretty objects.

Hydrophane is an opal which may be dirty white, yellowish, brownish, reddish, or greenish in colour, and which in its natural condition has little lustre and translucency. In mass it is almost opaque, and very little light passes through even the thinnest of splinters. Hydrophane possesses, however, one very remarkable property on which depends its occasional application as a gem. By the absorption of water it becomes almost perfectly transparent; some specimens even acquire the play of colours characteristic of precious opal, and are then known as *oculus mundi*. The capacity of hydrophane for absorbing large quantities of water is due to the great porosity of the substance; so eagerly does it suck up water that it will adhere to the tongue; moreover, its immersion in water is often accompanied by a hissing sound due to the rapid expulsion of bubbles of air. The transparency of hydrophane, acquired in the way described, is not permanent, however, and on drying the stone gradually becomes again cloudy and opaque, any play of colours it may have acquired being lost. So long as the water in which the stone is immersed is pure the phenomenon may be repeatedly observed. Hydrophane is sometimes used as a gem, and when this is the case it is cut with a rounded surface in the form of a lenticle and set *à jour* in rings or as a pin, so that there is nothing to prevent the stone being immersed in water at will and its peculiar property exhibited. It is not surprising that the behaviour of hydrophane, under the circumstances described, inspires considerable awe and wonder in the minds of Eastern people, and especially of the natives of Java and other East Indian islands, by whom the stone is much worn as an amulet. It is said that a large number of stones are every year exported from Europe, and especially from Oberstein, to these islands and there sold to the natives.

The transparency acquired by hydrophane after immersion in water is very fleeting; a more permanent effect, lasting perhaps as long as a year, is obtained by placing the stone in hot oil. A somewhat different effect again is obtained by allowing the porous mass to become impregnated with pure wax or spermaceti; the stone is then cloudy when cold, but when slightly warmed and the wax melted it assumes a brown or grey colour and becomes highly translucent or almost transparent. For this reason the mineral is sometimes known as pyrophane. It can be coloured by immersion in coloured liquids, and it is said that in former times it was brought into the market dyed red or purple.

Hydrophane is not particularly abundant, and as there is a certain appreciable demand for it it commands a fair price, the value of any given stone depending upon the size of the stone and the degree of transparency or play of colours it acquires when placed in water. The most important locality is probably Hubertusburg in Saxony; the mineral occurs here in a porphyry, either as thin strings or in nodules of chalcedony, with amethyst, rock-crystal, and common opal. When found in the mother-rock the siliceous masses are often still soft and gelatinous, the material gradually assuming the characters of hydrophane as it dries up on exposure to the air. It also occurs with the precious opal of Hungary, with the fire-opal of Mexico, with the various kinds of opal found in the Faroe Islands and Iceland, and at some other localities where opal is found, but always sparingly and in small masses.   The majority of the stones which come into the market do not much exceed lentils in size.

Cacholong (*Kascholong* (G erman), mother-of-pearl-opal, or mother-of-pearl-agate) is an opal with very little translucency, a feeble lustre of the mother-of-pearl type, and a milk-white, reddish, or yellowish colour.   It breaks with a large conchoidal fracture with very smooth surfaces.   Like hydrophane, it is very porous and adheres to the tongue ; but, unlike this, does not become transparent on immersion in water.   It is fashioned into all kinds of small articles and fancy goods, and is sometimes cut *en cabochon* for gems, some specimens when polished having quite a pretty appearance.   In some cases the stone is built up of alternate light and dark bands of material, as in onyx, or it may be interbanded with thin layers of bluish or greenish chalcedony, specimens of this description being sometimes used for cameos.

Fine specimens of good size are not very frequently met with, and consequently command rather high prices.   Cacholong is found in small amount at various localities, usually in thin layers 1 to 4 lines thick, these layers alternating with chalcedony.   The name is said to be derived from the Cach river in Bucharia, Central Asia, in which it occurs in the form of loose pebbles ; it has been also derived from the Tartar word *kaschtschilon*, meaning beautiful stone.   The above-named river has long been mentioned in connection with the occurrence of cacholong, but nothing further about the locality is known.   The stone was formerly known to the inhabitants of the region as kalmuck-opal, or, as it was thought to be a kind of agate, as kalmuck-agate.   It is also found in the basalts of the Faroe Islands and of Iceland, and as reniform and botryoidal incrustations on the limonite of Hüttenberg in Carinthia ; also on the shores of the Bay of Fundy in Nova Scotia.   The mineral, which at no time has had any great importance, occurs at all these places in association with opal and chalcedony.

TURQUOISE.

This stone is referred to as oriental turquoise, true or mineral turquoise, and *turquoise de la vieille roche*; its little-used mineralogical name, calaite, is derived from a name used by Pliny for a green stone supposed to be identical with the precious stone now under consideration. Turquoise is always opaque and is usually of a green colour, the best qualities only being blue. It never occurs in distinct crystals, in this respect differing from all the valuable precious stones hitherto considered, with the exception of opal. It is unique also in its chemical composition, since it belongs to the phosphates, a group of minerals which includes no other precious stone of the first rank.

Turquoise is a hydrous phosphate of aluminium, having the formula $2Al_2O_3.P_2O_5.5H_2O$, which corresponds to a percentage composition of: alumina (Al_2O_3), 47·0; phosphorus pentoxide (P_2O_5), 32·5; water (H_2O), 20·5 = 100·0. In the analysis of actual specimens it is found that these proportions are not invariable, a circumstance often observed in the case of substances which do not occur as crystals. Besides these constituents turquoise always contains small quantities of other substances. There is always from 1 to 4 per cent. of iron oxide and from 2 to 8 per cent. of copper oxide, the presence of which is important, since it is to this that the fine colour of the mineral is due. A blue oriental turquoise, probably from Persia, was found on analysis by Hermann to contain:

		Per cent.
Alumina (Al_2O_3)	. . . . . .	47·45
Ferric oxide (Fe_2O_3)	. . . . .	1·10
Cupric oxide (CuO)	. . . . .	2·02
Lime (CaO)	. . . . . .	1·85
Manganous oxide (MnO)	. . . .	0·50
Phosphorus pentoxide (P_2O_5)	. . . .	28·90
Water (H_2O)	. . . . . .	18·18
		100·00

Professor S. L. Penfield has recently (1900) investigated the chemical composition of turquoise, and has arrived at the conclusion that the copper and iron, which are always present, are not accidental impurities but are essential constituents of the mineral. The new formula he proposes, namely $[Al(OH)_2, Fe(OH)_2, Cu(OH), H]_3 PO_4$, represents turquoise as a derivative of ortho-phosphoric acid, in which the hydrogen atoms of the acid are largely replaced by the univalent radicals, $Al(OH)_2$, $Fe(OH)_2$, and $Cu(OH)$ in variable amounts.

When a fragment of turquoise is heated over a flame in a narrow tube closed at one end, it decrepitates, that is to say, it flies into small fragments with a loud crackling noise; at the same time water is expelled and condenses on the cool parts of the tube. When heated more strongly, for example, when ignited in a platinum crucible, a brownish-black mass results, usually so incoherent that it falls to powder at the slightest touch; in some cases the fragment of mineral is directly converted by the intense heat into such a brown powder.

Turquoise by itself is infusible in the blowpipe flame; by virtue of the phosphoric acid and copper oxide the mineral contains, it colours this, or any other colourless flame, green. It is usually soluble both in hydrochloric acid and in nitric acid, but specimens from different localities behave somewhat differently in this respect, some being unattacked by either of these acids.

It has already been stated, that turquoise has not hitherto been found in crystals. It occurs as irregular masses, completely or partly filling cracks and crevices and other cavities in the mother-rock. When a rock cavity is completely filled up, the turquoise usually takes the form of a plate, the thickness of which is rarely more than a few millimetres, while its area may be considerable. When, on the other hand, the cavity is only partially filled, the turquoise forms a lining of greater or less thickness to its walls, the surface of the lining layer of turquoise being frequently mammillated, botryoidal, or stalactitic.

In accordance with the absence of any definite crystalline form, there is a complete absence of cleavage in the mass. The fracture is sub-conchoidal to uneven : a clean, fractured surface shows but little brilliancy of lustre, and has usually a wax-like aspect, though at times the lustre may incline to the glassy or vitreous type. The lustre is greater when the stone is cut and polished, but is never very brilliant, the beauty of the turquoise depending for the most part upon its colour alone. The mineral is opaque, except in the thinnest of splinters through which a certain amount of light only is transmitted.

The naked-eye appearance of a freshly fractured, or cut and polished, surface of turquoise suggests a perfect continuity of structure. If, however, a thin section of the mineral be examined under the microscope, it is found to be built up of innumerable grains of irregular form arranged in an irregular manner. An examination of these grains in polarised light proves them to be doubly refracting, which demonstrates the fact that turquoise, in spite of the absence of any external crystalline form, is not amorphous, but is a compact aggregate of microscopically small crystalline individuals. In examining thin sections of turquoise under the microscope, less transparent portions with circular outlines are sometimes observed ; such appearances are probably due to the beginning of weathering n the turquoise substance. Small foreign bodies, which may possibly be chalcedony, are also visible sometimes under the microscope.

Turquoise is either green or blue, the former colour being much more frequently seen than the latter. The colour in both cases is due to the presence of copper phosphate, and probably also of iron phosphate, in small amount, intermixed with the colourless aluminium phosphate, which constitutes the greater proportion of the substance of the stone. In examining sections under the microscope it is only rarely, especially in blue Persian turquoise, that the pigment is seen to be located in definite strings or in cloudy patches with ill-defined boundaries. In most cases the colouring matter is in an extremely fine state of division, and is uniformly distributed throughout the turquoise substance. Thin sections of the mineral are almost colourless, perhaps faintly yellow ; the blue or green colour appears only in slices of some thickness. It is not at all unusual, however, for very thin sections of a deeply-coloured and almost opaque mineral to appear colourless and transparent.

The colour of turquoise ranges from sky-blue to mountain-green, the latter being not a pure green, but a green containing both grey and blue tones. Turquoise of a very intense and deep colour rarely occurs, but pale shades in great variety and forming a complete series from blue to green are to be met with. Of all these shades, the pure sky-blue of the deepest possible shade is most prized, and it is only turquoises of this colour which in Europe and the East are valued. The more the colour of a turquoise inclines towards green the less valuable does it become, and specimens of a distinct green colour are used as gems nowhere in the Old World, except in some parts of Arabia. It appears, however, that in former times turquoise of a green colour was in certain cases, for example by the ancient Mexicans, thought as much of as the blue variety, and we shall see later on that at the present day the natives of these regions frequently wear green turquoise in preference to blue.

As a rule, the turquoise is of one uniform shade of colour over its whole surface. In the case, however, of stones from certain localities, especially the Sinai Peninsula, there is sometimes visible a network of fine streaks of a paler shade of colour, which in cut stones shows up sharply against the deeper-coloured background. It is characteristic of true turquoise that its blue colour is just as beautiful by candlelight as by daylight, whereas other blue substances resembling turquoise appear of a dingy grey colour.

The colour of some turquoise is very unstable. Many stones, for example, from the Meghara valley in the Sinai Peninsula and from New Mexico begin to grow dull and pale directly they are taken out of the mine, and after a short time their colour completely disappears. It is stated as a general rule that the blue colour of turquoise is unstable and is gradually bleached by sunlight, the blue at the same time assuming a greenish hue. This is not always the case, however, and many turquoises retain their colour unaltered for a very long period. Sir Richard Burton testifies to having seen set in the musket of a Bedouin a very fine blue stone which, in spite of its exposure to sun, wind and weather for at least fifty years, had retained all its beauty of colour. The colour of turquoise is also said to be easily affected by perspiration from the body.

The original fine colour of a stone which has been bleached by wear or by prolonged exposure to sunlight can sometimes be restored by immersing it in ammonia, or by the application of grease ; even, it is said, by wearing it in such a way that it comes in contact with the natural grease of the hand. The restoration is not permanent, however, the blue soon disappearing, so that it is advisable to guard against the possibility of fraud in this direction. Since the alteration in colour usually proceeds gradually from without inwards, it is possible to improve the appearance of a stone by repolishing ; the operation would, of course, need repeating from time to time.

By the weathering of the mother-rock turquoise completely loses its colour and lustre. Thus rough specimens are often met with, the centre portion of which is of a fine blue colour, but is enclosed by an outer layer of dull white weathered material, which must be removed before the stone is cut. Sometimes the weathering process has proceeded so far that the whole mass is altered into a loose, crumbly material in which there may be found here and there grains of blue turquoise still unchanged.

Since the beauty and value of the turquoise depends almost entirely on its colour, attempts have been made to improve stones which are lacking in this respect by artificial means. A certain amount of success has attended these efforts, the method usually adopted being to impregnate the stone with Berlin-blue after the manner described in detail below under agate. The artificial colouring matter does not penetrate the stone deeply and the coloured layer can be scratched off with a knife. Moreover, stones so treated appear of a dingy grey colour by candlelight, and when immersed in ammonia either become green or lose their colour altogether, which is not the case with stones of a natural colour.

The specific gravity of turquoise is rather variable, values ranging from 2·6 to 2·8 having been observed. It is the least hard of any of the valuable precious stones ; its hardness, which is the same as that of felspar, being represented by 6 on the scale. Turquoise is therefore easily scratched by quartz or by a file, but is just hard enough to scratch ordinary window-glass. On account of its softness it requires special care when worn as a gem, although, being opaque, small scratches are less noticeable than in transparent stones.

Whether the ancients were acquainted with turquoise is doubtful, but the stone was certainly known in the Middle Ages. Its use as a gem is very general at the present time, the stone being as much prized in Eastern as in Western countries. Especially in the East, in Turkey, Egypt, Arabia, and Persia, it is much worn, being regarded by Orientals as a lucky stone. Thus it is to be met with everywhere in these countries, if only as a fragment of poor quality set in tin. It is much used in the decoration of the handles and scabbards of daggers and swords and of the trappings of horses and for other similar purposes. The name turquoise is said to signify Turkish gem. In Western countries large turquoises are frequently mounted with a border of small diamonds, while small stones form an effective frame for certain other precious stones of large size.

As is usually the case with opaque stones, turquoise is nearly always cut *en cabochon* with a plane undersurface of circular or oval outline (Plate XX., Fig. 2). Exceptionally large and fine specimens are said to be sometimes cut as table-stones or as thick-stones ; but since the existence of facets does nothing, on account of the opacity of the stone, towards enhancing its beauty these forms of cutting are but rarely adopted. The turquoise is often engraved with various devices ; in the Orient, for example, with quotations from the Koran, the letters being filled in with gold. Stones which are intended to be engraved are often cut with a flat instead of a rounded surface.

This precious stone, which is so generally prized, has a very considerable value, and in the Middle Ages was worth even more than now. The price of single stones varies with their size and their colour, the most valuable being of a pure sky-blue colour uniformly distributed and free from patches. As the colour inclines more and more to green the stone becomes less and less valuable ; while with regard to size, small turquoises are abundant and consequently cheap. Pieces of turquoise the size of a pea are rare, and when of a good colour command a high price. Small stones are bought and sold in thousands, rather larger specimens in dozens, while those above a certain size are sold singly. A carat stone of the best quality may be worth about fifty shillings, but the price of larger stones, owing to their rarity, is not in the same proportion to their weight. In the case of rough specimens it is very essential that they have a certain thickness, so that when cut *en cabochon* they shall not be too thin and flat, as is the case when the turquoise forms only a thin layer on the matrix.

Large turquoises of fine quality are few in number. Among such may be mentioned a heart-shaped stone 2 inches in length, which some time ago was in the possession of a Moscow jeweller, and which had been formerly worn as an amulet by Nadir Shah : an inscription from the Koran is engraved on it in gold, and it is valued at 5000 roubles. A turquoise in the collection of the Imperial Academy at Moscow measures more than 3 inches in length and 1 inch in breadth. The largest and finest stone in existence is said to be one in the treasury of the Shah of Persia. The most important turquoise mines known are situated in the dominions of this monarch, by whom the finest stones were formerly appropriated.

With regard to the occurrence of turquoise in nature it has been stated already that the mineral is found in veins of greater or less extent (Plate XX., Fig. 3) in certain rocks, having been deposited from aqueous solution in the cracks and crevices of these rocks. When the cavity is incompletely filled the turquoise forms a thin crust on its walls, and the surface of the incrustation may be mammillated, botryoidal, stalactitic, &c. The mother-rock of turquoise differs at different localities ; thus at one place the matrix may be quartzose slate, at another sandstone, and at a third trachytic rocks, the latter being remarkable as the bearer of the finest qualities of turquoise. The mineral appears never to have been found in limestone; statements as to its occurrence in this rock have been shown to be based on error.

The occurrence of turquoise in **Europe** is only sparing, and what has been found hitherto is almost entirely of the green variety, which is unsuitable for cutting as gems. The colour of European stones may in some few cases incline to blue, but is never of a pure sky-blue. At most localities the turquoise veins appear to be in a matrix of quartzose slate, as, for example, at Oelsnitz in Saxon Voigtland, and at Steine and Domsdorf near Jordansmühl in Silesia.

By far the most important localities for fine blue turquoise are in Asia. Of these the most famous are in **Persia**; hence the finest stones are referred to as "Persian turquoise." The name given to this their favourite stone by the inhabitants of the country is *piruzeh* (Arabic, *firuzeh*); and in the opinion of C. Ritter the word turquoise is a corruption of this.

The most important Persian turquoise mines, and those which yield precious material almost exclusively, are situated in the district of Nishapur, fifteen geographical miles west of Meshed in the province of Khorassan. In recent times details concerning this locality have been given by Tietze, Bogdanovitch, and the Persian General C. Houtum Schindler, who, at the beginning of the eighties, was for some time governor of the mining district and acting manager of the mines.

The mountains in the neighbourhood consist of nummulitic limestone and sandstone associated with clay-slates and interbedded with large masses of gypsum and rock-salt. All these beds have been broken through by younger volcanic rocks belonging to the Tertiary period, and consisting of porphyritic trachytes, or, according to some observers, of porphyry (felsite-porphyry). They form a chain of mountains extending from west to east between Kotshan and Nishapur. The occurrence of turquoise in this district is confined to the southern slopes of Ali-Mirsai, a peak in the chain with a height of 6655 feet. In this limited area are situated all the turquoise mines, not only those at present open, but also many now abandoned which were formerly worked, some being of great antiquity. The mountain is penetrated at a height of 4540 feet by a valley in which is situated the village of Maaden, 5100 feet above sea-level, and in latitude 36° 28′ 15″ N. and longitude 58° 20′ E. of Greenwich. This village is the centre of the area in which turquoise mining is carried on; the mines lie in the immediate vicinity of this to the north-west, and range in altitude from 4800 to 5800 feet above sea-level. All the inhabitants of Maaden earn their livelihood by work connected with the mining, cutting, and selling of this precious stone.

The original mother-rock here consists exclusively of porphyritic trachyte, which occurs in a weathered condition, and in brecciated masses consisting of blocks of the same rock cemented together by brown iron-ore (limonite). The turquoise fills up cracks and crevices in the trachyte and between the blocks forming the breccia; and being the latest formed mineral is deposited in and on the limonite. The latter frequently fills the rock cavities to only a partial extent, and the remaining spaces are filled by turquoise. This is found in layers of greater or less extent and only moderately thick, usually from 2 to 6 and never more than 13 millimetres in thickness. As a rule, it is found between layers of limonite of greater or less thickness, but this is not invariably the case. At other times the turquoise occurs in small masses of irregular shape and ranging in size from that of a pea to that of a bean. These small masses are either distributed irregularly through the rock or collected together so as to form plate-like masses within the limonite. The occurrence of turquoise in small veins running obliquely through larger veins of limonite, and sometimes extending into the surrounding blocks of trachyte, is also to be mentioned. Only in rare cases is turquoise found filling up cavities in the interior of blocks of trachyte forming the breccia. This mode of occurrence is of interest to mineralogists, since it shows that the formation of turquoise follows on the decomposition of felspar crystals of which it often takes the external form ; in other words, we have a pseudomorph of turquoise after felspar.

Turquoise is found, moreover, not only in the compact trachyte and in the trachyte-breccia, but also in the masses of débris formed by the weathering of these rocks and slowly accumulated at the foot of the mountain. The precious stone lies loose in the detritus, and is frequently coated with a white crust of weathered material, which must be removed before the fine blue colour of the stone can be seen. Sometimes the whole mass of turquoise is weathered to a white, crumbly material, which is, of course, useless as a gem. The turquoise-bearing deposits of alluvial débris have a thickness of from 2 to 20 metres : close to the foot of the mountain these deposits are less thick, while at some little distance away low hills have been carved out of the originally continuous mass by the action of the weather. Turquoise of good quality is to be found only in the uppermost portion of these secondary deposits to a depth of about 2 metres ; at a depth of 6 metres greenish and whitish stones of poor quality only are found, while below this turquoise is completely absent.

There are several hundreds of mines in this neighbourhood ; in the year 1876 there were 266 being worked, but the majority of them have since been abandoned. Some of these mines have been worked for centuries and are mentioned in the treatise on mineralogy written by the Arab Mohamed-ibn-Mansur in the year 1300. According to this work there was a legend to the effect that the richest of these mines were opened by Isaac the son of Abraham, and they are consequently known as the Isaac mines. For a long period they have been worked according to the best methods ; shafts to the depth of 150 feet were sunk, and levels and galleries driven to a length of 100 feet or more, though of small height and width. Pillars were left to give the necessary support to the roof, and, where necessary, ventilation shafts were sunk, so that the whole working was designed and carried out in a systematic manner.

According to the opinion of General Schindler, mining operations were probably carried on up to 1725 by the Persian Government ; and it is to this authority that the adoption of the methods described above was due. The management of the mines was subsequently transferred to the inhabitants of Maaden, and from this time the industry began to decline. Systematic methods were gradually given up ; the supporting pillars were removed in order to obtain the turquoise they enclosed, all precautions were neglected, the deposits were worked only with a view to rapid gain, and there was no thought for the future. Consequently the work became very dangerous and the yield decreased ; many of the workings became inaccessible, and at some places, where formerly existed properly constructed mines, there are now funnel-shaped depressions 60 to 80 feet across and as much as 250 feet deep, which have been formed by the falling in of the shafts and galleries.

These old mines are not in all cases completely abandoned ; the excavation of turquoise-bearing rock is sometimes still carried on, and both the loose rock lying in the workings and the refuse-heaps outside are worked over for turquoise, usually by women and children. New mines are always being sunk, and these are in almost all cases successful, since the deposit of turquoise extends throughout the whole mountainside.

The alluvial deposit also is worked for turquoise ; the detrital material is excavated, and after the larger blocks have been sorted out the remainder is washed in order to render the turquoise distinguishable. These washings at one time were of little importance, but came more and more into prominence as mining in the mother-rock became more and more neglected, until now they are by no means insignificant.

About 200 persons were engaged in the 'eighties in the mining of turquoise; of these about 130 were employed in mining operations in the mother-rock and the remainder in the alluvial deposits.

The stones collected here in these various ways are usually roughly cut *en cabochon* on the spot and then taken by the elders of the village, fifteen to twenty in number, into Meshed for sale. Owing to this fact, Meshed is sometimes incorrectly supposed to be a locality for Persian turquoise. From this place the stone travels, usually through the hands of Bucharian merchants, to Russia, especially to Moscow and to Nizhniy-Novgorod, and is sold at the fairs held at the latter place to dealers, by whom it is carried to all parts of the world. Nishapur has likewise been supposed to be a locality for the fine Persian turquoise, but because the stone is rarely to be seen or bought at that place its actual occurrence in the near neighbourhood has frequently been doubted by travellers.

The yield of the turquoise mines at the end of the seventies was about 25,000 tomans, or £8300, per annum, one-third of this sum being paid into the State treasury. According to other reports the value of the annual yield is much higher. General Schindler was informed by the turquoise merchants at Meshed that turquoises to the value of £12,000 were exported to Russia annually, while the smaller sales in Meshed itself amounted to £4000. This latter item is made up, for the most part, of turquoises mounted in tin or silver, but never in gold, and sold to pilgrims as lucky stones. Many stones are also exported through Yezd on the Persian Gulf to Constantinople.

In the year 1882 a determined effort was made by the Persian Minister of Mines to reorganise the management of the turquoise mines and thus to increase their yield. For four years but little improvement was noted, then General Schindler was placed at the head of affairs, and it was hoped that as much as 800,000 francs' worth of turquoise might be exported to Paris every year. These expectations were not entirely realised, as in the first year of the new régime turquoises to the value of 300,000 francs only were obtained; the employment of European methods, however, soon led to a substantial improvement in the yield, and the future of the mines became more and more hopeful.

A short time ago Mr. Streeter, a London jeweller, offered to rent and work the mines. As, after a thorough examination of the property, it was found that an outlay of from £50,000 to £60,000 would be necessary to set the mines in good working order, the idea was abandoned. It is said that the work is now likely to be undertaken by an American company.

Persian turquoises from this locality are often of a beautiful dark blue colour, but pale blue and green stones are also frequently met with. Stones found in the alluvial detritus and having a white external crust of weathered material are said to be of a specially fine colour. The colour, as a rule, is permanent, but in some of the newly opened mines turquoises have been found which, in a very short time after being taken out of the mine, lose their colour and become perfectly white. These stones are preserved in damp earth until they are sold to some unsuspecting person, who, in a short time, receives an unpleasant surprise. It is natural that such occurrences should have given rise to a certain distrust and suspicion of Persian turquoise in the minds of dealers in this stone.

Stones from different mines differ in quality, and they are classified on the spot according to size and shape, and especially according to colour, into three groups. Those of a uniform deep sky-blue colour and of a shape suitable for cutting *en cabochon* are classified as ring-stones; these are of the best quality and are not very abundant; they are found most frequently in the alluvial detritus. Stones of medium quality are divided into four sub-classes; the best of these are sent to Europe, the remainder being used in Persia and elsewhere in the Orient. Stones of the poorest quality, that is to say, of a pale blue or green colour, are sent only to Arabia, since in this country size and not colour or quality is the chief consideration. A pound's weight of stones of the first quality is worth at the mines about £90, while the same weight of stones of the third quality is worth only about £5. In Europe the price is far higher : it has been calculated that 25s. would be paid in Europe for a stone which could be got at the mines for 10s. A carat-stone is worth at the mines from 5s. to 10s. according to quality, the higher price being paid only for stones of the best quality.

There are other localities for turquoise in Persia besides Maaden, but all are little known and apparently much poorer. Turquoise has been found recently at Tabbas in the province Khorassan, but not of good quality. Bogdanovitch mentions a deposit of turquoise discovered not long ago, somewhere to the south of Meshed, about eighteen days' journey from this town. Another locality which has been known, though imperfectly, for a longer period is the province of Kerman in the interior of Persia. The mineral occurs at several places north-east of the town of Kerman, in the great range of mountains composed of volcanic rocks which stretches from north-west to south-east. At Chemen-i-Mô-Aspan, four fersakhs (eighteen miles) from Pâriz and opposite Gôd-i-Ahmer there are turquoise mines, which were worked until quite recently ; the stones found here have a greenish tinge. At Kârîk, north-east of Shehr-i-Bâbek, there is an old mine with two shafts, one of which was destroyed by an earthquake only a few years ago, while the other has not been worked for many years. A few veins of pale-coloured turquoise were found some years ago near Mashîz, on the slopes of the Cheheltan mountains, the highest peak of which has an altitude of over 12,000 feet. Turquoise is said to occur, and formerly to have been mined, in the neighbourhood of Taft, near Yezd, on the Persian Gulf.

It is stated that there are turquoise mines, yielding mostly green stones, further to the north-west, beyond the Persian frontier between Herat and Western Turkestan. According to the statements of ancient Arabian writers, the precious stone was found at Chodshent, from whence came also the green *callais* (*callaina*) of Pliny, now considered to be identical with turquoise. Other localities in the same region have also been recorded ; for example, in 1887 in the mountain range Kara-Tube, fifty kilometres from Samarkand. The turquoise occurs here in limonite and quartzose slate, and the place was, at some unknown time, the scene of mining operations. Finds of turquoise have been made in the same region in our own time ; for example, in the Syr Daria country in the Kuraminsk district (in the Kara Mazar mountains), and also in the Karkaralinsk district in the Kirghiz Steppes (Semipalatinsk territory of Siberia). These and other occurrences in the same region have no commercial importance and need no further consideration.

The next most important locality for turquoise in the old world is the **Sinai Peninsula** : the mineral is found for the most part in the neighbourhood of Serbâl, near the west coast. The best known mines are situated in the Wadi Meghâra or Maghâra (meaning hollow valley) ; these are very ancient and were worked on a large scale in ancient Egyptian times, according to H. Brugsch, as early as the period of the Third Dynasty in the reign of King Snefru, 4000 B.C. The discovery of numerous inscriptions and implements of various kinds proves that a garrison was maintained here by the Egyptians for the protection of the turquoise mines and of an important copper-mining industry. The existence of these turquoise mines was for a long period completely forgotten ; they were at length rediscovered by Major C. K. Macdonald. Work was at once recommenced, and some of the finest and largest stones found were shown in London at the Exhibition of 1851. One in the possession of Major Macdonald was as large as a pigeon's egg, but in a very short time lost its colour and became greenish-white and, compared with its original value, quite worthless. The same fate overtook many of the stones exhibited in 1851 ; one which had been sold for a high price became, in the course of a year, perfectly colourless. Specimens presented by Major Macdonald to the British Museum in 1862 still retain their fine blue colour although they have been exposed to a strong light for many years.

These ancient mines are situated on the northern slope of the Meghâra valley, 150 feet above its floor. This side of the valley is of red sandstone, while in the porphyry, of which the opposite side is formed, no turquoise is found. The precious stone fills up crevices and fissures in the rock, and is found in tabular pieces of about the same dimensions as in Persia; the mode of occurrence is therefore similar to that of the stone in the latter country.

The turquoise of the Sinai Peninsula is not, however, confined to the sandstone of the Wadi Meghâra, but is found in porphyry outside the valley. In the form of thin plates it penetrates the porphyry, which forms part of the Serbâl, and differs from the turquoise found in the sandstone in that its beautiful blue colour is permanent. These stones are collected and sent to market by the Bedouins ; moreover, some of the stones sent to Europe by Major Macdonald are said to have come not from the mines in the Meghâra valley but from the porphyry of Serbâl. The exact situation of the mines in this region is carefully hidden by the natives, and no details concerning the occurrence are known. A locality apparently of special importance is Moses' Well ; also Neseb or Nasaiph Well between Suez and Sinai. The occurrence of turquoise at this locality is probably distinct and definite in character, since the stones show under the microscope a peculiar structure different from that of turquoise from other localities. It is impossible to definitely locate the place, since every well which supplies drinking water is called Moses' Well by the Bedouins ; according to H. Fischer, however, it is situated in latitude 29° N., about five miles from the Serbâl. Adhering to the stones which come from this place is a brownish-red, ferruginous powder formed of friable, granular quartz ; it is probable, therefore, that the stones occur in sandstone as at Wadi Meghâra.

The best turquoises from the Sinai Peninsula are quite equal to Persian stones, and some even surpass these in beauty and depth of colouring. As a rule, however, the colour of these stones is of a more whitish-blue, the lustre is more glassy, and the material rather more brittle. Fine stones from this locality appear on the markets as Egyptian or Alexandrian turquoises; they were formerly regarded as artificial products, but detailed examination has shown them to be the natural mineral.

Arabia proper is another turquoise locality; at least three mines are stated to be situated in the " Midianite country "; two of these are supposed to be still worked, but the stones found there very soon lose their colour.

Despite certain statements to the contrary, turquoise has not hitherto been found in any of those Asiatic countries which are remarkable for their wealth of precious stones ; that is to say, neither in India, nor in Burma, nor in the Island of Ceylon.

In the new world the principal turquoise deposits are situated in the south-western states of the North American Union. These are not unimportant even now, but in former centuries they were of much more prominent interest. The most important mines are in the State of **New Mexico**, which formed a portion of the ancient kingdom of the Aztecs. The precious stone was much admired by the ancient Mexicans ; they prized it more highly than gold, and used it in the decoration of all kinds of objects as well as for a gem. It appears, however, that it was green and not blue turquoise which was held in such peculiar esteem. The green precious stone *chalchihuitl*, so much esteemed by this ancient people, is considered by some authorities to be identical with green turquoise; others, however, suppose it to have been emerald, jade, green jasper, or some other green mineral.

After the fall of the Kingdom of Mexico the turquoise still continued to be the favourite stone of the inhabitants of the region, that is to say, of the Pueblo and Navajo Indians. W. P. Blake states that the stone is known to these people as *chal-che-we-te*, which is supposed to be a corruption of the old name *chalchihuitl*. So greatly was the stone prized by these Indians that only with the greatest difficulty could they be induced to part with their turquoise-decorated tools and implements to white men. Such objects, moreover, were almost always buried with the dead, as recent excavations in the Indian burial-grounds of that region have shown.

The best known of the ancient turquoise mines is situated on the mountain named after this stone, Mount Chalchihuitl (or Mount Chalchuitl); this mine was the first to be rediscovered and was found by William P. Blake in the 'fifties. Mount Chalchihuitl forms part of the conical mountain group Los Cerillos, about twenty-two miles south of Santa Fé, the capital of the State of New Mexico; it is situated on the northern bank of the Galisteo river, which flows westward into the Rio Grande and separates the Los Cerillos district from the important mining district of the Placer or Gold Mountains.

This mountain group, and in particular the turquoise-bearing Mount Chalchihuitl, consists of sandstone, probably of Carboniferous age, intersected by dykes of augite-andesites. These andesites, and the volcanic tuffs with which they are associated, contain in various parts of the mountains ores of lead, copper, silver, and gold in no inconsiderable amount. They are usually much decomposed and completely bleached by the action of volcanic gases and vapours. The alteration and weathering of the rocks of this range has resulted in the formation of new minerals of various kinds. Thus by the alteration of the felspar in the volcanic rocks kaolin is formed; from this mineral the turquoise, subsequently formed, derives its alumina, the phosphoric acid being derived from the apatite in the same rocks, and the copper, to which the colour of the stone is due, from the copper ores also embedded in these rocks. All the turquoise appears to have originated in this way, the formation of turquoise always following that of kaolin. The mineral is found here as elsewhere in small nodules and thin veins, with a mammillated or botryoidal surface, in the andesite or andesitic tuff, which is altered to a whitish or yellowish clayey mass. The turquoise is so generally distributed through the rock that patches of it may be seen almost everywhere on the walls of the mine.

One of the ancient Mexican mines, which was without doubt worked before the discovery of America by Europeans, has been described by W. P. Blake as an enormous funnel-shaped pit, the sides of which are steep and precipitous. At one place there are even overhanging rocks forming a kind of cave, while at another the slopes are more gentle, owing to the falling in of waste material from above. An idea of the great age of this artificial excavation may be derived from the fact that on its sides are growing pines, cedars, and other trees hundreds of years old. It is about 200 feet deep and 300 feet wide, and out of it many thousands of tons of solid rock must have been excavated. In its neighbourhood are to be seen similar but smaller pits, and it would seem that the whole surface of the turquoise-bearing mountain was turned over in the search for the precious stone. Beside these surface-workings there exist also underground mines excavated at the same time, some of which are of no inconsiderable extent. These were discovered when attempts were made to rework this old deposit; and in the old mines were found relics of this long-past age in the shape of miners' tools of various kinds. Everything points to the fact that these ancient mines had been carefully covered up before being left, in order, no doubt, to conceal their exact whereabouts from strangers and unauthorised persons. How extensive must have been these ancient workings is shown by the one fact, among many others, that the heaps of barren rock thrown out from the mines cover an area of no less than twenty acres. Here again are to be found numbers of large growing trees, proving the great age of the heaps.

The abandonment of the mines was due to a great national disaster which befell the Indians in the year 1680. Owing to the undermining of the ground by the Indian miners a large section of the mountainside suddenly fell in, killing a number of workers on the spot. This accident was the immediate cause of the uprising of the Pueblos, which resulted in the expulsion of the Spaniards from the country.

At the beginning of the eighties of last century, after the opening up of the valley of the Rio Grande by the construction of a railway, a company was formed to undertake again the mining of turquoise, and at the same time the metallic ores of the region. It was soon found that though turquoise of a fine blue colour does exist, yet the greater part of the material is green or bluish-green in colour, and that to obtain a single stone suitable for a gem, however small, it was necessary to work through many tons of rock. Throughout the whole deposit the stone is poor in quality, and the company was soon obliged to give up work ; nevertheless, between 1883 and 1886 stones to the value of 3000 dollars were found. At the present time work on a small scale is carried on by a few poor white men and Indians, who, by lighting fires on the rock, make it friable, thus rendering the work of excavation less difficult. As a result of the adoption of this method the greater part of the turquoise is destroyed. What little is saved is roughly worked into rounded or heart-shaped ornaments pierced with a hole. These are sold at Santa Fé or to travellers at the railway stations as objects of local interest. The price at present is very low, only 25 cents (about 1s.) being asked by Indian dealers for a mouthful of such stones. But few of these stones find their way into the hands of the jewellers, for it is only seldom that the Indian dealers have really good stones to offer, and, moreover, because of a fraud attempted some time ago, confidence in them has been greatly shaken. These men placed on the market turquoises of a specially fine dark-blue colour, which was found by Mr. G. F. Kunz to be due to a surface application of Berlin-blue.

Another occurrence of turquoise in the same neighbourhood, from which the ancient Mexicans obtained a rich yield, was rediscovered at the beginning of 1890, and is now known as the " Castilian Turquoise Mine." It is situated seven miles from Los Cerillos, on the road to Santa Fé, and one and a half miles from Bonanza. The mother-rock is the same as at Mount Chalchihuitl, but the turquoise is of a better colour than that found at the latter locality. Several thousand stones with an aggregate value of 100,000 dollars have already been found, some being of a very fine blue colour, though not equal to Persian turquoise.

Also in New Mexico, in the south-west corner of the State, are situated the newly-discovered deposits of turquoise in the Burro Mountains, fifteen miles south of Silver City in Grant County. They are now being worked by a company, and some good stones have already been found. The existence of ancient mine barrows shows that this deposit, like others in the region, had been worked at some period now long past. Here also the turquoise occurs in the rock in the form of strings and veins. There was once found here a plate of turquoise 8 inches across and $\frac{1}{8}$ to $\frac{1}{4}$ inch in thickness ; and it is said that the mines have yielded as much as 10 kilograms of fine turquoise in a month. It occurs for the most part in kidney-shaped masses encrusted with a thin layer of siliceous material. Turquoise occurs in trachyte in the Cow Springs district of this region, and there is yet another occurrence at Hachita in the same county.

Another recently discovered locality for turquoise is situated in the Jarilla Mountains (Doña Anna County, New Mexico), 150 miles east of the Burro Mountains and 200 miles south of Los Cerillos. Here, again, ancient surface-workings reaching down to the solid rock are met with; these, judging from the character of the vessels and tools found there, must be centuries old. The turquoise is found in thin almost vertical cracks and crevices in trachyte, and is sometimes accompanied by copper-pyrites; the mode of occurrence is, therefore, the same as elsewhere in this region. In the Shoo-ar-mé mine in the Jarillas, which has a shaft 70 feet deep, turquoise has been found in abundance. It occurs usually in hemispherical or kidney-shaped masses, but also in irregular masses, completely filling up the cavities in the rock: a slab of turquoise was once found here which was 3 square feet in area and ¾ inch thick. Stones found at some depth are usually blue in colour; those which lie nearer the surface are frequently green, probably owing to weathering. Stones which when first taken from the ground are of a fine blue colour, sometimes almost indigo-blue, on drying or exposure to air lose their colour and will then adhere to the tongue. Nevertheless good specimens of permanent colour are frequently found, more than 50 kilograms of good marketable turquoise having been obtained in six months from one of the newly reopened mines.

Turquoise has been discovered at many other places in New Mexico. Many of these deposits were worked by the ancient Mexicans until the solid rock was reached, when, owing to the inadequacy of their primitive methods, the work had to be abandoned.

Other places in the United States have also yielded turquoise, and these will be briefly mentioned.

A large amount of turquoise was obtained by the ancients from a deposit about twenty miles from Tombstone in Cochise County, **Arizona**; it is situated in a spur of the Dragoon Mountains, not far from the former Apachan capital, Cochise, and south-east of the present capital, Tucson. The mountain is now known as Turquois Mountain, and as silver ores also occur in the neighbourhood there has sprung up quite a mining industry, the district in which it is carried on being known as the Turquois District. Several large excavations have been made in the mountain, but for some time now the work has everywhere been given up. The deposit is not as rich as in Mount Chalchihuitl, nor are the ancient workings as extensive. The colour of the precious stone is for the most part some shade of green.

In Mohave County, also in Arizona, is situated the turquoise locality known as Mineral Park. Here finely coloured turquoise has been found in three veins, 1 to 4 inches in thickness and about 100 yards apart; these have been followed up for almost half a mile. Large quantities of turquoise were obtained from this deposit first by the ancient Mexicans and subsequently by the Spaniards.

In the Columbus district of southern **Nevada**, about five miles north of Columbus and half a mile south of the Northern Bell mine, turquoise occurs in a brown sandstone in the form of veins and small grains. The single stones found here, though small, are finer in quality than any occurring elsewhere in North America, some few being of unsurpassed beauty and great value. Most of the turquoises found here are sent to San Francisco; besides these isolated stones the sandstone containing small pieces of turquoise embedded in it is also mined and affords an effectively coloured ornamental stone.

There are two other American localities of which mention must be made, namely, Holy Cross Mountain in Colorado, and Taylor's Ranch, Chowchillas river, Fresno County, California. Though neither are important commercially, yet the latter has a certain interest from a mineralogical point of view, since here turquoise occurs in hexagonal prisms some inches in length as a pseudomorph after apatite.

In Australia turquoise has recently been discovered in the State of Victoria; no details as to the occurrence are known, but turquoise has been obtained from a mine called the " New Discovery."

ARTIFICIAL PRODUCTION.—As in the case of all valuable precious stones, attempts have been made to produce artificially a substance resembling the naturally occurring stone, but saleable at a lower price. A certain measure of success has attended these efforts, since there has been produced a mass of which the chemical composition does not differ essentially from that of naturally occurring turquoise, and of which the colour, lustre, hardness, specific gravity, fracture, and general appearance are the same as in that stone. The manufacture is said to be carried on chiefly in Vienna, France, and England: the method adopted is to submit to great pressure a chemical precipitate of the same composition and colour as turquoise. The details of the method are not exactly known, but it would seem that hitherto it has been impossible to produce stones of any considerable size.

These artificially made turquoises are put on the market together with natural stones; they are of very good colour, and it is impossible by mere inspection to distinguish them from the turquoise of nature. This being so, the appearance on the market of an unusually large supply of good stones naturally creates suspicion in the minds of possible buyers; this was the case with a large supply of Persian stones and of the Egyptian or Alexandrian turquoises mentioned above, all of which, however, proved to be genuine.

Owing to a difference in their behaviour when heated, it is possible to distinguish artificially produced turquoise from the natural stone. The latter, when heated, decrepitates violently and is reduced by ignition to a brownish-black powder, or to a loose mass which easily falls to a powder and does not fuse. The artificial product, on the other hand, does not decrepitate, and on ignition is not reduced to powder, but fuses or runs together into a hard mass, which at least in the interior retains its blue or bluish-green colour; some or the artificial stones even fuse with moderate ease to a black bead. It is obvious that this method of detection can be adopted only when the complete destruction of the turquoise is of no consequence, as in buying a large parcel of stones, or when it is possible to detach a small splinter from the back of a large stone. It is said to be possible to recognise an artificially produced turquoise from the fact that after lying in water it assumes a darker shade of blue, and on the wet surface can be made out a network of cracks; moreover, it is said that such stones become softer after immersion in alcohol. There are always to be found adhering to genuine stones particles of the mother-rock, and especially of brown limonite, which is so frequently associated with turquoise; the presence of such particles was for a long time a sure indication of the genuineness of a stone; this is not now the case, however, for artificial stones are at present furnished with brown specks of hydrated iron oxide in order to imitate natural stones still more closely.

In the manufacture of artificial turquoise the reproduction of all the essential characters of the natural mineral is attempted, and so similar is the artificial product to the natural stone that it is frequently impossible to distinguish between them except by the employment of special tests. Besides these artificial stones other substances are substituted for turquoise, especially a glass paste, which is made by adding to a mass of glass 3 per cent. of copper oxide, $1\frac{1}{2}$ per cent. of manganese oxide, and a trace of cobalt oxide. These paste imitations are easily distinguished from genuine stones; their lustre is of the ordinary vitreous type, which is not difficult to recognise, especially along the margin of the stone, which is almost invariably splintered during the process of grinding, and where also the conchoidal fracture characteristic of glass may be detected. Moreover, unless the paste is made with the greatest possible care, it is almost certain to contain air-bubbles, which are never present in genuine stones. A few other substances, sometimes substituted for turquoise, will now be considered below.

BONE-TURQUOISE.

Bone-turquoise is also known by various other names, such as tooth-turquoise, occidental turquoise, *turquoise de la nouvelle roche*, odontolite, and fossil turquoise. The name is applied to the bones, and especially to the teeth, of certain extinct vertebrates, such as the *Mastodon* and *Dinotherium*, which during the process of fossilisation have taken up phosphate of iron, the so-called blue iron-earth, and so acquired a fine sky-blue colour, or by taking up copper salts have assumed a green colour. The latter case is rarely met with, and as bone turquoise of a green colour is not prized it need not be considered further.

These blue-coloured fossil teeth and bones are found at various localities, and are especially abundant in the Miocene beds of Simorre, Auch, and other places in the department Gers (province Gascogne) in France, where they are in most cases the remains of *Mastodon angustidens*. At Simorre the deposit was for a short time even systematically mined. When first taken out of the ground the teeth are of a dingy greyish-blue colour, which changes when they are heated to a fine sky-blue. The teeth of the mammoth, so often found in Siberia, are also sometimes of a fine blue colour.

Bone-turquoise, like true turquoise, is frequently cut *en cabochon*, and, though much less valuable than the latter, commands a fair price, especially in fine pieces of considerable size. Its colour by daylight is scarcely distinguishable from that of true turquoise, but in candlelight it assumes a dull grey hue, and when immersed in alcohol or water gradually fades. On cut and polished surfaces paler stripes on a darker background may frequently be detected, an appearance which is due to the structure of the tooth. This structure can be clearly made out by examining thin sections of the material under the microscope ; and at the same time it will be seen that the colouring matter is collected in small tubular cavities. Specimens showing dendritic markings, that is to say, moss-like patches of a brown or black colour, are met with not infrequently ; they are much less valuable than material which is not so marked.

Bone-turquoise may be easily distinguished from the true mineral turquoise. In the first place the former contains up to 11 per cent. of calcium carbonate, hence when a small fragment of bone-turquoise is placed in hydrochloric acid, or a drop of acid placed on a large piece of the same substance, there is a brisk effervescence. Moreover, when heated, a distinct smell of burning becomes perceptible, owing to the presence of organic matter in the substance. When rubbed it becomes electrified and retains its charge for a long time. It is less hard than true turquoise, and is not, therefore, susceptible of as high a degree of polish. The specific gravity, on the other hand, is greater, ranging from 3 to $3\frac{1}{2}$, so that a piece of bone-turquoise sinks in liquid No. 3 (sp. gr. $= 3\cdot0$), while mineral turquoise floats. The dull grey colour assumed by bone-turquoise in artificial light also serves to distinguish it from mineral turquoise.

An imitation of bone-turquoise may be produced by allowing calcined ivory to remain for a week immersed in a warm solution of copper sulphate, to which excess of ammonia has been added. After this treatment the ivory acquires the same fine blue colour as naturally occurring bone-turquoise.

LAZULITE.

There is, perhaps, but one mineral which, by any possibility, could be mistaken for turquoise, and which may sometimes be substituted for this stone. The mineral lazulite (not to be confused with lapis-lazuli, which will be considered below) probably never finds application as a gem under this name. It occurs in nature as monoclinic crystals of a sky-blue colour, and also in compact finely granular masses known as blue-spar. In chemical composition it is very similar to turquoise, being a hydrous aluminium phosphate, but containing in addition some magnesium and iron. Its hardness is rather less than that of turquoise, being $5\frac{1}{2}$; the specific gravity, $3\cdot1$, is considerably higher, so that lazulite sinks in liquid No. 3 (sp. gr. $=3\cdot0$). The lustre is of the common vitreous type, showing no inclination to waxiness, so that in this respect also lazulite differs from turquoise. Fine crystals of lazulite are found in a friable sandstone at Graves Mountain, Lincoln County, Georgia, U.S.A.; also on quartz which, in the form of thin veins, penetrates clay-slates at Werfen in Salzburg, where the massive blue-spar occurs. Massive material of a fine deep blue colour is also found in Brazil. Lazulite as a precious stone is met with extremely rarely in the trade.

CALLAINITE.

Callainite, or callais, is a mineral which was used as a precious stone in prehistoric times. It is found exclusively in an ancient Celtic grave at Mané-er-H'rock, near Lockmariaquer, in Brittany, as rounded fragments ranging in size from that of a flea seed to that of a pigeon's egg. In many respects it resembles turquoise, but differs from this stone in being beautifully translucent. It is almost always green in colour, the shades ranging from apple-green to emerald-green; it is sometimes veined and spotted with white, blue, black, or brown. In chemical composition callainite is very similar to turquoise, being a hydrous phosphate of alumina; the constituents, however, are present in rather different proportions in the two stones. Like turquoise, this mineral is not distinctly crystallised, being an aggregate of microscopically small grains. The hardness ranges from $3\frac{1}{2}$ to 4, and the specific gravity from $2\cdot50$ to $2\cdot52$. Callainite was obviously used by the ancient Celts as an ornamental stone, but where the specimens found in these ancient graves were originally brought from has not yet been discovered. The mineral is probably a variety of variscite, but is more transparent and of a finer colour than typical specimens of this mineral.

OLIVINE.

The olivine group embraces a large number of minerals, of which, however, only the one from which the group takes its name has any application as a gem. Precious olivine is known as *chrysolite* and to French jewellers as *peridot*. The name chrysolite is given to this stone on account of its fine yellowish-green colour, but the term in its original meaning is more correctly applied to stones of a pronounced golden-yellow, such, for example, as yellow topaz, which is referred to by Pliny and other ancient writers under this name. The name olivine is also descriptive of the colour of the stone, which is always a green showing shades of yellow and brown like that of the fruit of the olive-tree. No other colour is ever shown by chrysolite of gem-quality, so that this stone differs from many of those hitherto considered, such as diamond, corundum, topaz, in which the range of colours is very large, and in this respect it is more like turquoise.

The chemical composition of chrysolite, and indeed of all the minerals of the olivine group, is comparatively simple. Common olivine, and the gem-variety chrysolite, is a silicate of magnesium having the formula $2MgO.SiO_2$; a portion of the magnesia is, however, always replaced by a corresponding amount of ferrous oxide. The following is the result of an analysis made by Stromeyer of fine, transparent, yellowish-green chrysolite of gem-quality from the Orient.

	Per cent.
Silica (SiO_2)	39·73
Magnesia (MgO)	50·15
Ferrous oxide (FeO)	9·19

Manganous oxide, nickelous oxide, and alumina were also present in minute amounts, but the presence of these constituents need not be further considered here. The amount of iron present varies to a certain extent, and it is on this variation that the differences between the shade of different specimens depend. The colour of chrysolite, as of green bottle-glass, is due solely to the presence of a small amount of ferrous oxide ; when this is present in larger amount the stone is darker in colour, and *vice versâ*. In connection with this subject may be mentioned the fact that a silicate of magnesium known as forsterite, a mineral of the olivine group, contains no iron and is perfectly colourless.

Olivine sometimes occurs as distinct crystals, which belong to the rhombic system. A form not infrequently taken by transparent chrysolite of gem-quality is that shown in Fig. 71. Here the faces of the rhombic prism are inclined to each other at an angle of 130° 3′, the side edges are truncated by a pair of small faces of the brachy-pinacoid, and the front and back edges by a pair of large faces of the macro-pinacoid. The combination of these three sets of faces gives an eight-sided prism flattened in the direction of the large pair of faces of the micro-pinacoid. The faces of the micro-pinacoid and of the prism are striated in the direction of their length, that is to say, parallel to the edges of the prism. Above and below the prism faces are triangular faces of a rhombic octahedron or pyramid, and above and below the pinacoid faces there are rectangular faces of a macro-dome and a brachy-dome, the former being represented in the figure as the larger. Finally, the crystal is bounded above and below by a pair of narrow rectangular faces or basal planes, which truncate the two sets of dome-faces and the pyramid. Other forms are sometimes present, as, for example, in Fig. 11 of Plate XIV., but they are very similar to those just described. Crystals of olivine are found not attached to the walls of drusy cavities but embedded in the mother-rock; they are therefore usually developed on all sides.

Olivine has no very distinct cleavage; the best is in the direction of the large macro-pinacoid, and there is a very poor one parallel to the basal plane. The fracture is always conchoidal. The hardness is slightly less than that of quartz, and is usually given as $H = 6\frac{3}{4}$; olivine is thus scratched by quartz, but easily scratches felspar and still more easily ordinary window-glass. Though the hardness of chrysolite compared with other precious stones is low, yet it is sufficient to admit of the stone receiving a good polish; the process of polishing is, however, not easy. On account of its comparative softness and its liability to become scratched and dull this stone is not very highly esteemed and is worn but rarely, especially as a ring-stone.

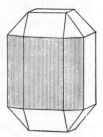

FIG. 71. Crystalline form of olivine (chrysolite).

The specific gravity of pure, transparent olivine ranges from 3·329 to 3·375. The greater the amount of iron present the higher is the specific gravity; hence the darker the colour the heavier the stone. Heavy, dark-coloured olivine, therefore, sinks in pure methylene iodide, but floats in liquid No. 4 (sp. gr. = 3·6). Stones of a light shade of colour and low specific gravity are often of about the same weight, bulk for bulk, as methylene iodide; they will therefore remain suspended at any point in the liquid. A stone having exactly this specific gravity will slowly sink to the bottom when the liquid is warmed, even when the vessel containing it is held in the hand, and will rise to the surface again when the liquid cools. Chrysolite may be readily distinguished from many other green stones by its specific gravity.

A certain amount of olivine is perfectly pure, clear, and transparent, and absolutely free from faults of any kind. Material of this description is termed " noble chrysolite," and this alone is cut for gems. The more abundant and more widely distributed " common olivine " is cloudy and translucent to opaque, and is unsuitable for this purpose. The lustre of olivine is of the ordinary vitreous kind, always, however, inclining to the greasy type; its brilliancy can be increased by polishing, as mentioned above. Being a rhombic mineral, olivine is doubly refracting; the index of refraction is not very high, but the strength of its double refraction is very great, the double refraction of this stone being stronger than that of any other precious stone except zircon. The least index of refraction for yellow light is 1·661 and the greatest 1·697; the difference between these two numbers, 0·036, is a measure of the double refraction; the mean index of refraction is 1·678. The refractive indices for light of other colours have not been determined; they would not vary much, however, from the values given above, since the dispersion is small and no brilliant play of prismatic colours is shown by cut stones.

We have already seen that the shade of colour shown by olivine depends upon the amount of ferrous oxide present. The colour inclines now to yellow, now to brown, and sometimes to green, but there is no marked difference between these shades, and they are never very deep or intense. The usual colouring of chrysolite is shown in Figs. 11 and 12 of Plate XIV., the one representing a crystal and the other a cut stone. Common olivine is not infrequently of a pronounced yellow colour. Specially named colour-varieties of this stone are sometimes, but not universally, recognised. Thus chrysolite proper is pale yellowish-green, peridot is deep olive-green, and olivine yellowish or light olive-green.

The dichroism of olivine is always feeble ; the two images seen in the dichroscope vary in colour between yellowish oil-green and pure grass-green with no appreciable yellow shade.

The other characters of this mineral, not affecting the appearance of the gem, but of use in the recognition of uncut stones, may be mentioned, namely, the fusibility before the blowpipe and the behaviour of the stone towards acids. Only olivine very rich in iron is fusible before the blowpipe ; this particular variety, which is never used as a gem, can be fused, but always with difficulty. All varieties of olivine when reduced to a fine powder are quickly decomposed, especially when warmed, by hydrochloric or by sulphuric acid, with separation of gelatinous silica. This character is utilised, probably unintentionally, in the process of polishing chrysolite, the final polish being given by the use of sulphuric acid instead of water.

The table-stone and the step-cut in their various modifications are the forms usually employed in the cutting of chrysolite ; one such form is shown in Fig. 12 of Plate XIV. Brilliant and rose forms are also sometimes employed. The table facet of table-stones and step-cuts is not infrequently cut with a curved surface, the result being a transition form between a faceted stone and the ordinary curved (cabochon) form which is sometimes used for chrysolite. The colour and lustre of olivine are often improved by the use of a gold foil, or in the case of very pale stones of a green foil.

Olivine is widely distributed throughout the rocks of the earth's crust. It is a constituent of basalt, and occurs most frequently in irregular grains or in large granular aggregates the size of a man's fist or head, or even larger, and consisting of small irregular grains of olivine intermixed with a few fragments of other minerals ; it is only rarely found in sharply defined crystals. Olivine occurs also in other igneous rocks, such as diabases and gabbros ; and rock-masses of large size are to be met with which consist wholly or largely of pure olivine. The mineral again is sometimes found interlaminated in gneiss or crystalline schists. Grains of olivine are sometimes to be found in the weathered débris of such rocks, but the mineral is usually the first to become altered by exposure to weather. Finally, it should be mentioned that olivine is an essential constituent of many meteorites.

The olivine found under the conditions mentioned above is scarcely ever suitable cutting as gems. That which occurs in diabase and gabbro is impure and opaque ; the material found in basalt and similar rocks, as, for example, that from Vesuvius, is frequently pure and transparent, but is almost always in grains of quite small size. The component grains of the larger aggregates also are nearly always of small size ; only rarely, as in the basalt of Mount Kosakow, near Semil on the Iser, in northern Bohemia, are they found of any considerable size. At this locality there sometimes occur transparent pieces the size of a hazel-nut and of a fine green colour from which gems can be cut. The olivine found in meteorites is also sometimes transparent and of a fine colour, but the grains or crystals in which it occurs are always of small size ; in only one or two extra-terrestrial bodies have fragments of transparent olivine been found sufficiently large to yield one-carat stones, and these are in very truth celestial gems.

The source of the chrysolite which is used in the trade for cutting as gems is somewhat obscure. It is probably identical with that of the transparent, finely coloured pebbles not infrequently seen in mineral collections. These pebbles are similar in all respects to cut chrysolites ; the largest are about the size of a walnut and they have obviously been collected in a river sand or gravel, the exact locality of which is, however, unknown.

Both "Pegu" and the "country of the Burmese" are mentioned as localities for chrysolite, but the occurrence of the stone in gem-quality here or in India is by no means well authenticated. The same is true for Ceylon, where chrysoberyl, so often confused with chrysolite, occurs; also for Brazil, the occurrence of chrysolite among the variously coloured stones of Minas Novas having been reported. In this case also it is highly probable that the stone referred to is chrysoberyl, since this stone is usually known to Brazilians as chrysolite.

Chrysolite of gem-quality is said, moreover, to come "from the Orient," "from Natolia" and "from the Levant," finding its way into Western markets by way of Constantinople and Austria. "Egypt" is also mentioned as a locality, especially for the transparent, green crystals, such as are represented in Fig. 11, Plate XIV., and often to be seen in mineral collections. More detailed statements mention Upper Egypt, and a locality east of Esneh between the Nile and the Red Sea, the mother-rock being supposed to be granite or syenite.

None of the occurrences mentioned above are well authenticated, and in the opinion of Mr. G. F. Kunz, the well-known American expert, an opinion based on a large experience of the precious-stone trade, no chrysolite suitable for cutting is at the present time found anywhere in nature. He considers that the material which now comes into the market is derived from old ornaments of various kinds, some dating back a couple of centuries. It is possible that the deposit, from which this old material was obtained, was exhausted or abandoned and that its exact position came to be forgotten. At the same time, however, Kunz records the occurrence of pebbles of chrysolite, suitable for cutting, with garnets in the sands of New Mexico and Arizona. Most of these, however, are small and not of very good colour, so that they are practically useless as gems.

True chrysolite, that is to say, noble olivine, and other precious stones similar in colour, are sometimes liable to be mistaken the one for the other. Thus chrysolite has occasionally been taken for emerald, though the two stones may be easily distinguished by means of the specific gravity, emerald being much lighter than chrysolite, which just sinks in pure methylene iodide, while emerald floats. The supposed emerald which ornaments the shrine of the Three Holy Kings in the cathedral of Cologne is in reality not emerald but chrysolite. The stone with which chrysolite is most frequently confused, however, is chrysoberyl. It has been stated above that this stone is known to the Brazilians as chrysolite; the chrysolite of French jewellers and the " oriental chrysolite " of the trade is also in many cases chrysoberyl. The two can be easily distinguished from the fact that chrysoberyl is both harder and heavier than chrysolite, the former sinking and the latter floating in the heaviest liquid (sp. gr. = 3·6).

Other pale green gems are occasionally referred to by the name chrysolite qualified by some prefix. Thus " Ceylonese chrysolite " is tourmaline of an olive-green colour; " oriental chrysolite" sometimes signifies yellowish-green corundum; the term "Saxon chrysolite " is applied to greenish-yellow topaz from Schneckenstein in Saxony; " false chrysolite " to green bottle-stone or moldavite, which will be considered later on; green specimens of the mineral prehnite are termed "chrysolite from the Cape." In Table 13, in the third part of this book, are given the methods by which these stones may be distinguished.

A yellowish-green glass recently much used for the commoner kinds of ornaments is also sometimes referred to as chrysolite or obsidian. It is very similar in appearance to true chrysolite, but, being singly refracting, is easily distinguished from this in the polariscope.

A good price can be obtained for large, pure chrysolites of a good deep colour, but large stones of average quality are not worth more than from 4s. to 7s. per carat. The mineral was more highly prized formerly than at present; it is now about equal in value to topaz.

CORDIERITE.

Cordierite is also known mineralogically as dichroite and as iolite, while jewellers refer to it as lynx-stone or as lynx-sapphire or water-sapphire (*sapphir d'eau*). The two latter terms are given to this stone on account of its blue colour, and specimens of cordierite have sometimes been mistaken for, or sold as, inferior sapphires. This feature, however, is the only one the two minerals have in common; in all other respects they differ very essentially.

The chemical composition of cordierite is expressed by the formula ·

$$H_2O.4(Mg,Fe)O.4Al_2O_3.10SiO_2 ;$$

it is thus a hydrous silicate of aluminium and magnesium with part of the magnesium replaced by ferrous iron. An analysis of a specimen from the "Orient," probably from Ceylon, gave the following results :

	Per cent.
Silica (SiO$_2$)	43·6
Alumina (Al$_2$O$_3$)	37·6
Ferric oxide (Fe$_2$O$_3$)	5·2
Magnesia (MgO)	9·7
Lime (CaO)	3·1
Water (H$_2$O)	1·0
	100·2

Later investigations have shown that the iron in cordierite is present, not as ferric oxide, as shown in the above analysis, but as ferrous oxide: it is doubtless to this constituent that the colour of the mineral is due.

Well-developed crystals of cordierite are not very commonly met with; the mineral crystallises in the rhombic system, and the crystals, as a rule, have the form of short prisms

FIG. 72. Crystalline form
of cordierite.

with basal planes, and with or without small pyramidal faces (Fig. 72). The faces of the crystals are usually rough and the edges somewhat rounded. There is no definite cleavage and the fracture is perfectly conchoidal. The hardness of cordierite is 7¼ on Mohs' scale; the mineral is thus slightly harder than quartz. Cordierite is brittle; it fuses before the blowpipe only with difficulty, and is not attacked by acids to any appreciable extent. Its specific gravity is rather variable, probably in correlation with the presence of a varying amount of iron; it ranges from 2·60 to 2·66, but both higher and lower values have been given. As a rule, the specific gravity does not differ widely from that of quartz, being in general a little lower.

Only perfectly transparent specimens of cordierite are suitable for use as gems, and these are less common than imperfectly transparent stones. The lustre is vitreous, but on fractured surfaces is inclined to greasy. After polishing, the stone becomes appreciably more lustrous, but is always far inferior to sapphire in this respect. The refraction is low, and the double refraction is feebler than in any other precious stone.

The range of colour shown by cordierite is rather extensive. Colourless, yellow, green, and brown stones are met with, but the most commonly occurring are of a moderately dark blue colour sometimes tinged with violet; only these latter are cut as gems. The most transparent and the most finely coloured stones are cut from pebbles found in Ceylon, the chemical composition of which has been given above. The bulk of the material used for cutting is derived from this source, that from other localities being, as a rule, cloudy and poor in colour. The colour of these pebbles varies between sky-blue and indigo-blue, the paler being sometimes distinguished as water-sapphire and the darker as lynx-sapphire.

The most salient feature of cordierite is its dichroism, that is to say, the difference in colour shown by the same stone when viewed in different directions. The phenomenon is more marked in this than in most other minerals, hence its name of dichroite. The maximum differences in colour are apparent in light which has travelled in directions parallel to the crystallographic axes of the stone, that is to say, in three directions perpendicular to each other, one of these being parallel to the edge of the prism shown in Fig. 72. The light transmitted in one direction is a fine dark blue, in another, perpendicular to this, a pale blue, while that travelling in the third direction is pale yellowish-grey or almost colourless. The light which travels in intermediate directions is intermediate in shade. When cordierite is examined with the dichroscope the images seen have very nearly the colours just mentioned, and this feature is of importance in distinguishing cordierite from other blue stones.

In cutting cordierite the dichroism of the stone must be taken into account. According to the orientation of the cut stone with respect to the crystal, there is produced a darker-coloured lynx-sapphire or a lighter-coloured water-sapphire. The step-cut and the table-cut are the forms most frequently adopted for cordierite; but on account of the depth of colour of this mineral the cut stone must not be too thick. The large front facet or table should be perpendicular to the direction along which light of the darkest blue colour travels, so that the stone will appear of the finest colour possible. As the plane of the table becomes more nearly parallel to this direction, the stone becomes of a more and more dingy pale blue or yellowish-grey colour. Cordierite is also cut en cabochon, and when this is the case the same care in the orientation of the cut stone with respect to the crystal is necessary. Cordierites cut en cabochon sometimes show a star of opalescent light similar to that exhibited by star-sapphires. The stone may also be cut in such a manner as to exhibit its dichroism to the fullest possible extent. This object is attained by cutting a cube, the faces of which are perpendicular to the three axes of the crystal. The cube mounted by one corner on a pivot, so as to show the three differently coloured faces, forms an interesting and remarkable object.

Cordierite is readily distinguished from sapphire, for the differences between the two stones are many and well marked. The dichroism of sapphire is much less pronounced than that of cordierite, while the hardness and specific gravity are much greater. Thus while sapphire sinks in the four heavy liquids, cordierite, as a rule, floats in all, though it may, in some cases, sink in liquid No. 1. The same features distinguish it from other blue stones, such as blue diamond, blue tourmaline, and kyanite, to be considered later, all of which are appreciably denser.

The occurrence of cordierite in nature is confined almost exclusively to granite and gneiss; it is found in some volcanic rocks, but only in small amount. Crystals regularly developed on all sides, but too dark in colour for gems, occur embedded in granite at

Bodenmais in Bavaria.   Irregular masses of material, which, in part, is of a fine blue colour and is sometimes transparent, occur in gneiss at Arendal, Kragerö, Tvedestrand, and other places in Norway, at Orijärfvi, near Åbo, in Finland, and elsewhere.   Fine stones are found in granite veins in gneiss at Haddam in Connecticut, U.S.A.   The pebbles found in Ceylon are sometimes as large as a nut ; they occur with other precious stones in the gem-gravels of the island.   Cordierite is also said to occur, together with blue and white topaz, in the district of Minas Novas in Brazil.

IDOCRASE.

Because of its occurrence on Mount Vesuvius idocrase is frequently referred to as vesuvian or vesuvianite.   It occurs at this locality in remarkably fine transparent brown crystals, which are sometimes cut at Naples as gems, and, on this account, are known in the trade as " Vesuvian gems."   The use of this stone as a gem is not extensive, and is mainly confined to Italy.   Crystals of gem-quality and of a green colour are found also in the Ala valley in the Piedmontese Alps ; a small number find their way into the gem markets through the neighbouring town of Turin.   Idocrase from other sources is scarcely ever cut as a gem, so that it may be regarded as an Italian precious stone.

Chemically, idocrase is a calcium-aluminium silicate containing small amounts of water, iron oxides, and other constituents.   The chemical composition was formerly considered to be the same as that of calcium-aluminium garnet ; this, however, has been shown to be incorrect, and the composition is now expressed by the complex formula $2H_2O.12CaO.3(Al,Fe)_2O_3.10SiO_2$.   Analyses of crystals from the two Italian localities mentioned above have given the following results :

	Brown, Vesuvius.	Green, Ala.
Silica (SiO_2)	36·98	37·36
Titanium oxide (TiO_2) . . .	—	0·18
Alumina (Al_2O_3)	16·70	16·30
Ferric oxide (Fe_2O_3)	2·99	4·02
Ferrous oxide (FeO)	2·01	0·39
Manganous oxide (MnO) . . .	0·57	—
Lime (CaO)	35·67	36·65
Magnesia (MgO)	2·62	3·02
Potash (K_2O)	0·08	—
Soda (Na_2O)	0·43	trace
Water (H_2O)	1·32	2·89
Fluorine (F)	1·08	—
	100·45	100·81

Idocrase crystallises in the tetragonal system ; well-developed crystals are of frequent occurrence.   They occur usually attached to the walls of cavities and crevices in rocks, and form beautiful druses.   The forms taken by two crystals of idocrase are represented in Figs. 73a and b ; the former represents a crystal from the Ala valley, and the latter one from Vesuvius.

The crystals in almost all cases take the form of prisms of greater or less length, with four, eight, or more prism-faces, which are distinctly striated parallel to the edges of the prism. The prism-faces are sometimes very numerous, and when this is the case the faces are narrow and the crystal almost cylindrical. The basal planes perpendicular to the prism-faces are usually moderately large ; between the basal plane and the prism-faces there are faces of a square pyramid (Fig. 73a), or of a square pyramid in combination with an eight-sided pyramid (Fig. 73b). These faces are usually very small in size, but sometimes extremely numerous ; indeed, crystals of idocrase have been met with which are richer in faces than crystals of any other mineral.

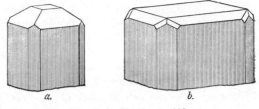

Idocrase shows only a suggestion of cleavage ; the fracture is imperfectly conchoidal to uneven. The mineral is brittle and the hardness (H $= 6\frac{1}{2}$) is

FIG. 73. Crystalline forms of idocrase.

rather less than that of quartz. It can be fused before the blowpipe with moderate ease, and, after being fused, is decomposed by acids ; before fusion it is unattacked by acids. The specific gravity varies between 3·3 and 3·5 according to the chemical composition. The brown crystals are rather heavier than the green ; thus the specific gravity of brown crystals from Vesuvius has been determined to be 3·45, and that of green crystals from the Ala valley to vary between 3·39 and 3·43. The mineral, therefore, sinks in pure methylene iodide, but floats in the heaviest liquid (sp. gr. $= 3\cdot6$).

In the matter of transparency different specimens of idocrase may differ widely. The majority of crystals are only translucent to semi-transparent, the free end of an attached crystal being frequently more transparent than the other. Only transparent or very translucent crystals are cut as gems : by polishing they acquire a very good lustre, which is vitreous in character, though on fractured surfaces it inclines to be greasy. The refraction is moderately strong, but the double refraction is feeble. The mineral presents a great range of colour, this being due to the iron and manganese which enter into its composition. Yellow, blue, and red, as well as almost colourless crystals, are met with ; but the colours most commonly seen are various shades of brown and green, and it is crystals of these shades which are used almost exclusively for cutting as gems. The dichroism of idocrase is moderately strong ; in the case of green idocrase from the Ala valley the two images seen in the dichroscope have the maximum colour difference when one is pure green and the other yellowish-green.

Idocrase occurs at numerous localities as a contact-mineral embedded in limestone, also in crystalline schists, gneiss, and serpentine. These modes of occurrence are well illustrated at the two localities in Italy already mentioned. The occurrence of the mineral under other conditions is rare, and for the present purpose need not be considered.

The green idocrase of the Ala valley is found on the Testa Ciarva, a bare, precipitous rock on the Mussa-Alp, in a band of serpentine more than a metre in thickness : the crystals are here associated with chlorite, and are attached to pale-green massive idocrase. It is at this same place that hessonite, the yellowish-brown variety of garnet, occurs ; the two minerals are not, however, found side by side. Fig. 1 of Plate XV. represents a specimen from this locality, having one large and several small crystals attached to the matrix. Fig. 2 of the same plate shows a stone cut from one of these crystals, the colour of which is a fine grass-green tinged with yellow. This has, therefore, a certain similarity in appearance to chrysolite, and is even sometimes known by this name. The yellowish tinge of the latter stone is, however, more pronounced, and its dichroism is scarcely comparable with that of idocrase. There is also a certain similarity between idocrase and other green stones, such as diopside, epidote, demantoid, &c.: the methods by which these may be distinguished are given in the third part of this book (Tables 13 and 14).

The idocrase of Vesuvius is found, together with other beautifully crystallised minerals, in the limestone blocks ejected from the old crater of Vesuvius, now represented by Monte Somma. The crystals are attached to the walls of cavities in this metamorphosed limestone, and range in colour from darkest brown to honey-yellow. A cut stone of brown idocrase from Vesuvius is represented in Plate XV., Fig. 3. Some of the pale brown stones found here are much admired; they are not unlike hyacinth (zircon) in colour, and, indeed, are sometimes mistaken for that stone. There is no difficulty, however, in distinguishing between the two minerals, owing to the density of zircon being so much greater than that of idocrase. The methods by which idocrase may be distinguished from other brown stones, such as smoky-quartz, brown tourmaline, axinite, &c., are given later (Tables 9 and 11).

The mineral known as xanthite and found at Amity, in Orange County, New York, is nothing more than dark yellowish-brown idocrase; it is sometimes cut as a gem, but is worn only in the United States.

For both green and brown idocrase the step-cut or the table-cut is employed; other forms are scarcely ever adopted. In correlation with the limited demand for idocrase as a gem, we find that both for brown and for green stones quite low prices are asked.

AXINITE.

In just a few cases transparent crystals of axinite are cut as gems, usually *en cabochon*; but, as a general rule, crystals of this mineral are only translucent, and therefore unsuitable

FIG. 74. Crystalline form of axinite.

for this purpose. Axinite has a considerable range of colour; it may be of a warm clove-brown, sometimes with a noticeable tinge of violet, or, on the other hand, of less pleasing colours containing a large proportion of grey. The dichroism is moderately strong; the images seen in the dichroscope vary in colour between violet-blue, cinnamon-brown, and olive-green. This feature distinguishes axinite from smoky-quartz, the commonest of brown gems, the dichroism of which is very slight; also from brown tourmaline, which, though strongly dichroic, shows different colours in the dichroscope.

Chemically, axinite is a silicate of aluminium and calcium, with water and boric acid, and with small quantities of iron and manganese, to the presence of which the colour of the mineral is due. Its composition is represented by the formula:

$$H_2O.6CaO.B_2O_3.2Al_2O_3.8SiO_2.$$

Axinite occurs not infrequently in fine crystals, but most usually in compact masses not suitable for cutting. The crystals belong to the triclinic system, and a frequently occurring form with striated faces is represented in Fig. 74. The crystals are peculiar in that the faces intersect in very acute angles; owing to this they often take an axe-like form, hence the name of the mineral. There is no distinct cleavage. The mineral is brittle. The hardness ($H = 6\frac{1}{2} - 7$) is approximately that of quartz, and the specific gravity varies between 3·29 and 3·30.   The lustre on natural crystal-faces is vitreous and frequently very brilliant, and is still further increased by polishing.

Axinite is found attached to the walls of cavities in ancient silicate rocks of various kinds.   The best crystals known occur in the gneiss of Le Bourg d'Oisans in Dauphiné (Department Isère).   The mineral is found at other places in the Western Alps, less frequently in the eastern (Tyrolese) Alps, at Botallack in Cornwall, and at other localities. The crystals are nearly always small in size, and have very little thickness, so that cut stones have only an insignificant appearance and a correspondingly low value.

KYANITE.

Kyanite, as its name implies, is usually of a blue colour, but may be white, pale yellow, grey, or black.   Crystals of a deep cornflower-blue are rare, the commonest are a pale sky-blue.   When sufficiently transparent, which is not often the case, this mineral forms very pretty gems.

Kyanite is comparable to a certain extent with sapphire.   It has, indeed, been occasionally mistaken and sold for this stone ; and the name sapparé, by which it is known to jewellers, recalls the same stone.   This latter name arose out of an error made by the Geneva mineralogist, Saussure, junior, who read the label attached to a supposed specimen of sapphire as sapparé ; the mistake has long been recognised, but the name remains, having become firmly established, especially among French jewellers.

Kyanite, like topaz, is a silicate of aluminium, but its composition differs from that of topaz in the absence of fluorine.   Its composition is represented by the formula $Al_2O_3.SiO_2$; a small part of the alumina is replaced by ferric oxide, and it is probably to this constituent that the colour of the mineral is due.   An analysis of fine blue kyanite from St. Gotthard gave :

	Per cent.
Silica (SiO_2)	36·67
Alumina (Al_2O_3)	63·11
Ferric oxide (F_2O_3)	1·19
	100·97

The mineral frequently occurs in definite crystals, which belong to the triclinic system ; the form most commonly met with is represented in Fig. 75.   The crystals are flat, elongated prisms, usually with a flattened six-sided cross-section, and not infrequently they are slightly bent.   They occur usually embedded in gneiss, crystalline schists, and other similar rocks. There is a very perfect cleavage parallel to the broad face, which in the figure is turned towards the front, and there is another less distinct cleavage parallel to one of the two narrower prism-faces next to the broad face.   These two narrower prism-faces are striated vertically, that is to say, parallel to the edges, while the broad cleavage-face is striated horizontally.   Terminal faces are usually absent, and when present are irregularly developed. At each end of the prism there is usually a plane face oblique to the length of the prism, along which the crystal has broken (Fig. 75) ; this plane of perfect parting (not true cleavage) gives rise to numerous fine horizontal cracks, which can be seen on the faces of the prism, especially on the large broad face, and are represented in the figure. Kyanite occurs in lamellar cleavage masses of considerable size, more frequently than in definite crystals. These masses are sometimes of a fine blue colour; they show the perfect cleavage surfaces, and are found in the same rocks as the well-developed crystals.

Kyanite is neither very hard nor very brittle. A curious feature of this mineral is the fact that its hardness is not the same on all faces of the same crystal, nor is it the same in different directions on the same face. Thus in certain directions it is easily scratched by felspar, while in others it is scarcely scratched by quartz; the hardness, therefore, varies between 5 and 7. In no other mineral has there been observed such a marked discrepancy between the hardness of different faces of the same crystal. In consequence of the comparative softness of kyanite in certain directions, special care should be taken in the wearing of cut stones.

The specific gravity varies between 3·56 and 3·60, the latter value being that of dark blue specimens of kyanite, which alone are used as gems. The density of the mineral does not thus differ very much from that of the heaviest liquid (sp. gr. = 3·6). Kyanite is infusible before the blowpipe, and

Fig. 75. Crystalline form of kyanite.

is unattacked by acids.

As a general rule kyanite can only be described as being very translucent; perfectly transparent stones are not at all common; and those which combine transparency with a fine dark-blue colour are still more uncommon. In cutting, portions which are pale coloured or not perfectly transparent are removed as completely as possible, and only material of a fine colour is utilised.

The dichroism of the mineral is distinctly observable, but not strong; it is best shown by stones of a dark colour. The images seen in the dichroscope vary in colour between pale and dark blue. The refraction of kyanite is considerable. The lustre of the crystal-faces is of the vitreous type, while that of the cleavage faces is pearly. The natural crystal-faces are not specially brilliant, but their lustre is increased to a certain extent by polishing.

Both the table-stone and the step-cut are employed for kyanite; but it is cut perhaps most frequently *en cabochon*, when its fine blue colour is almost comparable to that of the sapphire. The two stones may be easily distinguished, however, for sapphire is both harder, heavier, and more lustrous than is kyanite. Moreover, on most cut kyanites there are to be seen fine, close-set lines, corresponding to the cracks mentioned above, running parallel to each other across the surface. These lines, especially in stones cut *en cabochon*, sometimes give rise to a band of feebly reflected light, similar to the appearance shown by cymophane. Small cleavage cracks in the direction of the perfect cleavage are also not infrequently observed. The most valuable stones are those which are pure and transparent and of a deep colour, but they are never worth very much, and their use, at any rate in Europe, is very limited.

Kyanite is moderately abundant and is distributed somewhat widely, but material of gem-quality is found at only a few localities, and then in small amount. Among others may be mentioned Monte Campione, near Faido, on the southern slopes of St. Gotthard, in Canton Tessin. Definite crystals, though only pale blue in colour, are found here in abundance; they have the chemical composition quoted above. Associated with them is reddish-brown staurolite, to be considered presently; both are found in a matrix of white, finely scaled mica-schist.

Massive kyanite, sometimes containing portions of gem-quality, occurs also in mica-schist in the Zillerthal and the Pfitschthal in the Tyrol. At the summit of Yellow Mountain, near Bakersville, in North Carolina, U.S.A., fine dark blue crystals occur in a white quartz-

vein in granite ; a few of these are of gem-quality. Stones suitable for cutting are sometimes to be found also in the river sands of Brazil ; and the mineral is found in association
with the diamond and with other minerals at Diamantina. The best crystals are said,
however, to come from some unknown localities in India, a country in which kyanite is
more extensively used than in Europe. There is no doubt but that kyanite occurs at many
places in India, but it has been suggested that the stones worn in that country have all
come from Europe.

STAUROLITE.

Staurolite consists essentially of silicate of magnesium and aluminium with a small
amount of iron ; it occurs always in well-developed crystals belonging to the rhombic system.
These are elongated prisms usually with six sides ; frequently two crystals are twinned
together, the angle between the prism-edges being approximately 60° or 90°. Staurolite is
usually of a dark reddish-brown colour very similar to that of some garnets ; some specimens,
however, are more like dark yellowish-brown topaz. It is usually cut in the same form as
is garnet, but may be distinguished from this stone by the fact that it is doubly refracting.
It is rarely transparent enough for cutting as a gem. Its hardness is $7\frac{1}{2}$ and the specific
gravity 3·7 to 3·8. The crystals usually occur embedded in mica schist, argillaceous schist,
or other similar rocks. They are found in large numbers together with kyanite in the
white mica-schist of Monte Campione, near Faido, in Canton Tessin, Switzerland ; a few are
sufficiently pure and transparent for cutting as gems. The mineral occurs also with
euclase and other precious stones in the gold-washings on the Sanarka river in the district
of the Orenburg Cossacks ; and a few grains of it are met with among the diamond-bearing
sands of Salobro in Brazil. Staurolite of a rich garnet-red colour, perfectly transparent and
fit for cutting as gems, has been found recently (1898) with ruby and rhodolite in Mason's
Branch, Macon County, North Carolina, U.S.A. The stone is used to a very limited extent
and is worth very little.

ANDALUSITE.

Andalusite is identical in chemical composition with kyanite, having the formula
$Al_2O_3.SiO_2$. It differs, however, in crystalline form, the crystals belonging, not to the triclinic but to the rhombic system. They take
the form of elongated, almost rectangular columns, which are usually
terminated by basal planes perpendicular to the prism-edges. They are
rarely modified by other faces, and are therefore very simple in form
(Fig. 76). There is a distinct cleavage parallel to the faces of the prism.

Such crystals are found in large numbers at many localities, especially
in gneiss and other crystalline schistose rocks. Though often of considerable
size they are useless as gems, for they are nearly always opaque and of a
dingy grey, green, or red colour. For cutting as gem-stones the transparent
pebbles found with white and blue topaz in the gem-gravels of the Minas
Novas district in Brazil are exclusively used. These are usually green, but specimens of a

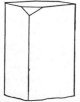

FIG. 76. Crystal-
line form of anda-
lusite.

yellowish-brown colour are sometimes met with. In reflected light the green pebbles usually appear of a dark shade of colour, but some are pale grass-green tinged with yellow. They are strongly dichroic, and this feature is distinctly observable even with the naked eye. When viewed in the direction of the two horizontal axes, indicated in the pebbles by the cleavage cracks, two shades of green are seen, one being rather more inclined to yellow than the other; when, on the other hand, the pebble is viewed through in a direction parallel to the prism-edge it appears of a pretty brownish-red colour. The pairs of images seen in the

dichroscope vary in colour between these principal shades. The two colours characteristic of andalusite are therefore the same as those shown by another dichroic precious stone, namely, alexandrite. Although the green of the alexandrite is more of an emerald-green and the brownish-red is deeper and more intense, yet the similarity between the two stones is close enough to admit of andalusite being passed off as the more costly alexandrite. The latter may readily be distinguished from andalusite by its greater hardness and higher specific gravity, the density of alexandrite being 3·64 while that of andalusite varies between 3·17 and 3·19.

In cutting andalusite, as in the case of other dichroic precious stones, the orientation of the cut stone relative to the crystal must be considered. It may be cut in such a way as to show one of the principal colours, or, on the other hand, so that both colours may be displayed. Cut stones are but little used and their price is low. The lustre of the mineral is of the vitreous type, and is but little increased by polishing. The hardness, H = $7\frac{1}{4}$, is slightly greater than that of quartz. Like kyanite, andalusite is infusible before the blowpipe, and is unattacked by acids.

The peculiar variety of andalusite distinguished as **chiastolite** (cross-stone) should not be forgotten. It has all the essential characters of andalusite, but occurs in elongated, irregular prisms up to an inch in thickness, which are always embedded in dark clay-slates. The peculiarity of the stone depends on the fact that the prisms are not only surrounded but also penetrated to a greater or lesser depth by the clay-slate. Thus, a rod of the clay-slate varying in thickness in different parts runs centrally throughout the whole length of the crystal. This is connected with the four portions which penetrate the crystal at its corners, so that a cross-section of the prism shows a dark cross on a white background (Fig. 77). The figure shows a number of cross-sections of a single prism arranged serially. From these it is apparent that the enclosed clay-slate increases in amount from below upwards, the white andalusite substance predominating in the lower, and the dark clay-slate in the upper portion of the prism.

FIG. 77.
Chiastolite

It is on the appearance presented by cross-sections of the prism that the value, such as it is, of chiastolite is due, for, in certain places, especially in the Pyrenees, such sections are worn as amulets and charms. Suitable crystals of sufficient size are to be met with at several localities in this neighbourhood; also at Salles de Rohan, near Brieux, in Brittany, and at many other places. Recently considerable numbers of fine, large crystals of chiastolite have been found in clay-slate at Mount Howden, ten miles north of Bimbowrie, in South Australia; these are of large size, reaching a length of 6 inches and a diameter of 2 inches; the cross-sections take a good polish and show cross-figures of various forms.

Chiastolite is a contact-mineral which occurs in abundance in clay-slates near their contact with granite. It is found, for example, at Hof, in the Fichtelgebirge, and in the slates round the Skiddaw granite in Cumberland, but it usually takes the form of thin needle-like prisms, which are not thick enough for the purpose mentioned above. Crystals as large as those found in the Pyrenees are rare.

EPIDOTE.

Epidote is another mineral which is used as a gem only occasionally when of specially fine quality. Almost the only locality at which transparent, finely-coloured crystals of gem-quality are to be found is the Knappenwand in the uppermost part of the Untersulzbachthal, in Pinzgau, Salzburg. There are many other localities, but they seldom yield stones which are sufficiently transparent and finely coloured for use as gems.

The chemical composition of epidote is expressed by the formula:

$$H_2O.4CaO.3(Al,Fe)_2O_3.6SiO_2.$$

The mineral is thus a hydrous silicate of calcium and aluminium, in which a variable proportion of the aluminium is replaced by ferric iron. An analysis of epidote from the Knappenwand is given below:

	Per cent.
Silica (SiO$_2$)	37·83
Alumina (Al$_2$O$_3$)	22·63
Ferric oxide (Fe$_2$O$_3$)	14·02
Ferrous oxide (FeO)	0·93
Lime (CaO)	23·27
Water	2·05
	100 70

Crystals of epidote are of very frequent occurrence. They belong to the monoclinic system and are usually prismatic in habit, the direction of elongation being parallel to the

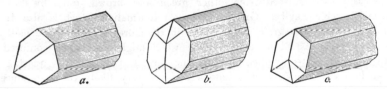

FIG. 78.   Crystalline form of epidote.

axis of symmetry, that is to say, perpendicular to the single plane of symmetry. Owing to the predominance of one pair of parallel faces of the prism the crystals usually present a flattened appearance. The elongated prism-faces are very often distinctly striated parallel to their length, that is to say, parallel to their mutual intersections, while the small terminal faces of the crystals are usually smooth. It frequently happens that two such prismatic crystals grow together in twin position, and when this is the case some of the terminal faces then form re-entrant angles with each other. As a rule, terminal faces are developed at only one end of the crystal, the other being attached to the matrix. A few of the forms most commonly taken by epidote are represented in Figs. 78a to c ; the first two of these, a and b, are simple crystals, the third, c, is a twin-crystal.

There is a definite cleavage parallel to one of the prism-faces, and one, rather less definite, parallel to another prism-face. The hardness is 6½, being thus rather less than that of quartz. The specific gravity ranges from 3·25 to 3·5 according to the amount of iron present. The specific gravity of crystals from the Knappenwand, which are comparatively rich in iron, ranges from 3·47 to 3·5 ; these therefore sink in pure methylene iodide, but float in the heaviest liquid (sp. gr. = 3·6). Epidote is fusible before the blowpipe, and the fused mass is decomposed by acids ; fresh, unfused material, however, is not attacked by acids.

Epidote of almost all degrees of transparency is to be met with, but perfectly transparent crystals of a dark colour are rare, except at the Knappenwand. Both the refraction and the double refraction are strong. The range of colour is somewhat extensive ; it depends upon the presence of iron, and the larger the amount of iron present the deeper the colour of the stone. Almost colourless crystals are rarely met with, pale yellow and red stones are more frequent, but the colours most commonly seen are more or less dark shades of pistachio-green. This colour, which may be described as a dark green with tones of yellow and brown, is so characteristic of epidote that this mineral is sometimes referred to as pistacite. This dark pistachio-green colour may be seen in Fig. 1 of Plate XIV., which represents a druse of crystals from the Knappenwand, while Fig. 2 of the same plate represents a cut stone. In reflected light the Knappenwand epidote appears dark green or, in thick crystals, almost black in colour. The light which is transmitted through the prism in a certain direction is, however, of a bright green colour, while that which travels in a direction perpendicular to this is yellowish-brown, sometimes tinged with red. Epidote is one of the most prominently dichroic of minerals ; the images seen in the dichroscope vary between green, yellow, and very dark brown.

The stone is cut in the forms usually employed for other darkly coloured stones, namely, in low step-cuts and table-cuts (Plate XIV., Fig. 2). Cut stones must not be too thick, otherwise the colour will be dark and unpleasing ; it is sometimes improved, however, by placing a burnished foil beneath the stone. According to the orientation of the large front facet or table with respect to the crystal the stone will appear more green or more brown in colour. The lustre of cut stones is vitreous in character and very brilliant.

Epidote is easily distinguished from other green and brown stones by its strong dichroism and high specific gravity. Green and brown tourmaline, which is also strongly dichroic, is much less dense (sp. gr. = 3·0 − 3·1) ; it therefore floats in pure methylene iodide. Diopside, chrysolite, and other green stones which might possibly be confused with epidote are all much more feebly dichroic, and the same is true of smoky-quartz. The confusion of epidote with brown stones, however, is not very likely to occur, since the former is usually cut in such a manner as to display its green and not its brown colour.

This mineral is generally found in ancient silicate-rocks ; the crystals occur either completely embedded or attached to the walls of cavities. At the Knappenwand, in the Untersulzbachthal, Salzburg, crystals occur attached to the walls of crevices in massive epidote, the so-called epidote-schist ; the exact spot is below the Poberg ridge, and was discovered in 1866. Here are found by far the most magnificent crystals of epidote known, and the locality affords one of the most beautiful examples of the occurrence of minerals. Thousands of crystals have been found there, many of which are distributed throughout the mineral collections of the world, while others have been cut as gems. Some are of considerable size, as much as 45 centimetres in length and 3 to 4 centimetres in thickness, but most are much smaller. In association with the epidote are to be found crystals of calcite, apatite, felspar, and certain rare minerals ; also fibres of asbestos, which sometimes form a thick felted mass round the point of attachment of the epidote prisms, as may be distinctly seen in Plate XIV., Fig. 1.

Epidote is found at many other places in Europe, but **nowhere** in crystals of such size and beauty as at the locality just mentioned. In America stones similar to those from the Knappenwand have been found at Rabun Gap, in Rabun County, Georgia; and fine dark green crystals have come from Roseville, in Sussex County, New Jersey, and from Haddam, in Connecticut; they are only occasionally, however, cut as gems. Again, a few crystals of green epidote are found with green tourmaline in Brazil. There must be many stones suitable for cutting to be found at the numerous localities for epidote, but since the use of this mineral as a gem is so limited and its value is so small, search for them is scarcely likely to be made.

PIEDMONTITE.

Piedmontite is a variety of epidote, in the composition of which alumina is replaced by manganese sesquioxide. This manganese-epidote is found principally in the manganese mines of San Marcello in the Piedmontese Alps, from whence the name piedmontite is derived. It sometimes occurs in magnificent cherry-red crystals, which, when they are of sufficient transparency, are cut as gems.

DIOPTASE.

Dioptase is a mineral of a deep green colour approaching, but always darker than, that of emerald. The two stones differ widely in chemical composition, however, for dioptase is a hydrous silicate of copper with the formula $H_2O.CuO.SiO_2$. The results of an analysis of a crystal from the best and longest known locality, namely, that in Siberia, are given below:

	Per cent.
Silica (SiO_2)	36·60
Cupric oxide (CuO)	48·89
Ferrous oxide (FeO)	2·00
Water (H_2O)	12·29
	99·78

The green colour of dioptase is due to the large amount of copper which is present, and for the same reason the mineral is sometimes referred to as emerald-copper.

Dioptase occurs usually in well-developed but small crystals. These are rarely larger than a pea, and belong, like crystals of emerald, to the hexagonal system. The six-sided prisms of dioptase, however, are terminated by the faces of a rhombohedron, and in rare cases the alternate edges between the striated rhombohedron-faces and the smooth prism-faces are replaced by narrow faces indicating a tetratohedral development of the crystal. This is shown in Fig. 79, which represents a Siberian crystal. There is a distinct cleavage parallel to the faces of a rhombo-hedron which truncate the polar edges of the primitive rhombohedron represented in the figure. On these edges there is therefore sometimes to be seen a pearly lustre, but otherwise the lustre of the crystals is of

Fig. 79. Crystalline form of dioptase.

the common vitreous type. Dioptase is not very hard; it is scratched even by felspar and is itself scarcely capable of scratching glass, so that its hardness is approximately represented on the scale by 5. While the hardness of dioptase is much less than that of emerald, the specific gravity is higher, being 3·28. The two minerals are therefore readily distinguished by means of these characters, but usually dioptase can be recognised at a glance on account of its dark colour, imperfect transparency, and the presence of numerous small cracks.

The principal locality for dioptase is the hill of Altyn-Tübe on the western slopes of the Altai Mountains in the Kirghiz Steppes in Siberia. This is a limestone hill, and is traversed by numerous crevices which are filled up for the most part by calcite; it is on this calcite that the dioptase in sparing amount is found: the mode of occurrence is illustrated in Plate XV., Fig. 4. A few comparatively large crystals of dioptase have also been found in some of the gold-washings in the Yeniseisk Government, where they occurred loose and in a more or less rounded condition. For a long time the occurrence of large crystals of dioptase was thought to be confined exclusively to Siberia, and the use of this mineral as a gem was limited to Russia, Persia, and countries in the near neighbourhood. This is still the case at the present day, although latterly fine crystals have been found in other parts of the world, especially in the French Congo.

In spite of the fine colour of dioptase, its deficient hardness and transparency prevent its extensive use as a gem; for the same reasons the price of the stone is always low.

CHRYSOCOLLA.

The amorphous mineral chrysocolla is closely allied to dioptase in chemical composition, being a hydrous silicate of copper, but containing more water than dioptase. Its hardness, $H = 2 - 4$, is low, sometimes less even than that of calcite; the mineral is lacking also in transparency, but in spite of these drawbacks it is sometimes cut as an ornamental stone on account of its fine green or blue colour. This is especially the case with stones from the copper mines of Nizhni-Tagilsk in the Urals. In the Allouez mine, near Houghton, in the copper region of Lake Superior, North America, the mineral occurs mixed with quartz; this affords a much harder material more suitable for cutting, and from it are obtained magnificent bluish-green stones half a square inch in area.

GARNIERITE.

Garnierite or noumeaite has certain characters in common with chrysocolla, and may therefore be mentioned here. Like chrysocolla it is opaque, amorphous, of a fine green colour, and softer than calcite. Chemically it is a hydrous silicate of nickel. It occurs as large masses in the island of New Caledonia, Australasia, where it is extensively worked as an ore of nickel, and where some few stones are cut for ornamental purposes.

SPHENE.

Sphene or titanite is a combined silicate and titanate of calcium having the formula $CaO.SiO_2.TiO_2$. The crystals belong to the monoclinic system, and are frequently finely developed. Sphene of a brown or yellow colour, opaque and unsuitable for cutting as a gem, occurs embedded in many silicate-rocks; while attached to the walls of crevices in the same rocks are found crystals of the same mineral, which are frequently very clear and transparent. These two varieties are sometimes respectively distinguished as titanite and as sphene: the latter alone is cut as a gem, and that only when perfectly transparent and of fine colour. Sphene is usually green, but may be yellow, brown, or even red: it is always distinctly dichroic. There are many stones which sphene of a green colour may resemble in appearance, such, for example, as chrysolite, idocrase, demantoid, and chrysoberyl. It never resembles emerald, however, since the colour has always a tinge of yellow and is never very deep and intense. Yellow sphene often resembles light yellow topaz in colour. The hardness of the mineral is low: $H = 5\frac{1}{2}$. Its specific gravity is somewhat high, ranging from 3·35 to 3·45; in pure methylene iodide, therefore, the mineral sinks.

Fine crystals, suitable for cutting, are found in crevices in gneisses and schists at various places in the Alps, especially in the Pfitschthal and the Zillerthal in the Tyrol, and also in the Swiss Alps. In America specially large and fine crystals occur at Bridgewater in Bucks County, Pennsylvania; these are often over an inch in length, and from those which are perfectly transparent fine stones, varying from 10 to 20 carats in weight, can be cut. Sphene is used as a gem to only a limited extent, and the price is always low.

PREHNITE.

This mineral has a fine green colour similar to that of chrysolite, and is therefore sometimes cut as a gem; but it is used for this purpose only to a very limited extent. Prehnite is a silicate of calcium and aluminium containing a small amount of water. It sometimes crystallises in tabular forms belonging to the rhombic system, but occurs most frequently as crystal aggregates which take the form of nodular, botryoidal masses with a radially fibrous structure. The mineral occurs, sometimes in masses of considerable size, in the amygdaloidal cavities of older volcanic rocks, such as basalt. It is met with at several places in the Alps, for example, at the Seisser-Alp in the Fassathal, Tyrol, at St. Gotthard, and elsewhere. Prehnite was first discovered at the Cape of Good Hope by Colonel Prehn, the Governor of the Colony, in the latter part of the eighteenth century; hence the names prehnite and "Cape chrysolite." The mineral is also found in the Lake Superior copper region and at many other places in North America, specially fine specimens being met with at Bergen Hill and Paterson in New Jersey.

The hardness of prehnite is rather over 6; the specific gravity ranges from 2·8 to 3·0. The mineral is usually translucent, rarely transparent; it has a vitreous lustre, and may be colourless, yellow, or green. The green variety is the only one suitable for use as a gem;

this is of a rich oil-green shade, and, when cut, may closely resemble chrysoprase in colour and lustre. The two stones may be distinguished, however, by the fact that chrysoprase is harder and much less dense than prehnite.

Chlorastrolite is a mineral which is sometimes referred to the species prehnite. It occurs as small rounded masses with a finely radiate or stellate internal structure, and of a bluish-green colour in the amygdaloidal trap rocks of the Isle Royale, Lake Superior, in the State of Michigan. These rounded masses, in many cases, have been weathered from the rock and are now collected as water-worn pebbles on the shores of the island. Owing to their fibrous structure they exhibit a chatoyant appearance similar to, but less perfect than, that of cat's-eye. When suitably cut they give gems of considerable beauty, which, however, are worn solely in North America. The largest nodule hitherto found has a diameter of 1½ inches, but the majority of specimens are much smaller.

Zonochlorite is a mineral of very similar character; it occurs in the amygdaloidal volcanic rocks of Neepigon Bay on Lake Superior in Canada. The rounded masses, which fill the amygdaloidal cavities, attain a diameter of 2 inches; they are built up of alternate layers of light green and of dark green material, and form rather pretty cut stones. The mineral, however, is rarely cut except in America, and even there is but little used.

THOMSONITE.

The essential characters of a precious stone are not, as a rule, present in thomsonite, but there is one variety of the mineral which is sometimes used for this purpose. This occurs filling amygdaloidal cavities in volcanic rocks at Good Harbour Bay on Lake Superior; these rounded masses are sometimes weathered out of the rock and found lying loose on the shore. Their internal structure shows a central mass of radially arranged fibres enclosed by a few concentric bands which follow the outer boundaries of the mass. In these concentric bands delicate shades of milk-white, yellow, and green alternate with each other, so that a section of the mass has an appearance which resembles that of agate. These water-worn pebbles, the largest of which are an inch in diameter, are collected on the shore and polished; no attempt is made to reach those still enclosed in the solid rock, and stones which are worked are only worn in the place of their origin.

Lintonite is a variety of thomsonite with alternate bands of green and of flesh-red; it is found in the above-mentioned locality.

NATROLITE.

Natrolite is another mineral belonging to the zeolite group; like thomsonite, it is a hydrated silicate of sodium and aluminium. It occurs sometimes as beautiful water-clear crystals of elongated, prismatic habit; these, however, have never been cut as gems. The mineral is also found in masses built up of radially arranged fibres marked by concentric bands of varied colours. Fine specimens of this variety of natrolite occur at Hohentwiel in Würtemberg; in these the concentric bands are coloured alternately isabel-yellow and pale yellow or white. The mineral is susceptible of a good polish, and is sometimes cut in such a way as to display the coloured bands to the best advantage. Even this variety of natrolite can scarcely, however, be regarded as a precious stone, as it is of but little value, and used only for inlaying and in the manufacture of small ornamental objects of various kinds.

HEMIMORPHITE.

Hemimorphite, otherwise known as electric calamine, is a hydrous silicate of zinc. It sometimes occurs as colourless, transparent, rhombic crystals, but more frequently in spherical or reniform shelly aggregates built up of radially arranged fibres and concentric layers. These rounded masses are frequently of a beautiful bright green or blue colour, which somewhat resembles that of turquoise, and, as in the latter mineral, is due to the presence of a small amount of copper. They are often cut with convex surfaces and used in ornamental work of various kinds. The mineral is not very durable, however, having a hardness of only from 4 to 5. Its specific gravity varies from 3·35 to 3·5, so that by means of these two characters hemimorphite may be readily distinguished from turquoise. Beautifully coloured specimens are found in the zinc mines at Laurion in Attica, at Santander in the north of Spain, at Nerchinsk in Siberia, and at other places; the application of the stone, however, is very limited.

CALAMINE.

Calamine, or zinc-spar, is carbonate of zinc, $ZnCO_3$, and is an important ore of this metal. It is found at the same localities as hemimorphite, and often occurs in the same kind of aggregates, which may be bright green, blue, or even violet in colour. It is used to a certain extent for ornamental purposes, the material from Laurion, in Greece, like the hemimorphite from the same locality, being worked for brooches, ring-stones, and as plates, &c.

THE FELSPAR GROUP.

The felspar group includes a number of minerals which are important constituents of the earth's crust, through which they are widely distributed.   Felspars in general are cloudy, opaque, and dull in colour, so that they possess none of the characters which would lead to their application as gems, or even as ornamental stones.   There are some few exceptions to the rule, however, and of these a detailed account, preceded by a general description of the characters of felspars, so far as they are of interest for our present purpose, is given below.

All the members of the group are silicates, that is to say, compounds of silicic acid and bases, and in all of them alumina is present.   Beside these constituents there is present also either potash, soda, or lime, or soda and lime together.   We may, therefore, distinguish potash-felspar, soda-felspar, lime-felspar, and lime-soda-felspar; in the latter, either the lime or the soda may predominate in different varieties.   These felspars of different chemical composition have received special names, some of which are mentioned below.

The felspars very frequently occur in well-developed crystals, often of considerable size.   Potash-felspar crystallises in the monoclinic, and the other kinds in the triclinic system.   All are very similar, however, in general form, and differ essentially only in the

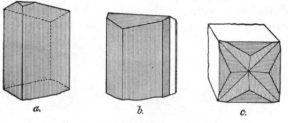

a.   b.   c.

FIG. 80.   Crystalline forms of felspar.

size of the angles between the faces, and this difference at most amounts to but a few degrees of arc.   Some of the forms taken by felspar are represented in Figs. 80a to c. In each case there is a rhombic prism, which in the simplest crystal (Fig. 80a) is terminated by two obliquely inclined faces.   Other faces are frequently developed, and there is nearly always a pair of parallel faces truncating the acute prism edges (Fig. 80b).   Very frequently two or more individuals grow together in accordance with various twin-laws, and give rise to complex groups (Fig. 80c).

Felspars frequently occur as constituents of rocks, such as granite, gneiss, trachyte, &c., when they usually take the form of irregular grains, which are embedded in the rock-mass. They also occur, however, as regularly developed crystals, which are attached to the walls of crevices or cavities in rocks of the same character, often giving rise to druses of great beauty.

There are certain physical characters which all the felspars possess in common.   All have two good and easily developable cleavages ; one, which is very perfect, is parallel to the obliquely disposed terminal face represented in Figs. 80a and b. On other parts of the crystal the lustre is vitreous, but on this particular face, usually known as the basal plane, the lustre is pearly; and from the cleavage cracks parallel to this face brilliant iridescent colours are sometimes reflected. The second cleavage, which, though good, is rather less perfect than the first, is parallel to the pair of faces which truncate the acute side edges of the prism (Fig. 80b); these faces together form what is known in crystallography as the clino-pinacoid.

The cleavages of monoclinic and of triclinic felspar are essentially the same, the differences between them being unimportant. In monoclinic crystals the two cleavage planes are exactly perpendicular to each other, and for this reason monoclinic felspar has received the name orthoclase, that is to say, cleaving at right angles. In triclinic felspar the two cleavage directions are not quite at right angles, and so this variety is distinguished as plagioclase, that is to say, obliquely cleaving. The two directions are not, however, very oblique; in one variety, for example, the angle is barely 93°. The cleavage of felspar is a feature which, taken in conjunction with its hardness, readily distinguishes it, even when in massive fragments, from other minerals. The hardness of felspar is 6; in fact, this mineral is the one which is chosen to stand sixth on Mohs' scale of hardness. The fracture of felspar in directions other than the cleavage planes is sub-conchoidal to uneven.

The specific gravity varies with the chemical composition between 2·5 and 2·7. The greater the amount of lime present the heavier is the felspar, pure lime-felspar or anorthite having the highest specific gravity of any. All felspars are fusible before the blowpipe, but only with difficulty. Neither pure potash- nor soda-felspar is attacked by hydrochloric acid, but anorthite and other varieties containing much lime are readily decomposed. The greater the amount of lime present the more easily is the felspar attacked by hydrochloric acid, and the more easily and completely is it decomposed.

As already mentioned, common felspar is in most cases dull in colour, cloudy, and imperfectly transparent. The colours most commonly seen are very pale shades of yellow, brown, and red, but more intense shades are also met with. The variety of potash-felspar, known as amazon-stone, has a fine green colour and is sometimes cut as an ornamental stone, but the yellow, brown, and red felspars are never used for this purpose. Quite colourless felspars, more or less perfectly transparent, are by no means rare; the only variety used for ornamental stones is that from the surface of which there is reflected a beautiful milky light.

The precious stones which are included in the felspar group are not much used nor are they of any great value. Nevertheless they have a certain importance. They will be dealt with below in some detail: first, amazon-stone, the use of which, for ornamental purposes, depends solely upon its colour; and secondly, those felspars which display a reflection of milky light or a play of colour, namely, moon-stone, sun-stone, labradorescent felspar, and labradorite. Other members of the felspar group are never used for this purpose.

AMAZON-STONE.

In colour amazon-stone is verdigris-green, sometimes tinged with blue. All shades of this particular colour, from the palest to the darkest, are met with, but only those stones which exhibit perfectly pure dark shades of colour are cut and polished. Stones showing white, yellow, or red patches and streaks are frequently met with, but are useless for ornamental purposes. Amazon-stone is opaque or only slightly translucent, and it acquires no specially fine lustre after polishing; its application as an ornamental stone depends, therefore, solely upon its colour, and the deeper and purer this colour is the more highly is the stone prized. This feature of amazon-stone is supposed by some to be due to the presence of a small amount of copper, but according to other authorities it depends on the presence of organic matter.

Amazon-stone is a potash-felspar. It is found in irregular masses as a constituent of granitic, syenitic, and similar rocks; also as regularly developed crystals, which may reach a length of 25 centimetres, in magnificent druses, in the crevices and fissures of rocks of the same kind. Fig. 1 of Plate XVI. illustrates the latter mode of occurrence, and a single crystal is represented in Fig. 81. The physical characters of amazon-stone are the same as those of other felspars. Its specific gravity, in correlation with the chemical composition, is somewhat low, varying between 2·55 and 2·66.

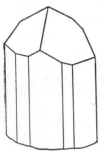

FIG. 81. Crystalline form of amazon-stone.

The name amazon-stone was first given in the middle of the eighteenth century to a green stone from the Amazon river in South America. It appears to be doubtful whether this was the same substance as that to which the name now refers; more probably it was nephrite, jadeite, or some other green mineral, for nothing is known at the present day of the occurrence of verdigris-green felspar in this region. It is known to occur in the Ural Mountains; here it is found in compact masses near Miask on the eastern side of Lake Ilmen as beautiful crystals associated with topaz and other minerals in granitic rocks. It was met with later at a few places in North America; of these occurrences the most beautiful is that of Pike's Peak in Colorado. Here it often takes the form of fine crystals, such as are represented in Plate XVI., Fig. 1; they are enclosed with grey quartz and flesh-red felspar in a coarse-grained granite. Amazon-stone of a fine colour also occurs in the coarse-grained granite of Allen's mica mine, near Amelia Court House in the State of Virginia, from whence have been obtained hundreds of tons of cleavage fragments, ranging up to 6 and 8 inches in length and of a magnificent green colour.

Amazon-stone is usually cut in the form of an oval or circular plate, the upper surface of which may be either plane or convex. These stones do not figure very largely in the precious stone trade, and are used most extensively in the countries in which they are found, namely, in Russia and North America. Several pounds may be paid for specially pure and fine stones of very intense colour or extraordinary size. The mineral is used not only as a gem but also in the manufacture of small objects of various kinds, such as bowls, vases, the stocks of seals, &c.

SUN-STONE.

The term sun-stone is applied to felspar of various kinds which have one feature in common, namely, the reflection of a brilliant red metallic glitter from a background which has little transparency and which is pale or almost white in colour. This metallic reflection is specially intense in direct sunlight or in a strong artificial light. The points which reflect the glittering light may be distributed singly and sparingly over the surface of the stone, or they may be numerous and closely aggregated; in the latter case there is over the whole surface a brilliant and glittering sheen. It is obvious that the name sun-stone is descriptive of this peculiar feature.

The explanation of the phenomenon is comparatively simple. Enclosed in the substance of the felspar are numerous minute and very thin scales of the mineral hæmatite (micaceous iron) arranged parallel to the direction of perfect cleavage in the felspar. The glittering reflection is therefore confined to the cleavage surfaces; other faces, parallel to which there are no inclusions, do not show it. The enclosed scales of hæmatite have a regular six-sided or rhomb-shaped outline, or they may be quite irregular. These scales, in consequence of their excessive tenuity, are transparent; they have a red colour, which may be distinctly seen when thin sections of the stone, cut parallel to the plane of reflection, are examined under the microscope. It is to the reflection of brilliant red metallic light from the surface of these scales that the glittering sheen of the stone is due. The more numerous and regularly distributed are these scales the more brilliant and uninterrupted is the reflection from the surface of the stone. If they are altogether absent there is no trace of the peculiar appearance characteristic of sun-stone; moreover, when the stone is turned about so that the reflecting surface is inclined towards the light at different angles the metallic sheen will be observed to appear and disappear. The metallic reflection is, of course, dull or absent in those portions of the stone in which the enclosures of hæmatite are few in number or altogether lacking, and in working the stone such parts are cut away and discarded. Sun-stone is very similar in appearance to that variety of quartz which is known as avanturine, and for this reason is sometimes referred to as avanturine-felspar.

At the beginning of the nineteenth century sun-stone was a great rarity and very costly. There were only a few small pieces known to be in existence, and the single locality given for it was Sattel Island (Setlovatoi Ostrov) in the White Sea near **Archangel**. The fragments of sun-stone found here were described as masses of cloudy, white, translucent felspar, in which were portions here and there showing a golden sheen. Subsequently the East Indies and Ceylon were mentioned as localities for sun-stone, but the occurrence of the stone there is probably not authentic.

In the year 1831 an occurrence of sun-stone was discovered at Verchne Udinsk on the Selenga, a river flowing into Lake Baikal in **Siberia**. The sun-stone is found here in a narrow vein of felspar, which runs vertically through a black rock. It is of a clove-brown colour, and in this case also the sheen is due to the enclosure of scales of hæmatite, which are arranged parallel to the plane of easier cleavage. The scales are present in large numbers, and when the cleavage face is inclined towards the light at a suitable angle a multitude of shining golden scales become visible. In all other positions the stone appears lustreless and of a uniform brown colour, but when turned in this one direction it appears as if suddenly gilded, producing a surprisingly beautiful effect. Even from broken fragments lying on the surface of the ground, and exposed to the action of weathering agencies, material was obtained superior to any which had been previously found. From a vein of considerable thickness below the surface were found bigger and more compact masses large enough to be worked into objects of some size, such as bowls and vases. Sun-stone also occurs as rounded pebbles in the Selenga river; these had been occasionally collected by passing merchants some time before the mineral was discovered in its original situation.

The most typical and beautiful sun-stone was discovered in the 'fifties at Tvedestrand in the south of **Norway**, where it occurs *in situ*. It is also found under similar conditions at Hitterö on the Christiania fjord. The sun-stone from the same region, for which Fredriksvärn was previously given as the locality, probably came in reality from Tvedestrand. The mineral never occurs in this region in regularly developed crystals, but always in masses embedded in white quartz, which forms a vein penetrating gneiss. This vein has on an average a thickness of a yard and has been worked for a length of at least six yards

It follows the vertical banding of the gneiss, which in the immediate neighbourhood passes into a mica-schist very rich in mica, or possibly into pure mica. Accompanying the sun-stone in this vein are a variety of minerals, including hæmatite, cordierite, hornblende, zircon, and probably also apatite. The felspar found near the margins of the vein is almost colourless, and does not show the metallic sheen characteristic of sun-stone. That found nearer the centre of the vein encloses more scales of hæmatite, and at the same time exhibits a marked metallic sheen. The finest material is usually obtained from the centre of the vein ; it occurs in irregular masses embedded in stuff of inferior quality which has to be cut away, the better portions only being kept for use.

Sun-stone occurs also in **North America,** for example, at Statesville in North Carolina. Some of the material found there is equal to Norwegian sun-stone, although the enclosed scales of hæmatite are much smaller than those enclosed in material from the latter locality. It is found at Fairfield in Pennsylvania and at other places in the same State. At Middletown, in Delaware County, Pennsylvania, very fine material, little inferior to Norwegian sun-stone, is found, with moon-stone, loose on the surface of the ground. At Media in the same county there is found, besides the brilliant red sun-stone, a variety of felspar which shows the same kind of sheen as sun-stone, but which is green instead of red in colour.

The sun-stone found at all the above-mentioned localities may be described as a triclinic lime-soda-felspar, in which the soda preponderates over the lime ; it is known mineralogically as oligoclase. The avanturine-felspar, so far considered, is thus a variety of the mineral species oligoclase, which is distinguished from ordinary oligoclase by the possession of a metallic sheen due to the enclosure of scales of hæmatite. Like all triclinic felspars, the sun-stone variety of oligoclase, or *oligoclase-sun-stone*, shows a series of fine striations on the principal cleavage face, that is to say, on the face which exhibits the metallic sheen. These striations have the appearance of straight lines running parallel to the plane of the less perfect cleavage, that is to say, parallel to the intersection of the two cleavage surfaces. They are a result of the peculiar repeated twinning of the crystal, and are not present in the avanturine-felspar to be presently considered. This difference, therefore, serves to distinguish the two kinds of avanturine-felspar when in the rough, but on cut and polished stones the striations are not seen.

The appearances characteristic of sun-stone are not limited entirely to oligoclase, but are exhibited also by other kinds of felspar, especially by certain specimens of potash-felspar or orthoclase. In these stones are enclosed thin scales of hæmatite, arranged parallel to the plane of perfect cleavage, which produce a glittering metallic sheen precisely similar to that of sun-stone already described. This *orthoclase-sun-stone* is found principally in North America, for example, at Glen Riddle, in Delaware County, Pennsylvania, where the felspar is salmon-coloured and partly transparent. Beautiful varieties of orthoclase-sun-stone have been discovered near Crown Point, in the State of New York. On the Horace Greeley farm at Chappaqua, New York, have been found small pieces of the same stone, almost equal to that found at Tvedestrand, Norway ; while in Virginia specimens have been met with at Amelia Court House in Amelia County.

Sun-stone is in every case cut with a quite flat or slightly convex surface parallel to the face showing the sheen. The stone is by no means extensively used, and moderately high prices are paid only for the finest specimens.

Avanturine-felspar can be readily distinguished even with the naked eye from avan-turine proper, that is avanturine-quartz, so that these stones are seldom mistaken one for another. In doubtful cases the hardness forms a distinguishing feature ; for while the hardness of quartz is 7, that of avanturine-felspar is only 6, so that the latter is readily scratched by quartz. Artificial avanturine-glass made in imitation of sun-stone will be considered under avanturine-quartz.

MOON-STONE.

The name moon-stone is applied to colourless, very translucent, or almost perfectly transparent felspar, which in a certain direction reflects a bluish, milky light, that has been compared to the light of the moon. This peculiar feature, like the metallic sheen, characteristic of sun-stone, is not confined to one member of the felspar group, but is exhibited by isolated examples of all the different varieties. It is shown to perfection in orthoclase, and for this reason moon-stone is frequently referred to as a variety of adularia, and the appearance characteristic of this stone as adularescence. It is by no means correct, however, to suppose that this feature is peculiar to adularia, for the same reflection of milky light is to be seen, though rarely, in the colourless and transparent soda-felspar, which is known to mineralogists as albite; moreover, the same is true of felspars having the chemical composition of oligoclase, which have been dealt with already under sun-stone. These adularescent felspars showing a reflection of soft, milky light, are variously referred to as girasol, fish-eye, wolf's-eye, " Ceylonese opal " or " water-opal."

Adularia is the felspar which most frequently shows the characteristics of moon-stone. It is a colourless and translucent to transparent variety of pure potash-felspar, for which the mineralogical term is orthoclase, and its chemical composition is represented by the formula $K_2O.Al_2O_3.6SiO_2$. It frequently occurs as fine crystals of large size and very simple form, such as are represented in Figs. 80a and b; it also takes the form of complex twin-crystals, as in Fig. 80c, which represents four individuals united together in twin position.

The specific gravity is 2·55, the same as for all orthoclase. The milky sheen characteristic of adularia is not exhibited over the whole surface of the stone, but only over a certain portion near a face which truncates the front and back edges of the prism (Figs. 80a and b), and in massive specimens over a corresponding part. The milky sheen is only visible when this part of the stone is in a certain position relative to the source of light and the eye of the observer. If the stone be turned out of this position the reflection of milky light ceases, but the sheen reappears when the stone is replaced in its original position. In Fig. 4 of Plate XVI. an attempt is made to reproduce the appearance characteristic of an irregular cleavage fragment of moon-stone.

Only specimens which show this sheen in a typical manner are cut as gems. To display the beauty of this stone to the best possible advantage, it is cut with a convex, polished surface of moderately strong curvature about the face which exhibits the sheen. The stone is very effective in cut and polished spheres, which may be strung together and worn as beads or for other ornamental purposes, when they have a certain resemblance to white pearls. Moon-stone is always cut either en cabochon or with a plane surface, never with facets. If the plane face be cut so that it is suitably orientated with respect to the crystal, there will be a milky sheen uniformly distributed over its surface; but if, on the other hand, the stone be cut with a convex surface, there will appear a band of bluish, milky light, crossing a colourless and almost transparent background (Plate XVI., Fig. 5). The area occupied by the band of light appears almost wholly lacking in transparency, while the area surrounding the band exhibits no milky sheen. There is no very sharp line of demarcation between the chatoyant band of light and the surrounding portions of the stone, but it is sufficiently abrupt to form an effective contrast. The more convex the polished surface of the stone the narrower and better defined is the band of light; this, however, is true only up to a certain point, for if the band is too narrow it will appear dull.   It is at no time very intense or strongly marked, but has always more of the nature of a soft, pearly lustre comparable to the chatoyancy of cymophane or cat's-eye, though the lustre of the band of light in the latter stone is more silky than pearly in character.   The effect of a cut moon-stone is heightened by mounting it in a closed, black setting.

The chatoyancy characteristic of moon-stone is exhibited to a variable extent in different specimens of adularia, but in the majority it is completely absent.   Moreover, this feature is often more prominent in one portion than in another of the same specimen, and in such cases the inferior portions are cut away and the best used for cutting as gems.   The more pronounced the chatoyancy of a stone the greater its value; one the size of a bean with a fine milky sheen is worth from 25 to 40 shillings, and the value increases considerably with the size.

The milky sheen is probably due to the presence of microscopically small, colourless, and brilliant crystal plates embedded in great numbers in the felspar, and arranged parallel to the surface from which the reflection of milky light takes place.   The presence of such plates can be observed in all specimens of adularia which show a milky sheen, and the more numerous are the plates the greater will be the prominence of this feature, which, however, is completely absent when there are no enclosures to be seen.   Chatoyant specimens of adularia are never perfectly clear and transparent but always cloudy, though possibly only slightly so, and this fact is also due to the same cause, namely, to the enclosure of small foreign bodies.

Splendid specimens of adularia are found at various places in the Tyrolese and Swiss **Alps.**   The mineral occurs here in fine crystals, of the forms shown in Fig. 80; these, together with quartz and other minerals, are attached to the walls of crevices in gneiss and other crystalline rocks.   This is the most important locality for the mineral, but it is rare for Alpine specimens to exhibit any chatoyancy at all, and still more rare to find the feature sufficiently well marked to make the stone useful for ornamental purposes.

The strongly chatoyant and beautiful moon-stone used for cutting is obtained almost exclusively from the island of **Ceylon.**   Its mode of occurrence there differs from that which obtains in the Alps.   Massive fragments of irregular shape, and as large, or larger than, a man's fist, are here embedded in a white kaolin-like clay, which has probably been derived from the weathering of a porphyritic igneous rock.   In all probability the moon-stone was one of the original constituents of this rock and the only one to resist the action of weathering.   It is found under these conditions among other places at Neura Ellia, near which is a spot, to the south-east of Adam's Peak, which is marked on some maps as " Moonstone Plain."   It is also of frequent occurrence nearer the centre of the island, and, indeed, is a constituent of many of the rocks of that neighbourhood.   It has also been observed to occur in close association with spinel, and in this case probably originated, together with this mineral and possibly also with the ruby, in the same mother-rock, namely, in a crystalline limestone.   Ruby and spinel are found under the same conditions at Mogok in Upper Burma, but here moon-stone is present in only small amount.   By the weathering of the mother-rock the moon-stone is set free, carried down the rivers and streams, and, as rounded pebbles, finally collected, together with other precious stones, out of the gem-gravels of the island.   It is said to be most abundant in gem-gravels at Bellingham, between the Point de Galle and Matura on the south coast of the island, but the greater part of the material imported into Europe from Ceylon for cutting is in the form of irregular masses, not in rounded pebbles, and has therefore been obtained from the original deposits. The stone is frequently cut in Ceylon in rounded forms, but so unskilfully that it usually has to be re-cut in Europe.

A small amount of material is derived from other sources, all of which are less important than that last mentioned. In Brazil fine crystals occur in gneiss in the neighbourhood of Rio de Janeiro, many of which are suitable for cutting as gems.

In **North America** moon-stone is found at various places, but the finest specimens come from Allen's Mica Mine at Amelia Court House in Virginia. These colourless and almost transparent fragments, the largest of which are half an inch across, occur embedded in a coarse-grained granite, from which they are won in the course of mining for mica. Moon-stone from this locality is quite comparable in quality to that from Ceylon.

Here in North America are to be found, besides adularia, other felspars which exhibit the chatoyancy characteristic of moon-stone, though only to a small extent. The external appearance of some of these differs in no wise from that of the *adularia-moon-stone* hitherto considered, but this is not true in all cases. The colourless and transparent soda-felspar, known as albite, sometimes exhibits the chatoyant appearance in question, and may then be distinguished as *albite-moon-stone*. A very beautiful example of this is afforded by the albite of Mineral Hill, near Media, in Delaware County, Pennsylvania, which is sometimes cut as a gem; also by that from Macomb in St. Lawrence County, New York, which is distinguished by the name of peristerite. It occurs with ordinary felspar and is sometimes cut as a gem. Many of the crystals of peristerite are chatoyant, and some of these are as fine as moon-stone from Ceylon; others are coloured with pale shades of green or yellow, and some specimens show different colours at the same time. Adularescent albite, which, as mentioned above, is distinguished by the name of peristerite, is found in crystals and in large massive pieces in veins of coarse-grained granite penetrating gneiss at Bathurst, near Perth, and at various other places in Canada.

Much rarer than moon-stones are transparent felspars with a reddish adularescence and a yellowish colour. They are sometimes referred to as sun-stone, but are quite distinct from the stone of this name described above. They occur at the same localities as moon-stone.

Very good imitations of moon-stone, which are frequently used in cheap jewellery, have recently been made in glass, and it is by no means easy to distinguish them by mere inspection from genuine stones. Glass imitations are, however, always denser and less hard than real stones; moreover, moon-stone is distinctly doubly refracting, while the glass imitations are singly refracting, so that these characters serve to distinguish the two.

LABRADORESCENT FELSPAR.

A beautiful coloured sheen is a noticeable feature of the potash-felspar or orthoclase found in the south of Norway between the Christiania and Langesund fjords. It occurs in masses in an augite-syenite, a rock which has sometimes been referred to as zircon-syenite, and is especially abundant in the veins of coarser-grained material by which the rock is penetrated. As more definite localities, Laurvik and Fredriksvärn, especially the latter, are often mentioned. The surface which shows the coloured sheen has a rather greasy lustre, and has the same orientation in the crystal as that of the chatoyant surface of a crystal of moon-stone. Contrasted with moon-stone, however, this felspar is grey and opaque, and its sheen, instead of being of a milky blue tinge, is of a very fine blue colour, or in rare cases green, yellow, or red. The sheen is also much more intense and brilliant than in moon-stone, and thus approaches more nearly to that of labradorite without ever quite equalling this stone in beauty. A polished plate of labradorescent felspar is represented in Plate XVI., Fig. 3 ; it is rarely used for gem purposes, since it is surpassed in this direction by the abundantly occurring labradorite, but the whole rock mass is frequently worked and utilised for such objects as gravestones and the ornamental facings of shops and public buildings.

LABRADORITE.

Labradorite, labrador-spar, or labrador-felspar, so named because of its occurrence on the coast of Labrador, is the most magnificent of all the felspars. It has an extraordinarily brilliant play of colours combined with an intense metallic lustre ; it thus forms an extremely effective material for many decorative purposes, although the stone in itself is of a dingy grey colour.

Labradorite from the locality mentioned, like oligoclase from Tvedestrand in Norway, the so-called sun-stone, is a lime-soda-felspar, in which, however, there is more lime than soda. The result of an analysis is given below :

		Per cent.
Silica (SiO_2)	. . . . . .	55·59
Alumina (Al_2O_3)	. . . . .	25·41
Ferric oxide (Fe_2O_3)	. . . . .	2·73
Lime (CaO)	. . . . . .	11·40
Soda (Na_2O)	. . . . . .	4·83
Potash (K_2O)	. . . . . .	0·30
		100·26

All lime-soda-felspars having a composition approaching that represented in the above analysis are referred to the mineralogical species labradorite, whether they come from Labrador or elsewhere.

The mineral is found but rarely in regular, well-developed crystals belonging to the triclinic system. More usually it occurs in irregular masses, and this is especially the case with that found on the coast of Labrador. As in all other felspars there are two well-defined cleavages, which in this species make an angle of 94 degrees, one of which is less perfect than the other. The more perfect cleavage surfaces have a distinctly pearly lustre, and moreover exhibit the same striations due to twinning as were described in connection with the sun-stone from Tvedestrand, only in labradorite the striations are usually wider apart and less regular and numerous. A very similar series of straight lines or striations, also due to twinning, is to be seen in labradorite on the less perfect cleavage surface.

The other characters of labradorite are essentially the same as in other felspars. Thus the hardness is 6 ; the mineral is fusible before the blowpipe, but with some difficulty ; the specific gravity is low, though rather higher than that of potash-felspar, being 2·70. Other features which distinguish labradorite from potash-felspar, or orthoclase, are the presence of the twin striations, and the fact that it is moderately easily decomposed by hydrochloric acid with the separation of gelatinous silica.

Labradorite is perfectly opaque, and of a rather dark smoke- or ash-grey colour. Naturally occurring specimens have little lustre, but take a good polish. When a fragment of this mineral is turned about in the hand, so that the angle at which light strikes it is varied, in one particular position the dull smoke-grey surface of the stone will be suddenly lit up with a magnificent display of colours. This happens when rays of light are reflected from the less perfect cleavage face, or from a certain other face, but in the latter case the play of colour is less striking. To get the best possible effect the direct rays of the sun or of a powerful artificial light must fall upon the second cleavage surface at a certain definite angle, and the eye of the observer placed in a suitable position to receive the reflected rays. The stone is cut, not with facets, which destroy its effect, but either perfectly plane, or with a very slightly convex surface, which must be parallel to the reflecting surface. If the cut and polished surface deviates far from this position no play of colour will be seen. It is evident, moreover, from what has been said, that the play of colour is only exhibited by a cut-stone when in a certain position relative to the incident light and the eye of the observer, and that it suddenly disappears with the smallest deviation from this position.

This sudden disappearance and reappearance of brilliant chatoyant colours is specially characteristic of labradorite, and constitutes one of its most striking features. On this account the stone is known to jewellers also as *changeant ;* and the play of colour as change of colour or labradorescence.

The sheen of labradorite is metallic, and at the same time brilliantly and intensely coloured. There is no art by which a reproduction or an imitation of it in any way comparable to the original may be produced. The colours recall those of iridescent objects and of precious opal, but in the latter case each surface shows a number of sharply defined and differently coloured patches, while in labradorite there are large areas presenting one uniform colour. The sheen of this stone may perhaps be best likened to certain other natural objects, such as the wings of tropical butterflies, which display the same kind of metallic colours, though even more lustrous and brilliant. As examples may be cited *Morpho cypris* and *Morpho achilles* of a beautiful blue colour, and *Apatura seraphina* which is green in colour, all being from South America.

The range of colour exhibited by labradorite is wonderfully extensive. There are blues of all tints from pure smalt-blue to violet; greens ranging from the purest emerald tint to blue and yellow; the most brilliant golden yellow, a bright lemon-yellow, a deep and intense orange, which shades off gradually in a strong copper-red or a warm tombac-brown. In some stones the metallic sheen varies in colour as the stone is turned about ; thus in one position it appears of a green colour and in another yellow; such a change is not, however, frequently met with. In Plate XVI., Fig. 2, an attempt has been made to reproduce the magnificent colour effects of labradorite.

Labradorite of different colours differs also in the frequency with which it is found ; thus blue and green are most commonly, and yellow and red most rarely, seen. It is not unusual for the reflecting surface of a specimen to show a sheen of one uniform colour over the whole area of that surface, as in labradorite with a blue sheen from Brisbane in Australia, the variety represented in Plate XVI., Fig. 2. More commonly, however, the same surface displays elongated streaks and irregular patches, which differ in colour and in intensity of colour, and are not very sharply defined. An interesting specimen of labradorite, not from Labrador but from Russia, has been described by the Parisian jeweller Caire. According to his description it displays a perfectly recognisable image of Louis XVI.: the head of the finest azure-blue colour stands out from a golden-green background, and is surmounted by a beautiful garnet-red crown with a border of rainbow colours, and a small, silvery, shining plume. The sum of a quarter of a million francs (£10,000) was demanded in 1799 by the owner of this remarkable object.

The metallic sheen of labradorite is not always displayed uninterruptedly over the whole reflecting surface, for there are often areas which show nothing but the dull body-colour of the stone. These areas often take the form of long, narrow strips with straight, parallel margins, and these may alternate with similar strips showing the most beautiful sheen. These uncoloured patches, especially those in the shape of strips, sometimes display a coloured sheen when the stone is moved into another position relative to the incident light and the eye of the observer, while at the same time the portions, which before were coloured, now assume the dull grey body-colour of labradorite. On replacing the stone in its original position these relations are reversed. Stones in which the sheen is interrupted in this way are much less valuable than those which display a uniform sheen over the whole reflecting surface : these latter are rare and always of small size.

The value of labradorite increases with the depth and brilliancy of its coloured sheen. Stones with a dusky sheen are referred to as " bull's-eyes " (*œil-de-bœuf*), and do not fetch very high prices. The value also depends to a certain extent upon the colour of the stone, since some colour-varieties are less common than others. The price of perfectly faultless stones is rather high, but is much less for less perfect examples. The best material is cut for use as gems ; larger pieces of inferior quality are utilised in the manufacture of small objects of various kinds, such as boxes, stick-handles, &c. It is customary, also, to make use of labradorite in the representation of objects with a metallic colour, such as the wings of butterflies in mosaics, and it is used for many other ornamental purposes. At the beginning of the nineteenth century small reliefs representing a mandrill baboon were much in vogue, and these were fashioned out of this stone in such a manner that only the snout and the other parts of the body, which are coloured in the living animal, showed the coloured sheen of the stone.

The chatoyant colours of labradorite have been explained in various ways, and it is possible that the yellows and greens and the blues are due to different causes. In the case of yellow and of green, the chatoyant colours are caused by the inclusion of minute, brownish translucent plates, of what appears to be hæmatite, magnetite, and ilmenite, in the substance of the labradorite. These plates are rhombic, hexagonal, or quite irregular in outline, and under the microscope are seen to be embedded in the felspar in great numbers and all parallel to the surface from which the coloured sheen is reflected. The blue colour, on the other hand, is not connected with the presence of enclosures, for it is sometimes very prominent when enclosures are completely absent. We are probably dealing here with a complicated optical phenomenon connected with the interference of light, for which a complete explanation remains still to be sought.

Labradorite was first discovered by Moravian missionaries among the Esquimaux of the **Labrador** coast towards the end of the eighteenth century. The first stones were brought to Europe in 1775, and specimens from Labrador were presented to the British Museum in 1777 by the Rev. Mr. Latrobe; the largest is a slab measuring 2 feet by 1 foot. The mineral forms one of the constituents of an igneous rock, pebbles and boulders of which are widely distributed in this region. The other constituent is hypersthene, a mineral of the augite group, with a fine copper-red, metallic reflection. The rock is very coarse-grained, and in consequence it is unusual to find pebbles which contain the two constituents side by side ; nearly always they are small and consist wholly of labradorite or of hypersthene. As to the conditions under which the rock occurs *in situ* but little is known. The bay of Nunaengoak, on the northern border of the mainland of Labrador, near Nain, is said to abound in the so-called " labrador-rock." East of the mainland is the small Isle of Paul (Tunnularsoak), which, especially in former times, was often referred to as a locality for labrador-spar, while the principal locality has been stated to be an inland lake west of Nain. Near this place the labradorite-hypersthene rock, which is known to petrologists as norite, is said to occur in a very coarse-grained hornblende-granite, portions of which may be seen attached to specimens preserved in collections. According, however, to other views, this supposed granite is in reality a coarse-grained gneiss, that is to say, a member of the crystalline schists.   It is stated by Mr. G. F. Kunz that labradorite has not been systematically sought for in this region for over a hundred years.

Hitherto we have considered only the characters of labradorite from the coast of Labrador.   The mineral has a very wide distribution as the constituent of many different kinds of rocks, but in by far the greater number of cases the coloured metallic sheen is absent, and the mineral appears of a dull grey or white, not in the least suitable for decorative purposes.   Nevertheless, chatoyant labradorite has from time to time been found at other localities, sometimes in such abundance and fine quality as to appreciably lower the price usually asked for good stones.   At other places the rock containing the labradorite is worked and used for the decoration of buildings or even as a building stone.

In 1781, soon after the first discovery of chatoyant labradorite in Labrador, specimens of similar character were discovered in **Russia** ; and this country has proved to be specially rich in this mineral.   It was found first in the form of boulders at Peterhof near St. Petersburg.   These boulders usually reflect a blue colour, but their play of colour is not equal to that of specimens from Labrador.   Boulders of uncommonly large size, measuring more than 4 feet by 2, were found on the banks of the Paulovka.   Numbers of labradorite boulders occur also at Miolö in Finland.   The mineral was first discovered in Finland in the twenties of the nineteenth century during the recommencement of work at a very old iron mine at Ojamo, in the parish of Lojo, in the neighbourhood of Åbo.   It differs somewhat from that found in Labrador, being very markedly translucent and almost colourless instead of grey.   Moreover, there are more colour reflections, and these are sometimes arranged with more or less regularity ; for example, there may be concentric chatoyant zones around a dark nucleus, the zones differing in colour, but each being uniformly coloured over the whole of its area.

The most important occurrence of chatoyant labradorite in Russia, or probably in any country, is, however, in Volhynia and the country extending to the neighbourhood of **Kiev**. Together with other minerals, especially diallage, it forms a rock known to petrologists as gabbro.   In places where this rock is very coarse-grained, single individuals of labradorite reach a length of 5 inches, but in other places they measure no more than a few lines.   The colour of the mineral is variable, it may be dark grey or green, and the same fragment sometimes displays several shades between pale and dark green.   There is a very fine play of colour over the face on which this phenomenon is usually observed : green, blue, yellow, and red are all to be seen, but the two first predominate ; the yellow usually occurs between stripes of green.

This gabbro with chatoyant labradorite is one of the features of the great south Russian granitic area.   It is not found in loose pebbles or boulders, but forms a great part of the solid rock of this region and is excavated in quarries at various spots.   It is worked, for example, on the banks of the Bystrievka stream, near Kamennoi Brod in the district of Radomysl, and it is this locality which furnished the material for the coloured columns which ornament the Church of the Saviour at Moscow.   Later on discoveries were made west of Kamennoi Brod, at Goroshki, and at many other points in the district of Zitomir, and now the rock has been traced into Government Kherson, where it was discovered in 1867 at Novo-Pavlovsk.

Very fine labradorite with a uniform blue sheen has recently come into the market ostensibly from Brisbane in Queensland.

The mineral is widely distributed throughout the **United States** of North America, and is found in great abundance in Lewis and Essex Counties in the State of New York, both *in situ* and also as boulders in glacial deposits. These boulders in the drift can be traced all the way down to Long Island and New Jersey, and they are so numerous in one of the rivers of Lewis County that it has been named Opalescent River. Large quantities of this labradorite are quarried at Keeseville, Essex County, New York, for monumental and building purposes. The mineral is also to be met with at various places in Pennsylvania, Arkansas, and North Carolina, but in no case is the material obtained from these localities cut as a gem, since labradorite from Labrador is not only more easily obtained but exhibits a finer play of colour, and is susceptible of a higher polish than is the case with material from the United States.

ELÆOLITE.

Elæolite is a variety of the mineral species nepheline. It is a silicate of sodium and aluminium, and crystallises in the hexagonal system. Its hardness lies between $5\frac{1}{2}$ and 6, and its specific gravity varies between 2·58 and 2·64. A characteristic feature is the ease with which it is decomposed by hydrochloric acid. Nepheline is found in nature in two forms, which differ widely in external appearance, but possess in common all the features characteristic of the species. Ordinary or " glassy " nepheline takes the form of colourless, or faintly coloured, crystals or irregular grains; it occurs as a constituent of many of the younger volcanic rocks, drusy cavities in which are often lined with crystals of glassy nepheline. The best crystals are perfectly colourless and transparent, and take the form of hexagonal prisms, which are usually terminated by the basal plane : they are found in the blocks ejected from the old crater of Vesuvius, now represented by Monte Somma. This " glassy " nepheline has none of the characters essential for a gem, and is, therefore, never cut as such.

The other variety of nepheline occurs as a constituent of ancient plutonic rocks, especially in the elæolite-syenites of certain districts; in this case it nearly always takes the form of irregular grains, and seldom occurs in definite crystals. This older nepheline differs essentially from the younger, glassy variety in the character of its lustre, which is distinctly greasy, hence the name elæolite (German, *Fettstein*). Moreover, instead of being transparent and colourless, it is cloudy, or, at most, only translucent, and of a colour ranging from bright bluish-green or brown to tile-red. On mere inspection we would assume elæolite to be a mineral perfectly distinct from nepheline; since, however, the two minerals are in complete agreement in such characters as crystalline form, chemical composition, specific gravity, &c., they must be included in the same mineral species.

The contrast between the external appearance of elæolite and that of glassy nepheline depends upon the presence in the former of numerous enclosures of microscopically small crystals, some of which may be augite or hornblende, while others belong to other mineral species. Both the colour and the greasy lustre of elæolite is due to these enclosures, the presence of which also gives rise to a wave of soft, milky light reflected with special distinctness from stones cut with a convex, polished surface. The broad band of light crossing the curved surface is very similar to that seen in cymophane and in cat's-eye, and travels over the surface as the stone is moved about before the observer. Stones which combine this appearance with an intense and pure body-colour are often very effective and comparatively valuable, since, though the mineral itself is abundant, specimens of this description are rare and seldom appear in the market. Usually it is stones of a green colour which are cut as gems, red stones being scarcely ever used for this purpose.

In external appearance the peculiar specimens of elæolite just described are apt to be confused with cymophane and cat's-eye. Both, however, are harder than elæolite, which can be scratched by quartz, while these cannot. Moreover, they are both heavier and sink in liquid No. 4 (sp. gr. = 2·65), in which nepheline floats.

Elæolite has been longest known from the south of Norway. It occurs in elæolite-syenite in pieces ranging in size up to that of a man's fist at several places in this region, among which may be mentioned Laurvik and Fredriksvärn, the material found at the former place being for the most part brown and green and that at the latter being red. Large fragments of red and of green elæolite are also found in a similar rock on the eastern shore of Lake Ilmen, near the smelting-works at Miask in the Ural Mountains. Greenland is another locality for this mineral; while the principal locality in the United States is Magnet Cove in Arkansas, where fine flesh-red, cinnamon-brown, and yellowish-brown elæolite of gem-quality occurs in abundance. At Gardiner and Litchfield, in Maine, elæolite of a fine green colour is found; and Salem, in Massachusetts, may be mentioned as another locality. In all cases the mineral occurs as a constituent of a rock similar to the one in which it is found in Norway and the Urals.

CANCRINITE.

The yellow cancrinite, which occurs in association with the elæolite of Litchfield, Maine, is sometimes cut and worn as a gem on account of its pretty colour. It is composed of the same chemical elements as elæolite, but contains in addition carbonic acid and water. It crystallises in the same hexagonal forms as nepheline. It is never perfectly transparent, and at best is only strongly translucent; its colour ranges from pale yellow to dark orange-yellow. The same mineral is also found at other localities; and further is not always of a yellow colour, but may be rose-red, green, &c. It is used as a gem only in the United States, and even there but rarely.

LAPIS-LAZULI.

Lapis-lazuli, or azure stone, also known as oriental lapis-lazuli, is an opaque mineral, usually of a magnificent blue colour.   It occurs in nature in extremely fine-grained to compact masses with an uneven fracture.   The crystals never exceed the size of a pea or bean, and are extremely rare ; they have the form of the rhombic dodecahedron, and belong to the cubic system.

The beauty of this stone depends entirely upon its colour.   In the finest and best specimens it is a dark azure-blue shading off to blackish-blue.   Plate XX., Fig. 1, represents a stone of a fine azure-blue with a slight tinge of black.   Deep blue specimens only are used as gems, and their colour is far more intense and beautiful than is that of any other opaque blue stone ; it is always deeper than that of turquoise, and on this account the two stones may be readily distinguished.   Very pale blue or almost colourless lapis-lazuli is also met with not infrequently ; such specimens might possibly be confused with turquoise were it not that they are so seldom used as gems.   As in similar cases stones of a pale colour are described as being " feminine," and those of a deep blue as " masculine."   The colour may be distributed with such perfect regularity that the stone is of one uniform tint throughout, but more frequently there are bands and patches which are white or some shade of blue differing from that of the mass of the stone.   Moreover, the blue colour is often flecked with yellow, shining, metallic specks of iron-pyrites, which are often supposed to be gold by the uninformed.   On the decomposition of the iron-pyrites the yellow specks are replaced by patches of rusty brown, which much disfigure the stone.   The precious stone known to the ancients as sapphire was probably the blue lapis-lazuli, and not the blue variety of corundum which now bears that name.

Lapis-lazuli, showing shades of colour other than the pure blue, is by no means uncommon.   At certain localities there frequently occurs a blue variety slightly tinged with green, but stones of a pure green are very rare, and the same is true for violet and reddish-violet stones, which also appear to be confined to certain localities.   It is unusual for a stone to exhibit more than one colour, but blue, green, and red or violet are occasionally seen in the same specimen.   When reduced to powder the mineral is always of the same tone of colour, but lighter in shade than it was before.

The colour of lapis-lazuli is not perfectly stable under all conditions ; it is altered, for example, by the action of heat.   If a pale blue stone be raised to a dull red-heat it frequently assumes a fine dark blue colour, the specimen which was comparatively worthless before is now fit for use as a gem, and has a considerable value.   In other cases the pure pale blue or dark green is transformed into a greenish-blue, which is not much admired ; and if the temperature is raised too high the stone is completely decolourised.   When green and violet stones are raised to a dull red-heat they frequently behave like the pale blue stones, and assume a deep blue colour, which in this case also greatly enhances their value.   A greenish-blue lapis-lazuli from Chile loses its colour on heating, but regains it during the process of cooling.

The finely granular structure of lapis-lazuli is apparent even to the naked eye.   In the mass the mineral shows no trace of cleavage, and the fracture is sub-conchoidal to uneven.   The lustre on a freshly fractured surface is of the vitreous type ; it is usually feeble, and though stronger in the material from certain localities, in other cases is quite dull. Even when cut and polished there is nothing approaching a brilliant lustre in this stone, and the more impure and patchy the specimen the more feeble will be its lustre. The lustre acquired by polishing is soon lost when the stone is in use, owing to its low degree of hardness, and it then becomes dull and less pleasing to the eye. Lapis-lazuli may practically be described as a perfectly opaque mineral; only in the thinnest of splinters is it somewhat translucent.

The hardness of lapis-lazuli is rather low, being about $5\frac{1}{2}$ on the scale; the mineral is thus readily scratched by quartz and is even scratched by felspar, though it is itself still capable of scratching window-glass. According to various determinations, the specific gravity of the mineral lies between 2·38 and 2·42; it will, therefore, float in the lightest of the test liquids, the specific gravity of which is 2·65, the same as that of quartz. It is perhaps the lightest of any of the minerals used as gems.

The mineral is decomposed by hydrochloric acid, and the white material, with which the blue is frequently intermixed, dissolves in the acid with effervescence, thus proving it to be calcite. At the same time the blue colour gradually disappears, and the whole process is accompanied by the evolution of a strong smell of hydrogen sulphide, a smell like that of rotten eggs. When heated before the blowpipe the mineral fuses with difficulty to a colourless and rather clear, bubbly glass.

When examined with a lens, or even with the naked eye, it is quite obvious that lapis-lazuli is not a homogeneous mineral, like diamond, ruby, and other precious stones, but a mixture of several substances; and this is demonstrated still more conclusively by chemical investigation, and by the examination of thin sections under the microscope.

Analyses show that though all specimens of lapis-lazuli contain the same chemical elements, yet these are not always present in the same proportion, and their relative amounts may vary between wide limits. Thus the percentage of silica, an important constituent and one which is always present, varies between 43 and 67, and there is just as much variation in the case of other constituents. The following analysis is that of a stone from the "Orient," 28·2 per cent. of calcium carbonate and 4·5 per cent. of magnesium carbonate having been deducted:

	Per cent.
Silica (SiO_2)	43·26
Alumina (Al_2O_3)	22·22
Ferric oxide (Fe_2O_3)	4·20
Lime (CaO)	14·73
Soda (Na_2O)	8·76
Sulphuric anhydride (SO_3)	5·67
Sulphur (S)	3·16
	100·00

A small amount of chlorine, up to about a half per cent., is also present in some specimens.

The existence of such wide variations in the chemical composition of lapis-lazuli leads one to suppose that the mineral is a mixture of different substances. The appearance of thin sections under the microscope confirms this assumption, for the single mineral constituents and their relation to each other are then distinctly observable. The ground-mass is usually white calcite or limestone of finely granular structure, and in this all the other minerals are embedded. The presence of calcite is the cause of the white patches and streaks in the mineral, and also of its effervescence, when placed in hydrochloric acid. It may be present in large or in small amount and is sometimes dolomitic ; that is to say, it contains magnesia, as is shown by the above analysis.

Embedded in the calcite are to be seen numerous grains of minerals of various kinds. A considerable number of these are quite or almost colourless, and consist of augite and of hornblende.   The remaining grains are constituted of the true lapis-lazuli substance ; they impart to the mineral its colour, and to a certain extent other of its characteristic features. At times they replace all other constituents, so that the whole mass is made up of them almost entirely, while at other times they are distributed singly through the calcite.   If these grains are present in large numbers the colour of the stone is deep and full, and according as they are distributed regularly or irregularly through the mass the latter is uniformly coloured, patchy, or streaky.   The colour of the stone as a whole depends upon the colour of these grains, and these may be blue, green, or violet, while the blue grains vary in shade from a deep, intense blue to one which is almost colourless.   In outline these coloured inclusions are nearly always rounded, angular, or ragged with numerous indentations and projections.   Sometimes, however, they take a regular crystalline form identical with that of the larger crystals mentioned above, that is to say, the form of a rhombic dodecahedron ; such crystals, therefore, also belong to the cubic system.   Their reference to the cubic system is supported by the fact that the majority of the grains are singly refracting ; a few indeed are doubly refracting, but these are no doubt cases of anomalous double refraction such as is often observed in cubic minerals.   Indications of cleavage parallel to the faces of the rhombic dodecahedron are also sometimes to be observed.   Not infrequently a certain number of the small blue grains are aggregated together in circular groups.

These constituents of lapis-lazuli, to which the colour of the mineral is due, do not appear to be all of the same character.   The two Swedish mineralogists, Bäckström and Brögger, during the course of an important investigation into the composition of lapis-lazuli, succeeded in isolating those pigment granules which differ in character.   It was found that one kind of these granules has the composition of the blue mineral haüynite, which will be considered later, since it is sometimes used as a gem.   A second kind has the composition of an artificial substance much used as a pigment and known as ultramarine.   Lapis-lazuli may, therefore, be considered to contain natural ultramarine, and, indeed, before the introduction of artificial ultramarine this pigment was derived exclusively from lapis-lazuli, and was naturally very expensive.   There are also, sometimes, a few granules of still another kind, differing again in chemical composition, but the two mentioned above are the most important.   Their composition is given by the following formulæ :

$$\text{Haüynite, } 3(Na_2,Ca)O.3Al_2O_3.6SiO_2.2(Na_2,Ca)SO_4.$$
$$\text{Ultramarine, } 3Na_2O.3Al_2O_3.6SiO_2.2Na_2S_3.$$

Each thus contains a molecule of sodium and aluminium silicate, $Na_2O.Al_2O_3 2SiO_2$, which in haüynite has some of the sodium replaced by an equivalent amount of calcium ; in addition to this a certain amount of sodium sulphate enters into the composition of haüynite, and in ultramarine some sodium sulphide.

From the analysis given above it can be calculated that in that particular specimen there was present 76·9 per cent. of haüynite, 15·7 per cent. of ultramarine, and 7·4 per cent. of other blue grains with the chemical composition of the mineral sodalite.   This mineral, like haüynite and ultramarine, contains sodium and aluminium silicate, but the sodium sulphate of the former mineral and the sodium sulphide of the latter is replaced in sodalite by sodium chloride.   The proportions in which these three pigments are present are very variable ;

haüynite is always present in the largest amount, ultramarine in small amount, or sometimes completely absent, while sodalite is present in still smaller proportion. The colouring properties of all three substances appear to be equal, so that the appearance of the stone as a whole is not affected by the proportions in which they chance to be present.

Each of these three silicates is decomposed by hydrochloric acid with the separation of gelatinous silica. The sodium sulphide contained in ultramarine is decomposed by hydrochloric acid with the evolution of hydrogen sulphide, a gas which has the smell characteristic of rotten eggs. When, therefore, lapis-lazuli is treated with hydrochloric acid the presence or absence of this smell indicates the presence or absence of ultramarine. The fact that these three silicates are all decomposed by hydrochloric acid accounts for the decolourisation of lapis-lazuli when subjected to the action of this acid.

It should be clear from the foregoing that lapis-lazuli, as it is brought into the market and applied to various ornamental purposes, is a limestone more or less richly impregnated with the pigments mentioned above.

It is probable that these, as well as the other mineral enclosures (augite, hornblende, &c.) contained in lapis-lazuli, have been formed by the action of an igneous magma, such, for example, as granite, on lime-stone. Lapis-lazuli is thus what is spoken of as a contact-product, as is the case with certain other minerals already described.

We must now consider the distribution of lapis-lazuli. The occurrence at many of the places stated to be localities for this mineral is by no means well au-thenticated. In some cases the regions mentioned as localities have

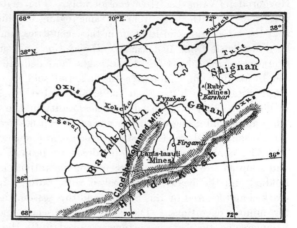

FIG. 82. Occurrence of lapis-lazuli in Badakshan.
(Scale, 1 : 6.000,000.)

never been thoroughly explored owing to their remoteness and inaccessibility, and in others they are merely places where the stone is marketed. Nevertheless, there are localities at which the occurrence of lapis-lazuli is well authenticated, and of which details concerning the conditions under which it occurs and the manner in which it is collected are known. The richest and most important localities are situated in Asia, and it is from this source that stones of the finest quality are derived. There is an occurrence of less importance in Chile, South America, and another, still less important, in the neighbourhood of Rome and Naples.

The longest known lapis-lazuli mines are situated in **Badakshan** in the north-east corner of Afghanistan on the upper reaches of the Amu Darja (river Oxus). The Central Asian occurrence of ruby and spinel already mentioned is in the same neighbourhood, and both localities were visited and described as far back as 1271 by the celebrated Venetian traveller, Marco Polo. The mode of occurrence of the mineral and the conditions under which it is collected have been described later by other investigators.

These ancient mines, which are still being worked, are situated in the upper part of the valley of the Kokcha (Fig. 82), a tributary on the left bank of the Oxus. The place lies to the north of the Hindu Kush, between these mountains and the Chodsha-Muhamed range.

in latitude about $36\frac{1}{2}°$ N. and longitude $70\frac{1}{2}°$ E. of Greenwich. There are probably other deposits of lapis-lazuli in this inaccessible region, especially in the Hindu Kush, but that in the upper Kokcha valley is the only one definitely known.

At the part where the mines are situated the valley is only about 200 yards wide and is shut in on both sides by precipitous walls of bare rock. The mines lie about 1500 feet above the bed of the river in a white and black limestone, the mother-rock of the lapis-lazuli. Three varieties of the mineral are distinguished: one of an indigo-blue colour, a second of light blue, and a third which is green in colour. The annual production amounts at the present time to about 36 poods, that is to say, about 5000 kilograms.

The greater part of the lapis-lazuli mined here, and specially the stones of finest quality, is sent to Bukhara, from thence they are sent to Russia, and being brought into the market at the fairs of Nizhniy Novgorod are distributed by the merchants assembled there to all parts of the world. By the time the mineral has reached the market of Nizhniy Novgorod its value has risen considerably. The material which is not sent to Bukhara goes, together with rubies from the same region, to China and to Persia, and it is probable that the lapis-lazuli said to occur in these countries, and in Little Bucharia and Tibet, has been imported from Badakshan. Descriptions of such occurrences to be found in the literature of the subject are always vague, and definite statements as to localities, &c., are sought for in vain. Moreover, the material sold in other parts of Asia, for example, in Afghanistan, Beluchistan, and India, and stated by travellers to occur in those regions, in all probability is imported from the locality in the neighbourhood of the Upper Oxus. The lapis-lazuli from which the ancient Egyptian scarabs were cut, probably came from Badakshan, as did also the material much used elsewhere in ancient times.

The mining methods used hundreds or, it may be said, thousands of years ago are still adopted. At the spot where mining operations are to be undertaken large fires are kindled and the heated rocks soaked with water. In winter the rocks are more easily cracked and fissured by this process, and this season, therefore, is considered most favourable for mining operations. The rock thus loosened and cracked is broken up with large hammers and the barren portions removed until the nests of lapis-lazuli are met with. Around each nest a deep groove is excavated, and the whole mass is then prized out with a crowbar. There are sometimes split off these masses, in a direction parallel to the bedding planes of the rock, large slabs equal to several Taurian maunds in weight, a maund being 30 or 40 pounds or more. It has been suggested that the deep blue colour characteristic of lapis-lazuli from Badakshan is due to heating by the fires employed for breaking up the mother-rock. This, however, is probably not the case, since it appears that in former times pale blue lapis-lazuli, which, it will be remembered, is sometimes changed into dark blue by the action of heat, was found side by side with the dark blue variety. Not infrequently the material from this locality is flecked with yellow iron-pyrites, while at other times this mineral is aggregated in large nests and bands.

Another group of lapis-lazuli mines is situated at the western end of **Lake Baikal** in Siberia (Fig. 83). Deposits are known on each of the streams Talaya, Malaya Bistraya, and Sludianka. According to Laxmann, the lapis-lazuli occurs here in a white granular limestone along the line of contact of this rock with granite. These details have been confirmed by later observers, and the mode of occurrence of lapis-lazuli at this locality is thus quite in accordance with what has been said above regarding the origin of this mineral. The material found in the neighbourhood of Lake Baikal is, in many cases, inferior in quality to most of that which comes from Badakshan, and it also contains much less iron-pyrites. Besides dark blue stones, violet, dark green, and pale red specimens are met with, and occasionally one which shows a bright red centre surrounded by a dull, dark blue border. The latter case would suggest that dark blue lapis-lazuli is derived by the weathering of red or violet material, and if this be so, the dark blue stone of Badakshan must be regarded as having undergone more alteration than has the violet and red material of Lake Baikal.

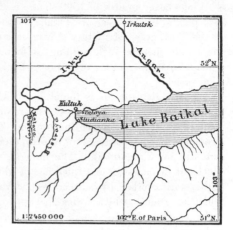

FIG. 83. Occurrence of lapis-lazuli in the neighbourhood of Lake Baikal. (Scale, 1 : 2,450,000.)

The output of these mines is very uncertain, and the working of the deposit is never a very profitable undertaking. Moreover, in new ground there is nothing to indicate the presence or absence of the mineral, and the choice of a favourable spot for fresh excavations is entirely fortuitous.

In the narrow valley of the small stream Talaya, which has a length of thirty versts (twenty miles) and flows into Lake Baikal two versts to the south of Kultuk. there are on the left side steep cliffs of white dolomitic lime-stone which in places are overlain by granitic rocks. In the limestone are veins filled with a marly limestone containing scales of mica, blocks of compact limestone, and nodules and fragments of lapis-lazuli (Fig. 84). Up to the year 1853 three mines were being worked in this deposit. The material obtained was poor in quality, however, and after the discovery of the richer deposits on the Malaya Bistraya the mines were abandoned.

The mines on the Sludianka stream are situated twelve versts south of the village of Kultuk ; they were worked for a very long period, were abandoned for a time on account of the poor quality of the lapis-lazuli and the hardness of the rock in which it was enclosed, but later were reopened. Here also the mineral is found embedded in white marble at the junction of that rock with granite and gneiss, the latter of which is frequently associated here with marble. Pebbles of lapis-lazuli are also found in the bed of the stream throughout its entire course of thirty-

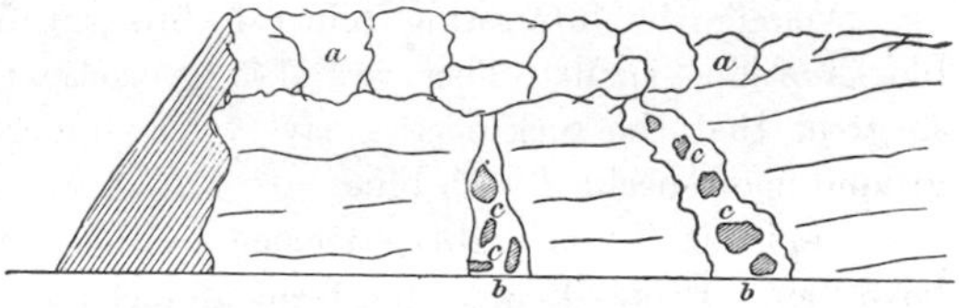

FIG. 84. Occurrence of lapis-lazuli on the Talaya River, Lake Baikal. (a, Granite ; b, Veins in limestone ; c, Nodules of lapis-lazuli.)

five versts. The material found in this region has a great range both of colour and quality ; the best is of the deepest and most beautiful ultramarine-blue ; the least valuable is pale and cloudy ; while the medium qualities sometimes show beautiful transition tints between blue and violet or celadon-green.

The richest deposits are those discovered in 1854 on the Malaya Bistraya (i.e. Little Bistraya) stream. All the best stones from this region have come from this particular spot, and the material found there as a whole is wonderfully uniform in quality. The mines are situated on the left bank of the stream, ten versts above its mouth, and no others in the neighbourhood have been worked for a long period. In this region granitic rocks predominate, and in the mountain ridge, which forms the right side of the valley, there are almost vertical beds of white, granular, dolomitic limestone, which has been altered by the granite on which it rests ; the lapis-lazuli occurs in the loose material which fills up the crevices and veins in this limestone.   The greater the depth of the excavation in this deposit the more abundant and finer in quality does the lapis-lazuli appear to be.   The workings cover an area of 7000 square feet, and, as nodules of lapis-lazuli, weighing as much as 3 poods (108 pounds) are found in the Malaya Bistraya, and pebbles of the same kind in the Turluntay, a tributary stream, it is probable that the deposit is still more extensive. Here and there a little sulphur is found in association with the mineral, and on the Malaya Bistraya stones of various shades of dark green are found in addition to the blue variety.

Lastly, lapis-lazuli is found in abundance in the **Chilian Andes**.   According to Philippi, the locality is in the Cordillera of Ovalle, only a few cuadras away from the main road leading to the Argentine provinces; it lies at the sources of the Cazadores and Vias, two small tributary streams of the Rio Grande, at only a little distance from the watershed, but still on Chilian soil.   The mineral, which is associated with iron-pyrites, occurs in blocks of various sizes in a thick bed of white and grey limestone, which rests on clay-slates and is itself overlain by another bedded rock rich in iron-ore and garnet.   This latter bed underlies granite, which forms the upper part of the mountain.   The weathering of these rocks has given rise to a small plain of secondary deposits consisting of pebbles of granite, slate, and iron-ore, and among these are found a few loose fragments of lapis-lazuli.   The Chilian lapis-lazuli is mined in some quantity; but it is pale blue in colour, often tinged with green and disfigured by white patches, so that it is much less valuable than material from the Asiatic deposits.

Small quantities of earthy lapis-lazuli are found in the ejected blocks of crystalline limestone of Monte Somma, the old crater of Vesuvius; also in the blocks of limestone included in the volcanic tuff of the Albanian Hills, near Rome.   Since the material is unsuitable for use as a gem these occurrences require no further comment beyond stating that here also, as in all other cases, the mineral is a product of the contact-metamorphism of limestone.

Among the interesting minerals brought from the ruby-earths of Burma are great blocks of lapis-lazuli.   These are of two varieties: in one the quantity of the blue mineral is so great that the rock-masses have a deep indigo tint, while in the other there is a white ground-mass speckled with blue.

Lapis-lazuli was highly esteemed by the ancients and was often engraved and cut in bas-relief.   It is still held in esteem and is used for ring-stones, brooches, and for similar purposes, not only in Western countries but also in the Orient and China.   Since the beauty of the stone depends upon its blue colour, portions for cutting are chosen with a view to obtaining stones as uniformly coloured as possible.   The stone is cut with a plane or slightly curved surface, facets being quite ineffective on account of the complete opacity of the mineral.

Lapis-lazuli is not now as valuable as it once was.   The price of a stone depends on its size, and on the purity, uniformity, and depth of its colour; a stone with none of these qualities is almost worthless.   Pure, azure-blue lapis-lazuli is the most valuable; the presence of white specks and patches detracts considerably from its value, but less so if they are distributed quite uniformly and regularly.   Pale blue and greenish-blue lapis-lazuli is worth but very little.

The rough material used for cutting is usually in small fragments, which are sold by the kilogram.   Pieces of the size of a nut are common; if of the best quality they are worth £15 per kilogram.   Larger pieces of equally fine quality will fetch in the European markets as much as £25 or £30 per kilogram.

Lapis-lazuli is used much more frequently for small articles, such as letter-weights, candlesticks, bowls, vases, and fancy articles of various kinds, than it is as a gem. Such articles were formerly very expensive, since each was cut out of a single piece of material, and rough blocks of good quality and sufficient size for the purpose are rare. Now such objects are made of metal and veneered with thin plates of lapis-lazuli. The mineral is used also as a decorative material in very ornate buildings, such as the Winter Palace at St. Petersburg and the Castle of Tsarkoe-Selo, in which there are rooms which are wainscotted with lapis-lazuli. The stone is also utilised in mosaics and for inlaying, the yellow, shining specks of iron-pyrites being made to represent stars in a blue sky. Pliny compared the stone to the star-bedecked firmament; and the shining metallic flecks of iron-pyrites do contrast in fact very effectively with the dark blue background of the stone itself.

Formerly lapis-lazuli had a very important technical application as a pigment. Dark-coloured fragments of the mineral were powdered up, the blue constituent separated as far as possible from the colourless portions, and the blue powder thus obtained worked up into a paint known to artists as ultramarine. This was the only fine blue pigment known, and was comparatively costly. It is now replaced by an artificial substance which very closely resembles the ultramarine of nature in colour, chemical composition, and other characters, and which is much lower in price.

There are certain opaque blue stones and artificial substances which it is possible to mistake for lapis-lazuli, and which are occasionally passed off as such. Very close imitations of lapis-lazuli can be made in glass, but the colour is less intense; the specific gravity is higher, and on the smallest broken surface the bright conchoidal fracture of glass can always be seen, as distinct from the dull uneven fracture of true lapis-lazuli. Again, agate is sometimes artificially coloured and sold in the trade as lapis-lazuli; the colour imparted to such stones is always a dark Berlin-blue and not the deep azure-blue of the genuine stone. Moreover, both the hardness and the specific gravity of agate are greater than those of lapis-lazuli. The blue mineral chessylite or azurite, a hydrated basic copper carbonate, is sometimes substituted for lapis-lazuli; this is softer and much heavier (sp. gr. = 3·8) than the latter, and is readily and completely dissolved with effervescence in hydrochloric acid. Turquoise is too light in colour to be mistaken for lapis-lazuli, and the blue mineral lazulite does not resemble it closely enough to admit of the one being substituted for the other.

HAÜYNITE.

This mineral has already been mentioned as one of the coloured constituents of lapis-lazuli. It occurs also in small irregular grains in certain volcanic rocks, and, far more rarely, as regular crystals which have the form of the rhombic dodecahedron. The principal localities are: the neighbourhood of the Laacher See, near Andernach on the Rhine (Niedermendig, &c.); the Albanian Hills, near Rome (San Marino, &c.); and the French Auvergne. Pure haüynite is sometimes of a beautiful blue colour, and almost perfectly transparent; when this is the case it may be cut as a gem, and then commands a fair price. It is said to be worn to a certain extent in France, but it has no importance whatever in the trade.

Haüynite, whether in crystals or in grains, has a definite cleavage parallel to the faces of the rhombic dodecahedron. It is translucent to transparent and singly refracting; its hardness is 5½ and its specific gravity 2·4. These characters serve to distinguish it from all other blue stones.

SODALITE.

This is another blue mineral belonging to the same group. Like haüynite, it occurs as grains enclosed in lapis-lazuli, and has a limited use as a gem. It occurs also as larger grains and as crystals belonging to the cubic system. It is usually dull or quite colourless, but specimens are sometimes met with of a pronounced blue colour and closely resembling lapis-lazuli in appearance. The material most frequently cut for gems is that found in a syenitic rock, loose blocks of which lie about on the surface of the ground at Litchfield, Maine, in the United States, where the stone is worn on account of its national origin. Similar material is yielded by rocks of the same kind at Ditro, in Transylvania, and at Miask, on Lake Ilmen, in the Ural Mountains. Fine large masses also occur on the Ice river, in British Columbia, and at Dungannon, in Hastings County, Ontario. Sodalite was used as an ornamental stone by the ancient inhabitants of the Bolivian tableland, beads of blue sodalite and of fluor-spar, together with arrow-heads of quartz and obsidian, having been found by the traveller, Alfons Stübel, in the ruins of Tiahuanaco, on Lake Titicaca, one of the most ancient cities of South American civilisation. Sodalite is not known to occur in the region, so that the material for these objects must have been brought from some locality unknown.

OBSIDIAN.

Obsidian is not a simple mineral, but a glassy lava or volcanic glass, belonging to the rhyolite group of rocks. It is sometimes worked for ornamental purposes, and is known to lapidaries by several names, among which are lava, black lava-glass, volcanic glass, and " glass-agate."

Obsidian, like artificial glass, is perfectly amorphous, and, therefore, optically singly refracting. It has a perfect conchoidal fracture, such as is seen in glass, and exhibits the vitreous lustre characteristic of that substance, though the lustre may sometimes incline to the greasy type. Typical obsidian is thus wonderfully similar to ordinary glass, and differs from it markedly only in colour and transparency. Obsidian may be black, grey, brown, yellow, red, green, and sometimes blue, but is always deeply coloured, and, because of this, almost perfectly opaque. Very thin splinters alone are transparent and at the same time colourless, or nearly so. Obsidian may be of one uniform colour, or it may be patchy and streaked with various colours. Thus a variety from North America, known as " mountain mahogany," is streaked with brown and grey, and when cut shows a grain like that of mahogany. Obsidian of a uniform black colour is more important and more widely distributed ; this variety, when perfectly homogeneous and uniformly coloured, is admired for the silkiness of its appearance, and is cut for ornamental purposes.

Very frequently obsidian is not uniformly coloured, nor even apparently homogeneous in structure, but contains embedded in it crystals, sometimes of appreciable size ; such specimens are useless for cutting. Those which appear to be homogeneous are not so in reality. When examined under the microscope they are seen to contain large numbers of minute, spherical, or elongated cavities (so-called vapour pores), minute crystals of all kinds, and other enclosures. These are too small to affect the beauty of the stone in the mass, but they sometimes give rise to the greasy lustre already mentioned, and in a few cases to a peculiarly beautiful, reddish, silver-white, greenish-yellow or golden-green sheen, which shows up very effectively against the dark body-colour of the stone. Obsidian showing a well-marked sheen is much valued, and is worth much more than the ordinary kind.

In the remaining characters, obsidian preserves its resemblance to artificial glass. Thus its specific gravity is low; it varies between 2·3 and 2·5, rarely reaching or exceeding 2·6. In correlation with the fact that the mineral is of the nature of a lava we find the chemical composition different for each occurrence. In every case, however, it contains silicates of the alkalies and of aluminium, and of these alkalies a small amount of potash is never absent. The amount of silica present varies between 60 and 80 per cent., and the same wide variation in amount is shown by the other constituents. These are the same as in ordinary glass, but are present in obsidian in different proportions. The following is an analysis by Abich of a fine black obsidian from the island of Lipari :

		Per cent.
Silica (SiO_2)		74·05
Alumina (Al_2O_3)		12·97
Ferric oxide (Fe_2O_3)		2·73
Lime (CaO)		0·12
Magnesia (MgO)		0·28
Potash (K_2O)		5·11
Soda (Na_2O)		3·88
Chlorine (Cl)		0·31
Loss on ignition (water)		0·22
		99·67

Again, obsidian is very brittle and breaks easily into sharp, angular pieces. On this account care should be exercised both in the wearing of obsidian ornaments and also in the cutting of the mineral. Its hardness is 5 to $5\frac{1}{2}$, the same as that of window-glass ; obsidian is therefore scratched by felspar, and still more easily by quartz, while it is itself scarcely capable of scratching window-glass. Acids, with the exception of hydrofluoric, have no action on this substance. It fuses before the blowpipe with intumescence, and then solidifies to a grey, porous mass.

Obsidian takes a high polish, and is worked as an ornamental stone in a variety of ways. The kind which shows a coloured sheen is cut en cabochon, the sheen being displayed to best advantage by this form of cutting. The ordinary black variety is used for mourning jewellery of all kinds, brooches, sleeve-links, necklaces, bracelets, &c., but it is not often used for ring-stones. It is usually cut with a plane or slightly convex surface, rarely with a pronounced convexity. Faceted stones in the form of rosettes are frequently seen, as are also faceted or spherical beads of obsidian.

Obsidian was at one time much more extensively used than it is now; at the present time it is replaced by artificial glasses, which in depth and uniformity of colour and silkiness of lustre are more than comparable with the naturally occurring glass. In the case of cut stones it is impossible to decide at once whether the material of which they consist is an artificial or a natural product. There is no difficulty, however, in distinguishing between these natural or artificial glasses and jet, another black mineral often used in mourning jewellery. As we shall see later, jet is a variety of coal, and, like all organic substances, is a bad conductor of heat. It therefore feels warm to the touch, while obsidian and glass, being better conductors, feel cold. Other black stones which may come into the market, black tourmaline and black spinel, for example, are heavier than obsidian, and sink in liquid No. 4 (sp. gr. = 2·65), in which obsidian floats. Black hæmatite is not infrequently used as a ring-stone, but its lustre is more distinctly metallic, and on unglazed porcelain it gives a red streak, while that of obsidian is colourless.

Obsidian of various kinds is widely distributed. It is found in some districts in extensive rock-masses built up of irregular, angular, or rounded blocks. A comparatively large proportion of these rock-masses is sufficiently pure and homogeneous to be cut for ornamental purposes. The rough material being thus very abundant, it is not surprising to find that the price of a cut stone is very little in excess of the sum paid for the labour of cutting.

The distribution of obsidian is so general that it would be impossible to mention all the localities for this rock by name; a few only of the most important will now be dealt with.

In Europe the island of Lipari is a locality at which fine obsidian occurs in abundance. A lava stream of obsidian, ranging in thickness up to 100 feet and having a breadth of an eighth of a mile, stretches from Monte Campo Bianco to the sea at Capo Castagno. The material also occurs in great abundance in the neighbouring island of Vulcano. On the island of Ponza there are dykes of black obsidian penetrating the volcanic tuffs. It is abundant in Hungary ; and in Iceland there is so much fine material suitable for cutting that obsidian is often referred to by lapidaries as " Icelandic agate."

The country in which in former times obsidian was most extensively used for all purposes is Mexico. Arrow-heads, spear-heads, knives, and other tools and weapons were fashioned out of obsidian by the ancient Mexicans, and, indeed, by some of the native Indians at the present day. They had learned the art of striking off a long, thin splinter of obsidian, with an edge so fine and sharp that it could be used as a knife, or even as a razor. Discoveries in the dwelling- and burial-places of these ancient people have shown that obsidian was fashioned into mirrors, masks, and all kinds of personal ornaments. The distribution of the mineral is general throughout Mexico, and extends northwards and southwards beyond the borders of that country. The ancient Mexicans appear to have derived most of their rough material from one particular spot, the so-called Cerro de las Navajas (Hill of Knives), first exactly described by Alexander von Humboldt. It is situated near Real del Monte, in the State of Hidalgo, north of the city of Mexico and near the town of Atotonilco. The ancient mines, which were worked long before the conquest of Mexico by the Spaniards, are still to be seen here, the marks of an ancient civilisation. The material found here shows a variety of colours, and specimens with a very fine sheen are met with, but the greater part of it is the ordinary black variety.

Among the localities in the United States at which fine obsidian occurs, may be mentioned Silver Peak, in the State of Nevada, and Obsidian Cliff, in the Yellowstone National Park. The brown and grey streaked " mountain mahogany " is found, together with other kinds of obsidian, along the Pitt river in California, and there are many other localities in this country which yield material suitable for cutting.

A locality of some importance in Asia is the Caucasus. Among the obsidian which is mined here is some of the variety which exhibits a coloured sheen. The material found at Mount Ararat has a very rich sheen, and is worked at Tiflis into personal ornaments, vases, bowls, and other large objects. The balls of obsidian found in the Marekanka river at Okhotsk in eastern Siberia are known to petrologists as marekanite, and are sometimes utilised for various ornamental purposes. Each ball may be uniformly coloured brown, grey, yellow, or red, or it may exhibit a number of colours. Similar material is found at other places, for example, in Mexico.

MOLDAVITE.

Moldavite is a glassy substance, the origin of which has not yet been definitely determined. It is known as bottle-stone, or as pseudo-chrysolite, on account of its resemblance to green bottle-glass and to green olivine (chrysolite), the resemblance being so close that faceted specimens of moldavite can only be distinguished from the substances mentioned by careful examination. To lapidaries the mineral is usually known as water-chrysolite.

Moldavite, like obsidian, has the chemical composition and physical characters of a glass. It can only be distinguished from obsidian with the naked eye by its perfect transparency and its green colour. The colour is never very deep, varying between leek-green and olive-green; specimens of a light brown colour are sometimes met with, but are rarely cut as gems. The mineral is amorphous and therefore optically singly refracting and not dichroic; it is brittle, breaks into sharp angular fragments, possesses a perfect conchoidal fracture and a strong vitreous lustre, all of which features it has in common with obsidian. Its hardness is about $5\frac{1}{2}$ and its specific gravity is 2·36, rather less than is usually the case with obsidian.

Although, externally, moldavite so closely resembles a piece of green bottle-glass or of transparent green obsidian, internally there are well-defined differences. Under the microscope moldavite is seen to contain vast numbers of minute air bubbles such as are seen neither in glass nor in obsidian, and the microscopic crystals always present in obsidian are absent in moldavite. Moreover, chemical analysis shows that the composition of moldavite is variable like obsidian, but that the substance contains more silica than is present either in artificial glass or in obsidian. Furthermore, moldavite contains no potash and very much less lime than is present in glass. The following is an analysis by C. von John of a specimen of moldavite from Trebitsch in Moravia:

	Per cent.
Silica (SiO_2)	81·21
Alumina (Al_2O_3)	10·23
Ferrous oxide (FeO)	2·45
Lime (CaO)	2·10
Magnesia (MgO)	1·08
Soda (Na_2O)	2·43
Loss on ignition	0·04
	99·54

The percentage of silica present may be as low as 76 and as high as 83. Moldavite fuses before the blowpipe only with great difficulty, and the fused mass after cooling is perfectly clear, so that in this respect also the mineral differs from obsidian and from glass.

It will be seen from the foregoing that though moldavite has many characters in common with green glass and obsidian, yet that there are important differences between moldavite and the two latter substances. It is, therefore, still uncertain whether moldavite is to be regarded as a natural glassy lava or as an artificial product. At the localities in Bohemia and Moravia, where the substance is found, there has flourished since ancient times a glass-making industry, and it is thus possible and even probable that moldavite is an artificial glass. Professor Suess has recently advanced the suggestion that moldavites are of cosmic origin and that they represent a hitherto unrecognised type of meteoric stone. He maintains that the peculiar surface markings cannot have been produced by attrition in water, and points to the resemblance these bear to the pittings of meteorites produced by the enormous resistance the latter encounter in their passage through the air.

The mode of occurrence of moldavite throws no light on the origin of the stone, for its presence *in situ* in any solid rock has never been unquestionably established. It is found loose in the ground in pieces which never reach the size of a man's fist, and which are more or less elliptical or flat and disc-like in form. The surface is wrinkled and scarred as if the material had been corroded, and so dark, dull, and rough that the transparency, delicate colour, and strong vitreous lustre of the substance would never be suspected.

The Bohemian localities for moldavite have been known for a long period. One of these is the district between Moldauthein and Budweis, in the south of the country, on the Moldau river (hence the name moldavite). Pebbles of the substance are found in the alluvial deposits of the river or are turned up with the soil in tilling the fields. The district between Prabsch, Klein-Horozek, and Zahoritsch is specially rich, but more so formerly than now. The stones found here are collected and sold to the lapidaries. Radomilitz, west of Budweis, is another locality, but the stones found there are lighter in colour than those met with elsewhere in Bohemia, and, moreover, are stated to occur in the ground-moraine of an ancient glacier of the glacial period. This mode of occurrence is by no means an unquestionable fact; but if it were, it would negative the theory that moldavite is an artificial product. The substance is found under the same conditions as in Bohemia at Kotschichowitz, near Trebitsch, in the Iglawa valley, and at other places in Moravia.

Rounded pebbles, measuring as much as an inch across and quite similar to the Bohemian bottle-stone, but less finely coloured, are found at Santa Fé in New Mexico, U.S.A. In this case the pebbles are, without doubt, a natural obsidian.

Moldavite is not extensively employed for ornamental purposes, and in former years was still less used. Cut stones are not worth more than sixpence a gram. The forms of cutting are those employed for olivine, namely, the table-stone and step-cut; the large front facet is often cut with a slightly convex instead of a plane surface.

In spite of the abundance and cheapness of moldavite, the substance is frequently imitated in an artificially prepared glass, which, when cut, can scarcely be distinguished from the true stone. It is easy, however, to distinguish moldavite itself from the green gem-stones it may resemble, namely, green tourmaline, chrysolite, idocrase, and demantoid, or even emerald. The specific gravity of each of these stones is appreciably higher than that of moldavite, and they all sink in liquid No. 4 (sp. gr. = 2·65), in which moldavite floats. Moreover, they are all harder than moldavite and, with the single exception of demantoid, are doubly refracting and dichroic, so that it is a simple matter to avoid mistaking bottle-stone for any of these precious stones.

THE PYROXENE AND AMPHIBOLE GROUPS.

HYPERSTHENE (with Bronzite, Bastite, and Diallage).

In this class are included those members of the pyroxene or augite group of minerals, which exhibit on one particular face a peculiar metallic sheen, and which depend on this feature for their application as ornamental stones.

HYPERSTHENE.

Hypersthene is remarkable for the display of a magnificent copper-red, metallic sheen, which shows up very effectively against the dark body-colour of the mineral. It is much esteemed as an ornamental stone, especially in France, but when the sheen is absent, as is sometimes the case, the mineral has no other feature of beauty, and is useless for decorative purposes.

The finest material comes from the coast of Labrador, and this is used, probably exclusively, for cutting as gems. The small island of St. Paul is often mentioned as a locality, and for this reason the mineral is also referred to as paulite, while lapidaries know it as "Labrador hornblende." It is associated here with labradorite, famous for its coloured sheen, and, as described above, the two together form rock-masses of considerable size. What has been said respecting the occurrence and distribution of labradorite in this region holds good also for hypersthene. The amount of material found here is considerable, and good specimens free from fissures can be easily cut out of the best portions of the larger rock-masses. Being, therefore, comparatively abundant, hypersthene is not a mineral of very great value.

Chemically the mineral consists essentially of a silicate of magnesium with the formula $MgO.SiO_2$, but a considerable portion of the magnesia is replaced by an equivalent amount of ferrous oxide. The following is an analysis of hypersthene from Labrador:

	Per cent.
Silica (SiO_2)	49·86
Alumina (Al_2O_3)	6·47
Ferric oxide (Fe_2O_3)	2·25
Ferrous oxide (FeO)	14·11
Manganese oxide (MnO)	0·67
Magnesia (MgO)	24·27
Lime (CaO)	2·37
	100·00

The mineral is one of the pyroxenes which crystallise in the rhombic system, but distinct crystals are not found in Labrador. The water-worn masses in which hypersthene occurs here are about the size of a man's fist, and exhibit more or less distinct cleavages and planes of separation. One of these separation planes is specially prominent by reason of the presence of innumerable crystalline plates embedded in the hypersthene. These plates or scales are all arranged parallel to the separation plane; they have a bright metallic lustre, and possibly consist of brookite, a crystallised modification of titanium dioxide. The tendency of the mineral to separate along this plane often gives rise to fissures in the stone, and special care must be taken during the operation of cutting to avoid the development of such fissures and cracks.

The coppery sheen of hypersthene is exhibited only on that face parallel to which the crystalline enclosures are arranged, that is to say, only on the plane of separation ; no trace of it is to be seen on other faces, and, moreover, when there are no enclosures the sheen is completely absent. It follows, therefore, that the minute enclosures embedded in the substance of hypersthene are not only the cause of the separation plane, but also of the coppery sheen characteristic of the stone. The beauty of hypersthene is best displayed by cutting a stone with a slightly convex surface and with the circular or oval base parallel to the plane of separation of the mineral. On polishing this surface it acquires a fine lustre and its sheen is also thereby intensified. If the polished surface is flat the metallic sheen is exhibited uniformly over the whole area, but when convex, the sheen is confined to that portion of the surface which is directed towards the source of light, and is then stronger and more intense. The effect of the sheen is diminished or entirely destroyed if the stone is not cut exactly in the manner described ; and the effect of facets is rather to detract from the appearance of a cut stone.

The body-colour of hypersthene is of a dingy brownish-black, but it forms an effective background for the strong coppery sheen. The mineral is perfectly opaque in mass but in very thin sections is transparent ; the tabular inclusions described above are to be distinctly seen when such sections are examined under the microscope.

The specific gravity of hypersthene is 3·4. Its hardness is 6, so that the mineral is scratched by quartz, but is itself capable of scratching glass. It is brittle and fuses without difficulty before the blowpipe, giving a black magnetic glass. It is not attacked by acids.

A few other minerals belonging to the pyroxene group, and more or less closely related to hypersthene, sometimes contain enclosures arranged parallel to a certain face, which also give rise to a metallic, or, in some cases, to a pearly, sheen. This is in all cases less beautiful than the coppery sheen of hypersthene, but the stones are sometimes cut and polished nevertheless. Their body-colour is not dark brown as in hypersthene, but some pale shade of brown, grey, or green. These minerals, which are known by the names of bronzite, bastite, and diallage, will be now briefly described and the features in which they differ from hypersthene indicated.

BRONZITE.

Bronzite is in reality a hypersthene which contains rather less iron than ordinarily. Owing to this difference in the chemical composition of the mineral, the specific gravity is lower, being only about 3·2, the colour is less dark, and the sheen paler and more bronze-yellow in character, though still very strong and with a brilliant metallic lustre. Bronzite sometimes shows indications of a fibrous structure, and when this is the case its sheen has a certain resemblance to that of cat's-eye. It occurs in association with felspar, and in masses sufficiently large for cutting, at Kupferberg in the Fichtelgebirge ; in serpentine at Gulsen, near Kraubat, in Styria ; on the Seefeld-Alp in the Ultenthal, Tyrol, and at other places. Bronzite is employed even less extensively than hypersthene.

BASTITE.

Bastite or schiller-spar is mineralogically identical with bronzite and differs from this latter only in external appearance. In colour it is of a pale greyish-green, and the sheen, which varies between the metallic and the pearly type, is also green. The principal locality is Baste in the Radauthal, near Harzburg, in the Harz. Single grains of the mineral, not infrequently rather large in size, occur here embedded in a serpentine, which varies in colour between dark green and black. The paler-coloured bastite with its metallic to pearly sheen contrasts very well with the dark serpentine in which it is embedded. The latter, flecked here and there with patches of sheeny bastite, forms a more decorative material than bastite or hypersthene alone, for when a fragment of serpentine containing bastite is moved about, the sheen does not disappear simultaneously from every patch of bastite but disappears from some only to reappear on others. The mineral is usually cut with a slightly convex or scutiform surface; its application as an ornamental stone is very limited, however, and it is used for the most part in the manufacture of small articles, such as snuff-boxes, letter-weights, &c.

DIALLAGE.

Diallage is rather less closely related mineralogically to hypersthene than either bronzite or bastite. In addition to the constituents of hypersthene it contains a large amount of calcium, and it crystallises, not in the rhombic, but in the monoclinic system. It resembles the other members of the pyroxene group so far considered in that it occurs most frequently in irregular masses and possesses a definite plane of separation, which displays a shining metallic sheen. In colour diallage may be dark brown, various shades of green including a very pale greenish tint and a pale greyish shade; it therefore shows more range of colour than other members of the group. The sheen is usually of the same colour as the stone itself; on the darker stones it is more metallic in character, and on the lighter it is more pearly. The rock known to petrologists as gabbro consists of diallage and felspar, and at many localities it occurs so coarsely grained that the diallage individuals are large enough for cutting; the application of the mineral as an ornamental stone is, however, extremely limited. Such coarse-grained gabbros are found, among other places, at Volpersdorf, near Neurode, in Silesia, at Le Prese in Veltlin, and at many other places in the Western Alps; at Prato, near Florence, and elsewhere in the Apennines; and in the island of Skye.

DIOPSIDE.

Among the minerals of the pyroxene group diopside is remarkable for its transparency and for the beauty of its colour, and on this account it is sometimes cut as a gem. It is a silicate of calcium and magnesium, the composition of which is represented by the chemical formula $CaMg(SiO_3)_2$, and in which a portion of the magnesia is replaced by ferrous oxide.

Diopside crystallises in the monoclinic system, and the crystals usually have the form of rather long prisms with an oblong cross section and with the edges replaced by narrow faces of a rhombic prism. They always occur attached at one end, and may be terminated quite irregularly at the free end, or there may be a greater or less number of obliquely placed terminal faces. Twin crystals are not infrequent, and irregular, columnar aggregates of crystals are often met with.

There is a moderately perfect cleavage parallel to the faces of the rhombic prism.   The crystals are brittle, and have a hardness of 6, almost the same as that of felspar; they are thus easily scratched by quartz but are themselves capable of scratching ordinary window-glass.   The specific gravity varies between 3·2 and nearly 3·3, and is higher the greater the amount of iron present.   The higher specific gravity is that possessed by the transparent, bottle-green crystals rich in iron from the Tyrol, which furnish most of the material cut as gems.

The lustre is strongly vitreous and is increased appreciably by polishing.   The transparency varies much in degree; in some crystals it is very nearly perfect, and those which are less perfectly transparent are not cut as gems.   Diopside is green in colour, but the depth of shade depends upon the amount of iron present.   Crystals which contain practically no iron are almost colourless; as the amount of ferrous oxide present increases the colour becomes deeper and deeper, and the crystals which are richest in iron are of a fine, deep bottle-green colour.   A characteristic feature noticeable even in the most deeply coloured stones is the existence of only a very slight dichroism.

There are but few localities from which material fit for cutting is obtained.   The pale greyish-green crystals which occur in association with hessonite in the Ala valley in Piedmont (Plate XIV., Fig. 7) are worked at Turin, and to a certain extent also at Chamounix, and are worn in rings and other pieces of jewellery, more especially in Italy. Similar crystals of a pale oil-green colour occur at De Kalb, in St. Lawrence County, New York, and are worn to a limited extent in North America.   The dark bottle-green crystals from the Schwarzenstein Alp in the Zillerthal, Tyrol, are still finer; they attain a diameter of 1 inch and a length of 5 inches, and occur attached at one end to chlorite-schist.   Some, especially the smaller crystals, are very transparent and of a fine colour, which, however, is frequently not uniform throughout the whole crystal.   Very often the attached end of the crystal is green while the free end is almost colourless, but the reverse is never the case. Such crystals were found formerly in some abundance; the green ends were used for cutting, and the cut stones, like those from Ala, were rather admired, especially in Italy.   At the present time they are less frequently found and are worth less than formerly.

The forms of cutting employed for diopside are those generally used for coloured, transparent stones, namely, the step-cut and the various modifications of the table-stone. In the case of dark-coloured stones the step-cut must not be too deep.

Diopside may be confused with other green stones or with green glass.   It can be distinguished from the latter by the fact that the glass is singly refracting.   It differs from emerald in colour and in specific gravity, being much heavier.   Chrysolite (olivine) is often very similar to diopside in colour, shows the same feeble dichroism, and has much the same specific gravity; it is appreciably harder, however, than diopside, which is easily scratched by this stone.   Green tourmaline, epidote, and alexandrite are readily distinguished from diopside by their strong dichroism; moreover, the last named is appreciably heavier, and the same is also true of idocrase.   Diopside and dioptase can scarcely be mistaken the one for the other, since the difference between them is very marked.

SPODUMENE.

Spodumene is a member of the pyroxene group which usually occurs as opaque, ash-grey crystals; only rarely is it transparent and suitable for gem purposes. A beautiful green, transparent variety is known as hiddenite. Chemically, spodumene is a silicate of lithium and aluminium with the formula $Li_2O.Al_2O_3.4SiO_2$; other elements, including iron, are also present, but only in small amount. From the fact that lithia is an essential constituent of hiddenite, this emerald-green variety of spodumene is sometimes referred to as "lithia-emerald." It resembles the emerald, however, only in the green colour, which is probably due to the presence of a small amount of iron or chromium.

Spodumene occurs in monoclinic crystals of prismatic habit and vitreous lustre. These crystals possess a very perfect cleavage parallel to the faces of the prism. In the transparent varieties the colour is usually pale yellow or yellowish-green, but some crystals (of the hiddenite variety) have an emerald-green tint which, however, is never as deep and pure as that of the finest emeralds. The two stones differ in that while spodumene is rather strongly dichroic, emerald is only feebly so.

Spodumene has a specific gravity of 3·17, and this is another feature which distinguishes the stone from emerald, for the specific gravity of the latter does not exceed 2·7. Moreover, the hardness of spodumene varies between $6\frac{1}{2}$ and 7, while in the case of emerald $H = 7\frac{1}{2}$ to 8. Hence the former mineral is, and the latter is not, scratched by quartz, and, conversely, while emerald is capable of scratching quartz spodumene is not.

Spodumene of gem-quality is exclusively an American product, having been found only at one locality in the United States, namely, at Stony Point, Alexandra County, North Carolina, and in Brazil.

In **North Carolina** the hiddenite variety is found at Stony Point, the name of which place has been altered to Hiddenite, associated with emerald, beryl, quartz, garnet, rutile, and other minerals, in drusy cavities in a gneissose granite. The first specimens of hiddenite were discovered in 1879 by Mr. W. E. Hidden; they had been weathered out of the mother-rock and were lying loosely in the ground. They were transparent and greenish-yellow in colour and were at first thought to be diopside, since at that time spodumene had never been found in fine, transparent crystals. A closer examination revealed their true nature, and later on, during systematic mining operations for emerald in the same locality, crystals of hiddenite were found in the original mother-rock. The size of the crystals varies considerably, the largest prisms having a length of 7 centimetres. Some crystals are remarkable for the possession of a peculiar corroded surface. Some, again, were of an emerald-green colour; these and the more ordinary yellowish-green crystals were all cut as gems. Very few of these cut stones have found their way into Europe, the majority having been kept in the country as objects of local interest; a fine cut stone is, however, to be seen in the Mineral Collection of the British Museum (Natural History). Fine, green, transparent hiddenite is worth from 50 to 100 dollars (£10 to £20) per carat, and the mineral is more likely to rise than to fall in value, since the deposit at Stony Point appears to be completely exhausted and no other has as yet been discovered. Altogether about 7000 dollars' worth of rough material has been obtained from this one deposit.

More recently fine, transparent specimens of spodumene of a pale yellow, rather than a green colour, have found their way from **Brazil** into the European markets under the name of chrysoberyl. This yellow, transparent spodumene occurs in Brazil in association with true chrysoberyl and was formerly mistaken for this stone, and it is probable that many supposed chrysoberyls from Brazil would turn out on closer examination to be spodumene. A few transparent pebbles of a beautiful blue colour were found formerly in the Rio de S. Francisco in the neighbourhood of Diamantina in Minas Geraes, and these for a long time were mistaken for blue lazulite.

Hiddenite can be distinguished from other green precious stones without much difficulty. The differences between this stone and emerald have been already pointed out. It is distinguished from chrysoberyl by the fact that it is much softer and much lighter, hiddenite floating in methylene iodide and chrysoberyl sinking heavily. Diopside differs from hiddenite in its specific gravity, which is higher, and its feeble dichroism; and the same is true also of chrysolite. Demantoid, the green variety of garnet, is often very similar in colour, but it is singly refracting, and therefore cannot be mistaken for the doubly refracting and dichroic hiddenite.

RHODONITE.

Rhodonite is a member of the pyroxene group, remarkable for its rose- or raspberry-red colour, which, however, in some cases may incline to a light chestnut-brown. The mineral is used as a gem and also is fashioned into all kinds of ornamental objects; the sarcophagus of Czar Alexander II., for example, is constructed entirely of rhodonite. The material used for such purposes is found at Ssedelnikova, in the neighbourhood of Ekaterinburg, in the Urals; it is finely granular to compact, and has a somewhat splintery fracture. It is worked for the most part at Ekaterinburg. The locality is situated on the eastern side of the Urals, south-west of Ekaterinburg and on the right bank of the Amarilka, a tributary on the right of the Isset; it is distant only a few versts from the gold-washings of Shabrovskoi. The rhodonite is here obtained from the quarries, which lie close together, in what appears to be a black clay-slate. The upper portion of the deposit contains a large admixture of quartz, and is therefore unfit for use. The rhodonite in places is very coarse-grained; such material, though unsuitable for cutting, is of interest mineralogically, since its nature can be easily made out. The whole deposit is penetrated by numerous cracks and fissures, the course of which is indicated by a black discolouration, due to weathering of the material along these cracks.

Rhodonite of the same description is also found, though less abundantly, in the manganese deposits of Wermland, Sweden; the material from this locality is not used for cutting. The American occurrence at Cummington, in Massachusetts, is more important. Here are found blocks of rhodonite, of a fine, rose-red colour and several hundred pounds in weight. The material is equal in quality to that from Russia, and is used for similar purposes.

Chemically, rhodonite is a silicate of manganese, which, when pure, has the formula $MnO.SiO_2$. Usually other constituents are present, especially calcium and iron, in greater or less amount. Crystals belonging to the triclinic system are met with not infrequently, but at localities other than the Urals. Very good ones are found, for example, in the manganese mines in Wermland. These are transparent and of a beautiful colour, but on account of their small size are scarcely ever cut as gems, the massive material already mentioned being the variety worked almost exclusively for decorative purposes. Massive rhodonite has a hardness which lies between 5 and 6, and a specific gravity which varies between 3·5 and 3·6. It is only slightly translucent, and its lustre is feeble, but it is susceptible of a moderately good polish.

Rhodonite is sometimes erroneously referred to as manganese-spar. This is a carbonate of manganese, of which the mineralogical name is rhodochrosite ; it is of a pretty rose-red colour, but is too soft to be worked for ornamental purposes.

On account of its colour, which is somewhat similar to that of rhodonite, we may here, in passing, briefly describe a mineral which belongs not to the pyroxene but to the mica group.

LEPIDOLITE.

The colour of this mineral is not a pure rose-red, but shows a tinge of blue or violet, and therefore inclines more to lilac. Lepidolite is a lithia-mica, a finely granular to compact variety of which is found in some quantity at Rozena, in Moravia. It is sometimes fashioned into small ornamental articles of various kinds for the sake of displaying its pretty lilac-red colour, but since the hardness is only 2, and it can be scratched with the finger-nail, it is useless for other purposes.

NEPHRITE. JADEITE. CHLOROMELANITE.

These three minerals are sometimes collectively referred to as nephritoids. They were used as precious stones in prehistoric times, and objects of various kinds fashioned out of them have been found in Europe among the remains of the Stone Age. At the present time their use for ornamental purposes is limited and mainly confined to a few countries outside Europe.

The first mentioned of these minerals, nephrite, belongs to the amphibole or hornblende group, and the other two, jadeite and chloromelanite, which only differ in unessential characters, belong to the pyroxene or augite group. In spite of the fact that these three minerals are classified mineralogically into two different, though closely related groups, they resemble each other very closely in external appearance, in certain other of their characters, and in the purposes to which they are applied. Structurally they may be described as being very finely fibrous to compact aggregates, the individual constituents of which are only recognisable when thin sections of the mineral are examined under the microscope. The substance, especially when polished, appears in each case, as far as can be made out with the naked eye, perfectly homogeneous, and the appearance is rather such as one associates with fused material. When thin sections are observed under a high magnification, however, it is seen that the substance in every case is built up of very numerous fibres matted together. Although the hardness of these minerals is not very considerable, being between 6 and 7 on the scale, yet, on account of their fibrous structure, they are exceptionally tough and more difficult to fracture than any other substances in the mineral world ; this character being specially conspicuous in nephrite. Because of their toughness, and also because of their pretty appearance, the nephritoids, even in prehistoric times, were used in the fashioning of idols, ornamental objects, and tools, such as axes and chisels. Such objects are now found in Europe and other countries, in ancient lake-dwellings, graves, &c., and also lying loosely in the surface soil.

Until comparatively recently, the three minerals under consideration were found both in Europe and America only in the worked condition, and the occurrence of rough material *in situ* was known only in Central Asia and in New Zealand. In accordance with these facts, a theory has been elaborated, mainly by Heinrich Fischer, of Freiburg (Baden), that the material, either in its worked or its rough condition, was carried probably from Central Asia to the places where it is now found.

Recently, however, nephritoids in the rough condition have been found in several districts where formerly only worked specimens had been known. Moreover, it has been observed that the microscopic structure of all such specimens is identical, and that it differs from that of the material which occurs in Central Asia. On these grounds, it is argued by F. Berwerth, of Vienna, and especially by A. B. Meyer, of Dresden, that the view held by Fischer is unsupported by fact, and that the objects found in any one region were worked out of rough material found in the same neighbourhood. This view is now very generally accepted, although, in some places where such remains are to be found, the source whence the rough material was derived has not yet been discovered. In these cases it is probable that, unless the deposit was exhausted by prehistoric man, a closer investigation of the district would reveal its location.

The nephritoids are opaque, or at most translucent; they are sometimes brightly coloured, but, as a rule, are an inconspicuous green or grey colour, or almost colourless. Their external appearance is thus widely different from that of gems proper, and their application in Europe is extremely limited. They are highly esteemed, however, by people in a primitive state of civilisation in other parts of the world. Thus the nephrite which occurs in New Zealand is worked by the Maoris, and the jadeite which occurs in Burma by the Burmese in the same way as these substances were worked by the ancient inhabitants of Europe in prehistoric times. The nephritoids are, however, most highly esteemed in China. Together with certain other minerals of similar appearance, they are there referred to as " yu," and are certainly more favoured by the Chinese as a nation than any other stone. There are several varieties of " yu," and they are worked in China not only for personal ornaments but also for plates, bowls, vases, sword-handles, idols, and such like objects, some of the work being executed with amazing skill and taste. Objects fashioned out of nephrite and jadeite are much valued elsewhere in the Orient, in Central Asia, Turkey, &c., but it is in China that the industry is most flourishing, and from this country that a considerable number of worked articles are exported.

NEPHRITE.

Nephrite is known as axe-stone, because it is frequently found fashioned into axe-heads; and also as kidney-stone, because it is often worn as a charm to prevent kidney diseases. Both in English and in French it is often referred to as jade. It belongs to the amphibole or hornblende group of minerals, and its chemical composition is represented by the formula $CaO.3MgO.4SiO_2$, in which a variable amount of magnesia is replaced by an equivalent proportion of ferrous oxide. The following is an analysis of nephrite from Eastern Turkestan :

	Per cent.
Silica (SiO_2)	58·00
Alumina (Al_2O_3)	1·30
Ferrous oxide (FeO)	,2·07
Magnesia (MgO)	24·18
Lime (CaO)	13·24
Soda (Na_2O)	1·28
	100·07

The composition of nephrite is exactly the same as that of another member of the amphibole group, namely, actinolite.   This mineral is found not infrequently in the form of long, thin rhombic prisms, belonging to the monoclinic system, which occur embedded singly in talc-schist, for example, in the Zillerthal in the Tyrol ; more often it occurs as radial aggregates of acicular crystals at many localities.   All the essential characters of nephrite, namely, the specific gravity, hardness, and cleavage, agree completely with those of actinolite, and in both minerals the colour is of a more or less intense green.   Nephrite is thus nothing more than a very finely fibrous to compact actinolite, the prismatic crystals of which are reduced to microscopically fine fibres.   Observation of thin sections of nephrite under the microscope shows that the fibres have the same characters as the larger prismatic crystals of actinolite, but that these fibres are woven and matted together in an altogether irregular fashion.   The disposition of the fibres with respect to each other is to a certain extent characteristic of each occurrence, specimens from different localities differing somewhat in their microscopic structure.

From what has been said regarding the structure of nephrite, it will be readily understood that the mineral never shows any external crystalline form.   It occurs always as irregular masses of larger or smaller size or in water-worn blocks.   There is never a cleavage through the whole mass of a specimen, but there is sometimes a distinct separation in one direction, which is due to the material having become schistose in character.   In this direction the mass can be broken up with comparative ease, but in others, owing to the toughness of the material, this can be accomplished only with great difficulty.   It is almost impossible to break up large blocks of nephrite with a hammer, especially when they are in the form of rounded boulders.   The method adopted in such cases is to develop cracks in the mass by subjecting it to sudden changes of temperature, for example, by heating it strongly and then plunging it into cold water.   The breaking up of the mass into its characteristic splintery fragments can then be completed by the help of a hammer.

Compared with the extraordinary toughness of nephrite its hardness is rather low, being not quite 6 on the scale (H = $5\frac{1}{2}$ − 6).   The mineral is therefore harder than glass but softer than quartz.   It is brittle, but with suitable tools can be worked on the lathe, the process requiring, however, special precautions and care.

The specific gravity is very nearly 3, but is slightly variable, probably because of the variable amount of iron present.   The usual limits are 2·91 and 3·01, but values up to 3·1 and even 3·2 have been given.   These high values are probably the results either of the presence of foreign matter (magnetite, &c.) or of inaccurate determination.   Hence most specimens of nephrite float in liquid No. 3 (sp. gr. = 3·0), but some specimens slowly sink. The specific gravity is of importance, since it affords a means whereby nephrite may be distinguished from the very similar but much heavier jadeite (sp. gr. = 3·3), which always sinks rapidly in liquid No. 3, and scarcely floats even in pure methylene iodide.

Nephrite is not attacked by acids.   Heated before the blowpipe it becomes white and cloudy and fuses with difficulty to a grey slag.   In contrast to this behaviour jadeite fuses easily, even in an ordinary flame, and, moreover, it colours the flame bright yellow, which is not the case with nephrite.

This mineral is never transparent, and is strongly translucent only in the thinnest of sections or along the sharp edges of splintered fragments.   In mass it is either perfectly opaque or feebly translucent.   Fractured surfaces are dull, but smooth polished surfaces have a good lustre of a somewhat greasy character.

The colour of nephrite depends upon the amount of iron present.   Like actinolite it is usually of some shade of green, brighter or paler according as there is much or little iron present.   In rare cases iron is completely absent and the mineral practically colourless. Various tints of green are met with, including grey-green, sea-green, leek-green, grass-green, &c.; moreover, yellow and brown nephrite is known, as well as grey nephrite showing a bluish, reddish, or greenish tinge.   Nephrite of a colour which is compared with that of whey is often much esteemed.   The mineral is usually coloured quite uniformly, but occasionally a specimen is met with which is streaked, spotted, veined, or marbled with several colours or shades of colour.   Nephrite from different localities usually differs in colour; thus at one locality nephrite of a certain colour predominates, while at another that of another colour is more abundant.   This feature, therefore, taken in conjunction with the microscopic structure and the chemical composition serves to determine the locality from whence any particular specimen has come.

With regard to the occurrence of nephrite in nature it may be stated that the mother-rock is in all cases a crystalline schist, the mineral being found with especial frequency in hornblende-schist.   It forms also more or less extensive bands in pyroxene-rock, serpentine, and other rocks of this class.   Nephrite occurs *in situ* principally in Eastern Turkestan and the regions east of this in China, in Transbaikalia, and in New Zealand.   A few years ago a small amount was found in Silesia, but elsewhere the occurrence of nephrite *in situ* is very sparing.   In regions where the mineral occurs in its primary situation, loose boulders are often found in alluvial deposits; boulders are also met with, among other places, in the lowlands of north Germany.

We will now consider the occurrence and distribution of nephrite, both in the worked and the unworked condition, in some detail.   Most of the nephrite found in **Europe** is in the form of axes, chisels, and other objects, dating back to prehistoric times, and the most famous and best known localities for these interesting remains are the ancient lake-dwellings in Lakes Constance, Zurich, Bienne, and Neuchâtel in Switzerland.   Similar objects are found in the neighbouring districts of southern Baden (on the Ueberlinger See) and Bavaria. The microscopic structure of the stones found at these places differs from that of all other known nephrites.   It is very probable, therefore, that the rough material was obtained from some deposit in the neighbourhood, perhaps from the Swiss Alps.   A few rounded pebbles of nephrite have been found on the shores of Lake Neuchâtel, and these are probably from some primary deposit in the neighbourhood, but in spite of systematic search no nephrite *in situ* has yet been discovered.

A few nephrite pebbles have also been found further to the east in the Sannthal above Cilli, and in the Murthal in Styria, the original deposit being doubtless somewhere in the neighbourhood, but at present undiscovered.   It is certain that nephrite has a wide distribution in the Alps, and as the geological investigation of these mountains proceeds, occurrences of nephrite *in situ* will no doubt be met with.   A few worked specimens of nephrite have been found between Switzerland and Styria, but no rough material.

Outside Switzerland and the neighbouring countries mentioned above, only a few prehistoric articles of nephrite have been met with in Europe.   Tools of jadeite, on the other hand, are comparatively common in Europe, and these also accompany the nephrite objects in Switzerland.   Many prehistoric axes have been found in France; all, however, are of jadeite, and not a single specimen has been conclusively proved to be nephrite.   In Italy worked articles of nephrite seem to be confined to Calabria and Sicily, but jadeite objects are distributed over the whole country.   A few discoveries have been made also in Greece, but in none of these countries has the rough material yet been found.

In Germany, leaving southern Baden and Bavaria out of consideration, only a few small nephrite axes have been found, most of them in the neighbourhood of Weimar, and in Silesia, for example, at Gnichwitz in the Breslau district.   Here in Germany, however, rough nephrite is more common, and is known to occur in several different ways.   A few erratic blocks and pebbles have been found in the glacial deposits left by the northern ice-sheet, among other places at Stubbenkammer, in the island of Rügen, at Potsdam, at Suckow near Prenzlau, and at Schwemmsal, north of Düben, in the Bitterfeld district.   The blocks of nephrite do not differ in their mode of occurrence from other boulders in the glacial deposits of the North German lowlands, and, like them, come doubtless from Scandinavia, having originated in the crystalline schists so widely distributed throughout that peninsula.

The occurrence of nephrite *in situ* in Germany is limited to Silesia.   This locality for nephrite was mentioned by Linnæus (1707–1778), but in course of time the occurrence was forgotten and only rediscovered in the eighties of the nineteenth century.   The places at which nephrite is found are the same as those described by Linnæus.   One of these is Jordansmühl in the Zobten mountains; here nephrite, usually of a dark green colour, forms a layer of considerable extent, and in places over a foot in thickness between granulite and serpentine.   The mineral occurs at the same place as rounded nodules, the largest of which measure 5 centimetres across, and in veins in the serpentine; the nephrite in the latter situation being white or of a pale green colour.   The nephrite of which the axes found at Gnichwitz, two hours' journey from Jordansmühl, are made, is very similar in character to the rough material found at this locality.   The other locality in Silesia, at which nephrite is found, is Reichenstein, a famous mining centre.   The material found here is compact and of a light greyish-green colour, sometimes tinged with red, and is indistinctly schistose; it occurs in layers, the thickest of which are 7 centimetres across, intercalated in a diopside-rock in the Prince adit.

The most important localities for nephrite, from whence alone the mineral is exported, are in Asia.   Chief among these is Eastern Turkestan (Little Bucharia) in the **Chinese Empire**; here, in the region south of Yarkand and Khotan, the Konakán and Karalá nephrite quarries are specially well known, but are not now systematically worked.   These quarries are situated on the right side of the valley of the Karakash, an upper tributary of the Khotan Daria, 500 feet above and one and a half miles from the stream, in the neigh-bourhood of Gulbashén and nine miles east of Shahidulla.   The latter place is situated on the above-mentioned river at a spot where there is a sharp bend towards the west; it lies in latitude about $36\frac{1}{2}°$ N. and longitude $78\frac{1}{2}°$ E. of Greenwich, in the region of the western termination of the Kuen-Lun mountains, and on the southern slopes of this range.   In these quarries there is a layer of nephrite 20 to 40 feet thick between gneiss and hornblende-schist.   The mineral occurs *in situ* at many spots on the northern as well as the southern slopes of the Kuen-Lun range, for example, at places further down the Khotan Daria, and on the Sirikia; and as pebbles in all the watercourses draining the northern and southern slopes of these mountains.   The nephrite of Eastern Turkestan is generally paler than that found elsewhere.

Another locality at which nephrite occcurs *in situ* lies further to the west in the Pamir region and on the Raskem Daria.   This river flows eastward, and after being joined by the Tash Kurgan from the south is known as the Yarkand Daria.   The nephrite mines lie on the right bank of the Raskem, a little north of the place where it bends suddenly from north-west to north-north-east, in about latitude 37° 4′ N. and longitude 76° E. of Greenwich.   The existence of this deposit was long suspected, on account of the presence of pebbles of nephrite in the lower reaches of the Yarkand Daria.   It was actually discovered in the year 1880, and the characters of the nephrite pebbles and of the material from this deposit were found to be in complete agreement.

The deposits of nephrite in this portion of the Chinese Empire are not, however, limited to the few spots mentioned above, but extend over an area which stretches from the Rasken Daria eastward over more than 5° of longitude as far as Kiria, in longitude about 82° E. of Greenwich, and probably still further.

Bogdanovitch has recently given a detailed account of the occurrence of nephrite in this region. According to this authority, between Mount Mustagat (longitude about 76° E. of Greenwich) and the meridian of Lob-Nor (longitude about 89° E.) there are no less than seven districts where primary deposits of nephrite are known to occur on the northern slopes of the Kuen-Lun and in the immediate neighbourhood of this range. The mineral is for the most part interbedded in pyroxene-rocks. Pebbles of nephrite are to be found in almost all the water-courses of Kashgar, and mining operations are systematically carried on in the valleys of the Jurunkash, Karakash, and the Tisnab. The workings on the first-named river at Kumat below Rhodan are specially well known. They are situated partly in the latest formed deposits of the river, the course of which is often altered to suit the exigences of mining operations, but more especially in the ancient terraces above the river, the detrital material of which belongs to the glacial period. Above Kumat the floor of the valley mentioned above is riddled with pits measuring from 1 to 1½ metres in depth. The Karangu-Tag hills on the Khotan Daria above Khotan, in longitude about 79½° E. of Greenwich, have been famed since ancient times for their richness in nephrite. The mineral is found here only in secondary deposits and not in its original situation, as at Balyktshi and the places mentioned above. The deposits recently discovered by the above-mentioned traveller at Shanut in the basin of the Tisnab, and at Lishei in the district of the Kiria Daria, are of the same character, as are also those in the famous Mount Mirdshai, or Midai, on the upper reaches of the Asgensal, a tributary of the Yarkand Daria. The primary deposits of nephrite have not been worked since the revolt of the Mahometans and the expulsion of the Chinese; they are assumed to be exhausted so far as concerns material lying near the surface, and the assumption is probably true. All the primary deposits of nephrite are in high mountain regions, many, indeed, above the limit of perpetual snow. From this situation much has been carried down to the valleys below by the agency of running water and glaciers.

According to the mode of occurrence and the manner in which it is obtained three classes of rough material are here distinguished: (1) material quarried out of primary deposits; (2) masses carried down from inaccessible heights by the ice of glaciers and still preserving their sharp edges and corners; (3) water-worn boulders and pebbles from the ancient glacial deposits or the later alluvial deposits of the rivers. These boulders and pebbles, having withstood successfully the jolting and grinding and blows of their journey down from the heights, are more likely to be free from internal cracks and fissures than are the other kinds; and for this reason boulders and pebbles of nephrite, apart from considerations of colour and such like, are more valuable than sharp-edged fragments. The work of quarrying the primary deposits of nephrite is made less arduous by lighting fires on the surface. This practice not only spoils the quality of the mineral, but also causes the mass to break up into fragments of comparatively small size, so that really large blocks are rarely seen. One such rarity is the gigantic monolith of the tomb of Tamerlane in the Gur-Emir mosque at Samarkand. Another was found by Bogdanovitch at the village of Ushaktal, between Karash and Toksun, having been left there in the middle of the eighteenth century on its way to Pekin. This is an irregular block, measuring 133·1 centimetres in length, 111·2 in breadth, and 94·6 in thickness. It probably came originally from the mines at Shanut; and though it has been exposed for a long period to both accidental and wilful damage, is still only one-third smaller than the monolith mentioned above.

Another occurrence of nephrite *in situ* has recently (1891) been discovered still further to the east. The locality lies to the north of the Kuen-Lun range, on the way from Kuku-Nor to the Nan-Shan mountains in the Chinese province of Kan-su. It is stated that the nephrite forms a vein in a soft rock, but the nature of this is not mentioned. It is probable that this deposit is the source of the numerous nephrite pebbles found in the streams and rivers of the district by former travellers. There is a brisk trade in nephrite carried on by the inhabitants of the northern slopes of the Nan-Shan range, round the villages of Kan-chu and In-chu, for example; and in almost all the villages in this region the stone is worked by the natives for the Chinese, while in the town of Su-chu-fu (a little south of latitude 40° N.) there are several workshops. The nephrite found in the Nan-Shan mountains is usually cloudy to translucent and of a light-green, milk-white, or sulphur-yellow colour.

It is not improbable that in the long ranges of the Kuen-Lun and the Nan-Shan there are other places where nephrite occurs *in situ*, since pebbles have been found by natives in the region; and the same is true for other parts of China, for example, in Yun-nan. The deposits actually known, however, have furnished not only a large proportion of the stone known as " yu," which is worked in China into the various objects already described, but also that which is applied to other purposes, and which is described as Central Asian nephrite.

A great many objects fashioned out of nephrite have been found in various parts of Asia, but it is doubtful whether the rough material came in every case from Eastern Turkestan and the other parts of the Chinese Empire mentioned above. Among other places discoveries have been made in Amur, Japan, East Cape, and the Chukchis peninsula; while in Syria and Asia Minor articles made of nephrite have been brought to light during the course of Schliemann's excavations. It is probable that the rough material was obtained in every case from some place in the neighbourhood of the spots where these articles are now found.

Of the occurrences of nephrite in other parts of Asia the first to be mentioned is that in **Siberia,** in the neighbourhood of Lake Baikal, and near the celebrated Alibert graphite mines worked by Faber, the Bavarian lead-pencil manufacturer. Nephrite and graphite are here very closely associated, and numerous scales of graphite are found embedded in the nephrite. Nothing is known regarding the origin of the latter, though here, as elsewhere, it occurs in a crystalline schist which constitutes the predominating rock of the region. Hitherto the mineral has been found only in blocks together with boulders of other material in the alluvial deposits of the water-courses of the region. These boulders, though not very numerous, are of considerable size, weighing as much as 1000 pounds or more. An immense polished block from this locality, weighing 1156 pounds, is exhibited in the Mineral Gallery of the British Museum of Natural History at South Kensington. The primary deposit is probably situated in the rocky mountainous district of Batugol, in the Sayan range to the west of Lake Baikal. The Soyots, or inhabitants of this district, wear the nephrite as ornaments, the women making necklaces and the men tobacco-boxes of it. Among the water-courses in which boulders of nephrite are found, may be mentioned the Byelaya and Kitoy rivers, both rising in the Sayan mountains and both tributaries on the left bank of the Angara, but the former emptying itself into the Angara further from Lake Baikal than does the latter. Also the Bistraya, a tributary on the right bank of the Irkut river which rises in the Chamar-Taban mountain, the Sludianka flowing northward into Lake Baikal (Fig. 83) and the Onot river.

Axe-heads of nephrite are also found not infrequently in eastern Siberia, both in the soil and in the ancient Tchudic graves at Tomsk and in the Altai Mountains. Rough nephrite is also said to occur in Amur.

The following details respecting the recent discovery of nephrite in Siberia are taken from Mr. G. F. Kunz. The search for nephrite in Siberia was greatly stimulated in the year 1897 by a command from the Imperial House of Russia that material be obtained for a sarcophagus to contain the remains of Czar Alexander III. Three expeditions were made by L. von Jascevski, the officer in charge of the Siberian division of the Geological Survey of Russia, to the eastern Urals for the purpose of discovering larger masses of nephrite than had been found before, and, if possible, of finding nephrite *in situ*. After thoroughly investigating the deposits and obtaining masses of the mineral in the region of the Onot, a region which had been visited by Alibert in 1850, and by Permikin in 1865, he visited the district of the Chara Jalga. In the bed of this river some masses of nephrite measuring 12 feet in length and 3 feet in width were discovered, but an even more important discovery was that of a ledge of nephrite of a magnificent green colour in its primary situation, this being the first observed occurrence *in situ* of nephrite in Siberia. The boulders in the stream furnished sufficient material for the purpose of the expedition, and for the past three years the Imperial Lapidary Works at St. Petersburg have been engaged upon the working of a small canopy to be placed over the the tombs of the Czar and Czarina. The canopy measures 13 feet in height and is constructed wholly of nephrite and rhodonite, of which latter material the entire sarcophagus for Czar Alexander II. had been made.

In **India** nephrite, or material similar to nephrite, has been found, but not in large amount. The occurrence is not important, and is confined to the southern part of the Mirzapur district in Bengal. The important occurrence of jadeite in Burma will be dealt with further on.

In **America,** as in Europe, jadeite is more common than nephrite. Objects fashioned out of nephrite are known to have been found in Central America, Venezuela, and Alaska, and primary deposits of nephrite, or at any rate rough nephrite, is said to occur on the Amazon river. Some of the material to which the name amazon-stone is applied has been supposed to be nephrite, but whether this is actually so is very uncertain. The occurrence of nephrite *in situ* at various places in Alaska and in the neighbouring parts of British Columbia is, however, well established. It is possible that both the Tchudis and the natives of the regions on either side of Behring Strait obtained their rough material from the deposits in Alaska. It is improbable, however, that the rough material for the nephrite objects found in South America, in Venezuela, Colombia, and Brazil, came also from Alaska. The nephrite out of which these objects are fashioned has definite characters of its own, and is more likely to have been found nearer at hand: where, however, is unknown.

The somewhat remarkable occurrence of nephrite in **New Zealand** was first discovered by Forster, who accompanied Captain Cook. The nephrite, which is usually of a fine green colour, occurs here partly *in situ* and partly as loose erratic boulders. From the earliest times this beautiful stone has been fashioned by the Maoris into weapons (battle-axes and clubs), chisels, axes, ear-pendants, idols, and other objects, and is still highly esteemed by them and known under the name " punamu." They recognise several varieties which are distinguished by special names. The conditions of the occurrence are not yet known in detail, but there appear to be three localities, all of which are on the west side of South Island. Fifteen miles from the mouth of the Arahaura (Spring) river occurs a band of " green schists" several feet in thickness; this is one locality. Another lies to the south of Mount Cook, in the vicinity of Jackson Bay or on Milford Sound; the occurrence here is probably in serpentine. The third locality is on Lake Punamu, in the province of Otago. New Zealand nephrite is exported to Europe and there fashioned into objects of all kinds. In these islands there are found two other green stones similar in appearance to nephrite but differing in chemical composition. These stones are known to the natives by the names "kawakawa" and "tangiwai"; they are often mistaken for nephrite and used instead of this stone.

Nephrite, both in the worked and in the rough condition, is found at many other localities in this part of the world. It occurs, for example, *in situ* in New Caledonia, New Guinea, the Marquesas Islands, the New Hebrides, the Society Islands, and Tasmania; but the occurrence in New Zealand is by far the most important.

An extraordinary number of minerals exhibited in collections are incorrectly labelled as nephrite. Almost every mineral which is compact, slightly translucent, and of a colour more or less resembling some shade of nephrite, has at one time or another been set down as nephrite, among other minerals so misnamed being different varieties of quartz, agate, serpentine, and zoisite. The hardness and specific gravity of true nephrite afford a means whereby it may be distinguished from other minerals with little difficulty. Nephrite is, or was, imitated in China by a glass paste called "pâte de riz," which is very similar to true nephrite, but appreciably harder. For a long time jadeite, chloromelanite, and nephrite were included under the common term jade; the differences between these minerals will be explained later.

It will be gathered from what has been said above that in remote bygone ages nephrite had a very extensive application, both for ornamental and utilitarian purposes. This is still the case in the Orient generally, and in China particularly, also among semi-civilised peoples; but in Europe the stone is but little used for any purpose. The fine green varieties, especially those from New Zealand, are occasionally cut *en cabochon* for ring-stones and pins, and whole rings are sometimes carved out of nephrite; but more frequently it is fashioned into small articles such as paper-knives and letter-weights. Nephrite ornaments are, generally, of little value, but very high prices are often paid in Europe for beautiful examples of Chinese art, the value of such articles lying not in the nephrite of which they are made, but in the marvellous workmanship they display.

In China the reverse is the case, and nephrite has an intrinsic value apart from the work expended upon it. "Yu" is an article of considerable commercial importance, and each colour and shade has a particular name and price. The most highly esteemed variety is of a pure milk-white colour and has the greasy lustre of hog's-lard. According to Bogdanovitch pebbles of this quality fetch as much as 200 roubles. The task of working such tough material is arduous indeed; for rough cutting double the weight of material removed is paid in silver; for finishing the expense depends upon the fineness of the work. It is not surprising, therefore, that articles carved out of nephrite are not cheap, even in China, and that they are still less so in Europe.

JADEITE. CHLOROMELANITE.

Jadeite and chloromelanite are very similar to nephrite in external appearance and in hardness, toughness, &c. By French mineralogists all three are included under the name jade, and this significance for the term has been often accepted elsewhere. Since the differences between the three minerals are now recognised the term jade has been given to nephrite, and is accepted as a synonym for nephrite. The word is derived from the Spanish *piedra de la hijada*, meaning kidney-stone; its use is always attended with confusion, and it would be an advantage to dispense with it altogether.

Although jadeite and chloromelanite are apparently so similar to nephrite, chemical analysis and microscopic examination show that they are really dissimilar, and that while nephrite belongs to the amphibole group of minerals, jadeite and chloromelanite must be included in the pyroxene group. ·All the essential characters of these two minerals are identical; they contain the same constituents, but whereas chloromelanite contains a considerable amount of iron, in jadeite there is very little of this element present. Chloromelanite may, therefore, be considered as a jadeite rich in iron and of a correspondingly dark colour.

Chemically, jadeite is very similar to spodumene, and the variety of that mineral known as hiddenite, the only chemical difference being that the lithium of spodumene is replaced in jadeite by sodium. It is essentially a silicate of sodium and aluminium with the formula $Na_2O.Al_2O_3.4SiO_2$; it always contains small amounts of other substances, however, and on this account no two analyses are ever identical. It must, however, be observed that the simple formula given above cannot be deduced directly from analyses of chloromelanite, and this is a point which requires further chemical investigation. A comparison of the two analyses quoted below, one of jadeite from Burma and the other of an axe-head of chloromelanite from Department Morbihan (Brittany), reveals a close correspondence in the chemical composition of the two substances. If, on the other hand, these two analyses are compared with that of nephrite, it will be observed that in nephrite aluminium and sodium are almost absent, that there is very little calcium either in jadeite or in chloromelanite, and that jadeite contains scarcely any magnesium.

	Jadeite. (Per cent.)	Chloromelanite. (Per cent.)
Silica (SiO_2)	58·24	56·12
Titanium dioxide (TiO_2) . . .	—	0·19
Alumina (Al_2O_3)	24·47	14·96
Ferric oxide (Fe_2O_3)	1·01	3·34
Ferrous oxide (FeO)	—	6·54
Manganese oxide (MnO) . . .	—	0·47
Lime (CaO)	0·69	5·17
Magnesia (MgO)	0·45	2·79
Soda (Na_2O)	14·70	10·99
Potash (K_2O)	1·55	trace
	101·11	100·57

Like nephrite, jadeite and chloromelanite may be described as very finely fibrous to compact aggregates. The minerals have no regular external form; a microscopic examination of thin sections shows them to consist of an irregularly interwoven mass of fine fibres. It is on this structure that the extreme toughness and compactness of the minerals, as well as their uneven splintery fracture, depends.

Transparent specimens of these minerals are never seen; very thin splinters are at most only translucent or feebly transparent. The lustre on fractured surfaces is slight, but is much heightened by polishing, being then somewhat inclined to be greasy; the same lustre may be seen on many rounded, water-worn pebbles. The substance of jadeite in itself is colourless, and many natural specimens are quite or almost white. More frequently the mineral shows a tinge of rose-red or some light shade of colour, such as pale grey, greenish-white, bluish-green, leek-green, or apple-green. Some varieties are white, with more or less sharply defined spots of a fine emerald-green colour. These latter are due to the presence of a small amount of chromium distributed through the stone in this irregular manner. The colour of the varieties which show a uniform shade of pale green is due to the presence of a small amount of iron. Chloromelanite, since it always contains a considerable amount of iron, is never colourless or pale in shade, but always dark green or almost black.

Having considered the features in which jadeite and chloromelanite are in agreement with nephrite, we must now pass to the consideration of their distinguishing characters.

The most important of these is the specific gravity. Jadeite is denser than nephrite, the specific gravity of the former varying between 3·30 and 3·35, while that of the chloromelanite variety, owing to the larger amount of iron present, reaches 3·4. Although in the majority of cases this is a valuable aid in discriminating between these minerals, it does occasionally happen that a specimen of jadeite, owing to the presence of inclusions of foreign matter, has a specific gravity as low as that of nephrite, and cannot be distinguished in this way. It is generally safe to assume, however, that a stone with a specific gravity of 3·3 is not nephrite. There is a slight difference between the hardness of the minerals, that of jadeite ($H = 6\frac{1}{2} - 7$) being rather higher than that of nephrite.

Another important difference between these minerals is the ease with which they can be fused. While nephrite fuses only with great difficulty, jadeite and chloromelanite are very easily fusible; indeed, splinters of typical chloromelanite or jadeite will fuse to a transparent, blebby glass even in the flame of a spirit-lamp without using a blowpipe at all. This is not invariably the case, however, for some specimens of jadeite and of chloromelanite are rather difficultly fusible, though never as much so as nephrite. In performing these blowpipe experiments it will be noticed that jadeite and chloromelanite, since they contain sodium, colour the flame an intense yellow, while nephrite does not do so.

Setting aside chemical analysis, the surest means whereby these minerals may be distinguished is by the examination of thin sections under the microscope. It is always possible to find some fibres which show the cleavage or optical features characteristic of augite or of hornblende, the mineral in the former case being jadeite or chloromelanite and in the latter nephrite. Although only a very small splinter of material is necessary for the investigation of the fusibility, chemical composition, and microscopic structure of any given stone, it may be inadmissible to remove even this from a worked object, and in such cases the only test applicable is the determination of the specific gravity.

Objects worked in jadeite of prehistoric age have been found frequently in Europe, Asia, America, and Africa, but the localities from whence the rough material was obtained are in most cases practically unknown. The only locality which has been thoroughly examined from a scientific point of view, and where large masses of jadeite are known to occur *in situ*, is in northern Burma. The material from this deposit, like nephrite from its Asiatic localities, is distributed over the whole of the Orient, where it is highly esteemed and used for the same purposes as nephrite; from which, indeed, it is often not distinguished, being included with this stone under the term " yu." Like nephrite, jadeite is not much esteemed in Europe, and is seldom applied to purposes of ornament.

Objects made of chloromelanite accompany jadeite articles in France, Switzerland, Mexico, and Colombia, while a large axe of the same material has been found at Humboldt Bay in New Guinea. Chloromelanite in the rough condition has never yet been met with; the articles which are found fashioned out of it all date back to remote antiquity, and are rarer than either nephrite or jadeite objects. Its existence as a distinct mineral was first recognised in an axe-head found in France.

Prehistoric objects of jadeite have a wide distribution in **Europe**. They usually have the form of axes, the so-called flat axes, in contradistinction to nephrite axes, which are generally thicker. In Switzerland, jadeite and chloromelanite objects are found together with articles of nephrite in the ancient lake-dwellings. In France, only jadeite and chloromelanite objects have hitherto been found. In Germany, articles made of jadeite are found all along the course of the Rhine and over the whole of the western portion of the country, namely, in Alsace, Baden, Würtemberg, Hesse, Nassau, Rhenish Prussia, and Westphalia (and in Belgium), and as far northward as Hanover and Oldenburg, and eastward as far as Brunswick and Thuringia. These objects are absent in the eastern part of Germany, but reappear in the Austrian Empire, namely, in Upper Austria, Carinthia, Carniola, southern Tyrol, and Dalmatia. Axes and other objects of jadeite are distributed throughout the whole of Italy. In Greece, and in the neighbouring portions of Asia Minor, numerous specimens have been found, Schliemann having found nephrite objects in excavations at Troy. In Egypt, jadeite often served as the material from which scarabs were cut.

Hitherto jadeite in the rough condition has been found only in small amount in the Swiss and Piedmontese Alps. Among the pebbles derived from the Alps, and which are met with on Lake Neuchâtel and at Ouchy, near Lausanne, on Lake Geneva, are some both of nephrite and of jadeite. Jadeite *in situ* undoubtedly exists at a few places in the crystalline schists of the Alps themselves, for example, at Monte Viso in the Aosta valley, and at San Marcel, both in Piedmont. These occurrences are only important inasmuch as they show that jadeite does actually occur *in situ* in Europe.

In **America**, articles fashioned out of jadeite and chloromelanite by the ancient inhabitants are found in large numbers in Mexico, in Central America, especially Costa Rica, and in the northern part of South America, especially in Venezuela. Jadeite is supposed by some to be the stone known to the ancient Mexicans as "chalchihuitl," but other authorities consider this term to signify turquoise. It is probable that the rough material for these objects was obtained near where they are now found, but no trace of a deposit has hitherto been met with. It has been suggested that jadeite is one of the green stones from the Amazon river to which the name amazon-stone was originally given, but this is very uncertain. Rough jadeite, as well as nephrite, is said to occur in Alaska.

The most important deposits of jadeite are in Asia. Those in **Eastern Turkestan** are of minor importance. The mineral here accompanies or occurs in the vicinity of nephrite, and, like this, is interfoliated with amphibole- and pyroxene-schists, but in less abundance. In the nephrite mines at Gulbashén, in the Karakash valley, jadeite has been found intergrown with nephrite. In the Pamir region jadeite occurs *in situ* in the valley of the Tunga, a tributary on the left bank of the Raskem Daria. The quarries in this deposit are situated about thirty or forty versts from the nephrite quarries in the Kaskem valley, and in latitude about 37° 40′ N. and longitude 76° E. of Greenwich. The deposit was at one time quarried by the Chinese, but since the latter were driven from Yarkand the quarries have been abandoned.

The occurrence of jadeite in **Upper Burma** is much more important. The deposit occupies a limited area on the upper course of the Uru river, and was first thoroughly investigated in 1892 by Dr. F. Noetling of the Geological Survey of India, from whom most of the following particulars are obtained :

The jadeite mines are situated in the sub-division of Mogoung, about 120 miles from this town, and near the river Uru, a tributary of the Chindwin (Maps, Figs. 54 and 55). The mineral is quarried out of the solid rock ; and is also obtained in the form of rounded boulders, which lie in the débris of the Uru river and are probably derived from some unknown deposit *in situ* in the neighbourhood of the river.

The mineral has been obtained from the débris of the Uru river for a much longer period than the primary deposit has been quarried. The boulders occur on the banks of a portion of the river stretching for fifteen or twenty miles down stream from the village and fort of Sanka. Above Sanka pebbles of jadeite are entirely absent, while below the stretch of river just mentioned they are met with so rarely that searchers are scarcely repaid for their trouble. The banks on each side of this stretch of the river have been worked probably for hundreds of years, and the stone is not even yet exhausted. Pits to the depth of 20 feet are sunk in the river alluvium at the foot of the hills which form the sides of the valley. The material excavated out of these pits contains pebbles of jadeite, quartzite, and other rocks. That in the bed of the river is brought up by divers. The pebbles of jadeite are all much rounded, but vary very much in size: three men are said to have been needed to remove one particular block, but boulders of such a size are found only exceptionally.

Good specimens of jadeite are also found, though rarely, in the red, clayey weathered product called laterite, which occurs in the Uru valley. These stones have a colourless nucleus enclosed in a red crust of a certain thickness. This red crust produces the effect of a fine sheen, and is due to the penetration of iron oxide from the clay in which the pebbles are embedded. Such stones, the colouring of which is quite permanent, are distinguished as " red jadeite," and are much esteemed, especially by the Chinese, who pay high prices for them.

The occurrence *in situ* of jadeite in this region has been known since the end of the 'seventies, having been discovered, probably by accident, about fifteen years before Dr. Noetling's visit. A quarry in the deposit at Tawmaw measures now about 100 yards across, and gives employment to between 500 and 600 people of the Katshin race.

The village of Tawmaw is habitable only during the dry season of the year; it is situated in latitude 25° 44′ N. and longitude 96° 14′ E. of Greenwich, six miles west of Sanka and 1600 feet above the Uru river. The jadeite forms a moderately thick bed in a little hill of dark green, or almost black, serpentine, which projects above the surrounding Miocene sandstones.

The labour of quarrying the jadeite is lightened by the practice of lighting fires on the surface. The jadeite, heated to a high temperature, rapidly cools during the night, and becomes cracked and fissured in all directions. The mass is thus broken up into blocks of manageable size, which can be dealt with by the quarryman's hammer. In consequence of this practice the jadeite obtained from the quarry is much inferior in quality to that collected out of pits in the river-banks; in spite of this fact, however, 90 per cent. of the total yield is derived from the quarry, and only 10 per cent. from the river pebbles.

The best of the material is sent on mules by the shortest overland route direct to China, the country in which jadeite is in most request. Another portion of the yield is sent to Mogoung, and from thence in boats to Bhamo, on the Irrawaddy, in consequence of which Bhamo is often incorrectly given as a locality for Burmese jadeite. From Bhamo the mineral is sent down the river to Mandalay, where it is carved in the large lapidary works or only roughly cut and sorted according to quality. What is not used up in Burma is exported through Rangoon by the sea route to China or to Europe.

The mineral found here is known both to the natives and to the Chinese by the Burmese name "kyauk-tsein." It is white, often with a marbled appearance, and contrasts sharply in the quarry with the dark green or black serpentine, by which it is surrounded,

Here and there in the white, translucent jadeite are sometimes patches of a fine emerald-green colour, the material of which differs in nowise from the ordinary white kind. Jadeite of this description is much valued and is carved for ring-stones, or the whole of a ring or bracelet may be fashioned out of it in such a way as to display the bright green spots in the most effective manner.

Fine white jadeite, as well as that spotted with green, is very valuable, even at the quarry. For a block measuring a cubic yard and containing a good deal of green material, £10,000 was demanded, and £8000 was actually offered by a Chinese merchant. A small green stone, large enough for a seal-stone, will fetch between 400 and 500 rupees there, but in Europe will be worth very much less. The value of the whole deposit, which seems inexhaustible, is thus enormous, although of course it must not be supposed that material like that described above is found every day, especially with the primitive methods of quarrying at present employed.

It may be remarked that though isolated specimens of "jadeite from Bhamo" may have the exceptionally low specific gravity mentioned above, yet, as a general rule, jadeite from Upper Burma has the normal specific gravity of 3·3.

This locality in Burma is the only one known to Europeans at which jadeite indisputably occurs *in situ*. It is very probable that there are several others in the same country, and in the region extending far into the Chinese province of Yun-nan, some of which may be worked by natives in the manner employed at Tawmaw. Moreover, in many of the rivers of these regions it probably exists as pebbles, and is sought for just as in the valley of the Uru river, but at present this is not indisputable. It is certain, however, that dark green and white pebbles, said to have been found in "Tibet in the northern Himalayas," are at present cut at Oberstein on the Rhine. They do not agree in character with those from the valley of the Uru, but more detailed statements as to the locality from whence they are brought are not forthcoming. There are often to be seen in mineral collections specimens labelled jadeite from Tay-hy-fu, or Talifu, in Yun-nan (latitude 26° N. and longitude 100° E. of Greenwich). This, however, is not in reality a locality for the mineral, but merely a stage on the journey from the Burmese deposits, described above, to Pekin; and on examination such specimens are found to agree in every particular with the jadeite known to have been found in Upper Burma.

QUARTZ.

No other mineral exists in such abundance, and at the same time affords so many different ornamental stones as does quartz in its several varieties. From one point of view quartz is one of the most important of precious stones, not on account of its value but because of its very wide distribution, and also because of the occurrence of even the most beautiful varieties in large masses. For these and other reasons all the ornamental varieties of quartz are classed with the so-called semi-precious stones.

In order to avoid repetition the characters common to all the varieties of this mineral will be first described, and afterwards the features which distinguish the varieties used as gems and for ornamental purposes.

Quartz is pure silica, that is to say, it is oxide of silicon with the chemical formula SiO_2, and consists, in the purest condition, of 46·7 per cent. of silicon and 53·3 per cent. of oxygen.

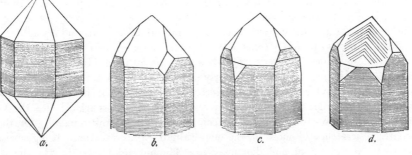

FIG. 85. Crystalline forms of quartz.

More frequently than not it contains impurities of various kinds, a fact to which the variety in colour of the mineral is due. Quartz differs from opal, another mineral which consists largely of silica, in that it is completely free from water.

Quartz differs also from opal in another important respect, namely, in that it is not amorphous but crystallised. Very finely developed crystals of quartz, usually with brilliant faces, are extremely common. They belong to the hexagonal system (Figs. 85a – d), and almost without exception take the form of regular six-sided prisms, the faces of which are characterised by the presence of very distinct striations perpendicular to the edges of the prism. The latter is terminated at one end (Figs. 85b – d), and in completely developed crystals (Fig. 85a) at both ends, by a six-sided pyramid, the faces of which intersect the prism-faces in horizontal edges. In addition to these common faces there are often others of small size (Figs. 53b – d), the arrangement of which indicates that the symmetry of quartz is that of the trapezohedral-tetartohedral division of the hexagonal system. The habit of quartz crystals is rather variable. In many cases the prism-faces are elongated, as in the forms represented in the figures; but sometimes they are short or even completely absent, and in the latter case the crystal becomes a double six-sided pyramid. The faces of this pyramid may be all of the same size, but more frequently large faces alternate with small ones, so that there are three large and three small faces (Figs. 85b – d). Owing to irregular growth it may happen that the faces of the pyramid all differ in size and are arranged in no particular order as to size. The small faces represented in Figs. 85*b* and *c*, between the prism- and the pyramid-faces are very often absent; when present they occur in regularly developed crystals on the alternate upper and lower corners (see Figs. 85*b* and *c*), but this arrangement is often disturbed by the twinning of the crystals (Fig. 85*d*).

Quartz crystals occur either embedded in the mother-rock, when they are developed on all sides, as in Fig. 85*a*, or attached to a rock surface, in which case only the free end is developed (Figs. 85*b* — *d*). The attached crystals sometimes form fine druses, as represented in Plate XVII. In this illustration the crystals are of an elongated prismatic habit, but sometimes, when only the terminal pyramid-faces are developed, the surface of the rock appears covered with close-set crystals.

Quartz occurs very commonly also in compact masses. As isolated irregular grains it forms an important constituent of many widely distributed rocks, such as granite, gneiss, &c., and, in more rounded grains, of sand, sandstone, grit, &c. Frequently, also, aggregations of microscopically small quartz grains are met with; these aggregations constitute the different varieties of compact quartz, such as hornstone (which includes the green chrysoprase), jasper, and others. Columnar aggregates are not uncommon, and when the columns or fibres are very slender the mass has a fibrous structure, as, for example, in tiger-eye. Altogether these compact varieties of quartz furnish quite a large number of ornamental stones, each of which will be described in its proper place.

The mineral has no distinct cleavage. The fracture of crystals and crystalline masses is conchoidal, almost as perfectly conchoidal as that of glass; in compact aggregates it is uneven to even, sometimes splintery. Quartz stands seventh on Mohs' scale of hardness; though not as hard as precious stones properly so-called, it is harder than any other widely distributed mineral, and scratches the majority with ease, as it does also window-glass. Steel or iron when struck against quartz gives rise to bright sparks, hence the German name "Feuerstein" (fire-stone) for flint, a compact variety of this mineral. On the other hand, quartz is scratched by most of the more valuable precious stones, being surpassed in hardness by topaz, corundum, and diamond, &c. The mineral is very brittle, and splinters may be broken off, at any rate, from large crystals, with little difficulty; some of the fine grained and compact varieties, however, are much less easily broken.

The specific gravity of pure quartz is 2·65, so that in this respect also quartz differs from opal, which is appreciably lighter. It should be observed that the value given above applies only to pure material, and that the specific gravity of impure varieties may be rather less or rather greater.

Quartz is infusible before the ordinary blowpipe, but can be fused in the oxyhydrogen flame. It is completely dissolved by hydrofluoric acid, but is unattacked by other acids; and is scarcely affected at all by a solution of caustic potash, in which, however, opal readily dissolves. On being rubbed it becomes electrified and retains its charge for nearly an hour.

The many varieties of quartz which occur in nature differ widely in external appearance, these differences depending on variations in structure, lustre, transparency, or colour.

Generally speaking, the lustre of quartz is of the vitreous type; some specimens exhibit a greasy lustre (greasy quartz), while finely fibrous aggregates sometimes possess a very good silky lustre. All varieties of quartz are susceptible of a high polish; the lustre of natural faces and fractured surfaces is often not particularly strong, but is much heightened by cutting and polishing.

In its purest condition quartz is perfectly transparent and colourless, and in many

Plate XVII

GROUP OF QUARTZ CRYSTALS (ROCK-CRYSTAL): FROM DAUPHINÉ, FRANCE

cases the coloured varieties also freely allow the passage of light. These transparent varieties of quartz are distinguished as precious quartz; between precious quartz and cloudy or opaque common quartz every degree of transparency is to be found.

Chemical examination of the coloured varieties of quartz shows that their colour is due to the presence of impurities intermixed with the colourless quartz substance. In some cases the pigment, the nature of which is unknown, is in the finest state of subdivision, and is distributed uniformly throughout the mass of the stone, so that it is impossible to distinguish individual particles even with the aid of the highest powers of the microscope. In other cases the colouring matter is seen under the microscope to be located in needles, fibres, grains, scales, &c., of other mineral substances, which are embedded in the quartz substance in such numbers as to impart to it their own colour. To the former class belong brown smoky-quartz, violet amethyst, yellow citrine, and rose-red rose-quartz, and to the latter green prase, blue sapphire-quartz, and others. The range of colour shown by quartz is very extensive ; no colour is entirely unrepresented, and the majority exist in a number of different tints and shades. There is very often a marked contrast between the colour, or the shade of colour, of different portions of the same specimen ; this is shown most typically in agate, that variety of quartz so much used for ornamental purposes. The most important of the coloured varieties of quartz have been enumerated above.

Optically, quartz, being a hexagonal mineral, is doubly refracting, but neither the refraction nor the double refraction is very strong. An inspection of the table given below will show how small is the difference between the ordinary and the extraordinary refractive indices for light of different colours :

	$o.$	$e.$
Red light	1·5409	1·5499
Yellow light	1·5442	1·5533
Green light	1·5471	1·5563
Blue light	1·5497	1·5589
Violet light	1·5582	1·5677

Although the double refraction of quartz is not very great, it is strong enough to make the images of any small object, such as a candle-flame, appear double when viewed through a transparent faceted stone (Fig. 26a). By this means quartz may be distinguished from glass of similar colour, since this will show only single images (Fig. 26b). A comparison of the numbers in the table given above shows further that there is little difference between the refractive indices of quartz for light of different colours, and consequently that the dispersion of this mineral is feeble, so that it exhibits no play of prismatic colours comparable to that of the diamond.

A peculiar optical feature possessed by quartz, and by only one other mineral, is the power of rotating the plane of polarisation of polarised light, but this being of purely physical interest need not be dwelt upon here.

The characters common to all varieties of quartz having been described, we must now direct our attention to the features which distinguish the ornamental from each other and from the common varieties. Of the ornamental varieties of quartz there are first those which occur in regular crystals, or in aggregates of crystals, the individuals of which are recognisable as such by the naked eye, though they may sometimes have irregular boundaries. Next we have the compact varieties, consisting of an aggregate of numbers of quartz crystals of microscopic size. The crystallised varieties may be further subdivided according to colour, and the compact aggregates according to structure and other characters. Thirdly, there is the division which includes chalcedony, the special characters of which will be described later.

The different varieties of quartz which are used for ornamental purposes may, therefore, be classified in the following manner, and will be dealt with below in the same order :

A. CRYSTALLISED QUARTZ.

Rock-crystal.

Smoky-quartz.

Amethyst.

Citrine.

Rose-quartz.

Prase.

Sapphire-quartz.

Quartz with enclosures.

Cat's-eye.

Tiger-eye.

B. COMPACT QUARTZ.

Hornstone (including wood-stone and chrysoprase).

Jasper.

Avanturine.

C. CHALCEDONY.

Common chalcedony.

Carnelian.

Plasma (including heliotrope).

Agate (including onyx).

A. CRYSTALLISED QUARTZ.

ROCK-CRYSTAL.

Perfectly limpid, colourless, and transparent quartz is known as rock-crystal. It stands out prominently among all other minerals by reason of its clearness and transparency, in which respects it often surpasses even the diamond, although it is not comparable with the latter in lustre or play of colours. An irregular mass of rock-crystal at first sight looks very like colourless glass or pure ice, and for this reason it is sometimes called glass-quartz. In ancient times, and indeed even in the Middle Ages, it was thought to be actually ice, which had been frozen so hard on the highest peaks of the Alps, where rock-crystal is very abundant, that it could not be thawed again. From this belief arose the use of the word crystal, a term first applied to rock-crystal.

This variety of quartz occurs very commonly in fine crystals, the prism-faces of which are almost without exception largely developed, so that the habit of the crystals is columnar (Figs. 85a – d). The small faces on the corners, between the prism- and pyramid-faces, sometimes present in this variety, are represented in Figs. 85b and c. Such crystals are, as a rule, attached at one end to the matrix, from which those represented in Figs. 85b – d are broken off, but crystals developed at both ends (Fig. 85a) are by no means uncommon. Crystals of this variety of quartz differ, therefore, in their development from those of common quartz, in which the prism-planes are short, or even altogether absent, and the pyramid-planes only present, while the small faces at the corners are of very rare occurrence. Rock-crystal often occurs in magnificent groups, like the one represented in Plate XVII. from the neighbourhood of Bourg d'Oisans, in the Dauphiné Alps in France.

There is considerable variation in the size of the crystals. The smallest measure a few millimetres in length and weigh no more than a few milligrams, while the largest, which are least common, measure some metres in length and weigh several hundredweights. The commonest are those of medium size, having the length and thickness of a finger or rather more.

Like all varieties of quartz, rock-crystal often contains enclosures of the most varied kinds, which, on account of the extreme clearness of the substance in which they are embedded, can be seen very distinctly. Beside enclosures of foreign matter, rock-crystal not infrequently contains cavities which are either vacuous or filled partially or completely with liquid. In the former case the remaining space is usually occupied by a bubble of gas, which moves about in the cavity as the crystal is moved. Sometimes such cavities containing liquid and an air-bubble are distinctly visible to the naked eye, but more often they are of microscopic size. In the latter case they are often crowded together in such vast numbers, especially at the attached end of the crystal, that it appears quite cloudy, while the free end is perfectly transparent. In many cases the liquid, which fills these cavities, may be easily proved to be carbon dioxide, but in other cases it is either water or a solution of common salt.

The enclosures of solid bodies, especially of crystals belonging to other mineral species, are more important. They may be extremely small in size but present in very large numbers, so that they impart to the whole crystal a uniform colour, as in the case of green prase and blue sapphire-quartz. On the other hand, these enclosures may be few in number and large enough to be seen with the naked eye, especially when they are conspicuously coloured. Scales of green chlorite, for example, are often seen either embedded in rock-crystals or attached to the surface in such numbers that but little of the colourless quartz substance can be seen, and the crystals appear of a green colour. Blades of white tremolite, needles of green actinolite (Plate XVIII., Fig. 2), and of red or yellowish rutile, are often enclosed in rock-crystal, and in many cases it is on the presence of these and other enclosures, both fluid and solid, that the application of some varieties of quartz depends.

Formerly rock-crystal was much used as a gem, and was cut as a brilliant, table-stone, or rosette. A bright vitreous lustre is imparted to it by polishing, and this can be slightly increased by igniting the stone. Under favourable conditions cut stones show a certain play of prismatic colours, and for this reason it was fashionable at one time to decorate chandeliers, hanging lamps, candlesticks, &c., with prisms and pendants of rock-crystal.

The mineral was at one time much used also for the fashioning of small articles of more or less utility, such as crystal-balls, letter-weights, seal-stocks, &c. In the Middle Ages all kinds of vessels, bowls, vases, and drinking cups, frequently so beautifully engraved with pictured figures as to be transformed into veritable works of art, were carved out of this substance. The art flourished most at the time when the manufacture of perfectly clear and colourless glass was still in its infancy, and rock-crystal was the source from whence material of this description could be most easily obtained. As the glass-making industry developed and reached perfection, the art of working rock-crystal became forgotten, for it was soon found that vessels equally transparent and finely finished could be made in glass with much less labour than in rock-crystal. Articles carved out of rock-crystal are of course more durable and not so readily scratched. Many art collections bear eloquent testimony to the perfection which the art of working rock-crystal had reached among the ancients, while in ancient writings there are to be found many records of historic vessels of this substance.

At the present day rock-crystal is little used for gems or for the purposes described above. On account of its comparatively great hardness and its resistance to chemical reagents it is devoted rather to utilitarian purposes, such as the construction of lenses for spectacles, telescopes, and other optical instruments, lenses of this material being less likely to be scratched than are those of glass. Its perfect transparency renders it useful for other optical purposes, while it is used also for the manufacture of very exact weights to be employed when extreme accuracy is required, for the pivot supports of various delicate instruments, and for other similar purposes.

The value of rock-crystal, like its use, has much diminished during recent times. It depends upon the purity, transparency, colourlessness, and freedom from faults of the material, such as enclosures of foreign matter, fissures, cloudy and coloured patches, and anything, in short, which detracts from the uniformity of quality of the stone. The value of any given stone depends also, of course, upon its size; small pieces of perfect quality are not uncommon, and for this reason the price of a cut ring-stone, even of the best quality, scarcely ever exceeds ten shillings. Masses of considerable size and of good quality, on the other hand, are not so easy to obtain, and the larger the size the higher relatively is the price.

Rock-crystal is very widely distributed, and is usually found, together with other minerals, in the cracks and crevices of various ancient rocks, in which it forms druses, often of enormous size. This mode of occurrence, taken in conjunction with the scarcely ever failing presence of fluid enclosures, indicates very clearly that the mineral for the most part has crystallised from an aqueous solution of silica. It would be impossible to mention severally all the localities at which rock-crystal occurs; a few typical occurrences only will be here enumerated.

Such in **Europe** are the high mountain peaks of the Tyrolese, Swiss, Italian, and French Alps. The crystals, which are attached to the walls of crevices in granite, gneiss, and other similar rocks, are collected and brought into the market by persons known in Switzerland as "Strahlern." Their task is anything but easy, since the crystals almost invariably occur in the highest and most inaccessible parts of these ranges. In the search for crystal-bearing druses a "Strahler" is guided by the quartz-veins, which extend like white bands across the faces of the rocks, and in which the druses are usually found. The character of the sound given out when the vein is struck with a hammer indicates whether there is likely to be a cavity, and therefore possibly a druse, at that particular spot. If there is, the rock is broken open with a pickaxe or by blasting with powder or dynamite and the crystals obtained.

These cavities and the crystals contained in them are not usually of very large size. Drusy cavities of enormous dimensions are, however, met with occasionally, and are known as crystal-caves or vaults. Several hundredweight of crystals have been found in one such cavity, and single crystals sometimes weigh a hundredweight or more.

A celebrated find of this kind was made in the year 1719 in a crystal-cave on the Zinkenstock, near Grimsel, in the Bernese Oberland. One crystal found here weighed eight hundredweight, many weighed a hundredweight or more, and altogether 1000 hundred-weight of crystals were taken from this gigantic druse. From a cave in the Vieschthal, between Münster and Laax, in Upper Valais, 1757 crystals, ranging in weight from 50 to 1400 pounds, were obtained. An occurrence of rock-crystal, only in comparatively small druses, but often mentioned, is in the slightly auriferous quartz-veins of La Gardette, near Bourg d'Oisans, in Dauphiné (Department Isère), in the French Alps. From this locality comes the group of crystals represented in Plate XVII.; they always possess peculiar oblique and unsymmetric terminations, which is characteristic, and distinguishes them from crystals from other localities. At the beginning of the nineteenth century material from here and elsewhere was cut in the lapidary works of Briancon on the Durance, Department Hautes

Alpes, and the water-clear stones cut here were called "Briançon diamonds." The working of rock-crystal at this place has long been abandoned; and, indeed, the Alpine material has now very little industrial significance, being replaced by Brazilian crystals, which are not only purer but also cheaper.

The weathered débris from the higher Alps, which contains rock-crystal detached from the parent rock, is transported to lower levels by glaciers and streams. The crystals are rolled along the beds of streams and rivers, becoming as they travel further down the valleys more and more water-worn, until at last they take the form of perfectly rounded pebbles, which, on account of surface scratches, appear to be cloudy, but which in reality are perfectly transparent inside. Such pebbles, the biggest of which are the size of a nut, are found in the Rhine, having reached this river through the Aar. At many places in Baden they were at one time cut under the name Rhine-quartz (Germ. *Rheinkiesel*) and constituted a by-product of the gold-washing industry. This *Rheinkiesel* was formerly, though incorrectly, considered to be finer and purer than the rock-crystal which had been brought down the Alps by other means. Similar pebbles were found and occasionally utilised in the same way at other places, for example, at Médoc and at Alençon (" Alencon diamonds ") in Normandy, at Fleurus in Belgium, and at Cayenne.

There are many extra-Alpine occurrences of rock-crystal, but only a few will be mentioned here. Magnificently clear, though not very large, crystals occur in the cavities of the world-famous statuary marble of Carrara in Italy. Very fine water-clear crystals, developed on all sides and often showing no distinct point of attachment, occur in the cavities of the dark Carpathian sandstone, or in the clay-slates with which it is interbedded, in the Marmaros Comitat, in north-east Hungary, bordering on Galicia. The crystals range in size from that of a pin's head to that of a nut; they are collected at the surface of the ground usually after heavy storms of rain. They have a wide distribution in the district, but Veretzke, in the valley of the Nagyag river, and Bocsko may be mentioned as typical localties. In reference to the region in which they occur these crystals are known as " Marmorosch diamonds." Small rock-crystals are found in crevices in Triassic (Lettenkohle) marl, in the Hessian county of Schaumburg, on the lower Weser. These were known as "Schaumburg diamonds," and were formerly used as gems, only a small proportion, however, being perfectly clear and fit for cutting.

In England fine, transparent rock-crystal is found in quartz-veins in slate at Tintagel and Delabole, in Cornwall, and elsewhere; they are known variously as " Cornish diamonds," " Bristol diamonds," &c.

Although the occurrence of fine rock-crystal in Europe is very abundant, it is far surpassed by that of other parts of the world. A large amount was being obtained from **Madagascar** at the end of the eighteenth century, and a considerable amount still comes from that island. The material found here is specially pure and clear, and often occurs in blocks of some size, the largest measuring as much as 8 metres in circumference. It occurs in special abundance in the form of isolated, partially water-worn blocks on the slopes of the Befoure mountains, where it is collected for the market. Large blocks of perfect quality, and weighing from 50 to 100 pounds, are not at all uncommon, and it is the abundance of excellent material in Madagascar which has caused the depreciation in prices.

Rock-crystal is common in **India** also. It was, and is still, worked there in various ways, but the objects now carved and worked in this material do not bear comparison with those of by-gone times. The industry is still in operation at Vellum, in the Tanjore district of the Madras Presidency, where brilliants, rosettes, spectacle-lenses, &c., are produced.

The material is obtained in the neighbourhood from a conglomerate of Tertiary age, in which it occurs in the form of pebbles. A once famous centre of this industry is Delhi, the bowls, vases, drinking-cups, and other objects carved there being renowned for beauty of design and skill in workmanship. The art is now forgotten, but the old mines at Aurangpur, fifteen miles south of Delhi, whence rough material was obtained, are still to be seen. There are many other localities for rock-crystal in India, but none appear to have any industrial importance.

Rock-crystal is very abundant in America, and specially so in **Brazil.** Not only are large amounts of very fine and clear rock-crystal obtained in this country, but also many of the coloured varieties of quartz (amethyst, citrine, &c.). The rock-crystal is much used for spectacle-lenses and other optical instruments, and for various other purposes. Brazilian rock-crystal ("Brazilian pebble") is easy to obtain, and is very cheap, and has therefore ousted other material from the market. A finished spectacle-lens cut in Brazil from Brazilian material, according to Mr. G. F. Kunz, of New York, fetches less than the cost of cutting a lens in that city. The States in which the mineral is most abundant are Minas Geraes, São Paulo, and Goyaz, especially in the last named in the Serra dos Cristaes, sixty-five miles from Santa Lucia and 200 miles from the town of Goyaz. The crystals here lie about loose on the ground, and some weighing sixty-four pounds have been found in the surface soil. Beside the rock-crystal there is quartz of every colour and quality. For a long time the occurrence gave employment to 200 persons, who in the space of two years collected 7000 tons of material. Later, however, the demand fell off, and finally almost ceased, mining operations becoming at the same time less active.

There are also many localities for rock-crystal in the **United States** of North America. At Chestnut Hill, in the State of North Carolina, are found pure crystals weighing a hundredweight or more; the largest American crystal known was met with there and weighed 131 kilograms (288 pounds). The next largest American crystal weighing 86 kilograms came from Alaska. On Lake George, in New York, and over a wide area in Herkimer County, in the same State, rock-crystals of small size but developed on all sides occur in cavities in the Calciferous sandstone. In lustre, transparency, and purity these crystals rival even those of Carrara, in Italy. They have been collected in large numbers for upwards of half a century, and both cut and rough stones are sold in the neighbourhood, mostly to tourists, under the name "Lake George diamonds" or some similar term. Some specimens enclose large drops of water and others black grains, often of considerable size, of a bituminous substance; in the former case the stones are of greater, and in the latter of less, value than ordinary specimens. Water-clear rock-crystal occurs in abundance in Crystal Mountain and elsewhere within a forty miles radius from Hot Springs, in Arkansas. Waggon-loads of these so-called "Arkansas diamonds," which are found in crevices in red sandstone, are brought by the farmers to Hot Springs and Little Rock, where thousands of dollars' worth are sold to the visitors at the baths. In the Washita river at Hot Springs there are quartz pebbles like those found in the Rhine, and so eagerly are these sought for by tourists that the supply has been artificially increased by grinding together fragments of rock-crystal in rotating barrels. Rock-crystal occurs abundantly also in Canada.

Rock-crystal of fine quality has recently been found at Mokelumne Hill, Calaveras County, California; an account of this occurrence has been given by Mr. G. F. Kunz in his report on precious stones for 1898. The large crystals are found embedded in the gravels and sands of an ancient river channel; they are only slightly water-worn, but are usually much stained on the exterior. One crystal measures 19 by 15 by 14 inches, and another

14 by 14 by 9 inches. From one crystal a perfectly flawless sphere measuring $5\frac{1}{2}$ inches in diameter has been cut; larger balls, with a diameter of $7\frac{1}{5}$ inches, are not entirely free from flaws. The rock-crystal of this locality thus rivals in size and transparency that of Japan, Brazil, Madagascar, and the Alps, which have hitherto been almost the only sources of such material.

Under some circumstances rock-crystal may be mistaken for other water-clear precious stones, and being cheaper is sometimes successfully substituted for one or other of these colourless gems. The resemblance to diamond in some of its features is indicated by the number of names given in different parts of the world to this stone in which the word diamond figures as an affix. Thus it is known as "false diamond" or "pseudo-diamond," and, according to the locality in which the crystals occurs, as "Marmorosch diamonds," "Schaumburg diamonds," "Arkansas diamonds," "Bohemian diamonds," "Irish diamonds," "Paphos or Baffa diamonds," "Fleurus diamonds," "Bristol diamonds," "Isle of Wight diamonds," "Quebec diamonds," and so on, and generally as "occidental diamonds." All these terms signify one and the same stone, namely, rock-crystal, and since even the finest specimens are not for a moment comparable either in brilliancy, lustre, or play of colours with diamond, any real confusion seems impossible.

In case of necessity rock-crystal and diamond can be distinguished by the help of the facts that the former is doubly and the latter singly refracting, that the diamond quickly sinks while rock-crystal floats in pure methylene-iodide, and also by the enormous difference in hardness between the two. The specific gravity of rock-crystal serves to distinguish it from all other colourless stones. These, arranged in order of density commencing with the heaviest, are: zircon, sapphire, topaz, spinel, tourmaline, and phenakite. Phenakite, the lightest of these, has a specific gravity of 2·99, so that it is considerably heavier than quartz, which has a density of 2·65 and remains suspended in the lightest test liquid, in which all the other stones sink.

Although rock-crystal is so cheap and abundant glass imitations are often substituted for it, and many of the so-called false diamonds are in reality nothing but glass. Genuine rock-crystal is always considerably harder, and often also lighter, than the glass imitations; the two can be readily distinguished by the help of the polariscope, rock-crystal being doubly, and glass singly refracting.

SMOKY-QUARTZ.

This is a transparent variety of quartz which is brown to almost black in colour. It is often called "smoky topaz," but has no connection whatever with topaz. Smoky-quartz when perfectly transparent is not infrequently cut, and with its deep, rich colour makes a rather effective gem.

The colour ranges from clove-brown to smoke-grey, passing by imperceptible stages to the perfect colourness of rock-crystal, and, on the other hand, through darker and darker shades till we reach specimens which, at any rate, in thick pieces, are almost perfectly black. Such dark smoky-quartz is distinguished as *morion*. Specimens of this stone are not always uniformly coloured, some portions being lighter and others darker. All darkly coloured smoky-quartz is distinctly, though feebly, dichroic, the two dichroscope images being coloured respectively yellowish-brown and clove-brown tinged with violet. The paler the colour of the stone the less the difference in colour between the two images, and the dichroism of very pale stones is scarcely observable at all. A crystal of smoky-quartz is represented in Plate XVIII., Fig. 3a, and a cut-stone of paler colour in Figs. 3b and c, of the same plate.

The colouring of smoky-quartz is due to the presence of a volatile organic substance containing carbon and nitrogen, which can be distilled off as a turbid liquid. The distillation is accompanied by a smell of burning, which is noticeable even when a dark coloured specimen, that is to say, one containing much colouring matter is broken, or when two fragments of the same are vigorously rubbed together. On igniting smoky-quartz, or on merely raising its temperature to 200° C., the organic colouring matter is completely destroyed, and the mineral becomes indistinguishable from rock-crystal. At a low temperature the brown colour becomes yellow like that of citrine, another variety of quartz. Probably not a little of the yellow quartz which comes into the market is none other than "burnt" smoky-quartz.

With the exception of colour, all the characters of smoky-quartz are identical with those of rock-crystal. The similarity in the habit of the crystals, and their mode of occurrence, is specially striking, and we are, therefore, justified in stating that smoky-quartz is simply rock-crystal of a brown colour. This is not always the case, for we shall see later that amethyst, a violet-coloured variety of precious quartz, differs from rock-crystal not only in colour but also in structure.

All that has been said with regard to the crystalline form (Fig. 85) of quartz, and its occurrence in crevices in the gneissic and granitic rocks of the Alps and at other localities, together with the existence of very large drusy cavities containing hundredweights of the finest crystals, applies equally well to smoky-quartz, and need not be repeated.

The most remarkable occurrence of smoky-quartz in large crystals of deep black morion was discovered in August 1868, in a crystal cave near the Tiefen glacier in Canton Uri, **Switzerland.** The cave was in weathered granite, and out of it was collected in a very short time 300 hundredweights of crystals, 200 hundredweights of which were beautifully transparent and suitable for cutting, and the remainder for museum specimens. Among the latter are some of remarkable size, the largest of which are now exhibited in the Berne Museum. Several of these crystals are distinguished by special names: the "Grandfather" is 69 centimetres long, 122 in circumference, and weighs 133½ kilograms. The "King" is rather thinner and less heavy, but longer and better preserved than any; it is 87 centimetres long, 100 centimetres in circumference, and weighs 127½ kilograms. The two smallest of the half-dozen crystals preserved at Berne are named "Castor" and "Pollux"; they are 72 and 71 centimetres in length and 65 and 62½ kilograms in weight. Each of these crystals was attached to the matrix at one end, but in the case of one large crystal, which has perfectly developed faces at both ends and on all sides, it is impossible to see the point of attachment, although it must originally have grown attached to the matrix just like the others; this is 82 centimetres long, 71 in circumference, and weighs 67 kilograms. Another crystal found at the same time and place is preserved in the Mineral Department of the British Museum (Natural History); it weighs 299 pounds and is just over a yard long.

Compared with this Alpine occurrence of smoky-quartz all others are unimportant. It is met with accompanying topaz, beryl, and specially amethyst, at Mursinka, in the Urals, and with beryl and topaz in the neighbourhood of Nerchinsk, in Transbaikalia. In the form of pebbles it is found with pebbles of rock-crystal in the Rhine; and in the same way at Alençon, in Normandy, and again in the gem-crystals of Ceylon.

In **Scotland** it is found on Cairngorm, a mountain on the borders of Banffshire and Inverness-shire, and on this account a brown transparent quartz is often referred to in

England as *cairngorm*. The search for crystals of smoky-quartz was formerly a very profitable industry in the districts contiguous to the great granite masses in this part of Scotland, but it has now been practically abandoned. The cairngorms were obtained by digging shallow pits and trenches in the decomposed granite and débris which covers most of the flat hill-tops, and also appears in many of the corries. The mineral occurs, together with large crystals of orthoclase and plates of muscovite-mica and sometimes beryl, lining cavities in veins of fine-grained granite which penetrate the coarse granite.

Material suitable for cutting is met with at several localities in the **United States.** Large quantities, associated with amazon-stone, are found in the coarse-grained granite of Pike's Peak, in Colorado, where several thousand dollars' worth of stones are collected and cut every year. The largest crystal found at this locality is over 4 feet long. At Mount Antero, in Colorado, at Magnet Cove, in Arkansas, and in Burke and Alexander Counties, North Carolina, smoky-quartz is obtained in not inconsiderable amount. Pebbles of smoky-quartz, and also of rock-crystal, are not uncommon on the coast at Long Branch, near Cape May, New Jersey, and are cut and bought by visitors as souvenirs. It would be superfluous, even if it were possible, to enumerate all the many localities for smoky-quartz in the United States. It may be mentioned, however, that here also occur crystals of very large size, weighing upwards of a hundredweight, of the most perfect transparency, and constituting some of the finest of gems.

Smoky-quartz is usually cut as a brilliant or table-stone, sometimes also in the step-cut, the form with elongated brilliant facets, or the Maltese cross (Plate XVIII., Figs. 3*b*, *c*). Under the name of cairngorm it is specially admired as a gem in Scotland. Besides being cut as a gem, smoky-quartz, like rock-crystal, is fashioned into a variety of semi-ornamental objects, its strong lustre and rich colour being displayed by such articles with good effect.

There are but few precious stones of a brown colour, and of these smoky-quartz is the most transparent, and also the most abundant. The brown precious stones for which it might conceivably be mistaken are axinite, idocrase, and brown tourmaline; others are of a more yellowish-brown colour, and therefore distinguishable at a glance from smoky-quartz. Each of the three minerals mentioned above is heavier than smoky-quartz, and the specific gravity of each being greater than 3·0 they all sink in liquid No. 3, in which smoky-quartz floats. Moreover, smoky-quartz is only feebly dichroic, while the other three are strongly so. Brown diamond, which may resemble smoky-quartz in colour, is readily distinguished by its strong and characteristic lustre, its single refraction, complete absence of dichroism and high specific gravity, the last named character causing it to sink heavily even in pure methylene iodide.

AMETHYST.

Amethyst, or occidental amethyst as it is called in order to distinguish it from "oriental amethyst," is quartz of a violet colour. The range of colour includes reddish-violet tints of pale or almost colourless shades, and deep, rich tones of pure violet. It is not uncommon for this stone to show patches of different shades of colour or colourless portions side by side with those of a violet colour, these differently coloured portions being sometimes arranged in regularly alternating bands or sectors. In a few rare cases crystals showing a second colour—yellow or green—have been met with.

Generally speaking, amethyst is cloudy: perfectly clear and transparent material is distinguished as precious amethyst; the latter alone is cut as gems, which are estimated according to their transparency and their depth and uniformity of colouring. Pale-coloured and patchy specimens are almost worthless as gems, but may be utilised in the manufacture of small semi-ornamental objects.

The colour of amethyst is very like that of violet corundum, the "oriental amethyst." It compares unfavourably with the latter in one respect, however, since in artificial light it appears of a dingy grey, while under the same conditions the "oriental amethyst" retains its beauty of colour unimpaired.

Amethyst is dichroic, but this feature is not equally prominent in all crystals, being well marked in some and scarcely appreciable in others. Of the two images seen in the dichroscope one is more reddish and the other more bluish in colour, the difference being more or less well marked according to the character of the crystal under examination.

The colour of this stone does not resist high temperatures. When subjected to the action of heat it changes first to a more or less pronounced yellow, gradually assumes a greenish shade, and finally disappears. The latter change takes place at 250° C., and the whole process of decolourisation takes place very rapidly. The change of colour from violet to yellow has a certain practical significance, since thereby the more abundant violet amethyst can be transformed into the rarer yellow citrine. As a matter of fact, many citrines, which are much prized as precious stones, are nothing other than "burnt" amethysts.

The colour of amethyst has been ascribed to the presence of various substances. The pigment is present in an extremely finely divided state and is mechanically intermixed with the colourless quartz substance, but the individual granules are not recognisable even under the highest powers of the microscope. In patchy specimens the colouring matter is irregularly distributed through the substance of the stone. The pigment has been variously supposed to be potassium ferrocyanide, ferric thiocyanide, or some organic substance, but is most commonly considered to be manganese, which is shown by analysis often to be present, though only in extremely small amount. For example, in a deeply coloured Brazilian amethyst only one-hundredth of a per cent. of manganese was present, while in a stone of a paler colour it was completely absent, so that this evidence cannot be taken as conclusive.

The form taken by crystals of amethyst agrees in all essentials with that of rock-crystal. Amethyst crystals (Plate XVIII., Fig. 1a), too, very often have the prismatic habit of rock-crystal, especially when they occur in the drusy crevices of gneiss and such like rocks. They usually differ, on the other hand, from rock-crystal in the absence of faces other than those of the prism and the terminal faces of the hexagonal pyramid, being in this respect more like common quartz. Crystals of amethyst with the pyramidal termination often occur so closely crowded together on the matrix as to form a columnar aggregate, the individuals of which have irregularly developed faces owing to overcrowding. Of the six terminal pyramid-faces it frequently happens that only three alternate ones are developed, the other three being very small, or even completely absent, so that the crystals sometimes have the appearance of cubes. Such cube-like amethysts are often met with amongst the material sent in large amount from South America for cutting.

Amethyst crystals are often built of numerous superimposed laminæ, of greater or less thickness, in twin position with respect to each other. This peculiar structure is indicated by a fine banding of alternately lighter and darker colouring, and also by the existence of two sets of delicate striations meeting at an angle on the terminal pyramid-faces (Fig. 85d). Delicate striations corresponding to this lamellar structure are also to be seen on fractured surfaces of amethyst, giving them a kind of rippled (thumb-marked) appearance. In this connection may be mentioned the so-called sceptre-quartz (Fig. 86), a peculiar form sometimes taken by rock-crystal, but much more frequently by amethyst. In sceptre-quartz then, there is a long, slender, usually colourless, transparent, or cloudy prism of quartz, attached to the end of which in parallel position is a thicker quartz crystal, which is usually transparent and of a violet colour.

Crystals of amethyst sometimes attain a considerable size, the largest known being over a foot in length. These very large crystals are rarely quite transparent or uniformly coloured, and are therefore unsuitable for gems; but there is an abundance of crystals which fulfil these conditions, and which are large enough to supply the market with any amount of rough material.

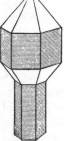

FIG. 86.
Sceptre-quartz.

With respect to the mode of occurrence of amethyst, we have already seen that crystals of elongated prismatic habit usually occur, like rock-crystal, on the walls of crevices and joints in granite, gneiss, and other rocks. Crystals in which only the hexagonal pyramid or three-faced cube-like termination is developed, have a different mode of occurrence. These usually line amygdaloidal, that is to say, almond-shaped, cavities in a black igneous rock, to which the name melaphyre is applied. The cavities were formed, in their efforts to escape, by the steam and gases imprisoned in the molten igneous rock, and have remained after the solidification of the rock. As long as the rock preserved its fresh, unaltered condition, these cavities remained empty, but with the weathering and alteration of the rock the cavities become wholly or partially filled with secondary minerals. The alteration process is set up by the percolation through the rock of water, which, becoming charged with carbon dioxide and other acids, is enabled to dissolve out many of the constituents of the rock, which are then redeposited in a new form in the cavities. Many and diverse are the secondary minerals formed in this way according to the conditions which prevail in different cases. Amethyst is an important member of this class, and together with it, in the same cavity, agate is frequently found. The secondary minerals thus formed, of course, take the shape of the cavity in which they are deposited, and are therefore referred to as amygdales, the different kinds being distinguished as amethyst amygdales, agate amygdales, &c. The largest may weigh a hundredweight, and the smallest are about the size of a pea. Rocks containing such amygdales are known as amygdaloids.

When the mother-rock becomes completely decomposed these nests of crystals are to be found among the weathered débris, and are often carried away by running water; the crystals they contain are finally deposited in the form of rounded grains and pebbles together with other pebbles in the alluvium of streams and rivers.

At one time the best known amethyst was that which occurs in the cavities of amygdaloidal rocks in the neighbourhood of **Oberstein** on the Nahe, a tributary of the Rhine. A large amount of it used to be cut in the famous lapidary works of Oberstein, but now the local supplies are for all practical purposes completely exhausted. The cutting of amethyst is still carried on at Oberstein on a large scale, but the rough material has now to be imported. A plentiful supply of fine material at low prices can be easily obtained from other parts of the world, so that the exhaustion of the localities in the Nahe valley has not affected the prosperity of the Oberstein lapidary works.

Most of the amethyst used for cutting is obtained from **Brazil** and the neighbouring country of **Uruguay.** The stones, packed in barrels or in skin sacks, reach Europe, and especially Oberstein, in hundreds of tons, and with them are sent other varieties of quartz, such as citrine, colourless rock-crystal, &c.

In the State of Rio Grande do Sul, in southern Brazil, and in Uruguay, amethyst occurs, together with citrine and a large amount of agate, in the amygdaloidal cavities of melaphyre

Quite recently (about 1900) there has been found in this region an enormous amethyst geode, a portion of which was shown by Herr C. W. Kessler, an agate merchant of Idar, at the Düsseldorf Exhibition of 1902. It was found by the agate-seekers at an elevation of 600 metres above sea-level in the Serra do Mar, in the State of Rio Grande do Sul, and about ninety miles north of the German settlement of Santa Cruz. The cavity was only about a metre below the surface in a reddish clay, the weathered product of melaphyre, while the lower portion of it still remains embedded in the solid rock. It measured 10 metres (33 feet) in length, $5\frac{1}{2}$ in breadth, and 3 in height, and was estimated to weigh thirty-five tons; about fifteen tons was shipped in ten pieces to Idar. The whole of the cavity is lined with brilliant crystals of amethyst of a deep bluish-violet colour and averaging about 4 centimetres across.

At a few localities in the State of Minas Geraes, in the northern portion of Brazil, where it is not associated with agate, the mineral has another mode of occurrence as groups of beautiful crystals. These are found on the Campos dos Cristaes in the neighbourhood of Diamantina, but the finest specimens come from the Ribeirão da Paciencia at Itaberava, near Cattas Altas, south of the town of Ouro Preto (Fig. 67), where they occur under the same conditions as the yellow topaz already described. In the gem-gravels of the district of Minas Novas are many pebbles of amethyst accompanying the pebbles of white and blue topaz, chrysoberyl, &c., already described, and which, like these, are probably derived from granite and gneiss.

In the **United States** of North America a certain amount of amethyst suitable for cutting is found, none of which, probably, leaves the country. It is most abundant at Deer Hill, near Staw, in Maine, but comparatively little of this is suitable for cutting. Good crystals of large size, transparent and finely coloured, are met with in the State of Pennsylvania, namely, in Chester County and other districts, but specially in Providence Township, Delaware County. Fine specimens have been found in Haywood County, North Carolina. The amethysts of Rabun County, Georgia, are remarkable in that they frequently contain large fluid enclosures, those in other amethysts being of microscopic size. The stone is widely distributed in the region of Lake Superior, especially on the Canadian side, Amethyst Harbour being a typical locality; but most of the material found in this district is unsuitable for cutting. The amethyst of Nova Scotia is not infrequently used as an ornamental stone, that from the Bay of Fundy and other places being often worked into objects of considerable size.

Another American locality still to be mentioned is Guanaxuato in **Mexico**. The crystals found here measure as much as a foot in length; the prism is usually pale in colour, the pyramidal termination alone being of a deep shade, and, moreover, the crystal as a whole is seldom transparent enough for cutting. Far finer stones than those found at Guanaxuato are frequently found in the ancient Aztec graves, so that there must be other localities in Mexico now unknown.

The amethyst pebbles found in the gem-gravels of **Ceylon**, being the finest known, superior even to Brazilian stones, are much in demand. They have been derived from the granitic and gneissic rocks of the neighbourhood, and occur with the other precious stones mentioned under ruby, with which they are collected in the manner already described. Amethyst occurs in the same way in a few of the rivers of Burma, also in small amount in India, but in neither country has the occurrence any importance.

The amethyst localities in the **Urals**, on the other hand, are extremely important, especially the neighbourhood of Mursinka in the Alapayev and Reshev mining districts, in the Ekaterinburg division of Government Perm, the village of Mursinka being in latitude

57° 40′ N. and longtitude 30° 37′ E. of Pulkova (the Russian observatory). Here amethyst is found, often just beneath the turf, in the drusy cavities of quartz-veins of no great size running through weathered granite. As already mentioned it accompanies beryl and topaz, which also occur in nests and veins in the granite, though never in the same cavity with amethyst, being always at a much greater depth, so that the collection of these stones is attended with more difficulty. In this district at the present time about 140 pounds of amethyst, 15 pounds of beryl and topaz, and more than 200 pounds of gold-quartz is annually collected.

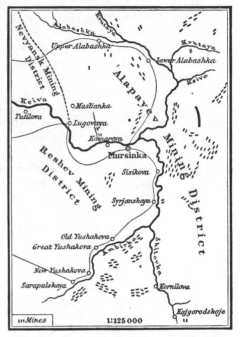

The industry is carried on preferably in winter, when it affords employment to 150 or 200 persons, the number of persons employed in the summer not exceeding twenty-five. The mines are the property of the Crown; there are about seventy-five of them, but only nine are at present worked. According to their position, which is shown in the accompanying map (Fig. 87), they fall into three groups, namely: (1) mines about Mursinka; (2) the Alabashka mines on the stream of this name; (3) the mines on the Ambirka stream, also known as the Sarapulskaya mines. The mines east of the rivers Alabashka and Shilovka, especially those near Sisikova and Mursinka, yield amethyst principally, while from those which lie between the villages of Upper and of Lower Alabashka and near Yushakova and Sarapulskaya topaz and beryl are more ex-

FIG. 87. Occurrence of amethyst near Mursinka in the Urals.

clusively obtained, the two last mentioned being localities for red tourmaline (rubellite) also. The precious stones found in this region are for the most part cut in Ekaterinburg, and many remain in the country. A certain proportion, however, find their way into western Europe by way of Nizhniy Novgorod, where great cosmopolitan fairs are held. Generally speaking, Uralian amethysts are pale in colour and patchy, but among them one sometimes meets with specimens of a deep violet colour comparable with the finest of stones from Brazil and Ceylon.

The chief sources for the supply of amethyst are the Urals, Brazil, Uruguay, and Ceylon. Other localities are of but little interest or importance; the mineral is found, for example, usually in crevices in gneiss, at several places in the Alps, among others in the Zillerthal, Tyrol, a locality which formerly yielded material fit for cutting. Also in veins of metallic ores at Schemnitz in Hungary; while material suitable for cutting often occurs in Spain with quartz of other colours, Carthagena in province Murcia, and Vich in Catalonia, being mentioned as localities. At the former locality amethyst occurs in water-worn pebbles as it does in Ceylon. Fine crystals have been found in Cornwall, also near Cork and on Achill island, Co. Mayo, translucent crystals up to 8 or 10 inches in length having been found at the last named locality. In the Auvergne, at a spot 40 kilometres from Clermont, amethyst was mined by the Spaniards about 150 years ago, and the mines here have recently been opened up again owing to the increased demand for amethyst: the best material, of a rich purple colour, now obtained at this place fetches 120 to 800 francs per kilogram, according to the size of the rough pieces. Nowhere, however, except in the countries mentioned above, is amethyst mined to any appreciable extent.

Amethyst is cut most frequently as a table-stone or in the step-cut (Plate XVIII., Fig. 1 b), the brilliant form being rarely adopted. Stones of a deep, uniform colour are mounted à jour without foils, but pale or patchy stones are mounted with a foil of the same colour. Fine large amethysts of a uniform colour now fetch from 10s. to 12s. per carat; they were formerly worth very much more, as can be gathered from the fact that at the beginning of the nineteenth century the celebrated amethyst necklace of Queen Charlotte of England was valued at £2000, while at the present day it would scarcely find a purchaser at £100. The large amount of fine material, discovered during the course of the nineteenth century in South America, is partly responsible for the depreciation in the value of amethyst. Previous to this it was a costly material for superior jewellery, now it is used principally for simpler and cheaper ornaments. Pale coloured or patchy stones are almost valueless, the rough material fetching only a few shillings a pound.

In ancient times amethyst was often used for seal-stones and engraved with various devices, beside being fashioned into larger objects, such, for example, as the bust of Trajan, carried off by Napoleon from Berlin. At the present day amethyst is but little used for such purposes.

On account of the similarity in colour between true amethyst and the far more valuable " oriental amethyst," it is possible to mistake the one for the other, unless it be remembered that the latter is far harder and heavier, and sinks in methylene iodide, while the former floats. True amethyst is distinguished from violet fluor-spar, the so-called " false amethyst," by its double refraction, greater hardness, and lower specific gravity. The double refraction of true amethyst also distinguishes it from violet coloured glass, which from its appearance alone is often not to be distinguished from the genuine stone. Artificially coloured violet quartz is obtained by strongly heating rock-crystal and then immersing it in a solution of some violet coloured substance. The colouring matter penetrates the cracks in the rock-crystal caused by the sudden change of temperature, and, on drying, imparts its colour to the stone. Stones which have been subjected to this treatment are easily recognisable on account of the cracks, and for the same reason they are seldom cut.

CITRINE.

The name citrine is applied to yellow quartz. The crystals of this variety are developed in the same way, exhibit the same two sets of striations meeting at an angle on the pyramid face, and in many specimens the fractured surface has the same rippled, striated character as in amethyst. It is obvious, therefore, that citrine differs from amethyst solely in colour, and even this can scarcely be regarded as a fundamental difference, seeing that amethyst assumes the colour of citrine when exposed to the action of heat. Some, indeed, have gone so far as to assert that yellow quartz seldom or never occurs in nature, and that the greater part of the yellow quartz now in existence is in reality " burnt" amethyst, or possibly " burnt" smoky-quartz, the brown colour of which is, as we have seen, changed to yellow under the action of heat. This assertion, however, is certainly incorrect, for citrine does undoubtedly occur in nature at the localities to be presently mentioned, sometimes in considerable amount.

Citrine is sometimes very pale or almost colourless. Among deeper coloured stones may be seen wine-yellow, honey-yellow, and saffron-yellow specimens, while others have quite a pronounced brown tinge. Stones of a deep brownish-yellow colour are very like topaz (Plate XIII., Fig. 2a), and those of a fine golden-yellow are quite equal in beauty to yellow topaz, and can scarcely be distinguished on mere inspection from the latter stone except by an expert.

This variety of quartz is, in fact, constantly passed off as topaz, with which stone it has nothing in common save colour. It is probably not going too far to state that this stone is never either bought or sold in the trade under its correct mineralogical name, but always under the name of topaz, perhaps with a qualifying prefix such as occidental, Indian, Bohemian, Spanish. Used in this sense then, " Indian topaz " signifies not the saffron-yellow topaz from Ceylon, already mentioned, but saffron-yellow citrine. By " Spanish topaz " is understood citrine of a deep brownish-yellow shade, while the term " golden topaz " is sometimes applied to golden-yellow citrine. The term " false topaz " is also used, but is usually applied to yellow fluor-spar.

It is not difficult to discriminate between citrine and true topaz, for topaz, with a hardness of 8, will scratch citrine. Topaz, moreover, is much heavier than citrine and sinks heavily in pure methylene iodide, in which citrine floats. Finally, while topaz is always rather strongly dichroic, citrine is scarcely dichroic at all.

Citrine is of course only cut as a gem-stone when perfectly clear and transparent, and its value as such depends upon the degree of its transparency, and the richness and purity of its colour. The finest specimens are at least as valuable as correspondingly fine amethysts ; and medium material, like amethyst, is sold for a few shillings a pound. The step-cut and the various modifications of the table-stone are the forms usually adopted for citrine, as also for amethyst, topaz, and coloured stones in general.

It was usual formerly to regard citrine as a rare mineral only to be found at a few scattered localities. Goatfell, in the island of Arran, where it occurs attached to the walls of crevices in granite, was one, Bourg d'Oisans, in Dauphiné, here associated with rock-crystal, was a second, while others were mentioned in Hungary and Croatia. It first reached the European markets in quantity in the thirties of the nineteenth century, having been discovered in Brazil and Uruguay, whence it was sent with amethyst to the lapidary works of Oberstein on the Nahe. The mineral occurs most abundantly in Brazil in the States of Minas Geraes and Goyaz, but a large amount of material is also yielded by Rio Grande do Sul, the most southern of the Brazilian States. The principal locality in Uruguay is the neighbourhood of Salto Grande on the Uruguay river. In both of these South American countries citrine is found, together with amethyst and agate, in amygdaloidal cavities in melaphyre. In the neighbourhood of Mursinka, in the Ural Mountains, it is met with in the gem mines together with amethyst, but in much less abundance. A variety of quartz, which when strongly heated assumes a fine yellow colour and which can then be sold as topaz, occurs in Spain at Hinojosa, in the province of Cordova, on the northern slopes of the Sierra Morena. Several hundredweights of this stone is mined annually, but the average quality is poor and it does not sell for more than four or five francs per kilogram. When cut it is bought and sold under the name of " Spanish topaz," but is not very much in favour. A few fine specimens of citrine suitable for cutting have been found in North Carolina, but here, as elsewhere in North America, the mineral occurs in very insignificant quantities.

ROSE-QUARTZ.

Rose-quartz (" Bohemian ruby ") is a translucent or semi-transparent, massive quartz with a somewhat greasy lustre and a fine rose-red colour.   This colour is not permanent, but quickly disappears on exposure to light or a high temperature.   It is probably due to the presence of some organic matter, but has also been ascribed to titanium dioxide, small amounts of which have been found in rose-quartz.   Specimens showing paler shades of the same colour merging imperceptibly into milk-white are not uncommon, and contrast well with those of the typical rose-red colour.   Stones of a deep shade of this colour are cut with a rounded surface, but are not in much favour and are very low in price.   In spite of this, rose-quartz is imitated in glass, so cleverly indeed that it is sometimes only possible to distinguish between the two by the lower specific gravity, greater hardness, and double refraction of the quartz.   The mineral is found in large irregular masses in granite in the neighbourhood of Bodenmais in Bavaria, and of Ekaterinburg in the Urals, and in Ceylon, India, Brazil, &c., but must always be classed with the less widely distributed varieties of quartz.

PRASE.

Prase is sometimes referred to by jewellers as " mother-of-emerald " in reference to the fact that it was formerly supposed to be the mother-rock of the emerald.   It is a translucent variety of quartz with a somewhat greasy lustre and a leek-green colour.   This last named character is due to the presence of innumerable minute fibres and needles of actinolite enclosed in the otherwise colourless and pure mass of quartz.   Prase was known to the ancients, and was used in the fashioning of ornamental stones and to represent foliage in mosaics, while as a gem it was frequently engraved.   In the latter form it is not uncommon among Roman remains, but the source from whence the Romans obtained their rough material is not now known.   There are many more recently discovered localities however ; thus crystals and crystalline masses, of the same quality as the material used by the ancients, occur in metalliferous veins with iron-pyrites and zinc-blende at Breitenbrunn between Schwarzenberg and Johanngeorgenstadt in the Saxon Erzgebirge.   Also in the Habachthal, in the Salzburg Alps, and at several places in Scotland, Finland, &c.   Prase is applied now to the same purposes as in ancient times, but to a very limited extent, and this, combined with its abundance in nature, makes it one of the lowest priced of ornamental stones.

SAPPHIRE-QUARTZ.

Sapphire-quartz (azure-quartz or siderite) is a blue, crystalline quartz, the colour of which is due to the enclosure of a large amount of a blue, fibrous to earthy substance belonging probably to the mineral species crocidolite.   It is only faintly translucent, has a slightly greasy lustre, and is not very suitable for cutting, hence it is used to a very small extent and is correspondingly low in price.   It occurs in moderately large amount in veins penetrating the gypsum of the Gypsberg at Mooseck, near Golling, in Salzburg.

QUARTZ WITH ENCLOSURES.

We have already seen that quartz frequently encloses other minerals and foreign substances of various kinds. These enclosures may be present singly in comparatively small numbers or in such multitudes as to impart an apparently uniform colour to the quartz substance, as is the case with prase and with sapphire-quartz just described. It is proposed to consider here those cases in which there are present single enclosures of a nature to contrast very markedly with the quartz in which they are embedded. It is obvious that enclosures in translucent or opaque quartz will be seen very indistinctly or not at all, so that it is only the enclosures in very transparent quartz which will engage our attention. On the other hand, the clearness with which enclosures in quartz are seen is not at all affected by the colour of the mineral, objects embedded in amethyst being quite as distinctly seen as those in rock-crystal. These two varieties of quartz are those which most commonly contain enclosures, the different kinds of which will be described below in some detail.

Hair-stone and **Needle-stone.**—These names are given to quartz with enclosures of isolated, needle-like or hair-like crystals of various substances, like that represented in Plate XVIII., Fig. 2, which encloses green needles of actinolite. In other cases the enclosures may be white fibres of asbestos, long, thin crystals of rutile ranging in colour from yellow to red and somewhat resembling straw in appearance, and so on. Quartz containing enclosures of this description is distinguished as needle-stone or as hair-stone according as the enclosed crystals approach more to the thickness and straightness of a needle, or to the fineness and sinuosity of a hair. Enclosures of rutile of a reddish-brown to yellow colour, are referred to as "Venus's hair," and the stone as a whole as "Venus's hair-stone." Enclosures of green hornblende, actinolite, or asbestos are often called "Thetis's hair." The fibres may be straight, bent, crumpled, or wound into a ball. The appearance of green fibres, probably of asbestos, embedded in quartz often resembles that of a piece of moss, and the quartz containing such an enclosure is known as *moss-stone*; such moss-like enclosures are also frequent in agate (moss-agate), to be considered further on.

Enclosures of various kinds are not at all uncommon in the rock-crystal from the Alps and other localities. Some specimens of rock-crystal from Madagascar enclose long grey crystals of manganite with a bright metallic lustre. From the Calumet Hill quarry, near Cumberland, in Rhode Island, U.S.A., is obtained a translucent, milk-white quartz containing numerous needles of black hornblende; a large amount of this material was formerly exported for cutting to Idar and Oberstein on the Nahe, but since 1883 the supplies have ceased. Similar material is obtained in Japan and Madagascar.

FIG. 88. Flèches d'amour from Wolf's Island, Lake Onega, Russia.

In the pale amethyst found in the amygdaloidal cavities of rocks or loose in the soil of Wolf's Island, in Lake Onega, north-east of St. Petersburg, there are long brown crystals of the mineral göthite ("needle-iron-ore") (Fig. 88); and there are many other similar occurrences. These stones are cut in St. Petersburg and Moscow under the name of "Cupid's darts" (*flèches d'amour*). The same name is used for other stones of similar appearance, such, for example, as the beautiful specimens which occur in North Carolina, U.S.A. Like all mineral objects of the kind they are provided with a slightly convex polished surface, and often cut with a heart-shaped outline. Whether such objects are cut and polished or not, depends upon whether they possess a sufficiently pretty or bizarre appearance, and they are sold more as curiosities than as ordinary precious or ornamental stones with a definite market value. Fine examples of the kind have sold for 50s. or more, but much depends upon the purchaser.

Quartz containing Fluid Enclosures.—As already mentioned, rock-crystal frequently encloses cavities which contain fluid of some kind and a bubble of gas, the movement of which follows every movement of the stone. Large inclusions of this kind are not common, but rock-crystal containing specially fine ones is met with in Madagascar, in the Alps, and elsewhere. The rock-crystal of Herkimer County, New York, and of Hot Springs, Arkansas, deserves special mention on this account, while the amethyst of Rabun County, Georgia, is frequently cut for the purpose of displaying its large fluid enclosures. Generally speaking, however, quartz which has nothing to recommend it but its fluid enclosures is cut even more rarely than is hair-stone or needle-stone, and is regarded more definitely as a curious natural object.

Gold-quartz.—This is transparent or highly translucent quartz enclosing veins or grains of native gold. In San Francisco and a few other large towns of western America it is often cut in the form of plates for brooches, or utilised for stick-handles, cuff-links, paper-weights, and other small objects of more or less utility. Some of the gold-mines of California, Oregon, Idaho, and Montana have furnished very beautiful specimens. The value of a stone depends upon the amount of gold it contains, which is estimated by the help of the specific gravity, and also upon the beauty of the specimen. A ring-stone is worth from two to upwards of ten dollars, according to its quality.

Gold-quartz is now much in favour as an ornament, as is testified by the fact that in some years from 40,000 to 50,000 dollars' worth of rough material suitable for cutting has been sold in this region; a single cutting works at Oakland, California, has used annually 10,000 dollars' worth of rough material, and a large firm of jewellers in San Francisco has sold cut stones to the value of 15,000 dollars in the same period. The stones need careful sorting out, and as they are very fragile and difficult to work, only about half of the material destined to be cut actually reaches the market.

White and cloudy gold-quartz is more common than the transparent variety, and recently perfectly black material has been met with. White gold-quartz is artificially coloured rose-red by immersing it in a solution of carmine. Attempts have been made to manufacture the substance by fusing together gold and quartz in an electric furnace, not, however, with very favourable results.

The gold-quartz of Australia, South Africa, and most of the auriferous regions, is just as suitable for cutting as is the Californian, but in none of these lands is it worn to the extent that it is in America. The auriferous quartz of La Gardette, near Bourg d'Oisans, in Dauphiné, which forms the matrix of the beautiful rock-crystals represented in Plate XVII., used to be cut as an ornamental stone. About 200 pounds of gold-quartz is obtained every year from mines in the neighbourhood of Mursinka in the Urals.

Rainbow-quartz (Iris).—This is the name given to specimens of rock-crystal which show brilliant iridescent or rainbow-colours over more or less large surfaces. The colour has no connection with the quartz substance, but is due to purely physical causes connected with the interference of light. The phenomenon is only shown by specimens of rock-crystal which are penetrated by numbers of fine irregular cracks. These cracks contain films of air, and to their presence is due the colours of rainbow-quartz, colours which are of exactly the same nature and arise in the same way as the brilliant rainbow tints of soap-bubbles. Specimens of iris are cut with a slightly convex surface, which should be as closely parallel as possible to the surface which displays the rainbow colours, this being not plane but more or less irregular and curved, corresponding to the conchoidal fracture of quartz. Rainbow-quartz is worth more than rock-crystal, and specimens which show a central portion with a play of colours surrounded by a colourless border have a considerable value. In most cases the iridescence is confined to certain portions of the mass of rock-crystal, and these must be carefully sawn out in order to be cut for ornamental stones. Large crystals are occasionally iridescent throughout their whole mass and can be used in the fashioning of articles of large size, such, for example, as the candelabra in the Vatican collection, which, however, may possibly be made up of several portions.

More or less fine specimens of rainbow-quartz are found occasionally at all the localities at which ordinary rock-crystal occurs. Iridescence can be produced in the latter by striking it with a hammer or by immersing it in cold water after it has been strongly heated, the object in both cases being to develop cracks in its substance. Specimens are, of course, often broken by subjection to this treatment, but this is of little consequence considering the abundance of the mineral. Many other transparent minerals display iridescent colours, and are cut for ornamental stones, all of which are known by the same name of "iris." This term is also applied to cut rock-crystal and also to paste, made to imitate an iridescent stone by the use of vari-coloured foils, and much used in the cheapest of jewellery.

CAT'S-EYE.

The term cat's-eye (quartz-cat's-eye, occidental cat's-eye, schiller quartz) is applied to massive quartz from the surface of which, especially when cut in a rounded form, there is reflected a wave of milky light. The appearance is exactly the same as that of cymophane, a variety of chrysoberyl, also known as cat's-eye, but distinguished from quartz-cat's-eye by the prefix oriental. Unfortunately the distinction between occidental and oriental cat's-eye is not always made, an omission which leads to much confusion, especially in statements respecting the occurrences of these stones. They differ very markedly in almost every character; thus cymophane or oriental cat's-eye is far more brilliant and beautiful and much less common than is quartz-cat's-eye; and the band of chatoyant light is more clearly defined in the former than in the latter. In correspondence with these differences is the fact that cymophane is a much more valuable stone than is occidental cat's eye. The two stones may be easily distinguished by their difference in hardness, chrysoberyl being harder than topaz, and therefore much harder than quartz, and in specific gravity. Chrysoberyl has a specific gravity of 3·7 and therefore sinks heavily in methylene iodide, while quartz with a specific gravity of 2·65 floats in that liquid.

Quartz showing the optical effect known as chatoyancy occurs in compact masses, which are not, however, aggregates of single grains, but of uniform crystalline structure throughout. It has a somewhat greasy lustre, is never transparent, but always more or less translucent. The colour is sometimes white, but more frequently olive-green to dark leek-green, these paler or darker shades of green always containing grey tones. The mineral also occurs of a pronounced brown or yellow colour and red tinged with various shades of brown and yellow. Blue stones occur as rarities. Green and yellowish-brown stones cut en cabochon are represented in Plate XVIII., Figs. 4a and b.

The chatoyancy of quartz-cat's-eye is due to the presence of large numbers of fibres of asbestos, which are embedded in perfectly parallel directions in its substance and which in some specimens can be distinctly seen with the aid of a simple lens. Specimens of quartz are frequently met with in which these fibres have been destroyed by weathering, their place being taken by fine, hollow canals, each canal corresponding to a single fibre of asbestos. The whole mass of quartz has in this case therefore a tubular structure, but the optical effect is the same as if the fibres were present.

The chatoyancy of quartz-cat's-eye is displayed to the best advantage when the stone is cut *en cabochon*, with a decidedly convex upper surface, and with the flat base parallel to the direction of the enclosed fibres. The rounded surface is then covered by a band of light of greater or less breadth in a direction perpendicular to that of the fibres, the other portions of the surface showing no chatoyancy. As the stone is turned about the band of light travels across its surface, finally disappearing at the edge of the stone as the light reaches a certain incidence.

The chatoyant band has a lustrous, silky sheen of a yellowish- or bluish-white colour, which has been compared with the shining eye of a cat. The stone is usually cut with a rather elongated oval outline, like that of a coffee-bean (Plate XVIII., Figs. 4a and b), in such a way that the band of light is coincident with its greatest diameter, thus producing the most favourable effect. The cats'-eyes of finest quality are those in which the chatoyant band is of uniform width, not too broad and very sharply defined. Inferior stones are those in which the band is interrupted, too wide, or possessed of ill-defined margins, so that it does not stand out in sharp contrast with the surrounding portions. Specimens in which the band is replaced by a patch of light are also considered to be of inferior quality. The stones most highly esteemed in Europe at the present time are those of a reddish-brown colour and with a delicate bluish-white sheen. They are worth as much as 50s., though at that price they must be very perfect and of some size. Generally speaking, quartz-cat's-eye, especially inferior material, is worth very little, while oriental cat's-eye, even the poorer qualities, always fetches a high price.

The finest quartz-cat's-eye is found in **India** and Ceylon. In this quarter of the world the stone is much admired, especially by the Malays, much more so than in Europe, where its popularity depends upon the vagaries of fashion. The reddish-brown stones are usually stated to come from the mainland of India, and the green and grey specimens from Ceylon, but this distribution may not always be quite constant.

Cat's-eye from India is supposed to come mainly from the Malabar coast, in the south-east, and Quilon and Cochin are mentioned as localities, but these statements are not indisputable, and the mode of occurrence is altogether unknown. Since most of the material reaches Europe in the cut condition it is impossible to decide whether the rough material is found *in situ* or in loose pebbles. At Ratanpur, in the district of Rajpipla, N.N.E. of Bombay, the stone occurs, together with agate, in the form of pebbles, which are undoubtedly derived from the basaltic rocks—the Deccan traps— of the region. Other localities are given in the vicinity of Madras, and in the valley of the Lower Kistna in the neighbourhood of the Palanatha mountains, north-east of Guntur; also in Burma a few stones are met with. Generally speaking the occurrence of cat's-eye in the mainland of India is insignificant, and individual stones never exceed two ounces in weight.

While on the mainland of India quartz-cat's-eye is scarce and oriental cat's-eye, as far as is known at present, is absent, in the island of **Ceylon** both occur in considerable amount. Quartz-cat's-eye occurs here in grains or pebbles, rarely larger than a hazel-nut, in the gem-gravels of Saffragam and Matura, which have been derived from the weathering of granitic rocks.   Most of the stones are green in colour, but there are also brownish-red and yellow cat's-eyes.   Like those found in India, they are cut *en cabochon* before being placed on the market.   A large number are sent to Europe, but many also are kept in the country, as cat's-eye is a favourite ring-stone both in India and Ceylon.   The pure olive-green cat's-eye with a narrow, sharply defined band of light is most highly esteemed by the Cingalese, who are extremely proud of the occurrence of cat's-eye in their island and firmly believe that it is found nowhere else.   This is of course an unfounded belief, for beside localities already mentioned there are several places in Europe where it occurs, though only in inferior quality.

In **Europe** cat's-eye of a pale green colour, but scarcely suitable for cutting, is found with asbestos in small brecciated veins in serpentine at Treseburg in the Harz mountains. Stones of rather better quality occur in the diabase of Hof and other places in the Fichtelgebirge in Bavaria; they are often cut but are much inferior to Indian stones. Cat's-eye of a quality suitable for cutting is not found in Hungary, in spite of the fact that this gem is sometimes referred to by jewellers as " Hungarian cat's-eye."

In Europe cat's-eye is worn principally in rings, pins, brooches, &c., the small size of the finest Indian and Ceylonese stones, indeed, precluding the possibility of any other application.   Larger objects cut from this mineral are rare; as an example may be mentioned a bowl of yellowish-brown cat's-eye in the Vienna treasury measuring 5 inches in length; the block of rough material from which this was cut must have been of considerable size.

A stone very similar in appearance to cat's-eye can be obtained by treating tiger-eye with hydrochloric acid.   The colouring matter is dissolved out by this treatment, and the greyish material which remains shows when cut the chatoyancy of cat's-eye.   Together with the brown cats'-eyes of Ceylon there are sometimes found specimens of satin-spar, a variety of calcite, which show a reflection of milky light very like that of true cat's-eye. This is much softer, however, and in contact with hydrochloric acid it effervesces, while true cat's-eye does not.

TIGER-EYE.

Tiger-eye is a variety of quartz with a finely fibrous structure.   It ranges in colour from yellow to brown, and when cut and polished in the direction of the fibres exhibits a magnificent golden lustre.   A polished piece of tiger-eye is represented in Plate XVIII., Fig. 5.

The mineral occurs in the form of thin plates or slabs, which are bounded by parallel surfaces, and are rarely more than a few centimetres in thickness.   The fibres are arranged parallel with respect to each other and perpendicular to the surface of the plate; they are not always perfectly straight; they may be curved, or each may have a sharp bend at a certain place.

Even on an ordinary fractured surface the silky lustre of the stone is very apparent, and this may be considerably increased by cutting and polishing.   A polished surface turned towards the light usually exhibits a series of magnificently lustrous yellow bands, arranged parallel to the surfaces of the original plate.   These lustrous yellow bands alternate with dull brown bands, which show little or no silky lustre.   A slight change in the position of the stone results in a reversal of these conditions, the dark bands becoming lustrous and the silky yellow bands dark.   Every movement of the stone, therefore, is attended by alternations of brightness and dulness in the bands.   This constant alternation in the appearance of the bands is due to the bends in the constituent fibres of the stone, and is one of its most characteristic features. The chief beauty of ornaments of tiger-eye, in point of fact, depends upon the aspect of the stone changing with every movement of the wearer.

In association with tiger-eye there often occurs another stone, which agrees with the former in every respect save that of colour. Thus, it has a finely fibrous structure, a silky lustre, and the same banded appearance of polished surfaces. The physical characters including the hardness are all the same, but the colour instead of being yellow is dark indigo-blue. This blue mineral is also cut as a gem and is known by the name of **hawk's-eye**. A detailed mineralogical examination demonstrates the fact that it is a colourless transparent quartz, embedded in which are innumerable fine blue fibres, all arranged parallel to each other and perpendicular to the surface of the plate. These fibres consist of the blue asbestos-like mineral crocidolite, a member of the amphibole group, and chemically a silicate of iron and sodium. The colour of sapphire-quartz is due to inclusions of the same mineral, but in this case the fibres are not parallel but arranged quite irregularly.

In their structure and mode of origin tiger-eye and hawk's-eye are very closely connected as will presently be seen. An examination of a piece of hawk's-eye often results in the observation that the stone is not of one uniform blue colour, but that portions of it are yellow. These blue and yellow portions may be present in equal proportions, or the blue may be present in larger proportion than the yellow, or the opposite conditions may prevail. Now these facts clearly indicate that the yellow substance, which in every respect is identical with tiger-eye, has been formed by the alteration of the blue hawk's-eye. Moreover, it appears that it is the crocidolite which has undergone this alteration, its constituents, with the exception of silica and iron, having been dissolved and carried away. The silica which remains behind retains the fibrous form of the original mineral, but assumes a yellow colour owing to the deposition between the fibres of a small amount of hydrated oxide of iron, this also being an alteration product of crocidolite. When in any specimen of hawk's-eye the alteration process has just begun, there will be little yellow patches scattered here and there, which will increase in extent as the alteration proceeds, until finally the blue hawk's-eye becomes wholly converted into yellow tiger-eye, the fibrous structure of the original stone remaining, however, unchanged.

Beside stones which show every stage of transition between hawk's-eye and tiger-eye, there are others which are intermediate in character between hawk's-eye and crocidolite, and tiger-eye and crocidolite. Like hawk's-eye and tiger-eye, crocidolite occurs in parallel-sided plates or slabs, with its asbestos-like fibres perpendicular to the surfaces of the plate. In the fresh condition this asbestiform mineral is of a blue colour, but when altered it is yellow, owing to the oxidation and hydration of the ferrous iron it contains to hydrated ferric oxide. This yellow alteration product of crocidolite, consisting of a mechanical mixture of silica and hydrated iron oxide, has been given the name of "griqualandite." By the infiltration of silica between the fibres of crocidolite tiger-eye and hawk's-eye are formed, so that it is possible to regard these minerals as silicified crocidolite or as pseudomorphs of quartz after crocidolite. The quartz assumes the structure of the original mineral, the fibres of which remain embedded in it. If these remaining fibres are fresh and unaltered, the resulting mineral is blue hawk's-eye; if, on the other hand, they consist of the yellow alteration product, it is yellow tiger-eye. It is also possible, however, that the alteration of the crocidolite, which is accompanied by the change in colour from blue to yellow, takes place in part at least after, instead of before, the silicification of the asbestiform mineral. Because of the relation which exists between crocidolite and tiger-eye the latter is frequently referred to in the trade as crocidolite.

Like the Cape diamond, the minerals now under consideration are found in Griqualand West, not at the same localities but in the neighbourhood of Griquatown, to the west of Kimberley, the centre of the diamond mining area. In former times asbestiform crocidolite and limonite (brown iron-ore) were stated to occur in association with tiger-eye at Lakatoo on the Orange River, and at Tulbagh, the material from the latter locality, according to earlier accounts, being famed for its beauty.

According to Professor E. Cohen, those occurrences of the minerals which are important from an industrial point of view are situated in the mountain range north of the Orange River which, a little west of Griquatown, extends first in a north to south, and then in a north-east to south-west, direction. The continuation of this range on the other side of the Orange River is formed by the Doorn Bergen. On the large official map of Cape Colony of 1876 this range is marked as the Asbestos mountains; but in ordinary maps the name is applied to a much shorter range, lying a little further to the east, and on these the former range is marked as the Lange Bergen.

Tiger-eye occurs at many places in this range, among others in the neighbourhood of Griquatown. The plates are embedded in a finely grained quartz-rock of a reddish-brown, coffee-brown, or ochre-yellow colour. The mountains, which do not stand very high above the plateau, are composed mainly of this rock, which is often very thinly bedded, and may be best described as a jasper-schist. The mineral is quarried here and sent in large quantities to Europe, especially to Oberstein on the Nahe and the neighbouring town of Idar, to be cut.

Tiger-eye was once a great rarity in Europe; not more than a quarter of a century ago it cost upwards of 25s. per carat. Now, owing to the underselling of two rival traders, the stone has been placed on the market in such large quantities that the price has fallen to little more than 1s. a pound.

All the tiger-eye and hawk's-eye which comes into the market is obtained from the Asbestos mountains, but the mineral is not by any means confined wholly to this district, and appears to have a wide distribution in South Africa. The traveller Mauch has found it, for example, much further to the east, on the upper Marico, a tributary of the Orange River. Outside South Africa, however, neither tiger-eye nor hawk's eye has hitherto been found.

While these stones were rare and costly they were cut with a flat or convex polished surface, and were used as gems for rings, brooches, and such like. Later on, when the price had fallen considerably, they began to be used for the manufacture of small articles of utility or ornament, such as the handles of umbrellas, and at the present time figures in bas relief, and intaglios are sometimes cut from fine specimens. Tiger-eye, indeed, has now a considerable application in the manufacture of ornamental and semi-ornamental objects, but hawk's eye, being less abundant, is much less extensively used. In cutting either of these stones care must be taken that the cut and polished surface is as nearly as possible parallel to the fibres; a slight deviation from this rule has a very prejudicial effect on the appearance of the stone.

It has been mentioned above that tiger-eye, after treatment with hydrochloric acid, assumes an appearance similar to that of grey cat's-eye; this is due to the fact that the hydrated iron oxide is extracted, and the fibrous silica left behind. The names tiger-eye and hawk's-eye have reference to the fact that these stones differ from cat's-eye in colour, but when cut *en cabochon* possess exactly the same chatoyant appearance.

B. COMPACT QUARTZ.

HORNSTONE.

Hornstone is a very fine-grained to perfectly compact quartz, consisting of an aggregate of microscopically small grains of the mineral.   The sharp edges of broken fragments are slightly translucent, and the stone is further characterised by its splintery fracture.   These two characters serve to distinguish hornstone from jasper, the latter having a smooth fracture, and being perfectly opaque, even in the thinnest of splinters.   Hornstone is usually of a dingy grey, brown, or yellow colour, so that it resembles horn in colour as well as in being slightly translucent, hence its name, which is an old mining term and not used by lapidaries at all.   As used by mineralogists, the term includes silicified wood and chrysoprase, the two varieties of hornstone which are cut as ornamental stones, as well as several other stones which are not used for this purpose, and which, therefore, need not be mentioned here.

Most hornstones, such as occur abundantly in many metalliferous veins and as nodules in limestone, clay, &c., do not possess the characters necessary for gems, but there are one or two varieties which are more attractive in colour, and which are therefore sometimes cut. One of these, known as chrysoprase, is green in colour ; and another, wood-stone, shows the structure of wood, and is in fact petrified wood.   The surface of wood-stone is sometimes prettily marked, and the substance is often fashioned into small ornamental objects of various kinds.   It is similar in many respects to wood-opal, already described, but differs from it in the fact that the petrifying material is quartz instead of opal.

Wood-stone (fossilised or silicified wood).—This stone has been formed by the impregnation with quartz of the woody substance of plants of former geological ages, and its gradual replacement by the same mineral.   The quartz, thus deposited, possesses characters which, on the whole, resemble most nearly those of hornstone.   Not only is the structure of the wood well preserved, but the identity of the original object can be easily recognised by reason of the perfect preservation of the characteristic external form of its trunk and branches.   When the material is cut, either longitudinally or transversely, and polished, the woody structure becomes still more apparent.   A longitudinal section along the axis of the trunk or branch displays the cells and vessels, alternating bundles of which are frequently coloured.   Transverse sections show these cells and vessels in section, and so perfectly are details of structure preserved that when thin sections are examined under the microscope it is often possible even to determine the nature of the original plant; many species of fossil palms, pines, &c. have, indeed, been so determined.

In wood-stone the walls of the vessels are usually of a dusky brown colour, while the material which fills up the cavities of the vessels and the spaces between neighbouring bundles is usually much paler in colour.   This only serves to show up still more clearly the woody structure, which, with the high lustre acquired after polishing, renders wood-stone very effective for ornamental purposes.   The markings of a cross-section often recall the speckled plumage of the starling, and such specimens have been referred to as starling-stone. The material is occasionally cut for mounting as a gem, but more frequently, though now less than formerly, it is fashioned into small semi-ornamental objects.   Wood-stone is the material which the ancient Babylonians used for some of their cylinder-seals.

The rough material is widely distributed, and the value of worked articles is little higher than the cost of the labour employed in their manufacture.   In Germany wood-stone

QUARTZ (CHRYSOPRASE) 497

is found principally in the Kyffhäuser mountains, the petrified trees being enclosed in sandstones and conglomerates of Permian (Rothliegende) age. Other localities for petrified wood are extremely numerous, but need not be enumerated here; it may, however, be mentioned that silicified wood, often in the form of gigantic trees, occurs in great abundance in the Western States of North America, for example, in Colorado, California, and Arizona, the petrified forest known as Chalcedony Park being situated in the last named State. The material is often cut in these States, more frequently as an ornamental stone for table-tops, pedestals, &c., than as a gem.

Under the mineralogical term beekite are included silicified corals, the petrifying material of which is quartz of much the same character as hornstone. Material in which the white coral is thrown up against a fine flesh-red background is at present cut under the name of "coral-agate"; according to the Oberstein lapidaries the rough material is obtained from the neighbourhood of Aden, Arabia, but similar material is also to be obtained elsewhere.

Chrysoprase.—Chrysoprase is a very fine-grained, moderately translucent hornstone of a beautiful apple-green colour. It possesses all the usual characters of hornstone, including the rough splintery fracture. The colour, which retains all its beauty in artificial illumination, is never very deep, and the palest shades are almost colourless.

The colour of chrysoprase, which can be seen on turning to the representation of a cut stone in Fig. 8 of Plate XX., is due to the presence of about 1 per cent. of nickel. This element is probably present in the form of a hydrated silicate, which, when heated, loses water and becomes decolourised, since the stone itself, when exposed to the action of heat becomes paler and paler, and finally quite white. The temperature necessary to produce this change is not very high, and a seal of chrysoprase, if frequently used, will gradually lose its colour. The same thing happens, moreover, when the stone is exposed to the direct rays of the sun, the decolourisation in the latter case being due not to the action of heat but to that of light.

That the decolourisation of chrysoprase is due to loss of water is demonstrated by the fact that the colour is restored by burying the stone in moist earth or cotton-wool. The colour of a bleached chrysoprase can be restored, or that of a pale stone improved, by immersion in a solution of nickel sulphate. It is even possible to produce chrysoprase artificially by immersing chalcedony for some time in a green solution of a nickel salt. This is sucked up into the pores of the chalcedony, and imparts to it the fine apple-green colour of true chrysoprase. It is sometimes almost impossible to distinguish these artificially coloured stones from true chrysoprase, and a great many are sold as genuine stones. The deception in such cases is not very serious, since artificially coloured chalcedony agrees in almost all its characters with natural chrysoprase, and, indeed, possesses some advantages over the latter, for its colour is usually finer and more uniform, and it is unaffected by exposure to light and heat.

Natural chrysoprase is rather difficult to work, being very liable, on account of its brittleness, to crack and splinter at the edges. Moreover, it must not be allowed to become over-heated during grinding since this has a prejudicial effect on the colour. A stone which has been over-heated not only loses its bright apple-green colour but also, to a great extent, its translucency. If reasonable care be taken in the process of grinding, the colour is retained unaltered, and after polishing the stone acquires a very fine lustre. Chrysoprase is usually cut with a convex or plane surface, which is frequently bordered with one or two series of facets (Plate XX., Fig. 8). It is suitable for ring-stones, pin-stones, &c., but not for seals and signet-rings, because of its tendency to be decolourised by heat. It was at one time much more highly esteemed than it is now; a fine stone of a bright green colour and considerable translucency was once worth upwards of £5, while now it would fetch scarcely half that sum, and stones of inferior quality are of course worth still less. Chrysoprase, however, is the most valuable of those compact varieties of quartz which are used as gems, and generally speaking is one of the most highly esteemed of the so-called semi-precious stones.

This beautiful mineral is used also as a decorative and ornamental material for furniture, mosaics, &c. For example, two tables inlaid with chrysoprase were ordered by Frederick the Great for the Sans Souci palace at Potsdam, and the same substance figures in the beautiful mosaic of the walls of the fourteenth-century Wenzel Chapel in the Hradschin at Prague.

Chrysoprase occurs in thin plates and veins, sometimes of considerable size, which are usually embedded in serpentine, of which it is a weathered product, and from which it derives its nickelous colouring constituent. Large pieces are seldom uniformly coloured; portions of darker and finer colour merge gradually into others which are pale or almost white, or into the yellow or brown common hornstone, which must be removed before the better portions are cut. Masses are frequently met with in which the hornstone passes in places into other varieties of compact quartz, such as chalcedony, and into opal. The latter mineral has been formed at the same time and in the same manner as chrysoprase, and is sometimes coloured green like the prase-opal already mentioned.

The most important localities for chrysoprase are in Silesia, where it occurs at various places in the neighbourhood of Frankenstein, to the south of Breslau. At Kosemütz chrysoprase of a deep, and sometimes also of a pale, colour occurs with chalcedony, opal, asbestos, and other minerals in veins in serpentine. Similar occurrences exist at Baumgarten and Grochau, while at Gläsendorf, Protzan, and Schrebsdorf the mineral lies in a yellowish-brown clayey earth, which overlies the serpentine and is a product of its decomposition. The chrysoprase frequently occurs so close to the surface that it is washed out by heavy rains and collected by peasants. Specimens of moderately large size can be obtained at Frankenstein, but they contain a good deal of impure and light-coloured material. The stones of a bright green colour are usually rare; the finest come from Gläsendorf.

According to an early account (1805) there exists in this district a vein of chrysoprase, three (German) miles in length, traversing the serpentine and associated rocks. This was accidentally discovered in 1740 by a Prussian officer at the northern end of the vein by the windmill of Kosemütz. Frederick the Great interested himself in this Silesian stone, and utilised it in the decoration of the Sans Souci palace. The occurrence was probably only rediscovered by the Prussian officer, for the chrysoprase in the Wenzel Chapel at Prague no doubt came from Silesia, and therefore must have been known in the fourteenth century. At the time of the rediscovery of this fine green stone the name chrysoprase was bestowed on it, although the name had been given by the ancients to an entirely different mineral.

There are no other European localities of any importance. Dark apple-green stones come from Wintergasse, in the Stubachthal, in Salzburg; and Ruda, in Transylvania, is perhaps another locality, but at both the mineral occurs only sparingly.

A certain amount of chrysoprase also occurs at a few places outside Europe. Very fine stones come from India, but exactly where they are found does not seem to be known. The mineral is also found in the nickel-ochre mine at Revdinsk, east of Ekaterinburg in the Urals, and at various places in North America. The most important of the latter is the nickel mine on Nickel Mount, near Riddle, in Douglas County, Oregon, where the mineral is found in veins of an inch in thickness in the nickel ores which occur there in serpentine. It is of a dark apple-green colour, and fine plates measuring some square inches in area can easily be obtained.

JASPER.

Jasper is a very impure variety of massive quartz, which is distinguished from hornstone by its large even conchoidal fracture, the dull lustre of the fractured surface, and by its perfect opacity and deep colour. There is no sharp distinction, however, between jasper and hornstone, nor is jasper definitely marked off from other impure varieties of massive quartz, such, for example, as ferruginous quartz (*Eisenkiesel*). Each of the compact varieties of quartz consists of an aggregate of microscopically small quartz grains ; the superficial characters of each depend upon the precise nature and amount of the impurities present. There are thus no hard and fast distinctions between these varieties ; one specimen will approach more nearly to the character of typical jasper perhaps, and another to typical hornstone, while one end of a third specimen will show the characters of one variety and the other end of another variety, the middle portion being intermediate between the two. In the same way, there is no sharp line of demarcation between jasper and chalcedony, the next variety of quartz to be considered, and there is every possible gradation between typical jasper and typical chalcedony, so that it is sometimes difficult to decide whether a particular specimen should be classed as one or the other. By lapidaries a stone which is perfectly opaque in the sharpest of splinters is referred to as jasper, and one which is slightly translucent as chalcedony. Mineralogically there is no essential difference between chalcedony on the one hand and jasper and hornstone on the other, as we shall see later.

It has been stated already that jasper is a very impure variety of quartz, containing as much as 20 per cent. of foreign matter or even more. This consists for the most part of clay and iron oxide, these substances in a more or less finely divided condition being disseminated throughout the quartz substance. Organic matter of various kinds has also been observed. It is the presence of such impurities in large amount which conditions the characters of typical jasper, namely, opacity, a large-conchoidal smooth fracture, a dull fractured surface, and a deep colour. A reduction in the amount of impurity present is attended by a lighter colour, a more uneven and splintery fracture, and some degree of translucency in the stone, which thus becomes more akin to hornstone.

The uses to which jasper is applied depend mainly upon its colour, which, in its turn, depends upon the amount of iron present. A stone which contains but little iron is practically colourless, but shows the other characters of jasper owing to the presence of clay. A white jasper of this description having the appearance of ivory is stated to occur as a great rarity in the Levant and to furnish beautiful gems. The colouring of coloured stones is seldom quite uniform, the different colours or shades of colour exhibited by the same specimen may be in irregular streaks and patches or arranged regularly in parallel or concentric alternating bands. Some stones are traversed by straight crevices and fissures filled with quartz of another colour and sometimes of quite different character.

The colours most commonly seen in jasper are brown, yellow, and red, green is fairly common, but blue and black more rare. Several colour-varieties of jasper are recognised some of which are distinguished by special names and will be dealt with in detail later on.

There is considerable variety in the manner in which jasper occurs in nature. It may occur in beds interstratified with other rocks, or in irregular nodules in various ore d eposits, especially of iron. Sometimes it is found in the joints and crevices of siliceous rocks, and at other times at spots where certain igneous rocks of the greenstone (diabase) family have intruded into clay-slates, jasper being an alteration or contact product of the latter. Irregular masses of jasper are referred to, independently of their colour, which is usually yellow, brown, or red, as common or German jasper, in contradistinction to the red or chestnut-brown ball-jasper, which occurs as regular nodules or balls in brown iron-ore, or loose on the surface of the ground. All kinds of jasper are of course very common as pebbles in the sands and gravels of rivers and streams and other alluvial deposits.

The different varieties of jasper have been much used as ornamental stones, especially in ancient times, when the stone was engraved, set in mosaics, and fashioned into objects of some size. Its application in the Middle Ages, and in comparatively recent times, was also extensive, but it has now fallen almost into disuse. Specially fine stones of uniform colour are sometimes cut up for gems, and in spite of the dulness of the fractured surface they take a fair polish, although the lustre is never very strong. The stone is principally used, however, for boxes, bowls, vases, table-tops, and small ornamental edifices. The construction of such objects may be accomplished with one large block, or it may be necessary to use several pieces, in which case they must be chosen with due regard to colour. Jasper is too abundant to be worth very much, and the value of all but exceptionally fine and uniformly coloured specimens is small.

Some of the colour-varieties of jasper occur singly and only at certain places, while at other places jaspers of several different colours are found. Owing to the wide distribution of this stone only the most important localities will be here enumerated.

Typical **red jasper** is represented by the ball-jasper of Auggen and Liel, near Mühlheim, in Breisgau. It occurs there as rounded nodules, ranging in size from that of a nut to that of a man's head, in brown iron-ore, with which it is mixed. The nodules are coated with white marl; in the interior they are of a dark, tile-red colour with white, yellow, or greenish stripes or markings.

Fine red jasper (or ferruginous quartz) sometimes traversed by veins of white quartz, occurs not infrequently to the west of Marburg, in Hesse, also in Nassau and other places in Germany, being there a contact product of clay-slate near the junction of this with diabase. The dark blood-red colour is very effective when the stone is cut. This red jasper occurs usually in rather small pieces, but blocks the size of a man's head, or larger, are sometimes found. Löhlbach, near Frankenberg, was formerly noted for beautiful jasper occurring in blocks of considerable size, which was known as "Löhlbach agate." It was once used to a considerable extent, and many artistic objects, worked in this material, are still preserved in the collection at Cassel.

Fine red jasper, as well as that of other colours, is found in the veins of iron-ore in many places in the Saxon Erzgebirge. Fine specimens of all colours are of frequent occurrence in the British Isles, particularly in Scotland.

Brown jasper is represented by the Egyptian jasper, sometimes called Nile jasper, although it does not come from the Nile. It is found in rather rough rounded nodules, on the smooth fracture of which there are concentric brownish-yellow bands alternating with the dark chestnut-brown colour of the rest of the stone. These nodules have been derived from the beds of the Nummulitic formation; they occur in large numbers on the sserir or stony areas of the Egyptian desert. There is a sserir to the east of Cairo, on the slopes of Jebel Mokattam, and over large areas of the Lybian desert there is nothing to be seen but rounded fragments of jasper. The rounding of these fragments has been effected not by water but by the action of wind-blown grains of sand.

A large amount of brown jasper associated with red and yellow jasper is found at Sioux Falls, in Dakota, North America. About 30,000 dollars' worth is cut every year in large works on the spot, the material being used specially for the ornamentation of buildings. It is known in America as "Sioux Falls jasper," and occurs in inexhaustible beds, which are excavated in quarries.

Yellow jasper is much employed in Florentine mosaics, but is otherwise unimportant. It has brownish and white streaks on an ochre-yellow ground, and is obtained from the island of Sicily, from Dauphiné, and elsewhere.

Green jasper is found principally in the Urals, where also it is worked. It occurs, among other places, at Orsk on the Ural river above Orenburg, where it forms a thick bed in gneiss. This bed has furnished blocks of sufficient size to be worked into vases and other large objects, all the cutting being done in the great works at Ekaterinburg. The colour is dark leek-green, and the stone is therefore known by the name of plasma, a variety of chalcedony, which it resembles in appearance. Green jasper is much esteemed in China, and is one of the stones to which the term "yu" is applied.

Blue jasper is always rather dull in colour, frequently showing a lavender or grey tinge, and is seldom used. We may mention here the so-called *porcelain-jasper*, which is not really jasper at all, but is clay baked and hardened by the burning of lignite (brown coal). It is usually lavender-blue in colour, but may be tile-red or yellow; it is widely distributed, especially in northern Bohemia, and a specimen is now and again used as an ornamental stone.

Riband-jasper has differently coloured riband-like bands which alternate regularly with each other. It is very impure and can scarcely be properly regarded as jasper, its chemical composition approximating more closely to that of felspar than to that of quartz. It closely resembles typical jasper in appearance, and, indeed, differs from it only in its fusibility before the blowpipe, and in the large amount of foreign matter it contains. Riband-jasper, in which the contrast between bands of colour is not sufficiently marked to be effective, is common enough; it is found in beds at Lautenthal, in the Harz, at Gnandstein, near Kohren, in Saxony, and other places. The Siberian riband-jasper, on the other hand, in which dark blood-red and leek-green bands alternate with great regularity, is most beautiful. It is said to occur near Verchne-Uralsk, at the junction of the Uralsda with the Ural river, but only in small loose pieces, so that only comparatively small articles can be made of it: larger objects, however, are often veneered with thin plates of riband-jasper. Jasper of exactly the same description is said to occur at Okhotsk, in Eastern Siberia, and good specimens are stated to be found also in Chutia Nagpur, in India. Fine riband-jasper with yellow and red, alternating with white stripes, is found in large amount at Collyer, in Trego County, Kansas. It is an excellent material for cameos, for which purpose onyx, or riband-agate, is also very well suited, owing to its regular banded structure.

The stones which are intermediate in character between jasper and chalcedony, and which may be termed **agate-jasper** or jasp-agate, usually show opaque, dark coloured portions intermixed in various ways with translucent, lighter coloured portions. Such stones are the once much talked of "jaspe fleuré" of jewellers, and they were at one time worked like jasper. The material occurred in large amount principally in Sicily, where a hundred varieties, distinguished by differences in colour and markings, were recognised. A fine agate-jasper occurs at various places in Texas, and is often called in America "Texas agate." In some of these stones the translucent agate predominates, and in others the opaque jasper, the former being sometimes distinguished on this account as agate-jasper and the latter as jasp-agate; they are scarcely ever used as gems, and none are of any importance.

AVANTURINE.

Avanturine or avanturine-quartz is a feebly translucent, fine-grained to compact quartz, with a conchoidal or usually splintery fracture; the surface of which has a speckled, metallic sheen, usually of a reddish-brown colour, but occasionally yellow, white, blue, or green. This appearance is caused by the presence, in the colourless quartz substance, of numbers of enclosures, which can always be seen with the aid of the microscope, and sometimes with a simple lens, or with the naked eye. In some cases these enclosures consist of small silvery, reddish-brown, or sometimes white, scales of mica; in others of minute plates of the green chrome-mica fuchsite, or of similar plates of a blue colour of some unknown mineral, while in other stones the sheen is caused by the existence of numbers of small cracks filled with hydrated oxide of iron. Each crack filled with this substance and every single scale of mica gives its own metallic reflection; the sheen of the mass is the sum total of these reflections, and is the more uninterrupted the more uniformly and closely set in the quartz are these enclosures. As a rule, reddish-brown avanturine closely resembles sun-stone, which for this reason is called avanturine-felspar, avanturine itself being often called avanturine-quartz. The two can be distinguished by the fact that the quartz is the harder and will scratch the felspar.

Avanturine is often set in rings, pins, brooches, cuff-links, &c. It is cut with a flat or slightly convex surface without facets, and acquires by polishing a fine, strong lustre. The more uniform and uninterrupted is the sheen of a specimen of avanturine the more valuable it becomes, and portions with a sheen of this discription are cut out of the large irregular masses of poorer quality which occur in nature. Avanturine was at one time much prized as a gem; now, however, it is more frequently used as a material for bowls, vases, and ornamental objects of various kinds, the sheen being of course less perfect and uniform over objects of this size than over the surface of a small gem. Reddish-brown avanturine with a coppery reflection is the most highly esteemed variety; other avanturine, such as the brown, reddish-yellow, white with a silvery white sheen, black with white spots, green and blue, though rarer, are less prized. The special beauty of avanturines is associated with the isolation of the glancing metallic scales, and the consequent appearance of so many starry, glittering points.

Avanturine is rather widely distributed, and specimens of considerable size are not by any means rare, although those possessing the most desirable characters and of a quality suitable for gems are not to be found at every locality.

The mineral occurs both in primary deposits and also as loose pebbles. The richest localities are in the Urals, where it is found, forming thick beds in mica-schist, at several places in the Taganai range, to the north of Zlatoust on the Ui, a tributary of the Ufa. Also at Kossulina, twenty-eight versts west-south-west of Ekaterinburg, the material found here being superior in colour and sheen, but traversed by cracks and therefore not to be obtained in large pieces. All the material found in this region is cut at Ekaterinburg. White and reddish-white avanturine is found in the Altai mountains at Beloretzkaya, thirty versts from the long-famous lapidary works at Kolivan (latitude about 51° N.) It is worked at Kolivan; and, together with the Uralian avanturine, has furnished material for the bowls, vases, and ornaments which have been from time to time presented by the Czar of Russia to European princes. A vase of this description was presented in 1843 by the Czar to Sir Roderick I. Murchison in recognition of his services in exploring the geology of part of the Russian Empire; it is 4 feet high and measures 6 feet in circumference at its largest part; it is preserved in the Museum of Practical Geology, London. Large blocks of avanturine of faultless quality are worth hundreds of pounds.

Fine specimens of this mineral are occasionally met with in India, but nothing definite is known as to their mode of occurrence or the exact locality. A very pretty green glistening variety from the Bellary district in Madras deserves mention; the scales of mica enclosed in it are of the green chromiferous variety known as fuchsite, and the mineral itself occurs in blocks from which slabs of considerable size can be cut.

Green avanturine of this description is very highly esteemed in China; it is one of the stones referred to as " yu " and is distinguished from the others as the imperial yu-stone. The imperial seal is said to be made of this material, which is esteemed far more highly than is nephrite. The locality from whence the Chinese obtain the stone is not known.

Avanturine is said to have been found at several localities in Europe; for example, in the neighbourhood of Aschaffenburg in Bavaria, at Mariazell in Styria, and at Veillane between Susa and Turin in Piedmont, where it is found as pebbles. Also at Nantes in France, Glen Fernate in Scotland, and in the neighbourhood of Madrid, where it is associated with pebbles of granite.

Avanturine-glass is an artificial product which resembles natural avanturine but possesses an even finer appearance. It is a colourless glass in which are embedded numerous small red octahedra, the faces of which are equilateral triangles. The chemical composition of the material, the well-defined crystalline form, together with the red colour and strong metallic lustre of the enclosures, points to the fact that the latter consist of metallic copper. The much lower hardness of avanturine-glass, together with the form of the enclosures, which can be readily made out with a lens, definitely distinguishes it in all cases from natural avanturine (avanturine-quartz) and from sun-stone (avanturine-felspar).

There is a rather improbable story to the effect that a glass-maker of Murano, near Venice, discovered the art of making this glass by accident (*par aventure*), by dropping some copper filings into molten glass, and that the name avanturine (or aventurine) originated thus, and was afterwards applied to the natural mineral. It is very likely that the fable was invented in order to preserve the secret of the art of manufacturing this magnificent glass by putting curious inquirers off the scent. The art came subsequently to be forgotten or lost, even in Murano, until the year 1827, when it was rediscovered by the glass-maker, Bibaglia, after many unsuccessful trials. The chief difficulty is to prevent the copper from separating out into clusters of crystals, and to obtain these crystals distributed regularly throughout the whole mass of glass as is necessary in order to obtain the most favourable appearance. The product of this art, which is still a trade secret, is a most beautiful glass, much finer than natural avanturine, and is much used in the manufacture of small ornamental objects. It is possible to obtain large blocks of the material so that it can be used for objects of considerable size. Avanturine-glass fetches from 40s. to 50s. per kilogram; it contains about $2\frac{1}{2}$ per cent. of copper and is made of a very easily fusible glass, which melts at a much lower temperature than does the copper.

C. CHALCEDONY.

The chalcedony group includes a number of siliceous minerals of compact structure and fine splintery fracture, which are characterised by the possession of a finely fibrous structure. The fibres are always recognisable under the microscope, and sometimes with a simple lens or with the naked eye; they are always very short and their optical characters differ somewhat from those of quartz. Moreover, the specific gravity and the hardness are both less than those of quartz, the former being $2 \cdot 59$–$2 \cdot 60$, and the latter $H = 6\frac{1}{2}$; and in addition the mineral is considerably more soluble in a caustic potash solution than quartz.

Although chemically chalcedony is, like quartz, pure silica, yet since its physical properties differ it must be regarded as another crystallised modification of this substance. Chalcedony was at one time supposed to consist of a mixture of quartz and opal, which would account for its greater solubility in caustic potash. Microscopic investigation, however, has shown this hypothesis to be untenable.

Chalcedony being a fibrous aggregate never assumes a regular external form, but it may, and often does, occur as a pseudomorph after some other mineral. That is to say, it assumes the crystalline form characteristic of that mineral by mere replacement of the substance of the crystal, and not in response to an inherent tendency in the molecules of chalcedony to arrange themselves in that particular form. Chalcedony does, however, occur very frequently in rounded, reniform, botryoidal, cylindrical, and stalactitic forms, in which the fibres are arranged perpendicular to the surface. Such masses of chalcedony are built up of very thin concentric layers arranged parallel to the rounded surface. This banding is never wholly absent, but may be indistinct ; it is shown more or less prominently on all surfaces, whether cut or fractured, but is displayed to greatest advantage by a polished surface cut perpendicular to the layers. According as the bands are more or less prominent, the mineral is distinguished as striped or unstriped chalcedony, the former being usually referred to as agate. No very sharp line of demarcation can be drawn, however, between agate and unstriped chalcedony, the one passing insensibly into the other.

A blow from a hammer will sometimes cause a piece of chalcedony to separate along a surface corresponding to the superimposed layers, the surface of separation being frequently smooth and bright and of course parallel to the original surface. As a rule, however, the layers are so firmly bound together that this separation does not take place, although it is very easy to fracture the mineral by a blow perpendicular to the rounded surface, that is to say, in the direction of the fibres. When crushed, chalcedony always breaks in the direction of the fibres ; this fracture is uneven and finely splintery, and with a feeble lustre inclined to the waxy type. After cutting and polishing the stone acquires a brilliant lustre of the vitreous type. Chalcedony is translucent to semi-transparent, never perfectly transparent. Light passes through it more readily in a direction parallel to the fibres than in one perpendicular to them ; a slab of the mineral cut in a direction perpendicular to the fibres will therefore be more translucent than one of equal thickness cut parallel to the fibres. The different layers of which a specimen of chalcedony is built up vary in translucency, some being almost transparent and others almost opaque. It frequently happens that clear and cloudy layers alternate with each other.

As a rule, chalcedony is colourless, milk-white, or some faint shade of grey, yellow, or blue, but black and pronounced shades of yellow, brown, red, and green are met with occasionally, and blue as a rarity. Some specimens are of one uniform colour throughout, but more frequently the different layers of the stone are differently coloured, the colour, however, being uniform throughout the same layer. This contrast in the colouring of the layers has the effect of bringing much into prominence the banded structure of the stone.

Chalcedony being somewhat porous it can be artificially coloured with ease simply by immersion in a fluid containing colouring matter in solution. The liquid is sucked up into the pores of the stone, and as the stone dries and the liquid evaporates the colouring matter is deposited in these pores and imparts its colour to the whole mass. So porous are some specimens of chalcedony that they will adhere to the tongue owing to the rapidity with which the moisture of that organ is absorbed, and when placed in water the expulsion of the air-bubbles from the pores is so rapid as to cause a hissing noise. Not infrequently the pores, in the form of round cavities or elongated canals, can be recognised under the microscope, or even with a simple lens. Different layers of the same specimen are not equally porous, some will absorb the colouring matter with great rapidity, while others will be scarcely coloured at all.

The intense colours shown by certain specimens of natural chalcedony have in many cases been assumed after the formation of the mineral, by a natural process analagous to the artificial method.

The artificial colouring of chalcedony has become quite a prominent feature of the agate-cutting industry. It is carried on to such an extent that there is actually more of the artificially coloured material sold than of the natural stone. The subject will be treated in more detail, however, when the working of the stone is described.

Chalcedony occurs in layers with a mammillated surface coating the surfaces of rocks in cracks and crevices, and in favourable positions in the form of stalactites. It is found usually in volcanic rocks, which, when chalcedony is present in any amount, are always in an advanced stage of decomposition. Like opal and amethyst, chalcedony is thus an alteration product of the rock in which it occurs, and when the last stages of decomposition have been reached the chalcedony is set free and lies loose in the soil in isolated fragments, which are often carried away by running water and finally deposited as pebbles in the beds of rivers and in other alluvial deposits.

Several varieties of chalcedony are recognised by lapidaries, the distinctions between them being based on differences in colour and the mode of its distribution, differences which are not, however, very sharply defined. It is, moreover, sometimes difficult to decide whether a given specimen should be referred to chalcedony, jasper, or hornstone. Mineralogically chalcedony is determined mainly by the finely fibrous structure, but lapidaries classify the compact quartz minerals according to translucency alone, all the perfectly opaque varieties being referred to as jasper, and all those which are more or less translucent as chalcedony.

The varieties of chalcedony here distinguished are as follows:

> *Common chalcedony* of a faint uniform colour.
> *Carnelian* of a uniform red colour, with which is included brown *Sard*.
> *Plasma* of a uniform dark green colour. This, when spotted with red, is known as *heliotrope*, or blood-stone.
> *Agate* with a prominent banded structure. The well-known *onyx* is included under this "striped chalcedony."

The sub-classification of these varieties will be given below.

COMMON CHALCEDONY.

This is chalcedony in the more restricted sense, and is known simply as chalcedony. In colour it is white or some very pale shade of grey, yellow, brown, blue, or green ; though, as a rule, it is uniformly coloured, specimens are sometimes met with which show cloudy patches (cloud-chalcedony) and an indistinct banding.   Common chalcedony is built up of layers just as agate is, but being all of the same colour the banded structure is inconspicuous ; specimens in which it is at all prominent are referred not to common chalcedony but to agate.   As examples of chalcedony with a pronounced colour may be mentioned the rare *sapphirine* of lapidaries found at Nerchinsk in Siberia, in Transylvania, and in India, which approaches in colour the blue of the sapphire, and the yellow *ceragate* (wax-agate or semi-carnelian), which has a waxy lustre.   Cloudy milk-white chalcedony is referred to as *white carnelian.*

The fibrous structure of common chalcedony is fairly apparent : the fracture is typically uneven and finely splintery, and the lustre of the fractured and of a cut and polished surface agrees with the description already given.   Though never perfectly clear the mineral is often very translucent, even in thick pieces, but in other cases, especially in milk-white stones, it is almost opaque.   Very translucent stones are termed " oriental chalcedony " to distinguish them from more opaque specimens, which are described as "occidental chalcedony."   The finest " oriental " specimens are not always perfectly translucent throughout, but show delicate cloudy patches, which are by no means prejudicial to their beauty, but rather the reverse.

Common chalcedony occurs as reniform, botryoidal, and stalactitic incrustations on the surface of rocks, and also as a filling in of the cracks and crevices of these same rocks.   It is of moderately wide distribution, but usually occurs only in small amount, the incrustations being too thin to admit of the material being cut.   Thicker masses of greater purity and beauty come from Iceland and the Faroe Islands, and a considerable amount also from India.   The Indian localities will be enumerated below under agate.

Chalcedony was more extensively used both formerly and in ancient times than now, its place having been taken by other stones.   It is cut for ring-stones and seal-stones, and it also furnishes a material for seal-stocks, cups, plates, bowls, vases, &c.   It is worked in Europe, principally at Oberstein and Idar on the Nahe, and at Waldkirch in Baden.   A considerable amount is done also in India, but the industry is of less importance there than in Germany.

It is only exceptionally large and fine specimens of chalcedony which have any considerable value.   The value of a stone depends principally on its translucency and on the uniformity of its colouring ; it is important also that there should be no cracks or faults of other kinds.   Cloudy, patchy, or fractured stones are practically worthless.   In spite of the fact that chalcedony is so abundant and so low in price, it has been closely imitated in glass ; such imitations, however, can be distinguished from the genuine stone by the fact that they are less hard, and that their specific gravity is higher.   A few varieties of common chalcedony are sometimes distinguished by special names, for example :

Spotted agate (spotted chalcedony, St. Stephen's stone) is a white or greyish chalcedony with small red spots.   In the finest specimens the spots are no bigger than dots, and these are distributed uniformly over the whole surface so closely that when viewed at some distance the stone appears to be of a uniform rose-red colour.

Mocha-stone (tree-stone or dendritic agate) is a white or grey chalcedony showing brown, red or black dendritic markings resembling trees and plants. These have been formed by the percolation of a solution containing iron or manganese through the fine cracks of the stone, and the subsequent deposition of the colouring matter originally held in solution. The brown and red marking are caused by oxides of iron, and the black by oxide of manganese. The fact that such tree-like markings are formed when a liquid travels in such a confined space can be easily shown experimentally by means of a coloured solution between two plates of glass. The dentritic markings of mocha-stone lie to a large extent in the same plane, and in cutting this stone the lapidary aims at displaying the marking with a coating of chalcedony of only just sufficient thickness to preserve it from injury. The surface of cut stones is either plane or slightly convex, and their outline may be round or oval. A mocha-stone suitable for mounting as a brooch is represented in its actual size in Fig. 89. Certain rare mocha-stones in which the white or grey chalcedony shows, besides dendritic markings, the red dots characteristic of St. Stephen's stone are specially beautiful.

FIG. 89. Mocha-stone.

Dendritic agates of this description are said to have come originally from the neighbourhood of Mocha in Arabia, a seaport at the entrance of the Red Sea; hence the name "mocha-stone." In later times it has been obtained principally from India, where it occurs in the Deccan traps (see under *agate*). Fine stones are met with as pebbles in the Jumna river, and the mineral also occurs in large amount north of Rajkot and the Kathiawar peninsula, and in the bed of the Majam river in the same district. Here are found blocks of spherical, botryoidal, or amygdaloidal form, weighing as much as 40 pounds, as well as rounded pebbles. Fine material suitable for cutting occurs also at various places in North America; for example, at Central City in the Rocky Mountains.

The most valuable mocha-stones are those in which the dendritic markings closely resemble the outlines of trees and plants; stones in which the markings take the form of irregular black and brown patches are worth but little. Very good imitations of mocha-stone are now produced, an Oberstein agate-dealer having succeeded in producing permanent markings of a kind similar to those of mocha-stone on the surface of cut chalcedony. These artificial productions are far more beautiful than the naturally occurring mocha-stone, but otherwise are difficult to distinguish from these. Both the natural and the artificial stones were at one time much used, and commanded a high price, but this is not the case at the present time.

Moss-agate is characterised by the presence of green enclosures, such as are found in many specimens of rock-crystal. These enclosures usually take the form of long hairs and fibres much intertwined, and have the general effect of a piece of moss; hence the term moss-agate. The stone is common in the volcanic rocks (trap) of western India, occurring in places with mocha-stone. It fills irregular veins in the decomposed trap at Rajkot, among other places, where blocks ranging in weight from $\frac{1}{2}$ to 30 pounds are obtained. It occurs also as pebbles in many Indian rivers, for example, in the Narbada, Jumna, and Godavari. A supply of natural green and of artificial yellow and red moss-agates has been obtained for some time from China, and these stones have largely replaced others on the market. Fine moss-agates occur in considerable numbers in the States of Utah, Wyoming, Colorado, and Montana, in North America.

Enhydros are stones which are more curious than beautiful, and are used but little as gems. They are flat, oval, and hollow nodules of strongly translucent, almost colourless chalcedony, partly filled with water. The water thus imprisoned can be distinctly seen through the thin wall of translucent chalcedony, and can be heard when the stone is shaken. These nodules are now found in masses of clay, but they were originally formed in solid rocks of volcanic origin, which in course of time have been decomposed and weathered to clay. The mode of occurrence of these enhydros is the same as that of ordinary chalcedony, and they must be regarded as chalcedony amygdales which have been only partially filled with mineral matter. Throughout the weathering of the rock they retain the form of the original rock-cavity, and when the process is complete can be extracted from the soft clay without injury, which would be impossible while they are embedded in solid rock. Such nodules of chalcedony containing water, and no bigger than a nut, were first obtained from Monte Tondo in the Colli Berici, near Vicenza, in northern Italy, where they occurred at rare intervals in weathered basalt. They were known to the ancients, and are mentioned by Pliny. At the present time they are obtained in considerable numbers from Uruguay, where they occur with agate in weathered amygdaloid, the largest being half the size of a man's fist.

When such enhydros are exposed to dry air, it sometimes happens that the water contained in them slowly evaporates, but is re-absorbed when the stones are immersed in water. This furnishes still another instance of the porous nature of chalcedony, and also throws much light on the origin of agate amygdales to be discussed below.

Enhydros are sometimes polished, although they are used so little as gem-stones. Stones of sizes less than a nut are chosen for this purpose ; their surfaces are smoothed and polished very carefully in order to avoid breaking the nodule, and the movement of the enclosed liquid can then be clearly seen. The stones thus prepared are mounted in rings, pins, &c., and are worn more as curiosities than for the sake of any intrinsic beauty they possess.

CARNELIAN.

Carnelian is red chalcedony. It may have every appearance of being uniformly coloured or, on the other hand, the different layers of which it is built up may differ slightly, though appreciably, in colour. Typical carnelian is of a deep flesh-red colour, hence its name, but every shade of colour between this and pure white and yellow is represented. It was mentioned above that white chalcedony is often termed white carnelian ; this term is most correctly applied to chalcedony with a faintly reddish or yellowish tinge, every possible shade between this and the deep flesh-red of typical carnelian being represented in natural stones. In some cases portions of the stone are paler than others, and specimens with a pale central portion passing gradually into a dark coloured exterior are not at all uncommon. Fine dark stones of uniform colour and free from faults are described as " carnelian de la vieille roche," or as " masculine carnelian." By transmitted light they appear of a deep blood-red colour, and in reflected light of a blackish-red shade. They are found in India, but are very rare, and on account of their great beauty are highly esteemed. Stones of a pale red, or of a yellowish-red colour, are described as " feminine carnelian," or simply as carnelian. Among several thousand stones there will be probably only a few to which the term " masculine " may be be applied ; the rest will be yellowish, brownish, greyish, or too pale in colour, patchy or disfigured by cracks and fissures. Compared with opaque red jasper, which is often of a carnelian colour, all carnelians, whatever be their colour, are strongly translucent.

The colour of carnelian is not due to the presence of organic compounds, as was once supposed, but to that of compounds of iron, the red colour being due to ferric oxide, and the yellow and brown to hydrated ferric oxide. Yellow or brown stones, when exposed to the action of heat, gradually assume the red colour characteristic of carnelian, owing to the hydrated ferric oxide losing water and becoming anhydrous. Many stones, which in the natural condition are of a dirty yellow colour and unsuitable for gems, on being heated acquire the fine carnelian colour, and with it a considerable increase in value.

Stones which contain too small a quantity of iron are always pale in colour and cannot be made to assume the deep flesh-red tint when heated. It is, therefore, desirable in some cases to introduce a little more iron. This is done by immersing the stones in a solution of iron nitrate, made by dissolving some iron needles in nitric acid, or in a solution of iron vitriol, which may be more easily obtained and is just as effectual. On heating, the stones assume the fine carnelian red colour ; care must be taken in all these operations, however, that the temperature does not rise too high, for on strongly igniting carnelian it becomes white and dull, and can then be easily crushed to powder.

The mode of occurrence of carnelian in nature is the same as that of common chalcedony and of agate, namely, as incrustations with nodular surfaces, and in the cracks and crevices, and especially in the amygdaloidal cavities, of volcanic rocks. On the disintegration of these rocks the rounded nodules or irregular fragments of carnelian remain loose in the ground, or are transported by running water, and finally deposited as rounded pebbles in the sands and gravels of rivers and streams. In this form carnelian is of moderately common occurrence. The material, which is cut and polished in the lapidary works, comes almost exclusively, however, from India, Brazil, and Uruguay, where it occurs and is collected with chalcedony of other kinds, especially agate, the occurrence being described below under agate. The localities at which chalcedony of finer quality occurs will now be enumerated.

In **India** blocks of carnelian, weighing as much as 3 pounds, are found in the Rajpipla Hills, at Ratanpur, on the lower Narbada river (Fig. 33). The material as it is found in the mines may be blackish, olive-green, milk-white, or, in fact, almost any colour except red. This tint is only acquired after the stone has been heated, the heating being effected partly by a long exposure to the sun's rays and partly by fire. Stones, of which the original colour was olive-green, assume an especially fine tint on heating ; they are much prized, and are largely cut in the neighbourhood of Cambay, near Baroda.

Deposits of carnelian are worked also on the Mahi river, north of Baroda, and the mineral is found at many other places in the volcanic district of western India, but is not everywhere collected and worked. Moreover, pebbles of carnelian are found together with jasper and other varieties of chalcedony in almost all the rivers. A very similar occurrence to those in western India is that in the volcanic rocks of the Rajmahal Hills on the Ganges, in Bengal, but this appears to be of little commercial importance.

In **South America** the best known locality for carnelian is Campo de Maia, fifty miles south of the Rio Pardo, which joins the sea at Porto Alegre, Brazil. The stones found in this district are remarkable for their regular spherical form. Wherever agate occurs in this region it is accompanied by carnelian, so that the latter is somewhat widely distributed.

Other places occasionally given as localities for fine specimens of carnelian are situated in Dutch Guiana and Siberia ; also at Warwick, in Queensland. Compared with Indian and Brazilian localities, however, all are unimportant. Numbers of carnelians were formerly found in Japan, where they were pierced and strung together as beads, and disposed of to the Dutch, who at one time traded extensively with this country.

For use as a gem carnelian is cut, like chalcedony, without facets, with a plane or convex surface and with a round, oval, or other outline. The other purposes to which it is applied are the same as already described under chalcedony. Carnelian, as a rule, is less brittle than chalcedony and is therefore a more suitable material to engrave upon. Gems of carnelian are often mounted upon a gold or silver foil, which considerably increases their colour and lustre; they are used rather extensively in cheap jewellery, more so than ordinary chalcedony.

It has been stated already that carnelian of a more or less pronounced brown colour is frequently met with. Brown carnelian, whether of a bright chestnut-brown or more of an orange shade, is distinguished as **sard**. This variety is not very sharply marked off from ordinary red carnelian, and it is often difficult to decide which name to apply to a given stone. The finest sard is considered to be that of an orange-brown colour, which in transmitted light appears of a fine red colour. The colour of sard has been compared to that of salted sardines, and from this the name is said to be derived. Some specimens of sard only acquire their characteristic colour after exposure to heat. Many are dotted with numerous opaque spots of darker colour; such stones are described as "sandy sard." Fine sard is uncommon and very valuable. It is found at the localities mentioned for carnelian, and is collected with this stone and with agate. A method is now known by which a fine deep and uniform brown colour can be imparted to chalcedony. This artificially coloured material is known as sarduine, and is in no respect inferior to natural sard in place of which it is often used.

PLASMA.

This is the name applied to green chalcedony, the colour of which ranges from dark leek-green, the commonest shade, through pale apple-green to almost white. The colouring substance is the so-called green-earth (a chloritic or micaceous substance), or possibly sometimes asbestos, like that enclosed in moss-agate. While, however, in the latter stone these enclosures are aggregated in patches conspicuously marked off from the quartz substance in which they are embedded, in plasma they occur in small grains and scales distributed throughout the whole stone and to which they impart a uniform green colour. Owing to its numerous enclosures plasma is much less translucent than the other varieties of chalcedony, and is more like jasper, resembling this stone also in the fact that its fracture is smooth rather than splintery. Microscopical examination shows that the stone possesses the fibrous structure of chalcedony; it is therefore essentially different from the fine-grained green jasper, from which, however, in particular cases it is almost impossible by mere inspection to distinguish it.

Plasma was at one time known only as the material of which objects found among the ruins of ancient Rome and in other Roman remains were made. The place whence the Romans obtained their rough material is still unknown, but at the present time a considerable amount of the mineral is obtained from India. It is especially fine in quality here, and occurs like carnelian in the volcanic rocks of the Deccan, especially in the district south of the Bhima river (Fig. 33) in Haidarabad, and as pebbles in the rivers Kistna and Godavari among others. Plasma of fine quality is said to have been found also at the first cataract of the Nile in Upper Egypt. Oil- or leek-green plasma, sometimes of rare beauty, is found also in the porphyry of the Hauskopf and Eckefels, near Oppenau, in the Black Forest. It occurs here in nodular masses built up of concentric shells of plasma, chalcedony, quartz, and other minerals, alternating with each other. Another locality for plasma in the same part of Germany is the Sauersberg in Baden-Baden, where also it occurs in nodules enclosed in porphyry. For gem purposes, however, plasma from the Black Forest is unimportant.

Heliotrope (" oriental jasper," " blood-jasper," or blood-stone), Plate XVIII, Fig. 6, is a green chalcedony differing from plasma only in that its green colour is spotted, patched, or streaked with a fine blood-red. These markings have been compared with drops of blood, hence the names " blood-jasper " and blood-stone. A famous sculpture executed in heliotrope is preserved in the national library at Paris. It represents the scourging of Christ, and the red marks of the heliotrope are skilfully utilised to represent drops of blood on the raiment. The most valuable heliotrope is that in which the spots are of a full red colour, uniform in size and distributed regularly over the surface of the stone. Material in which the red marking is in streaks or large patches is considered to be inferior. The ground-mass, moreover, which is always appreciably less translucent than the red portions, should be of a uniform deep green shade. Stones in which the red markings are replaced by yellow spots and patches are sometimes met with ; they are far less beautiful than ordinary heliotrope, and are scarcely ever used as gems. Red spotted heliotrope is worn, like plasma, in rings, pins, brooches, and similar ornaments, and also is fashioned into small objects of various kinds. The rough material is obtained almost exclusively from India, where it occurs with agate, carnelian, plasma, &c. As localities are mentioned the district north of Rajkot in the Kathiawar peninsula, west of Cambay, where masses weighing as much as 40 pounds are found, and Puna, south-east of Bombay. Heliotrope and other similar stones are often stated to be exported to Europe from Calcutta ; it is scarcely probable, however, that material found in the west of India should be exported from an eastern port. The material sent from Calcutta may possibly be obtained somewhere in the east, in the Rajmahal Hills on the Ganges, but nothing is definitely known with regard to this. Compared with the Indian occurrence of heliotrope that of Europe is quite unimportant ; fine specimens are found at several places in Scotland, especially in the basalt of the Isle of Rum. Fine heliotrope has been found recently in Australia, and numerous specimens of the same stone come from Brazil where they occur with carnelian, agate, &c.

AGATE.

Agate is the most important variety of chalcedony, and is used far more extensively than any other. It is built up of layers which differ conspicuously from each other in colour and translucency, and on this account has a banded or striped appearance. At times these different layers are very similar in colour and transparency, the banded appearance being then less prominent, but typical agate always shows these sharply contrasting bands.

The width of the bands is usually the same throughout their whole course, and is often, though not always, extremely small. The small size of some may be realised from the fact that in a stone examined under the microscope by Sir David Brewster there were no less than 17,000 definitely marked bands in the space of an inch. A thin plate cut perpendicular to the layers of a very finely banded agate, when held up to the light, shows rainbow colours; in other words, the plate forms a grating which gives a diffraction spectrum. Stones which are finely banded enough to show this phenomenon are known as *rainbow-agates ;* they have no importance as gems.

Different layers vary very much in translucency, some being almost transparent, and others practically opaque. The colours of the different layers are, generally speaking, the same as those of common chalcedony. Thus some layers are almost colourless, or show pale shades of grey, blue, yellow, or brown, while the colours of others may be pronounced shades of yellow, red, brown, or grey, green and blue being always rare.   Each single layer is usually uniformly coloured throughout its whole extent, and can be identified with one or other of the colour-varieties of chalcedony already considered.   Thus the pale-coloured layers agree in character with common chalcedony, the red with carnelian, the brown with sard, and so on.   According to the predominant colour of a stone it may be distinguished as chalcedony-agate, carnelian-agate, &c.; and here also may be mentioned agate-jasper (jasp-agate) already described, in which layers of translucent chalcedony alternate with bands of opaque jasper.

The beauty of agate and its application to ornamental purposes depend upon the contrast in colour and translucency between the bands of which it is built up.   Finely marked and very translucent agates are sometimes distinguished by the prefix " oriental," stones less perfect in these respects being referred to as " occidental " agates.   The majority of agates as they occur in nature are light coloured, and but few show strongly marked colour contrasts, such as deep shades of red, yellow, or brown.   These stones, however, like ordinary chalcedony, can be artificially coloured, as we shall see later on when the methods of working agate are under consideration.

The two figures of Plate XIX. illustrate the varied courses taken by the bands shown on a cut and polished surface of agate.   According to the direction and arrangement of the bands, different kinds of agate are distinguished.

In *riband-agate* the different layers are parallel to each other, their surfaces being plane or uniformly curved, without indentations or prominences.   A surface cut perpendicular to the layers will show straight or curved bandings.   A riband-agate in which milk-white cloudy bands alternate with bands of another colour, the two sets being sharply marked off from each other, is known by the general term *onyx*.   According to the shade of the coloured bands several sub-varieties are recognised.   Thus, the term onyx in its more restricted sense is applied to agate in which the second set of bands are black; when these coloured bands are one of the pale shades characteristic of common chalcedony, the stone is described as *chalcedony-onyx*, when they are red as *carnelian-onyx*, and when brown as *sard-onyx*.   From the point of view of the working of agate, the different varieties of onyx have a definite importance, and will be referred to again.   In a modification of riband-agate, known as *ring-agate*, the differently coloured bands are disposed in concentric circles.   A stone with a dark coloured central spot surrounded by a series of concentric rings has a certain resemblance to an eye, and is therefore referred to as an *eye-agate*.   A kind of ring-agate can be artificially produced from ordinary agate or chalcedony: the point of a steel rod is placed on a cut surface of agate and is then smartly struck with a hammer.   There is then formed around the point a system of concentric rings, which give the stone a very pretty appearance.

When the banding is disposed so as to form re-entrant and salient angles, its outline being the same as that of the plan of a bastion or fortress, the agate is called *fortification-agate*.   The markings of *landscape-agate* suggest the outlines of a landscape picture, and when the banding recalls the outlines of a ruin, as is the case in brecciated-agate, to be described below, the stone is described as *ruin-agate*.   In *cloud-agate* cloudy patches contrast with a more translucent background; *star-agate* shows star-shaped figures; *shell-agate* or *coral-agate* resembles in appearance fossilised shells and corals, the petrification of these objects in agate actually at times taking place.   A number of other names, most of which are descriptive of the sub-variety to which they are applied, are recognised, but need not be enumerated here.

PLATE XIX

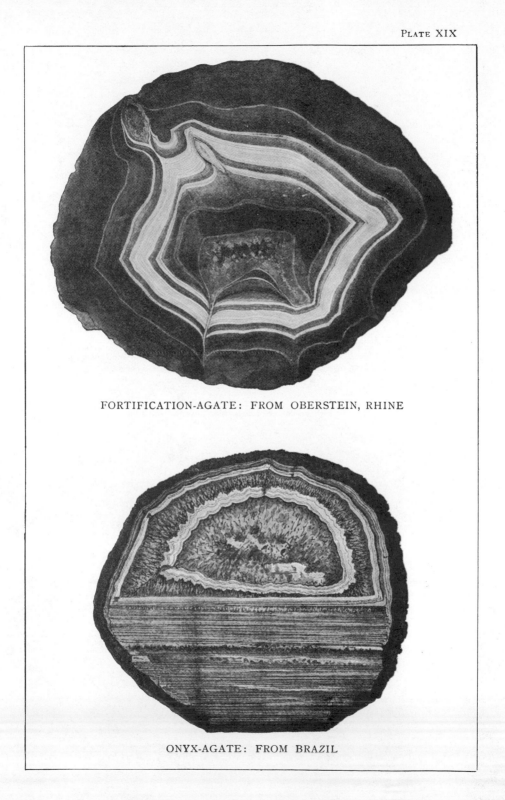

FORTIFICATION-AGATE: FROM OBERSTEIN, RHINE

ONYX-AGATE: FROM BRAZIL

Agate occurs only rarely as veins filling the fissures of rocks. There are typical examples of this mode of occurrence at Halsbach, near Freiberg, in Saxony, where the veins consist for the most part of coral-agate, and at Schlottwitz, near Wesenstein, in the Müglitz valley, in Saxony. The excellent riband-agate found at the latter locality has narrow, brightly-coloured bands arranged parallel to the surfaces of the vein; the latter contains, besides agate, common chalcedony, jasper, quartz, and amethyst. At a certain spot the material of one-half of the vein has been completely broken up by earth movements. The angular fragments of agate thus produced were afterwards cemented together by amethyst, the contrast in colour between these two minerals rendering the resulting stone, the well-known *brecciated-agate*, very effective for decorative purposes. The fragments of agate in brecciated-agate sometimes suggest a representation of ruined buildings, this variety being on that account referred to as ruin-agate. This brecciated-agate was discovered in 1750 and mined in some quantity. Like other Saxony agates, for example those found in porphyry at Altendorf and Rochlitz, it was fashioned into all kinds of articles. A beautiful rose-red ornamental sand was at one time prepared from the coral-agate of Halsbach, but at the present time the mines are almost always filled with water and therefore inaccessible.

Agate occurs much more commonly, however, in the vesicles of certain volcanic rocks, such as porphyries and basalts, and specially in amygdaloidal melaphyres. These almond-shaped or amygdaloidal cavities have already been mentioned under amethyst; besides this mineral most of them contain agate, and this, to distinguish it from the agate which occurs in veins, is sometimes termed amygdaloidal agate. Almost the whole of the material used for cutting is amygdaloidal agate. The cavities which are filled principally with agate are called agate-amygdales; their external surface is usually very rough and pitted.

The layers of agate of which these amygdales are built up usually follow more or less closely the surface of the amygdale. When exactly parallel to the external surface we get riband-agate, and when not exactly parallel we get one or other of the sub-varieties already described. Many South American amygdales are remarkable in that the bands follow the outline of the cavity for part of their course, and then, leaving the wall of the cavity, take a short cut across it (Plate XIX, Fig. *b*). Usually in agates from all other localities, and indeed in many South American amygdales also, the layers of chalcedony closely line the walls of the cavity, as represented in Fig. *a* of Plate XIX.

The agate of the amygdale is only rarely in immediate contact with the rock in which it is embedded; more frequently a thin layer of a green, earthy, chloritic, or micaceous mineral intervenes, the same substance as the so-called green-earth, which occurs enclosed in moss-agate and which imparts to plasma and heliotrope their green colour. The layers of agate which succeed the layer of green-earth do not, as a rule, completely fill up the cavity of the amygdale, and such amygdales containing a central space are known as geodes. The innermost layer of agate enclosing this central space may have a reniform or botryoidal surface, as is common in all varieties of chalcedony, or it may hang in stalactites from the roof of the cavity. Sometimes such a cavity is completely filled up by a later deposition of agate, which encloses the stalactites of the former stage of growth and gives rise to the so-called *pipe-agate*.

Amygdales consisting entirely of agate are, however, very rare; the innermost layer surrounding the central space usually consists of crystallised quartz—often of amethyst—the pyramidal terminations of the crystals projecting into the space. On the other hand, the whole of the central cavity may be filled with crystalline quartz showing no crystal faces (Plate XIX, Fig. *b*). In amygdales of this description there may be only a small central nucleus of amethyst, the bulk of the structure being of agate, or, again, the agate may form only a thin layer enclosing a nodule of amethyst, as already described under amethyst.   One accordingly distinguishes agate-amygdales and amethyst-amygdales, though the difference is only one of proportion.   In many cases the innermost layer of an amygdale is constituted by crystals of calcite or other minerals, especially the hydrated silicates belonging to the zeolite group, which rest upon the crystals of amethyst.

Sections of agate-amygdales often disclose other peculiarities which are important, inasmuch as they throw light upon the mode of formation of the amygdale.   Thus it has been observed that the central cavity of the amygdale is placed in communication with the exterior by a narrow canal, formed by the layers of agate bending sharply outwards at this particular point.   At least one such canal, and very frequently several, exist in every amygdale.

The external opening of the channel is frequently to be seen on the surface of the amygdale as a funnel-shaped depression, but is sometimes obscured and only to be made out in sections.   These canals may be empty or, on the other hand, completely filled up with agate of the same description as that of the innermost layer of the amygdale, being indeed continuous with this.

An explanation of the origin of agate-amygdales, sufficient to account for all their details of structure, can be deduced without much difficulty from the general character and the mode of occurrence of these structures.   First, however, must be noted that the amygdaloidal cavities, in which the agate occurs, are found in the rocks already mentioned only when these have been much weathered and their constituents partly removed.   The more advanced the decomposition of the rock—which may indeed proceed as far as almost complete disintegration—the more likely is it that the amygdaloidal cavities will be filled with mineral.   On the other hand, the less decomposed is the rock the less likely is agate to be found in its cavities, these cavities in fresh unaltered rock being always quite empty.

It follows from this, therefore, that the material which fills the cavities, namely, the silica, has been derived from the rock in which they are enclosed.   During the process of weathering, the silica and other constituents of the rock have been dissolved out by water and subsequently redeposited in the cavities.   The banded structure of the amygdale, however, indicates that the deposition did not proceed continuously and uninterruptedly, but that between the formation of two successive layers there was an interval of greater or less duration.

To explain this phenomenon the existence is assumed of hot intermittent springs, such as are now seen to perfection in the geysers of Iceland and in the United States National Park on the Yellowstone river.   The essential condition for the formation of amygdales is that the hot or warm water rising up from the depths shall saturate the rocks, and that it shall sink again, leaving the rocks dry for a period.   The hot water dissolves out the silica and other constituents of the rock, and the solution fills up the amygdaloidal cavities. When the waters sink these cavities are emptied, only a film of water covering their walls being left behind.   On the evaporation of this film, which readily takes place at such a high temperature, a thin layer of silica is deposited on the walls of the cavity.   When the hot spring again rises the same thing takes place and a second layer of silica is deposited, and so on, until the cavity is more or less completely filled up.   Each time the fluid passes into and out of the cavity by the canals mentioned above, which are for this reason often referred to as tubes of entry or escape, the fluid also perhaps passes to a certain extent through the porous agate itself.   Crevices, fissures, and other cavities in the rock are, of course, filled with agate in the same way.

The size of the agate amygdales of course depends upon that of the amygdaloidal cavities of the rock.   The smallest of these are about the size of a pea or a hazel-nut, while the largest are of considerable dimensions.   The weight of an amygdale depends not only on its size but also on the size of the central cavity.   The heaviest known is from Brazil and weighs about 40 hundredweights.

The localities at which amygdale-agate is found are very numerous, but at only a few does it occur in sufficient quantity and suitable quality for ornamental purposes.

At one time the most important occurrence was in the district of the Nahe, a tributary of the Rhine, in what is now the Oldenburg principality of **Birkenfeld**, and in the neighbouring portions of Prussia.   A portion of this district was formerly included in the Rhenish Palatinate, and for this reason the Palatinate or Zweibrücken are sometimes given as the localities for these agates even now.   Melaphyre and amygdaloidal rocks are distributed very widely in this region, and almost everywhere these contain agate, though not always of the best quality.   Very fine material is obtained from the Galgenberg, near Idar, and from the Struth, near Oberstein, both localities being situated in Birkenfeld, the latter on the Nahe and the former close to it in the small side valley of the Idarbach ; also from the Rosengarten on the Weisselstein, near St. Wendel, in Prussian territory.   The agate of this region has been systematically collected for centuries, and is cut and polished in the numerous lapidary works established there, especially at Oberstein and Idar, which even now turn out a supply sufficient for the whole world.

Other European localities though numerous are unimportant.   The mineral occurs in amygdaloidal rock at several places in northern **Bohemia** ; for example, on the Jeschkenberg near Friedstein, in the Kosakow mountains near Semil, in the Tabor and Morzinow mountains near Lomnitz, and in the Lewin mountains near Neu-Paka.   At these localities many of the amygdales have been weathered out of the rock and lie loose in the ground, or are found as pebbles in the rivers, for example, in the Iser and Elbe.   Besides agate, carnelian and other varieties of chalcedony, as well as jasper, are found here and cut at the lapidary works at Turnau, Liebenau, and Gablonz.

Mention should also be made of the so-called Scotch pebbles, very pretty agates, which occur in considerable abundance at Montrose in Forfarshire, in Perthshire, and at many other localities in **Scotland**.

The occurrence of agate in countries other than Europe, especially in South America and to a certain extent also in India, is far more important.   From the former country since the exhaustion of the German localities the lapidary works at Idar and Oberstein have obtained an abundant supply of excellent material.

In **South America** the principal localities for agate are situated in the Brazilian State of Rio Grande do Sul and the neighbouring country of Uruguay, the material from this region being termed, irrespective of locality, Brazilian agate.   In association with it are found the crystallised varieties of quartz already mentioned, namely, amethyst and citrine, and several varieties of chalcedony other than agate, especially carnelian.   In this region, just as on the Nahe, melaphyres and amygdaloidal rocks extend over a wide area, and here also agates and other quartz minerals have been formed in their cavities.   Very frequently the amygdales have been weathered out of their mother-rock, which is usually completely decomposed and altered to a red or brown, highly ferruginous clay.   When this clay has been washed away by rains the agate amygdales are left behind loose on the ground, and they are often found, too, as rounded pebbles in the alluvial deposits of streams and rivers.   The amygdales are usually shaped like a cake, flat on one side and rounded on the other, and are sometimes of considerable size, the one weighing 40 hundredweights, mentioned above, having been found in Brazil.

The principal locality for agate, carnelian, and other varieties of chalcedony is a mountain chain consisting mainly of decomposed melaphyre, which is about 400 miles in length and extends from Porto Alegre, in Rio Grande do Sul, in the east to the district of Salto on the Uruguay river, in Uruguay, in the west. Cutting through this chain in the north are the Rio Pardo and Taquaire, flowing into the Gulf of Alegre, and in the valleys and beds of these rivers carnelian is found, while in the heights above striped agate is more common. From the Campo de Maia, fifty miles south of the Rio Pardo, come sardonyxes weighing as much as a hundredweight and often of magnificent colour. In the tributaries of the Uruguay, in the districts of Tres Cruces and Meta Perro, bluish agate as well as the striped variety is found.

For a long time this locality was the only place from whence bluish-grey agate was obtained. Though dingy and unattractive in themselves they possess two distinct advantages over the coloured agates of Oberstein. In the first place, they are easily coloured artificially; and in the second place, the layers of which they are built up are perfectly straight, which is a great assistance to the lapidary in the cutting of onyxes. Natural stones of a black colour are very rare; among several thousand hundredweights of material there will perhaps be scarcely a single black specimen. Layers of a fine emerald-green colour are also very rare; when they do occur, they are always situated immediately under the amethyst which rests on the agate. Rose-red agate is uncommon, but the deep flesh-red colour of carnelian is rather frequent in Brazilian agates.

The occurrence of agate and the quartz minerals with which it is associated in Brazil was discovered in 1827 by a native of Oberstein. These stores of mineral wealth, which up till that time had been lying idle, were then collected from the surface and from the loose clayey ground and were sent in great amount to Oberstein, these minerals soon becoming important articles of export. In spite of the abundance of material its collection is not very easy, for the localities are situated for the most part in inaccessible regions, and the task of transporting the stones to the coast is extremely arduous. In spite of these difficulties, however, large quantities are sent every year to Oberstein, where but little agate from other localities is now cut.

In order to give an idea of the importance of this export, and incidentally of the agate-cutting industry, the following figures representing the weight of agate, including carnelian and other varieties of chalcedony, obtained from Rio Grande do Sul alone are given:

	Cwts.			Cwts.
1872–3 . . .	3100		1877–8 . . .	1825
1873–4 . . .	3850		1878–9 . . .	1530
1874–5 . . .	1200		1879–80 . . .	1950
1875–6 . . .	1900		1880–1 . . .	380
1876–7 . . .	1720		1881–2 . . .	700

It will be seen that the output varies a good deal; the price of rough material is also subject to fluctuation, varying between 5000 and 10,000 reis per arroba (£2 to £4 per hundredweight). For exceptional material as much as 100,000 to 200,000 reis per arroba may be paid. There are no cutting works in Brazil itself; an attempt to establish such was once made by a few emigrants from Oberstein, but was unsuccessful; articles of cut agate, therefore, have to be imported from Oberstein.

In **India** the mother-rocks of agate and the other varieties of chalcedony, already mentioned, are the amygdaloidal rocks or so called traps of the Deccan plateau. These Deccan traps cover an area of thousands of square miles and extend into the surrounding districts, namely, the ancient kingdom of Gujarat, with its capital of Surat, the district to the west of the Gulf of Cambay, now known as the Kathiawar peninsular, and a portion of Rajputana. Another occurrence which should be mentioned is much further to the east in the Rajmahal Hills, which are situated in latitude 25° N., at the point where the Ganges bends to the west. These hills, as well as the neighbouring portions of Bengal, consist of a volcanic rock similiar to those of the Deccan plateau.

The crevices and amygdaloidal cavities in these rocks are everywhere filled with agate, carnelian, &c., which are set free by the weathering of the mother rock and are found loose in the ground or as pebbles in the Godavari, Wanda, Kistna, Bhima, and several other rivers rising in or flowing through the Deccan. In some places the angular or rounded blocks of chalcedony are more or less firmly bound together by a ferruginous cementing material, thus forming extensive and sometimes thick beds of conglomerate. These beds of conglomerate or secondary deposits of agate constitute the main source of the material used for cutting.

Although the several varieties of chalcedony are widely distributed throughout India, they are not everywhere of a quality suitable for cutting. Among the numerous localities at which material of good quality is obtained there are two of special importance, which have been mentioned before as localities for moss-agate, mocha-stone, heliotrope, carnelian, &c. These are the neighbourhood of Ratanpur, on the lower Narbada river, in the State of Rajpipla, and the country north of Rajkot, in the Kathiawar peninsula, where the varieties of chalcedony just mentioned are everywhere accompanied by agate.

The best known deposits are at Ratanpur. whence for more than 2000 years the lapidaries at Broach have obtained their supplies of rough carnelian, agate, &c. The best specimens all come from a thin bed of conglomerate, the fine colour of the stones being doubtless due to the ferruginous cement by which they are bound together. This bed is made accessible by the excavation of pits, which measure 4 feet across and about 30 feet in depth, the deepest measuring not more than 50 feet. From these pits extend, in all directions along the agate-bearing bed, horizontal galleries, for distances up to 100 yards. The stones found in this bed are usually under a pound in weight, and among them are a few cat's-eyes. Some need to have their fine colour developed by exposure to heat while others already possess it. Part of the material collected is sent in boats to Broach and Cambay, the sites of the principal native lapidary works, part to Europe — principally to Oberstein which it reaches through London—and the rest to China, a considerable amount of Indian chalcedony, especially carnelian, being cut in that country.

Native lapidaries practice their trade at many places, where suitable rough material is obtained, for example, at Jabalpur, in the Central Provinces (Fig. 33), and at Banda on the Khan, a tributary of the Jumna, in which a large number of pebbles of chalcedony are found. Very important at one time were the lapidary works at Broach, near Baroda, and not far from the mouth of the Narbada river, and at Almadabad a little further to the north. The only lapidary works which are of importance at the present day are those at Cambay, on the gulf of the same name, north of Bombay. Articles worked here are designed and finished in accordance with European, Indian, and Arabian tastes, and together with rough material are exported to these countries.

The murrhine vases, famous in ancient times, are said to have come from Ulein (or Ouzein), in latitude 23° 10′ N. and longitude 74° 14′ E. of Greenwich, in this same part of

India.   These, therefore, must have been fashioned out of agate or some variety of chalcedony and not as has been assumed of fluor-spar, which is not known to occur in this region, and, moreover, is rare throughout the whole country.

Applications of Agate.—Agate and other varieties of chalcedony not only furnish a material for ornamental objects of the most varied kinds; but are also applied to many other uses ; in fact, no other stone is applied to purposes so diverse.   The character and design of the ornaments manufactured of these stones change with the fashion of the hour, and certain articles which one year are made and sold in their thousands, in the next year will be unsaleable.   The favour with which the different varieties of agate and chalcedony are regarded also depends on the caprice of fashion ; at one time red carnelian is most in demand, at another green heliotrope or plasma, and at a third black agate or onyx.   Agate ornaments of all kinds were specially popular from 1848 to the middle of the fifties, and this period was the harvest time of the agate industry.   It is only specially large objects or articles of great artistic merit for which high prices are demanded ; as a rule, agate ornaments are extraordinarily cheap, but in spite of this the material is imitated very closely in glass.   This so-called agate-glass can be easily distinguished from genuine agate by the much greater hardness of the latter.

The objects most freqently fashioned out of agate are small articles of personal ornament.   These, which are of the most varied description and are manufactured in enormous numbers, include sleeve-links, breast-pins, hair-pins, ear-rings, pendants for watch-chains, necklaces, bracelets, buckles, rings, ring-stones—often worked as seal-stones, sometimes with raised figures—seals, and signets.   The mineral is also largely used as a material for articles of more or less utility, such as stick- and umbrella-handles, children's toys (marbles), match-boxes, toilet-cases, snuff-boxes, seal-stocks, pen-holders, knife-handles, chessmen, counters, bowls and vases of every size and form, holy-water founts, cups, dessert plates, sauce-bowls, candlesticks, &c.   The variously coloured chalcedonies are utilised in mosaic work, which is applied to all kinds of decorative purposes.   The objects of technical importance for which agate is used include mortars, burnishing tools for gold-workers and bookbinders, smooth stones for paper and card manufacturers, rollers for the use of ribbon manufacturers, and pivot supports for balances and other delicate mechanical instruments.

Since the year 1850 there has sprung up with Central Africa a peculiar trade in charms of brown and black agate, the so-called olives.   These objects have the form of cylinders from $\frac{1}{2}$ to 3 inches long, they are pierced in the direction of their length, and must display a white central band.   In the middle of the sixties several hundred thousand thalers' worth of these charms were manufactured at Oberstein and exported to the Soudan ; some firms exported these goods to the value of 40,000 thalers (£6000).   The demand reached its height in 1866 ; after that date it began to fall off and is now very small.   Red carnelian charms having the form of a perforated triangle are exported to Senegal.

During the manufacture of the articles named above and of others the agate undergoes many and varied processes, which include cutting, grinding, polishing, boring, engraving, and artificially colouring.   It is essentially a German industry, and is carried on mainly at Oberstein, on the Nahe, and at Idar and other places in the neighbourhood ; also in Waldkirch in the Black Forest, where, however, fewer agates than other precious and semi-precious stones are cut.   At Oberstein and Idar, on the other hand, it is agate and the other varieties of chalcedony and quartz minerals which are principally worked ; other stones including malachite, lapis-lazuli, and in recent times even the diamond, are cut there, but in quite insignificant numbers.   The whole world is supplied with agate goods from these works, and although the agate sold at watering-places, tourist resorts, and such like places is described to be of local origin this is not actually the case. The agate-cutting industry has been monopolised by Oberstein for centuries, and in no part of the world, not even in India, has it anything approaching a rival. Wherever a cut agate is seen it may be pretty safely stated that it was worked at Oberstein. The mineral is not only cut, bored, engraved, and coloured in these works, but frequently also mounted, usually in gilded brass. Specialisation in these works has been carried out to such an extent that each workman executes his part with perfect skill and the greatest possible rapidity. The consequence is

Fig. 90. Agate-grinding and polishing workshop at Oberstein.

that the agate goods produced here are more perfect in workmanship and lower in price than those manufactured anywhere else. We will now consider the methods employed in rather more detail.

Agate-cutting.—The primary cause of the establishment of agate-cutting works at Oberstein was the presence of abundance of rough material in the neighbourhood. The earliest authentic record of the existence of the works dates back to 1497, but they were doubtless in existence before that date. After flourishing for some centuries, the industry gradually declined owing to the exhaustion of the native stores of rough material. The discovery of the Brazilian deposits about the year 1830 gave a fresh impetus to the trade, all the greater because the fresh material was specially suitable for the application of artificial colouring methods, which had been discovered a short time previously. At the present time scarcely any agate of local origin is cut at Oberstein, the rough material being obtained principally from Brazil, though some Indian chalcedony, especially certain varieties such as carnelian, moss-agate, mocha-stone, &c., is also worked. Material from other parts of the world, not only agate and other varieties of chalcedony, but amethyst, citrine, rock-crystal, and indeed almost all precious stones, including diamond, are also cut there to a certain extent. The rough material is imported from its place of origin by merchants who, following an old custom, dispose of it to the lapidaries by public auction, which is held from time to time in Oberstein or Idar.

The workshops in which the agate-cutting is performed are now fitted with every modern convenience, including grinding discs rotating in a horizontal plane, and steam-power is used throughout. The arrangement of, and the methods adopted in, one establishment do not differ essentially from those of any other.

Besides the workshops fitted with modern appliances, there are still retained the mills driven by water-power, which, with various modifications and improvements, have been in use since early times. There are altogether about 200 of these mills; a series of them are situated on the Idar stream, which flows into the Nahe close to Oberstein, and on which the village of Idar is situated, and others on neighbouring streams. These are reserved mainly for grinding work proper, boring and engraving being performed in special shops. A grinding mill of this description is represented in Fig. 90. In these mills three or five (in the figure three) grindstones of sandstone, which measure up to 5 feet in diameter, are sunk in a pit and are made to revolve three times a second by means of a water-wheel. The front surfaces of the grindstones are provided at either side with ridges and furrows so that the agates can be ground to any desired form.

The rough agate is first broken up with hammer and chisel; the more valuable pieces are roughly cut into the shape they are to assume by a metal disc charged with emery powder or, in recent times, with diamond powder. After this it is ground by being pressed with considerable force against the cylindrical surface or in the groove of the rotating grindstone, which is kept constantly wet by a stream of water. The workman, who performs the operation, lies with his body in a hollowed-out, trough-like bench, the cuirass, which stands close to the grindstone. He is enabled to apply the requisite pressure to the stone by pushing with his feet against two posts firmly fixed at the end of the bench. The stone is held in the hands and turned about on the grindstone until it has assumed the desired form. At each of the grindstones, which are about a foot in width, two workmen can work when necessary, one on either side. During the process of grinding the agate becomes phosphorescent and emits a brilliant reddish-white light.

Fig. 90 is a diagrammatic representation of a grinding-mill, while Fig. 91 is a picture of an actual workshop, namely, that of Herr August Wintermantel, at Waldkirch, in the Black Forest.

Vases, bowls, and such like objects are hollowed out by other workmen on a small grindstone; each process, in fact, is carried out by special workmen with the help of special appliances.

The polishing machine imparts a high polish to cut stones, and leaves them completely finished and ready for sale. It consists of a cylinder of hard wood on a disc of lead or tin, and the polishing material is tripolite applied in a moist condition. This work is so simple that it can be left in charge of children. The man and the two women represented to the left in Fig. 90 are engaged in polishing.

For some purposes stones require to be pierced, and the operation is performed by rapidly rotating steel points charged with emery or diamond powder, or provided with a fine diamond splinter.

Engraved Agates were regarded with great favour by the ancient Romans, and although they are not esteemed as highly at the present day, yet the number of stones treated in this way is not inconsiderable.

The varieties best suited for engraving upon, but not devoted exclusively to this purpose, are the onyxes, including onyx proper, chalcedony-onyx, carnelian-onyx, and sard-onyx. These can be so cut as to display a white or light figure against a darker-coloured background. The flatter an onyx is, and the more regular are its layers (Plate XX., Figs. 5a, b), the better suited it is for this purpose; hence Brazilian agates with their plane layers (Plate XIX., Fig. b) are specially suited for onyx stones, and on that account

FIG. 91.   Agate-grinding workshop of Herr August Wintermantel, at Waldkirch (Baden).

more valuable. Stones which are to be engraved have the form of plates, and are cut out of even-layered agates, so that the plane of the plate is parallel to the banding of the stone and is of one uniform colour. The uppermost layer is then partly cut away, so that the portion allowed to remain has the outlines of some figure, usually a portrait. It is generally arranged that the figure cut out of the uppermost layer shall be white or light coloured, this being well thrown up by the background afforded by the coloured lower layer. A *cameo* cut in carnelian-onyx is represented in Plate XX., Fig. 7, and others in Figs. 93 and 94. Sometimes the uppermost black layer is cut through and the figure worked in the underlying white layer, as in the *intaglio* of Fig. 92; in this case also one gets a white figure on a dark background. A seal-stone of carnelian with a letter engraved upon it is represented in Plate XX., Fig. 6. A kind of stone much sought after for cameos is one in which a white layer is partly covered by a red layer. It is then possible to produce a figure showing two colours, the red layer being utilised to represent hair, clothing, or such like. The work is performed, as we have already seen, by a tool known as the style.   The actual work of agate-engraving is performed principally· in Paris and Italy, but the plates of onyx used by the engravers are prepared at Oberstein.   In Italy onyx is largely replaced by plates cut from the thick shells of certain marine molluscs, these shells, like carnelian-onyx, being built up of red and white layers.   They are very much softer than agate, and can be cut with much less labour.

It is not only stones with plane surfaces, however, which are cut as cameos, for there are a few onyx vessels in existence decorated in this way, the body of the vessel being formed by a layer of one colour and the raised figures cut in a layer of another colour. The famous onyx vase preserved in the collection at Brunswick is a monument to the skill of ancient artists in this direction.

The **artificial colouring** of stones is an important branch of the agate industry at Oberstein, especially since the supplies of rough material have been obtained principally from Brazil.   Most of the agate found in nature is of a dingy grey colour, and, until the discovery of the methods by which it may be coloured, was totally unfit for cutting.   The first step was to colour the stones black, this art being stated to have been imparted to an agate merchant at Idar by a customer at Rome, who paid an annual visit to Idar for the purpose of buying onyx.   The art, which had been long known in Rome, was afterwards extensively practised at Oberstein, and in course of time was developed and improved. Not only was a black colour imparted to stones but also brown, yellow, blue, green, and red.

The artificial colouring of agate is possible because of the porous nature, not only of this stone but also of all other varieties of chalcedony, by virtue of which they absorb any coloured liquid in which they are immersed.   All specimens of agate are not equally porous, neither are all layers of the same piece ; some are very porous, readily absorbing a large quantity of coloured liquid and thus acquiring an intense colour, while others are quite or almost non-porous, absorbing colouring matter very slowly and with great difficulty, so that they are only faintly coloured.   Specimens answering the former description are described by lapidaries as " soft " and the latter as " hard."   In this sense the agate found at Oberstein is " harder " than that occurring in Brazil, which is specially suitable for artificial colouring.   In Brazilian agates, however, the outer portion of the amygdale, the so-called skin, is difficult to colour, and the milk-white layers are practically non-porous and cannot be coloured at all.   If the stone is allowed to remain long enough in the colouring liquid, this will penetrate to the innermost parts of the stone, so that the whole mass, and not merely a superficial layer, is coloured.   It has been observed that the absorption of liquid takes place much more rapidly radially, that is to say, perpendicular to the layers, than along the layers, this being due to the radial extension and arrangement of the pores already described.

The principle of the colouring process is thus quite simple, although in practice there are several points which require attention if a good result is to be obtained, and the work is entrusted only to experienced workmen.   The most essential qualification for the work is a familiarity with the appearance and character of different types of stones, but even with this qualification it is not always possible to foretell the result of immersing a given stone in a given liquid.   It may happen, for example, that after the immersion in the colouring liquid of a number of stones apparently equally suitable for the purpose, some will be green and others blue.   The complete process is still in many cases jealously preserved as a trade secret, as was the case at first with the process of staining the stones black.

The latter process has been known in Oberstein since 1819.   The stone to be coloured black is first washed clean and dried without the application of heat, and then immersed in an aqueous solution of honey or sugar in a perfectly clean and new pot. The whole is kept at a temperature below boiling-point for two or three weeks, during which time the loss of water by evaporation must be made good in order to keep the stones covered with liquid. The stones are then taken out, washed, and placed in another pot with commercial sulphuric acid (oil of vitriol) and again warmed. The honey or sugar absorbed by the stone during its immersion in the first liquid is decomposed by the sulphuric acid with

FIG. 92. Antique intaglio.

FIG. 94. Antique cameo.
(*See* page 521.)

FIG. 93. Antique cameo.

separation of carbon, which imparts its black colour to the stone. The " softer " stones, even after a few hours immersion in the acid, acquire a deep black colour, others require an immersion lasting for a day or several days, and some pieces only assume the black colour after being immersed for a considerable time. When the stone is as black as it is possible for it to be, it is taken out of the liquid and quickly dried in an oven. It is then cut and polished, and finally rubbed with oil or allowed to soak in this liquid for a day, the oil not absorbed being wiped off with bran. This final treatment hides small cracks and improves the lustre of the stone.

The fine black agates now bought and sold in the markets have been artificially coloured in this way, as have also the onyxes (Plate XX., Figs. 5a, b), in which the black layers, now alternating with the white, were originally greyish or bluish in colour. Not only do highly porous layers become a deep velvety black, but less porous bands acquire a more or less dark shade of brown. The shade of colour imparted artificially to a stone depends, to some extent, on its original colour ; a red layer, for example, after being artificially coloured will have a reddish tinge, and so on. It is a significant fact that, according to the way in which the stone will acquire the black colour, a hundredweight of agate may be worth £5 or £250. When offering such stones for sale, it is therefore customary to submit small fragments as samples for the purpose of testing whether they will acquire a good colour.

The tinting of agates with colours other than black is comparatively unimportant, but a few details of the methods adopted are given below.

A fine lemon-yellow, such as agate in its natural state never possesses, may be imparted by placing well dried agates in a pot of hydrochloric acid and gently warming in an oven. The colouring will be completed in a fortnight.

Blue, ranging in shade from the finest and deepest indigo and azure to a delicate sky-blue, can also be imparted to agate; natural stones of this colour are, however, never found in nature. The agate is first impregnated with yellow prussiate of potash (potassium ferrocyanide), and afterwards warmed in a solution of iron vitriol (ferrous sulphate). By the interaction and oxidation of these salts, a Berlin-blue compound is deposited in the pores of the stone, to which it gives its colour. This method, however, is only one out of several. Artificially coloured blue agates sometimes closely resemble lapis-lazuli, and for this reason are known as "false lapis-lazuli." They may be distinguished from genuine material by a difference in the shade of colour, and by the fact that the latter is softer than is agate.

A green colour is imparted by impregnating the stone with chromic acid, and afterwards exposing it to a high temperature. An apple-green shade like that of chrysoprase can be imparted with a solution of a nickel salt.

Agate is coloured red in the same way as is carnelian, namely, by impregnating it with a solution of iron vitriol and then exposing it to heat. Agate is also frequently coloured brown, as was mentioned above under sard.

Some agates undergo an advantageous colour change by being simply heated or " burned "; the bluish or greyish tints, for example, may thereby be changed to milk-white, and the yellow and brown to a fine red.

MALACHITE.

Malachite is a mineral sometimes used as an ornamental stone, although it has nothing to recommend it for this purpose save a fine green colour. Chemically it is a hydrated carbonate of copper, having the formula $H_2O.2CuO.CO_2$. The percentage composition of the purest specimens is:

Cupric oxide (CuO)	71·95
Carbon dioxide (CO_2)	19·90
Water (H_2O)	8·15

The water entering into the composition of malachite is easily driven off by heat, the mineral, at the same time, losing its green colour and becoming black. When a fragment of malachite is placed in hydrochloric acid, or a little acid dropped on the surface of the stone, there is a brisk effervescence owing to the evolution of carbon dioxide. This reaction serves to distinguish malachite in its rough condition from other green minerals of similar appearance. When completely dissolved in hydrochloric acid a green solution of copper chloride is obtained, and drops of this impart a magnificent blue colour to a colourless flame.

Malachite sometimes occurs distinctly crystallised, usually taking the form of small needles belonging to the monoclinic system. The mineral occurs much more commonly, however, in the form of compact masses of greater or less size, the crystalline nature of which is indicated by the radially fibrous structure. Compact masses, which are apparently not crystalline, and often even quite earthy in character, are also met with.

Malachite of the finest quality suitable for ornamental purposes occurs as nodular masses often of considerable size, the surface of which may be rounded, reniform, botryoidal, or stalactitic. These nodules have a radially fibrous structure, and are built up of concentric shells. The external rounded surface is often blackish and dull, but fractured surfaces are always green in colour, and show a slight silkiness of lustre owing to the fibrous structure of the mineral. The green colour of such a fractured surface is not quite uniform, and narrow bands, alternately lighter and darker, are often to be seen following the external outlines of the stone, in the same manner as in agate. The presence of these bands together with the fibrous structure gives the stone a kind of grained appearance, which is often very effective for decorative purposes. A nodule of malachite in its natural condition is represented in Plate XX., Fig. 4a, while Fig. 4b, of the same plate, shows a polished slab of the same material.

The physical characters of malachite must now receive our attention. Owing to the large amount of copper present the specific gravity is high; values ranging from 3·5 to 4·0 have been given for different specimens, but the mean value is usually between 3·7 and 3·8. The hardness is only about $3\frac{1}{2}$, so that malachite is scratched even by fluor-spar. Being opaque the liability to become scratched is not as serious a disadvantage as it would be in the case of a transparent stone, but nevertheless it is advisable to protect malachite ornaments as far as possible from injury in this direction. Because of the softness of this mineral it is impossible to obtain a very strong lustre on a cut surface, although the polish is decidedly good. Malachite is not brittle, and can therefore be worked on the lathe.

Malachite is a widely distributed mineral, but generally occurs in small masses only, intergrown with other copper minerals. These impure masses are not suitable for cutting, but they form at many localities not unimportant ores of copper. The wide distribution of malachite is explained by the fact that other cupriferous minerals and ores—copper-pyrites, cuprite, &c.—are readily altered to malachite; in fact, the mineral always originates in this way. Although malachite of inferior quality is abundant at many localities, large masses of pure material suitable for ornamental purposes are by no means common.

The largest amount of malachite suitable for cutting has been found in the **Ural Mountains**, where also the largest and purest masses of material hitherto found have been met with. In a sense, therefore, malachite is a Russian mineral, for in no other country is it found in such abundance. The Urals are very rich in copper, but only a few of the deposits are of importance from our present point of view, the majority yielding only material fit for smelting. Malachite suitable for cutting has been obtained from the copper mines at Nizhni-Tagilsk, at Bogoslovsk in the northern part of the mountains, and at Gumeshevsk further to the south. At the beginning of the nineteenth century there was a considerable yield of material suitable for cutting; this has gradually fallen off, until at the present day stones of this description are obtained only from the Medno-Rudiansk mine, near Nizhni-Tagilsk, the other mines being exhausted or yielding impure material only. The malachite usually occurs in nest-like masses in veins in limestone, from which it must be extracted by the ordinary operations of mining. The material, which is not sold in the market in the rough condition, is cut in the lapidary works at Ekaterinburg.

The mines which formerly yielded the finest material in the greatest abundance were those of Gumeshevsk, situated fifty-six versts south-east of Ekaterinburg and very near the 58th meridian of longitude east of Paris, in the district of the Chussovaya. Reniform, stalactitic, and tube-shaped masses of malachite, of various sizes and of a quality never seen elsewhere, were found in these mines embedded in red clay. The larger of these masses reached a weight of 10 poods (360 pounds), but the majority were smaller than this. The largest single mass of malachite found in these mines is preserved in the collection of the Institute of Mines in St. Petersburg. It is a smooth reniform mass having a height of 3 feet 6 inches, almost the same breadth, and a weight of about 90 poods (3240 pounds); it is valued at 525,000 roubles.

At Nizhni-Tagilsk, in the northern Urals, a little south of latitude 58° N. and in longitude about $57\frac{1}{2}$° E. of Paris, copper ores of a similar nature occur in the same manner as at Gumeshevsk, but fibrous malachite in reniform masses suitable for cutting is much rarer and less fine than at the latter place. In the year 1835 there was found here, however, a mass of malachite of such a size as had never been seen before nor has been since, and which far exceeded the large mass from Gumeshevsk, mentioned above. It was $17\frac{1}{2}$ feet long, 8 feet broad, and $3\frac{1}{2}$ feet high; it was quite solid in the interior, and was estimated to weigh not less than 25 to 30 tons. The whole mass was of a fine emerald-green colour and was thus quite suitable for cutting. The Medno-Rudiansk mine, which alone yields any considerable amount of fine malachite, is situated in the Nizhni-Tagilsk district; the other mines, as stated above, are now exhausted.

This is true also of the mines of Bogoslovsk, situated further to the north, in latitude about $59\frac{2}{3}$° N. and longitude $57\frac{2}{3}$° E. of Paris, on the upper Turya river. From the fact that they stand on this river, fifteen to eighteen versts further to the east, these workings are also known as the Turyinsk mines. Here also malachite occurs in reniform masses, which, however, are poorer in quality than those found at Gumeshevsk, and never of specially large size.

The locality for malachite suitable for decorative purposes next in importance to the Urals is **Australia.** The mineral occurs in this continent most frequently in small masses, valuable only for the copper they contain. The larger masses of malachite of superior quality, which are occasionally met with, are in no way inferior to Uralian material, and resemble this in every particular. The mineral is specially abundant in Queensland, and magnificent specimens have been obtained from the Peak Downs copper mine in that State. Fine malachite occurs also in New South Wales, and at Wallaroo and Burra-Burra in South Australia.

Malachite is worn more frequently set in brooches and ear-rings than as a ring or pin stone. It is usually cut with a plane or slightly convex surface and no facets, these in an opaque stone being quite ineffective. Table-stones and step-cuts, however, are sometimes seen, while the stones to be used as ear-rings are given a club-shaped form.

This mineral is far more extensively used as a material for letter-weights, inkstands, candlesticks, ornamental bowls and vases, and even for objects of considerable size, such as mantel-pieces and table-tops. These larger objects, however, are usually made of copper or some other material and only veneered with thin plates of malachite. The art of veneering lies principally in so piecing these plates together that the joins shall be as inconspicuous as possible, this end being attained by a skilful utilisation of the grained structure of the mineral. The industry flourishes most in Russia, where it is common to meet with large and beautiful objects made of malachite; and those displayed in the palaces of European princes have been in many cases presented by the Czar of Russia. The Isaac Church of St. Petersburg is famous for the beauty of its massive columns of malachite. A number of columns of the same material were found in the temple of Diana at Ephesus, and now adorn the Sophia Church at Constantinople.

The uses to which malachite is applied are thus very similar to those of lapis-lazuli, but malachite in large masses is less rare, and has not more than a tenth of the value of lapis-lazuli. The supply of malachite in small pieces is always more than equal to the demand ; the price of such material is therefore low, but large compact masses of fine colour are rare and command much higher prices.

The general appearance of malachite is so characteristic and peculiar to itself that the mineral can scarcely be mistaken for any other stone. Green chrysocolla, the so-called " siliceous malachite," sometimes resembles malachite, but can be distinguished from it by the fact that it does not effervesce when a drop of hydrochloric acid is placed upon it, while malachite, being a carbonate, does do so. This test may be applied to a cut stone with no serious damage, provided the drop of acid is placed on an inconspicuous part and quickly wiped off.

CHESSYLITE.

The chemical composition of this mineral is very similar to that of malachite, but the constituents are present in different proportions, the chemical formula being $H_2O.3CuO.2CO_2$. It has a fine dark blue colour like that of lapis-lazuli, a character to which its other name, azurite, has reference. Its specific gravity is 3·8, and its hardness $3\frac{3}{4}$, so that it is both denser and softer than lapis-lazuli, from which it may also be distinguished from the fact that it effervesces with acid like malachite. In mass, chessylite is scarcely ever perfectly transparent, being at best only translucent. Its lustre is vitreous and the mineral takes a good polish. It is used to only a very limited extent.

SATIN-SPAR.

The term satin-spar embraces certain finely fibrous varieties of three distinct mineral species, which are usually white in colour and possess in common a satiny or silky lustre.

Calcite.—This mineral sometimes occurs as veins with a finely fibrous structure, the fibres being arranged perpendicularly to the two parallel walls of the vein. When the plate-like masses from such a vein are broken across, the fractured surface shows a fine silky lustre which can be increased by polishing. This variety of satin-spar is, therefore, sometimes cut for beads, or ear-rings, or similar ornaments. A cut and polished convex surface shows a band of chatoyant light, as in cat's-eye, which is due to the same cause, namely, the finely fibrous structure of the mineral. Polished objects very soon become scratched and disfigured, for the mineral has a hardness represented by 3 only. Some of the best material is obtained from near Alston in Cumberland, where it occurs in straight regular veins measuring 2 or 3 inches across in a black shale of Carboniferous age. It has a snow-white colour with sometimes a delicate rosy tinge, and the satiny lustre is very conspicuous on polished surfaces cut parallel to the fibres.

Marble is a granular, crystalline aggregate of calcite, but as this variety has no application as a precious stone it need not be considered further. Onyx-marble from Tecati in Mexico and onyx-alabaster from Egypt are varieties of stalactitic marble, which in the majority of cases also consist of calcite.

Aragonite.—This mineral, like calcite, consists of calcium carbonate ($CaCO_3$), but differs from the latter in crystalline form and in physical characters. It also occurs sometimes in finely fibrous aggregates, when it cannot be distinguished on mere inspection from finely fibrous calcite, for which in its varied uses it may be substituted. The calcareous deposits (Sprudelstein) of the hot springs of Carlsbad in Bohemia also consist of aragonite; this material is often banded in red, white, and brown, and is fashioned into all kinds of small articles, which are sold as souvenirs to visitors at the baths.

Gypsum.—This mineral also occurs like calcite, but more frequently, in fibrous veins, the fibres of which are perpendicular to the walls of the veins. When the mineral has a very finely fibrous structure it exhibits a fine satiny lustre, and may be cut for ornaments of various kinds just as calcite is. The hardness is still less, however ($H = 2$), and the stone can be scratched even with the finger-nail. Material of this character is abundant in the extensive deposits of gypsum, largely quarried for the manufacture of plaster of Paris, in the Triassic rocks of Derbyshire and Nottinghamshire. Massive gypsum with a granular structure is known as *alabaster*, but, like marble, is never used as a gem.

FLUOR-SPAR.

Fluor-spar or fluorite is much too soft for use as a precious stone, but notwithstanding this is sometimes worn as a ring-stone or other ornament for the sake of its fine colour. It is often substituted for more valuable precious stones which it happens to resemble in colour, but its chief use is as a material for ornamental objects of various kinds. When perfectly clear and colourless fluor-spar has an important application in the construction of apochromatic lenses, and the mineral has several other technical applications.

Fluor-spar is an abundant and widely distributed mineral. It occurs in a massive and compact condition, filling veins and crevices in rocks, and is associated with barytes and metallic ores, especially of lead. Magnificent groups of regularly developed crystals resting on the matrix are sometimes found in cavities in these veins, the crystals being finely coloured and transparent enough for use as gems. The finest specimens are found in England, in the counties of Derbyshire, Cumberland, Cornwall, and Devonshire, the two first named being most famed for these crystals. Good specimens occur also in the metalliferous veins of the Harz, in the Erzgebirge, and in the Black Forest. The mineral is so widely distributed that it is only possible here to enumerate the most important localities.

In its purest condition fluor-spar contains 48·72 per cent. of calcium and 51·28 per cent. of fluorine, and is thus fluoride of calcium with the formula CaF_2. The mineral frequently occurs in fine crystals belonging to the cubic system. The cube is the form most commonly assumed, but the octahedron and other simple forms occur alone or in combination with each other. Twin crystals are not uncommon, two cubes interpenetrating in such a way that they are symmetrical with respect to each other about a face of the octahedron. Massive crystalline aggregates of granular or columnar structure, and compact masses of use only for technical purposes, are also common.

Fluor-spar has a perfect cleavage in four directions parallel to the faces of the octahedron, which truncates the corners of the cube. It is brittle, and is the mineral which stands fourth on Mohs' scale of hardness; it is scratched, therefore, by ordinary window-glass. The specific gravity varies between 3·1 and 3·2, so that the mineral sinks in test liquid No. 3 but floats in No. 2. When heated before the blowpipe, fluor-spar generally decrepitates violently and falls into small splinters, which are scattered about with some violence. This is explained by the fact that it usually contains a number of small cavities which are either vacuous or filled with liquid, and this on being exposed to heat expands and reduces the stone to splinters. Fluor-spar alone is not very fusible, but heated with other minerals it melts readily, and for this reason is used as a flux in the reduction of metallic ores, hence its name. Some specimens are remarkable in that when raised to a temperature below red-heat they phosphoresce, that is to say, they emit a beautiful bluish or greenish light. Fluor-spar is completely decomposed by sulphuric acid with the formation of hydrofluoric acid, a gas much used for etching glass and certain precious stones.

The behaviour of fluor-spar with respect to light has an important connection with the uses of the mineral as an ornamental stone. It has a characteristic moist-looking vitreous lustre, and ranges from perfect transparency through all degrees of translucency, to complete opacity. The most important character of the mineral is its colour, which is extremely variable; no other mineral, in fact, displays such a wide range of beautiful colours, and no colour represented in the mineral kingdom is absent. Perfectly pure fluor-spar, which is rare, is limpid and colourless. By the mechanical intermixture of very small amounts of foreign substances, for the most part of organic matter, colour is produced, and this is either destroyed or altered when the fluor-spar is heated. The colour of fluor-spar may be very pale and delicate in shade, or, on the other hand, it may be so deep and intense as to be only recognisable in thin sections of the stone, thick pieces appearing almost black. The powder or, in other words, the streak of the mineral is always white, or at least very pale. Owing to the irregular distribution of the colouring matter fluor-spar often has a patchy appearance. Specimens showing several colours alternating with each other in regular bands are common, this being especially the case in massive, crystalline aggregates. Crystals may have a nucleus which differs in colour from the external portions, for example, the central portion may be yellow and the outer violet.

The fluor-spar used for ornamental purposes must be transparent and of a fine uniform colour. Material of this description is known in the trade by the name of the precious stone it most resembles in colour with the prefix false, " false topaz," " false ruby," " false emerald," " false sapphire," " false amethyst," &c., being some of the terms used for stones of different colour. Coloured fluor-spar may be passed off for stones other than those just mentioned : thus yellow fluor-spar may resemble yellow quartz (citrine) as well as yellow topaz, and red fluor-spar either red tourmaline (rubellite) or red corundum (ruby). It is to be noted that the inaccurate term " false topaz " is applied to yellow quartz as well as to yellow fluor-spar.

Yellow fluor-spar is very common; it is found at Freiberg, Gersdorf, and other places in the Saxon Erzgebirge, in various shades of yellow, including wine-yellow, honey-yellow, and brownish-yellow.

Red octahedra, usually more or less corroded on the surface, occur in crevices in gneiss in the Swiss Alps, but are not abundant. They are found, for example, on the St. Gotthard, where Göschenen is often named as a locality; on the Zinkenstock, near the Grimsel; in the Tavetsch valley in Graubünden; in Wallis; Tessin, &c. The crystals are usually of a light rose-red colour, darker shades being rare.

The colour of green fluor-spar or "false emerald" is sometimes very fine and approaches to that of the true emerald. Crystals from certain English localities, from the porphyry of Petersburg near Halle, and from the metalliferous veins of Badenweiler, are remarkable in this respect. A recently discovered occurrence is that at Macomb, in St. Lawrence County, New York, where thousands of beautiful green crystals, with a total weight of fifteen tons, were found in a single large cavity.

Blue fluor-spar, or "false sapphire," of a very dark colour, is found in the tin mines of the Erzgebirge, and also in salt mines at Hall in the Tyrol. The cubes of fluor-spar from Alston Moor in Cumberland have the peculiar property of appearing by transmitted light of a fine green, and by reflected light of a dark blue colour. The phenomenon from its occurrence in fluor-spar is known as fluorescence. Stones of this description are sometimes mounted à jour in pins, rings, &c., in order to display their fluorescence. Other blue crystals, especially when of a dark shade of colour, have a perceptible violet tinge, while others are lighter in shade and of as pronounced a violet colour as amethyst itself. Magnificent crystals of violet fluor-spar, or "false amethyst," are found in the lead mines of Weardale, in Durham.

These various colour-varieties of fluor-spar are cut in the forms adopted for the precious stones they respectively resemble. They acquire a good polish, but require great care both in cutting and in wear, since the lack of hardness and the perfect cleavage renders them liable to be scratched or cracked and splintered. They are never worth very much, and can always be distinguished from the valuable precious stones they may resemble in colour by their lack of hardness, their specific gravity, single refraction, and by the fact that, in correspondence with their crystalline form, they are not dichroic.

Though little worn as a gem, fluor-spar is somewhat extensively used as a material for ornamental objects such as bowls, vases, candlesticks, letter-weights, or even for columns, mantel-pieces, &c. The so-called spar-ornaments of this description are manufactured chiefly in England of the variety of fluor-spar known as "blue john," which is found in large amount and of fine quality at Tray Cliff, near Castleton, in Derbyshire. This is massive, coarsely-grained material of a dark blue colour tinged with violet, frequently intersected with white and yellow bands. The stones are ground into the desired form and can be worked on the lathe; the process, however, needs great care, for, owing to the brittleness and easy cleavage of the mineral, it is very liable to splinter. Since the commencement of the industry in 1765 many matters of technique have been learnt by experience. The fluor-spar is now, for example, impregnated with resin in order to make it tougher and less liable to splinter. Material so treated can be fashioned into vessels the walls of which are not more than 1 or $1\frac{1}{2}$ lines in thickness, this excessive tenuity being necessary because of the depth of colour of the mineral. When "blue john" is exposed to a temperature just below a red-heat the dark violet-blue colour changes to a beautiful amethystine violet, which is not shown by stones in their natural condition. The operation needs to be performed with great care, for not only is the stone liable to crack, but if the temperature is too high it will be completely decolourised. Owing to the comparative abundance of the mineral and the ease with which rough material is obtained, articles fashioned of fluor-spar do not cost much more than the value of the labour expended upon their production; this is not inconsiderable, owing to the difficulty of the work.

It has been suggested that the murrhine vases of the ancient Romans were made of fluor-spar, but there is no conclusive evidence to support this, and it is more probable, as suggested under agate, that the material was some mineral other than fluor-spar. Pale yellow and rose-red beads cut out of fluor-spar have been found, together with beads of sodalite, in the ancient ruins of Tiahuanaco, near Lake Titicaca, on the high Bolivian plateau.

APATITE.

Apatite is another mineral which occurs of various colours closely resembling those of certain precious stones, and for this reason transparent stones are sometimes cut. Apatite is better suited than fluor-spar for this purpose, since it has a hardness of 5 and it possesses no distinct cleavage. Chemically it is a phosphate of calcium containing chlorine and fluorine ; it crystallises in the hexagonal system, the crystals, which are often very beautiful, usually taking the form of a six-sided prism with a basal plane perpendicular to the prism-faces, or with other terminal faces.

In a pure condition apatite is perfectly colourless and limpid, but by the intermixture of foreign substances it becomes variously coloured. Lilac, violet, or pale green crystals of apatite are found with the tin ores of the Erzgebirge, for example, at Ehrenfriedersdorf in Saxony, and in the old copper mine of Kiräbinsk, near Miask, in the Urals. Pale yellow apatite, the so-called asparagus-stone, occurs in talc-schist in the Tyrolese Alps. The deep green variety, known as moroxite, occurs embedded in crystalline silicate-rocks and in marble at many places ; for example, in North America, especially in Canada ; on the Sludianka, a river flowing into Lake Baikal, in Siberia ; at Arendal, in Norway. Sky-blue crystals are found at certain localities in Australia and Ceylon. Certain green, rose-red, and violet apatites, remarkable for their transparency and the beauty of their colouring, occur with tourmaline in the crevices of the granite of Mount Apatite, near Auburn, in Androscoggin County, in the State of Maine, U.S.A., and were formerly mistaken for tourmaline.

The variously coloured apatites just described, when sufficiently transparent, which is not usually the case, may be cut as gems. The variety most frequently cut, perhaps, is the green moroxite from Canada, but even this is used to only a very limited extent and is very low in price.

The mineral may be distinguished from other similarly coloured stones by its hardness and its specific gravity, the latter feature approaching very closely to the specific gravity of fluor-spar, which is $3\cdot2$. Apatite is doubly refracting and slightly dichroic, characters which serve to distinguish it from fluor-spar. From beryl and emerald, to which some cut apatites are very similar in appearance, the mineral may be distinguished, as pointed out above, by its hardness and its specific gravity, apatite sinking in test liquid No. 3 (sp. gr. $= 3\cdot0$), in which beryl and emerald float.

IRON-PYRITES.

Iron-pyrites, or simply pyrites, is often known to jewellers as marcasite, though by this term mineralogists refer to another species. It is disulphide of iron, its composition being represented by the formula FeS_2, and is the only mineral sulphide with a metallic lustre which is used for ornamental purposes. It crystallises in the cubic system, has a specific gravity of 5·0, and a hardness ($H = 6\frac{1}{2}$) a little less than that of quartz. The mineral is brittle, and when heated before the blowpipe it burns with a blue flame and gives off sulphur dioxide, easily recognised by its penetrating odour. When struck with a steel, iron-pyrites gives out brilliant sparks and a smell of sulphur dioxide, due to the ignition of the particles struck off. It is unattacked by hydrochloric acid, but completely decomposed by nitric acid.

This mineral is of a pale brass-yellow colour and possesses a brilliant metallic lustre, well shown in cut stones, which are usually given the form of a flat rosette. On account of the comparative hardness of the stone the lustre is retained for a long time, and the edges of a cut stone do not very soon lose their sharpness.

Up to the eighteenth century iron-pyrites was much esteemed, especially in France, as an ornamental stone on account of its brilliant lustre and pretty colour ; it was used for the decoration of shoe-buckles, garters, snuff-boxes, &c., besides being set in brooches, bracelets, and other articles of personal ornament. Afterwards the mineral grew more and more out of favour until its use was practically given up. Attempts have been made to reinstate it in its former position ; thus in the year 1846 a large amount of iron-pyrites, from Geneva and the Jura Mountains, set in an old-world and once admired fashion, was sent to Paris. The articles were readily bought up at first, but being dear and the setting not in accordance with the taste of the period there was no sustained demand for them. There is sometimes used in cheap jewellery at the present time a material, occurring in the neighbourhood of Dublin, which consists of a thin incrustation of small and brilliant crystals of iron-pyrites on the surface of black shales of Carboniferous age. Large polished plates of iron-pyrites, which were probably used as mirrors, have been found in the ancient graves of the Incas. These discoveries first drew attention to the mineral, which was known for a time as Inca-stone. In reference to its supposed health-giving properties it was also known as health-stone, and was much worn in the shape of amulets, as well as set in necklaces, pins, ear-rings, &c., and was possessed of considerable value. Iron-pyrites exists in nature in very great abundance, and is one of the most widely distributed constituents of the earth's crust.

HÆMATITE.

Hæmatite is an opaque mineral with a metallic lustre and dark steel-grey to iron-black colour. Chemically it consists of ferrous oxide, Fe_2O_3, and in its purest form contains 70·0 per cent. of iron and 30·0 per cent. of oxygen. Specially fine rhombohedral crystals occur not infrequently in the extensive and important deposits of iron-ore in the island of Elba, and in veins and crevices in the gneiss of the Alps, and at other places. Only those of some thickness are black and opaque, and exhibit a brilliant metallic lustre; minute crystals of extreme thinness appear by transmitted light transparent and of a fine blood-red colour, hence the Greek name hæmatite (blood-stone). The streak of this mineral on unglazed porcelain is dull and of a dark cherry-red colour, a feature which distinguishes it from all other black stones possessing a metallic lustre.

Hæmatite occurs more abundantly than in crystals as compact irregular masses of a black colour, with a shining metallic lustre, and with the same specific gravity (sp. gr. = 4·7) and the same hardness, namely, that of felspar (H = 6), as the crystals possess. The latter are scarcely ever cut, but the massive material is worked in the lapidary works at Oberstein and elsewhere for the manufacture of ornamental objects. The material used for this purpose, which is obtained in masses of considerable size, is said to come from India, and similar material occurs also in Brazil, in the States of São Paulo and Minas Geraes.

Finely fibrous hæmatite, the application of which is perhaps still more extensive, does not, as a rule, possess the brilliant metallic lustre of the crystals and of the compact crystalline masses just mentioned, and its colour is reddish and more like that of the streak. Specimens are sometimes met with, however, which, while retaining their fibrous structure possess a brilliant metallic lustre and a steel-grey to black colour. Material of this description is as suitable for cutting as the crystalline masses described above, but fibrous hæmatite when distinctly red in colour is useful only as an ore of iron. The latter variety usually occurs in rounded, botryoidal, or reniform masses and is then known as kidney-iron-ore; this occurs lining cavities in massive hæmatite of a pronounced red colour. Hæmatite of this character suitable for cutting was at one time obtained in the veins of iron-ore at Kamsdorf, near Saalfeld, in Thuringia, but these deposits are now practically exhausted; also in the famous old iron-mines in the island of Elba, where magnificent druses of beautiful crystals are often found. Material suitable for cutting is found also in the deposits of iron-ore in Scotland, and in the iron-mines of west Cumberland, but the most important occurrence is in the iron-mines of the north of Spain, near Bilbao, in the Basque Provinces, and at Santiago de Compostela in Coruña. Rough material of suitable quality for cutting is found also at many other localities beside those enumerated above.

The deeper the black colour and the more perfect the metallic lustre of hæmatite the more effective is it for the various ornamental purposes to which it is applied. For ring-stones it is cut with a flat surface, on which is engraved a figure or a letter, so that it may serve as a seal-stone, and this is one of its most general uses. Stones cut *en cabochon* for rings and other articles of personal ornament sometimes display a band of chatoyant light, which, though cloudy and dull, is similar to the characteristic feature of star-sapphire, and

is due to the fibrous structure of the mineral.   Stones of this description are cut with a slightly convex surface, and are set in brooches, bracelets, medallions, &c., as well as in rings.   The material is sometimes used for beads, which are strung together and worn as necklaces or bracelets.   Not infrequently such beads are chatoyant, and when this is the case they resemble black pearls in colour and lustre.   Small cubes of hæmatite are mounted as pins.   Generally speaking, however, the mineral has a very limited application as an ornamental material, for example, in mourning jewellery, and is very low in price.

Hæmatite has been used as an ornamental stone from the very earliest times. Numerous cylinder-seals, some of them engraved, have been found in the ruins of Babylon, and ornamental objects of the same material are found in the ancient Egyptian tombs. In classical times hæmatite was extensively used for intaglios and for other similar purposes, being a stone of fine appearance and easy to work.

The fine powder of hæmatite is much used as a polishing material in the working of precious stones, especially some of the softer kinds, although another material is available and is also used for this purpose, namely, the artificially prepared oxide of iron known as jewellers' rouge.   The massive fibrous variety of hæmatite, which is cut as an ornamental stone, constitutes the material of which polishing tools for burnishing gold and silver jewellery are made.   The city of Santiago de Compostela is stated to supply almost the whole of the world with these tools.

ILMENITE.

Ilmenite is very similar to hæmatite in appearance and is sometimes used for ornamental purposes.   The chief difference between the two minerals lies in the fact that ilmenite contains titanium dioxide as well as iron oxide.   A variety of ilmenite known as iserine occurs as black, rounded grains in association with sapphire in the sands of the Iserwiese in Bohemia.   Except for the absence of a fibrous structure, ilmenite possesses essentially the same characters as hæmatite, but is susceptible of a higher degree of polish ; it is distinguished from hæmatite by the fact that its streak is black instead of red, and that it is occasionally magnetic, which is never the case with hæmatite.   The application of the mineral as an ornamental stone is so limited that further consideration is for the present purpose unnecessary, but it may be mentioned in conclusion that an effective ornamental stone occurs at Cumberland, in Rhode Island, U.S.A., which is composed of grains of white quartz embedded in black ilmenite.

RUTILE.

Rutile is a mineral which consists of titanium dioxide (TiO_2), and is frequently found as crystals belonging to the tetragonal system.   It ranges in colour from red through brown to black, and possesses a strong metallic to adamantine lustre.   Some specimens are sufficiently beautiful to be cut as gems, and may then be mistaken at a first glance for black diamond, from which the stone is distinguished, however, by its lower degree of hardness (H = $7\frac{1}{2}$), its greater density (sp. gr. = $4\cdot2 - 4\cdot3$), and, in transparent specimens, by the strong double refraction and marked dichroism. Rutile is of common occurrence, but does not, as a rule, possess the characters essential for a gem ; it is, therefore, only rarely cut.

AMBER.

This material, so much used for personal ornaments, is not strictly speaking a mineral at all, being of vegetable origin, and consisting of the more or less considerably altered resin of extinct trees.   It resembles minerals in its occurrence in the beds of the earth's crust, and for that reason may be considered, like other varieties of fossil resin, of which it is the most important, as an appendix to minerals.

NATURE AND CHARACTERS.—It is proposed to deal first with true amber or amber proper, the succinite of mineralogists, the principal locality for which is the district of Samland, in the province of East Prussia, and to conclude with a consideration of other resins, which are similar to amber and used for the same purposes, but which are of much rarer occurrence.

In respect to its **chemical composition,** amber, like other resins, consists essentially of carbon, hydrogen, and oxygen.   These elements are combined together in somewhat variable proportions, but on an average the material contains 79 per cent. of carbon, 10·5 per cent. of oxygen, and 10·5 per cent. of hydrogen, a composition which is represented by the chemical formula $C_{10}H_{16}O$. A small amount of sulphur is sometimes present, and a little inorganic material, which remains behind as ash when the amber is burnt.   Pure amber contains only about $\frac{1}{5}$ of a per cent. of ash, but when enclosures of foreign substances are present this percentage rises considerably.

Amber is not homogeneous nor of simple constitution.   Apart from the inorganic ash it contains several substances which can be separated by chemical means.   By distillation is obtained a small quantity of oil of amber (an ethereal oil) and some succinic acid ; the latter is an especially characteristic constituent of true Prussian amber, in which it is always present, though in variable amount.   Perfectly clear and transparent specimens contain 3 to 4 per cent. of succinic acid, cloudy specimens contain more, and in frothy amber there may be as much as 8 per cent.   By treating the fine powder with alcohol, ether, and other solvents, four resinous substances, differing from each other in chemical composition and in melting-point, may be separated.   The insoluble residue, amounting to from 44 to 60 per cent. of the whole, is a bituminous substance, the so-called amber-bitumen.

Amber is perfectly insoluble in water, and is only slightly attacked after a contact of some duration with alcohol, sulphuric ether, acetic ether, and other solvents.   This affords an important distinction between true amber and the many similar resins, which are so often substituted for it, the latter being quickly attacked by alcohol and the other solvents mentioned above.   In concentrated sulphuric acid finely powdered amber is perfectly soluble even in the cold, and it is completely decomposed by boiling nitric acid.

When amber is heated it softens, swells, and gives off a characteristic and pleasant odour.   Between 280° and 290° C., that is to say at a higher temperature than is the case with other resins, it melts, and is at the same time decomposed with the evolution of the two volatile constituents, oil of amber and succinic acid.   These substances are given off in the form of white fumes and are sometimes accompanied by a small amount of water vapour.   The fumes have the same peculiar aromatic odour, and owing to the presence of succinic acid in them act as an irritant when inhaled into the respiratory passages, causing violent coughing. The non-volatile residue, amounting to about 70 per cent. of the weight of the original material, is a shining black substance known as colophony of amber. It is soluble in oil of turpentine and in linseed-oil, and its solution in either of these substances is known as amber-lac or amber-varnish, and is much used for varnishing. This lac is characterised by its great hardness on drying and only its dark colour stands in the way of a more extensive application. When heated in linseed-oil amber becomes soft and pliable, a property of which practical use is made, as we shall see later.

When heated in the air amber ignites and burns with a bright sooty flame, hence the German name *Bernstein* for amber. During combustion the same aromatic odour becomes apparent, this being so characteristic of amber as to be sufficient to distinguish it from other resins of similar appearance. It also leads to the use of a limited amount of amber as frankincense.

The **physical characters** of amber are in every respect those of a resin. It is perfectly amorphous and shows no indication of crystalline structure. It occurs in irregular rounded nodules having the form of rods, drops, plates, &c., and never in masses bounded by plane surfaces. There is no cleavage and the fracture is conchoidal, but masses of amber are often penetrated by irregular cracks. Sometimes a shelly, concentric separation is observable, the mass being built up of thin layers one within another, and only loosely held together. Specimens of this kind are known as *shelly amber*, while those which are as compact as if they had been cast are known as *massive amber*. In extreme cases these two kinds differ from each other very widely, but there are specimens to represent every possible gradation between the two. The structure is closely connected with the mode of formation of the amber, and has an important bearing on the practical application of the stone.

The specific gravity of amber ranges from 1·05 to 1·10; the substance is thus only slightly heavier than water, and approximates still more closely to the density of sea-water. Its hardness, $H = 2\frac{1}{2}$, is a little greater than that of gypsum, and it can therefore be scarcely scratched with the finger-nail. It is harder than most other resins, a difference which affords another useful distinguishing feature. It is not very brittle and can therefore be carved, worked on the lathe, or bored with little difficulty—important practical considerations. When amber is cut with a knife parings are not obtained but only powder.

When rubbed on cloth amber becomes strongly charged with negative electricity, and attracts to itself scraps of paper and other light objects; it is, indeed, from the ancient name for amber, electron, that the word electricity is derived. When rubbed very vigorously the characteristic odour becomes perceptible, but the amber does not become sticky as do other resins under similar circumstances, since the temperature of its melting-point is far higher than that induced by the vigorous rubbing. Amber is a bad conductor of heat, and in consequence feels warm when in contact with the hand. This character alone serves to distinguish it from minerals and from glass of similar appearance, since these feel cold to the touch.

In most cases amber possesses a fine resinous lustre, which is considerably heightened by polishing, but some specimens are dull and are not improved by polishing, these being therefore unsuitable for ornamental purposes.

Amber ranges from perfect transparency to complete opacity. In the same specimen there may be clear and cloudy portions side by side, these merging gradually into each other and never sharply defined. This is a characteristic feature of genuine amber, and one which distinguishes it from certain similar substances, to be mentioned below. In transparent specimens it may be observed that amber is singly refracting, this being in accordance with its amorphous character. A feeble double refraction is sometimes observed, especially in the neighbourhood of enclosures of foreign bodies, these no doubt setting up internal strains.

The colouring of amber is very uniform in character, no colour but yellow having been met with in the large quantity of Baltic amber hitherto collected. It varies in shade, however, from the palest yellow to dark yellow and brown. Material which has undergone a surface alteration is often red in colour, but fresh specimens never show this colour. Green and blue amber is very rare and will be considered below.

In spite of the uniformity of its colour several **varieties** of amber are recognised, the distinctions between them being based principally on the association of different shades of colour with different degrees of transparency. These different varieties differ in the capacity for acquiring a polish, and some by reason of their general appearance are more suitable for ornamental purposes than others, and are therefore of greater commercial importance.

Transparent amber is described in the trade as *clear*. Shelly amber is nearly always clear ; it is never cloudy throughout and rarely so in alternate layers. Massive amber, on the other hand, is nearly always more or less cloudy ; perfectly transparent specimens of massive amber are rare, though more frequent than cloudy specimens of shelly amber. The clear massive variety occurs in masses ranging in colour from almost perfect colourlessness to dark reddish-yellow. Water-clear amber is very rare, and is described as " yellow clear " ; the reddish-yellow, or " red clear," is more frequent.

Several varieties of *cloudy* amber are recognised in the trade, namely, "flohmig," " bastard," " semi-bastard," " osseous," and " frothy amber." Material which possesses characters intermediate between those of these varieties is distinguished by compound terms of a descriptive kind, such as, " clear-flohmig," " flohmig-clear," " flohmig-bastard," osseous-bastard," &c.

Flohmig amber is slightly turbid, with the appearance of having been clouded by a fine dust ; like the clear variety it is susceptible of a fine polish. The term " flohmig " is derived from the East Prussian word " Flohmfett," which signifies the semi-transparent, yellowish fat of the goose or duck, which this variety of amber is supposed to resemble in appearance.

Bastard amber is more turbid, but is still susceptible of a good polish. Various terms are used to signify the extent of the turbidity : thus, material which is cloudy throughout is termed *bastard proper*, while that in which cloudy portions are dotted about in a clear ground-mass is known as *clouded bastard*. Colour distinctions are also recognised : pure white to greyish-yellow shades of bastard-amber are described as *pearl-coloured*, the material with paler tones being known in the trade as " blue amber," not to be confused with the rare amber which is actually blue in colour. Yellow and brownish-yellow bastard amber is described as *kumst-coloured*, from the East Prussian name " Kumst " for cabbage (Sauerkraut), the former being described as pale and the latter as dark " kumst "-coloured. The specimen of amber represented in Plate XX., Fig. 9, is " kumst "-coloured ; part of its surface is polished and the rest in its natural condition.

Semi-bastard amber is intermediate in character between bastard and osseous amber, combining the appearance of the latter with the capacity for receiving a polish of the former.

Osseous amber, or more briefly *bone*, is opaque, softer than the varieties described above and inferior to them in its susceptibility of polish. In colour it ranges from white to brown, and, as the name implies, it has the general appearance of bone or ivory.

By the combination of the characters of the different varieties enumerated above, there arises a number of different colour-varieties of amber, which are classified into two groups under the descriptions " variegated osseous clear " and " variegated osseous bastard."

Frothy amber is opaque, very soft, and incapable of receiving a polish; it often encloses crystals of iron-pyrites in large numbers.

Leaving out of consideration those colour-varieties of amber which are rare or unusual, the pearl-coloured, and next to this the " kumst "-coloured, is the rarest of the varieties of every-day occurrence. These two kinds of amber are in more general favour in Europe than is the clear amber, and for this reason command the higher prices, but we shall see later on that the fashion and taste prevailing in different countries is not uniform.

The different degrees of turbidity in amber, with which are connected the differences between the numerous varieties of this substance, were at one time thought to be due to the presence of varying amounts of water. It is now known, however, that the turbidity is due to the enclosure of vast numbers of bubbles of various sizes, but always too small to be seen with the naked eye, or with a simple lens, their identification requiring the examination of thin sections under a high power of the microscope. These cavities are distributed throughout the ground-mass of the amber, which always consists of a pure clear resin, almost water-clear to reddish-yellow in colour. The differences in the appearance of the various kinds of amber depend upon the number and the size of these cavities, the smallest of which have a diameter of 0·0008 and the largest 0·02 millimetre. The cavities are the smallest and the most numerous in osseous amber, having a diameter of from 0·0008 to 0·004 millimetre and a distribution of 900,000 per square millimeter. " Flohmig " amber contains the smallest number of bubbles, 600 per square millimetre, but these, with a diameter of 0·02 millimetre, have the maximum size. Between these two extremes lie all the other varieties, the study of which has shown that the more numerous and the smaller are the bubbles the amber contains the more turbid it becomes. When the cavities are less numerous and larger in size the amber in which they are embedded is clearer, and when it contains no bubbles is perfectly transparent.

The small bubbles of air hinder the passage of light through the amber and thus produce the appearance of turbidity. A large portion of the light which enters a mass of amber is reflected from the surface of the air bubbles and thus fails to reach the eye. Could these air-spaces be filled with a transparent substance of the same, or nearly the same, refractive index as amber, which varies between 1·530 and 1·547, the turbidity would disappear and the substance would become perfectly transparent.

This, in fact, can be accomplished without any great difficulty by an operation which is known as **clarifying** the amber, one which is frequently performed in the trade for the purpose of making cloudy amber transparent and thereby increasing its value. The rough material is completely immersed in rape-seed oil in an iron vessel, and then very slowly heated to about the temperature at which the oil boils and begins to decompose. It is then allowed to cool and this must take place just as slowly and gradually as the preliminary heating, otherwise the clarified amber will become cracked and possibly fractured. The smaller the fragments of amber operated upon the quicker is the process completed; the heating of large pieces must be continued for a considerable period, and not infrequently needs to be several times repeated. The time required for the operation depends also upon the character of the material, for different pieces of amber of the same size will not require the same length of time to complete the operation. The clarifying process begins on the surface and spreads gradually inwards.

The rationale of the method is the penetration of the fine cracks in the amber and the filling up of the cavities with rape-seed oil, the index of refraction of which is 1·475, only slightly different from that of amber itself. Light is thus enabled to pass through the mass of amber without hindrance and the substance appears in consequence clear and transparent. If the oil is mixed with some colouring matter, this also penetrates the amber and imparts its colour to it.

Unless the greatest care be taken the operation of clarifying amber results in the development of peculiar cracks, which in some aspects resemble fishes' scales. These are at first so small as to be scarcely noticeable, but gradually become more and more conspicuous and begin to show iridescent colours, until towards the end of the operation they become quite obvious as shining, golden cracks. These are known to the amber workers as " sun-spangles," and their presence often serves to distinguish a clarified from a naturally transparent specimen of amber.

Having dealt with the yellow varieties of amber in some detail, we may now briefly consider the rarer kinds of a blue or green colour. *Green amber* ranges in colour from a pale to a blackish-green and from olive-green to the apple-green of chrysoprase, sometimes exhibiting white clouds. The colour of *blue amber* may be azure-, sky-, or steel-blue. Both blue and green amber is turbid, and, in fact, the turbidity has a close connection with the colouring of the stone, for this is due not to the presence of pigment, but to a peculiar modification of the rays of light caused by the presence of the numerous air-bubbles, which, as we have seen above, is the cause also of the turbidity of amber. These minute air-bubbles, to which both the colour and the turbidity of blue amber is due, are arranged in it in layers of some thinness, just as in bastard or osseous amber; and the colour phenomenon for which they are responsible is of the same nature as that observed in other turbid media. When subjected to the clarifying process blue amber loses both its turbidity and its colour, assuming the yellow colour of ordinary amber.

The phenomenon of *fluorescence* is very conspicuous in some specimens of amber, these appearing of a yellow to brown colour in transmitted light, but of a dark bluish or greenish colour in reflected light. Fluorescent specimens of Prussian amber are rarely seen, but are less uncommon among the amber-like resins of Sicily, Burma, and other localities. Not being suitable for ornamental purposes they are scarcely ever worked, and are worth less than specimens of ordinary amber.

One serious drawback to the use of amber as an ornamental material is its tendency to **change in colour** with the lapse of time. This change in colour is due to a chemical alteration which takes place gradually from without inwards. Pale-coloured specimens become darker, and those which were yellow become red or brownish-red. The change is noticeable after the lapse of only a few years, but differs in character in different varieties of amber. Clear amber becomes slightly darker and redder in colour and numerous cracks develop in its substance. In bastard amber an external layer becomes brown in colour and assumes a waxy lustre. Osseous amber acquires a porcelain-like lustre, and in frothy amber an external layer, sharply marked off from the remainder, becomes quite clear but brittle.

These alteration processes proceed gradually, especially along cracks in the material, until the whole mass has undergone the change. They were at one time attributed to the action of light, but have since been observed to take place in the dark. When a piece of amber is kept in water or otherwise excluded from contact with air, the change which takes place in it is much less in extent, so that the process is simply a case of atmospheric weathering.

Under natural conditions the weathering process has sometimes proceeded so far that only a small nucleus of the mass remains fresh and unaltered, the outer shell being much cracked and fissured, and its surface honeycombed with a shallow sculpturing, such as is represented in Plate XX., Fig. 9. The external weathered layer is easily detachable from the nucleus of fresh, unaltered material, which is often pitted with close-set, shallow, conical depressions. These effects of weathering are only to be seen in specimens which have lain in dry earth, those which have been embedded in perfectly dry sand being often changed and altered throughout their whole mass, and the surface deeply honeycombed. Material, on the other hand, which has lain in water or in moist earth, thus being preserved from contact with air, is often scarcely altered at all and does not show even the surface sculpturing described above.

Beside air-bubbles, the importance of which has been explained already, other **enclosures** of various kinds are found in amber, some of which are of special significance. Drops of water occur not infrequently, but enclosures of solid matter, either organic or inorganic, are more frequent.

The inorganic substance most commonly enclosed in amber is iron-pyrites, which, in the form of quite thin lamellæ, fills up cracks and crevices in many specimens, especially of shelly amber. It is also frequently enclosed in frothy amber as already stated. The presence of iron-pyrites in a specimen of amber naturally interferes with the working of it, and such material is of little value to the turner.

The enclosures of organic material, partly of vegetable and partly of animal origin are of great importance. The vegetable enclosures consist mostly of finely divided particles of carbonised wood, which are black in colour and are present in many specimens in greater or less abundance. The amber retains its usual yellow colour, and does not appear black in consequence of these black enclosures. Black amber does not actually exist, what is known as such being in reality quite another substance, namely, jet, which will be considered later on. The black resins which sometimes occur in company with amber are quite distinct from this and, further, are not suitable for ornamental purposes. The particles of carbonised wood now found enclosed in amber are the remains of the pine-trees from which the amber-resin was exuded, the so-called amber-pine (*Pinites succinifer* of Göppert). Larger fragments of wood, needles, and other parts of the tree, are also found as enclosures, but more rarely. Recognisable remains of other plants also occur though still less commonly, these often consist of flowers and leaves, the structure and form of which are perfectly preserved by the amber in which they are embedded.

This is also the case with animal remains, which are found in amber in great number and variety. These include insects of various kinds, such as ants, moths, and especially flies ; also spiders, and as great rarities snails and other small animals. Inclusions are confined almost entirely to the clear shelly amber and are scarcely ever found in turbid massive amber. The animal and plant remains found embedded in amber belong to extinct species, but these do not differ widely from present-day organisms. They lived and flourished at the time known to geologists as the Tertiary period, and were imprisoned in the resin exuded from the pine-trees living at the same time. So perfectly has their form and structure been retained, that these fossil plants and animals are almost as suitable for biological investigation as are living forms, often throwing light on the character of the fauna and flora of the period. A piece of amber containing some organism is often cut for ornamental purposes in such a way as to bring the enclosure into prominence.

WINNING OF AMBER.—The first amber to be collected would naturally be that which had been washed up on to the beach from amber-bearing strata beneath the sea. A certain amount is still obtained in this way, and is known as *sea-amber* or as *sea-stone*. It is distinguished from ordinary amber by the absence of a crust of weathered material. This was, no doubt, present while the substance remained in its original situation, but by the action of the sea was subsequently entirely removed, the only trace of its existence being the presence of a few surface pittings on portions protected from the action of the waves. Being exposed to all the disintegrating forces of littoral waters, masses of amber containing incipient cracks and fissures would be very unlikely to escape fracture, and for this reason sea-amber is usually free from such flaws.

The collection of sea-amber is an easy matter, for it is simply picked up on the flat sea-beaches. During landward wind-storms large quantities of amber are loosened from the floor of the sea and thrown up upon the beach, often entangled with masses of sea-weed. This is sorted over, and the larger pieces of amber are collected, but the smaller fragments, which do not repay the trouble of collecting, are left, and often accumulate at spots high up on the shore. Besides collecting the masses of amber thrown up on the shore the searchers also wade as far as possible into the sea, and by means of long-handled nets drag ashore the floating masses of sea-weed in order to secure the amber entangled in it. The amber thus drawn from the sea is known as *drawn-amber* (German, Schöpfstein), while that which is thrown up upon the beach is referred to as *strand-amber*.

Amber is collected in these ways all along the coast of Samland, but specially on the west coast, north of Pillau; also along the whole of the east coast of the Russian Baltic Provinces, Livonia and Courland; through East and West Prussia; in Pomerania and Mecklenburg; and along the whole coast of the Jutland peninsula in Schleswig-Holstein and Jutland; at some places in greater, at others in smaller amounts. After Samland, the Jutland peninsula with Schleswig-Holstein is the most important locality for amber. It occurs in greater abundance on the west coast, which is washed by the North Sea, than on the east. Stavning peninsula and Fanö Island are stated to be specially rich localities. The substance appears to be less abundant on the North Friesian Islands, Ramö, Sylt, Föhr, &c., but to be very plentiful on the shore of the Eiderstedt peninsula, where at ebb-tide large quantities of amber are exposed on the fore-shore and collected. The richest locality in this neighbourhood is the Hitzbank, a sand-bank which extends from the last-named peninsula far out to sea. The amber-seekers are therefore known here as the Hitz runners. The mouth of the Eider is another prolific spot, but amber is collected along a strip of coast stretching as far south as Büsum, although the yield is small in South Dithmarschen, at the mouth of the Elbe, and on the coasts of Hanover, Oldenburg, and Holland. The search for amber on the fore-shore is attended with great danger owing to the risk of being caught by the returning tide. In North Dithmarschen the seekers, who are known as amber-riders, follow the outflowing tide, often on horses and after collecting as much amber as possible hurry back to the shore as soon as the tide turns. Amber is also often collected from boats in this district. It is said that the yield here has considerably fallen off.

The islands on the coast of Holland and East and North Friesland were referred to by Pliny as " insulæ glessaridæ," that is to say the amber islands, and they have been called also the Electrides. Later on the much richer district of Samland became known to the Romans, who, at the beginning of the time of the Frankish emperors had entered into commercial relations with East Prussia for the supply of the highly esteemed amber, most of the material so supplied being probably strand-amber, with perhaps a little drawn-amber.

Other methods are now adopted in Prussia for the obtaining of sea-amber besides the primitive ones already described, the material being collected from the floor of the sea by the help of appliances of various kinds.

The amber lying at the bottom of shallow waters is scooped and raked up by persons in boats, with the aid of small hoop-nets provided with long handles. The boulders which are strewn over the sea-floor and hinder this operation are sometimes hauled up, and utilised as building stone, a desideratum in this part of the country.   The amount of amber wedged in between such-boulders makes their removal quite worth while.   A large amount of amber has been obtained in this way among other places in the neighbourhood of Brüsterort, on the north-west corner of Samland, where, after the removal of the boulders, the amber was raked up with drag-nets.   Amber is obtained in this way on the coast of Samland only; the method has been tried, but unsuccessfully, on the coasts of West Prussia; the amber obtained here consists exclusively of that picked up on the shore or drawn from the sea.

The methods for the winning of amber hitherto described are all somewhat primitive, but the more rational method of **diving** has been adopted more recently and has yielded a rich harvest.   The pioneers of this method were Stantien and Becker, a large firm of amber merchants at Königsberg, and since 1869 the amber lying loose on or embedded in the sea-floor has been collected by divers furnished with every modern appliance.   The sea-floor in the neighbourhood of Brüsterort and the village of Gross-Dirschkeim to the east was first explored, and after the exhaustion of these supplies, a spot further south near Palmnicken was worked.   At the present day, however, even this method is given up by reason of the vastly richer yield afforded by mining operations, but before passing to the consideration of the latter, the operations of dredging and of surface digging for amber must be described.

Dredging for amber is performed not in the open sea, but solely in the Kurisches Haff, and is undertaken only by the firm just mentioned.   The floor of this lagoon near the village of Schwarzort, a little to the south of Memel on the Kurische Nehrung consists of a bed of alluvium very rich in amber.   Towards the east the same bed rises above sea-level and near Prökuls is extensively worked in diggings.   Dredging for amber was commenced in 1860, and this year marked a turning-point in the amber-winning industry, for whereas, up to that date the amber markets were supplied mainly with sea-amber, the supplies subsequently consisted almost entirely of dredged amber.   The latter also is free from cracks and fissures and from the weathered crust, and, indeed, differs in no essential respect from sea-amber.   The work was commenced with three small hand-dredgers and was not at first very successful; when the right spot was found, the undertaking developed to quite an unexpected extent, and so large steam-dredgers provided with powerful machinery were brought into requisition.   The floor of the lagoon was excavated to a depth of from 7 to 11 metres, and about half the annual production of East Prussian amber was obtained in this way.   The industry gave employment to about 1000 workpeople, and incidentally led to a great development of the small fishing village of Schwarzort.   The deposits in the lagoon are now exhausted and no dredging has been done since the end of November 1890.

Besides being picked up on the shore, drawn from the sea, and raked or dredged up from the sea-floor, amber has been obtained since ancient times in **diggings** both on the shore and inland.   The amber thus obtained, the so-called *pit-amber*, differs from sea-amber in that it is enveloped in a thick weathered crust and is much cracked and fissured, though these flaws are not visible on the exterior owing to the crust.   Not only in East Prussia but also in all parts of the district where it occurs, digging for amber in the glacial and alluvial deposits and in the Tertiary strata is carried on.   Formerly the amount of material obtained in this way was small compared with that collected from the sea, but now these conditions are reversed.   Since the year 1873 specially large supplies of amber have been obtained from a greyish-green sandy clay, the so-called " blue earth," an amber-bearing stratum of Lower Tertiary age, which at the present day is the most important source of the material.

The surface **alluvial deposits** were first worked for amber in the south-east of East Prussia, south of the railway line between Ortelsburg and Johannesburg (on the Allenstein and Lyk railway), in a district bordered on the east by the Pissek and on the west by the Omulew river, and extending into Poland as far as the neighbourhood of Ostrolenka on the Narew. A large amount of amber has been found in former centuries in the surface diggings here and elsewhere in Poland, and in West Prussia, especially at Steegen on the Danzig Nehrung. At the last named place and at Prökuls, where are situated the most important of these alluvial deposits, the firm of Stantien and Becker commenced their digging operations; these workings were at first on quite a small scale, but have, although at other places, now developed to an enormous extent, the whole world being supplied with amber from the diggings of this firm. Prökuls is situated on the mainland and on the railway line between Memel and Tilsit, to the south of Memel and opposite Schwarzort. The deposit is identical with that which lies beneath the Kurisches Haff, which was formerly dredged for amber in the neighbourhood of Schwarzort. The diggings at this spot did not, however, yield nearly the amount of amber obtained by dredging, and have therefore long been abandoned.

In **glacial deposits** amber is present everywhere in the North German lowlands, but usually in small amount. At places where the glacial deposits overlie " blue earth " *in situ* amber occurs in greater abundance, but since the discovery of these richer deposits depends upon chance or accident the occurrence has no economic importance. Such deposits may be met with in the excavations necessitated by alterations and improvements, in the digging of sand and gravel pits, or in peat cutting. Small deposits and pockets of amber occur and are worked at many places in East and West Prussia, Pomerania, Mecklenburg, Schleswig-Holstein, Denmark, and further to the west, also in the province and kingdom of Saxony, in Silesia, and further eastward in Russia. A few examples of the richness of the deposits in East and West Prussia and Pomerania are given below. From a small pocket at Krebswalde, near Elbing, 700 pounds of amber was obtained. As rent for the diggings at Schillehnen, near Braunsberg, 400 ducats was formerly paid. The deposits at Gluckau, near Danzig, have been worked for at least 170 years, and as late as 1858 a fine piece of amber weighing 11 pounds 13 ounces was found there. At Karthaus, a large amount of amber was met with in pockets in loam, and at Berent, Konitz, Czersk, Tuchel, and Polnisch-Crone in West Prussia, and Treten and Rohr, north of Rummelsburg in Pomerania, amber has been obtained for more than 100 years from loamy veins, which penetrate the glacial sand to a depth of 23 metres.

The total amount of amber obtained from alluvial and glacial deposits is as nothing compared with that derived from **Tertiary strata,** namely, the banded sands of the lignite formation, and especially the " blue earth." Practically the whole of the genuine amber which now comes into the market is obtained from these deposits. The workings of the latter, both surface and underground, are confined to the coast east and south of Brüsterort on the north-west corner of Samland, and are entirely absent from the interior of the country, from the remaining portion of the Baltic coast, and from the coast of the North Sea.

Wherever the amber-bearing Tertiary beds are accessible, whether above or below sea-level, they have been worked for many centuries in **open workings.** Important workings are situated at Kraxtepellen, Gross-Kuhren, Klein-Kuhren, Georgswalde, Rauschen, Sassen, Wannenkrug, &c. A great wealth of material was found in the " blue earth " at Loppehnen, four-horsed waggons being required for its removal.

When the working of the "blue earth" in open diggings was commenced is not definitely known, but the diggings were probably in full operation as early as 1836. To reach the amber bed, which is only about one and a half spades in thickness, it is necessary to remove a great weight of superimposed material, consisting of later Tertiary strata (the lignite formation), and glacial deposits, often several metres thick, besides banking up the pits to prevent the inrush of the sea. The cost of working is thus considerable, and the fact that the deposit can be worked at a profit bears witness to its richness. In the "banded sands" of the lignite formation which overlie the "blue earth" are irregularly distributed pockets of amber. Though a considerable amount of material has been obtained in open workings, the sands are not now systematically worked because of the expense attending the process, and the bulk of the material is obtained from underground mines in the "blue earth."

Mining operations in the "banded sands" were undertaken by the Government as far back as the end of the eighteenth century (1781), but after working for twenty-four years these were abandoned. To the firm of Stantien and Becker belongs the credit of having successfully initiated the system of underground mining in the "blue earth," a system which has been followed with increasing profit up to the present day.

At the beginning of the seventies two attempts were simultaneously made in this direction. The Royal Prussian Mining Administration opened workings at Nortycken, near Rauschen, some distance inland from the north coast of Samland, but these had to be abandoned because of the impossibility of preventing the entry of water into the pit from the water-bearing blown sands lying above the "blue earth." The workings opened by Stantien and Becker at Palmnicken, between Pillau and Brüsterort, on the west coast of Samland, were a brilliant success. The success of this attempt led, in 1870, to the construction of a large open working on the shore itself, the output of which for a period of five years showed a steady increase. The amber-bearing stratum lies here 6 to 8 metres below sea-level, and dams of massive wood-work were required to protect the open diggings from the sea. The underground workings are free from this objection, as also from the interruption of work in the cold season of the year, and, moreover, they do not necessitate the lying fallow of a large area of fruitful land. At this place the whole of the amber-bearing "blue earth" is excavated from shafts, levels, and galleries, the amber so obtained being freed from the earth which adheres to it by washing in specially constructed appliances.

The amber thus obtained and cleaned, the so-called *dam-stone*, is enclosed in a thick crust of weathered material, which is removed in the cask-washers. In this operation the amber is placed with water and sharp sand in rotating barrels, which are kept in motion until the last trace of the opaque crust is removed. Another operation, the so-called Klebs' washing, completes the preparation of the rough material, which cannot then be distinguished from sea-amber. It is next examined with regard to colour, transparency, and the presence or absence of cracks, with the object of deciding its value and the class of work for which it is best suited, an impossibility in the case of material still enclosed in its opaque crust. The stones thus prepared are then sorted, the different qualities being placed on the market separately.

The increased yield of the mines has more than compensated for the exhaustion of the deposits in the Kurisches Haff, which at one time was dredged for amber. In the year 1893 the mines yielded 6000 hundredweights of amber, half of which was suitable only for the manufacture of varnish, while the other half consisted of pieces of medium size suitable for ornamental purposes. Beside the 600 men engaged in the mines, 400 more find employment in the sorting-rooms at Königsberg, so that the production of amber by the firm of Stantien and Becker alone provides a livelihood for upwards of 1000 men and their families. In 1884 the total production of amber in East Prussia was 3000 hundredweights, of which 1000 hundredweights was obtained by dredging at Schwarzort, where 1000 workpeople were employed; 1700 hundredweights from the mine at Palmnicken, employing 700 workers; 200 hundredweights were obtained by divers out of the sea at Palmnicken; and the remaining 100 hundredweights were picked up on the shore, drawn from the water, raked up from the sea-bottom, or dug up at various spots. In 1874, ten years before, the total production was only 1100 hundredweights, from which the rapid increase in yield can be seen.

The winning of amber has occupied the inhabitants of the Baltic since very early times. Articles of amber have been found in graves of the Stone age in East Prussia, showing that the material was highly esteemed even at that time. It is not surprising to find that from the earliest times efforts have been made by the ruling powers to obtain control of the amber deposits. Thus, amber was once declared to be the property of the German Crown, the existence of some ancient right being possibly the basis of the declaration. The material has remained Crown property up to the present day in all the districts which were not taken by the Poles, that is to say, in East Prussia. In West Prussia and other districts, though the amber-winning industry is subject to certain imposts, the material itself was not, nor is it now, the absolute property of the Crown.

Up to the year 1811 the rights of the State were strictly enforced, and the privilege of collecting and selling amber from the sea-shore was granted only to persons who paid for it. The impossibility of preventing unlicensed persons from collecting amber, and the demoralisation of the villages on the coast in consequence of the continued evasion of the law, led the Government in 1811 to lease their rights, first to a company and afterwards to a contractor. Under these conditions, which lasted until 1837, unlicensed persons were strictly forbidden to pick up even the smallest fragments of amber, and any infringement of these regulations was visited with the severest punishment.

These restrictions did not entirely achieve their object, for at certain favourable spots along the coast the inhabitants of the villages still succeeded in smuggling a certain amount of amber. In 1837 the irksome restrictions were wholly removed, and the shore from Nimmersatt on the Russian border as far as Polsk, east of Danzig, was leased to the shore communities themselves, who then had the right to pick up amber on the shore, draw it out of the sea, rake it up from the sea-bottom, dig for it in the land bordering on the sea, and sell it to whomsoever they would. Further to the west, as far as the mouth of the Weichsel, the town of Danzig had long before that time rights to work in the same manner. In 1868, however, when it had been found that a profitable working of the littoral deposits of amber required the expenditure of a considerable amount of capital and skill, not in the power of the villagers to give, the mining rights were withdrawn from the inhabitants of the shore villages in favour of persons better fitted to exercise them. The subsequent increase in the total production of amber, consequent on the introduction by Stantien and Becker of dredging in 1860 and mining in 1873, has been already pointed out. Dues are paid by this firm and by others engaged in similar operations to the owners of the rights.

An interesting light is thrown on the development of the amber-producing industry by a comparison of the yearly revenues derived by the State from this source. Up to the year 1811, when the industry was State-managed, the yearly income was £1100. From 1811 to 1837, when the rights were leased to a company and then to a contractor, the yearly income amounted to £1500; the taxation of the village communities brought in £1700;

and in later times the amber revenues have amounted to about £35,000, the exact sum received by the Prussian treasury in the year 1894–5 being £35,500. Of this, £33,850 was contributed by the firm of Stantien and Becker alone for the mining rights of the Palmnicken mine and other places. The remaining £1650 was paid for the right to collect amber along certain stretches of shore, which for some years had been knocked down to the highest bidder, and many of these workers also were under the control of the firm just mentioned.

A more or less systematic winning of amber also exists in the interior of the country, but it is impossible to single out localities for special mention. Pieces of amber picked up casually must be yielded up to the State officials, who are authorised to pay the finder a reward fixed by law. The material accumulated in this manner is sold periodically by the State.

WORKING OF AMBER.—Every piece of amber of a suitable size and quality is worked for ornaments or for articles used by smokers, the latter application being far more extensive than any other. The material suitable for this purpose is known as *work-stone*. Pieces which are impure or too small, together with the fragments detached in the working of larger stones, are melted down and used for the manufacture of lac and varnish, material of this kind being known as " varnish." The same kind of material has been used to a certain extent in recent years for the preparation of pressed amber, of which more will be said below. About half of the total production of amber must be placed in the " varnish " category, its value being about one-tenth that of the annual yield.

The application of the so-called work-stone must now be considered more in detail. The material is mostly turned on the lathe, but it may be fashioned into the desired form also by cutting and grinding, and a certain amount is cut up into thin plates. By immersion in hot linseed-oil, amber is not only clarified, as stated above, but is also softened and made flexible, a fact of great importance in the amber-working industry.

Nearly one-half of the total production of amber is devoted to the manufacture of articles for the use of smokers, namely, cigar- and cigarette-holders, mouthpieces for the same, and for pipes, &c. This manufacture is one of the staple industries of Vienna, the articles made there being famous all the world over for their excellence. A quantity of material worth 40 per cent. of the value of the annual yield of rough material goes to Vienna, and this includes only the best qualities of amber. The German towns in competition with Vienna for this trade are Nürnberg, Königsberg, Ruhla near Eisenach, and Erbach in the Oldenwald. Nürnberg and Erbach are engaged exclusively in the manufacture of articles for the use of smokers, which are all exported, while Ruhla manufactures the same class of goods, but for home use. In France, Paris and St. Claude in the Jura, are engaged in the same manufacture, the articles being for home use, and the value of the rough material employed being about 10 per cent. of that of the total yield. In England there is no important manufacture of amber goods; in Holland and Belgium ornamental articles of amber are manufactured, but not for export. In Russia both smokers' requisites and ornamental articles are manufactured at Polangen and Krottingen to the north of Memel on the Prussian border, the industry affording a livelihood to the whole of the population of the district. At Zhitomir, in Volhynia, are made only articles used by smokers, including the cigar-holders mounted with Tula silver, supplies of which are sent to Warsaw, St. Petersburg, Riga, Ostrolenka, and Odessa. In North America amber is worked for the same class of goods, but only for home use. After a period of depression, due to the pressure of Viennese competition, the industry in Turkey, especially at Constantinople, is now in a flourishing condition, and is still growing.

Thin plates of amber for use in inlaid work, mosaics, &c., were once much in request, but are manufactured now only in small numbers.

Ornamental objects of amber are manufactured in great variety, and in accordance with the taste of different peoples. Perhaps the commonest of such articles are beads, rounded or faceted, and perforated so that they can be strung together and worn as necklaces and bracelets, or used as rosaries by Roman Catholics and Mohammedans. Also pieces of special form for necklaces and bracelets, brooches, and other articles, these being often delicately carved.

A certain number of amber ornaments are manufactured, as we have seen, in Russia (at Polangen and Krottingen), at Constantinople, and in England. The industry flourishes also in China and Korea, where round beads for mandarins' chains are made in great numbers. China now uses annually for this purpose from £7500 to £10,000 worth of rough amber, while formerly the beads were imported ready-made from Germany.

The total production of amber ornaments in each of the countries enumerated is, however, small, and Germany in this respect comes before all, while for smokers' requisites Austria stands first. Danzig, Berlin, Stolp in Pomerania, and Worms manufacture almost exclusively ornamental articles, with which they supply the whole world, with the exception of the countries mentioned above. Some of these articles remain in Europe, but the remainder are exported from Hamburg, London, Marseilles, Bordeaux, Livorno, Trieste, and Genoa, some passing through Moscow and the fairs of Odessa and Nizhniy Novgorod to Turkey, Persia, Armenia, the Caucasus, Siberia, various parts of Africa far into the interior, China, India, Arabia (where numerous rosaries are disposed of to pilgrims going to Mecca), the West Indies, North and South America, &c. The articles must be made to suit the special requirements of each country, and certain classes of goods, which are more in request than any others, are manufactured in large quantities. This is the case, for instance, with beads, and in the wholesale trade six varieties are distinguished, namely:

1. *Olives*, elongated, elliptical beads.
2. *Zotten* (German), cylindrical, slightly rounded, almost plane, at the two ends.
3. *Grecken* (German), like " Zotten " but shorter.
4. *Beads proper*, spherical.
5. *Corals*, beads with facets.
6. *Horse-corals*, flat, clear beads faceted at the two ends.

These different beads are made of various sizes and in different qualities of amber, those of bastard being the most valuable. " Bastard olives " and " bastard beads " constitute at the present time the most valuable article of export of the German amber-industry. According to the size and colour of the beads, the price varies in

" Bastard olives " from £2 10s. to £25 per kilogram.
" Bastard beads " „ £3 12s. „ £15 „ „

Another article manufactured in large numbers is the *manelle*. This is a flat, polished disc of amber, in the middle of which is cemented an amber bead, the latter being of clear amber and the disc of bastard, or *vice versâ*. The bead is not infrequently set upon a piece of tin foil to increase the lustre, and the underside engraved with flowers or other devices. These " manelles " serve as the centre-pieces of necklaces and bracelets, and are much appreciated in Persia, Armenia, and Turkey.

Cylinders of amber were for a long time much sought after by certain races in Central Africa and South America to be used as ornaments for the ears. To meet this demand, now to a large extent fallen off, these " cylinders " were manufactured in large quantities, and of various sizes, the largest being about 5 centimetres in length and 2 in diameter, but with a rather broader base.

Other amber ornaments need not be described in detail since there are none of a particular type which remain long enough in favour to justify their manufacture in large numbers. New designs and devices must be constantly brought out to meet the ever-changing demands of popular taste. In the same way the taste of the people of different countries must also be studied by the manufacturer of amber goods, not only with regard to the design of these goods, but also with respect to the variety of amber in which they are executed. Thus for articles exported to Russia, fine bastard amber only must be used ; while in Holland the most admired variety is clear amber ; in Germany, both clear and bastard amber is favoured ; in France, bastard amber ; in China, clear amber ; in West Africa, the semi-osseous varieties with a brownish tinge ; and so forth.

The value of the whole of the rough amber annually used up lies between £100,000 and £150,000.   Of this, 40 per cent. is taken by Austria, 20 per cent. by Germany, 10 per cent. by Russia, 10 per cent. by France, 10 per cent. is used in the manufacture of varnishes, and the remaining 10 per cent. is used in the home manufactures of North America, China, Turkey, and other countries already mentioned.

THE AMBER TRADE.—We have already seen that the amber trade is almost completely controlled by the firm of Stantien and Becker, since they are responsible for the production and marketing of almost the whole of the yield. The initiation of mining operations by this firm effected a complete revolution in the trade, for whereas in former times almost the whole of the annual yield consisted of sea-amber, and only a very small proportion of dug-amber, these conditions are now reversed. The supposed superiority of sea-amber over dug-amber lay in the fact that the former was free from the enveloping crust of weathered material, so that its colour and quality could be seen at a glance, and, moreover, was, as a rule, sound and free from cracks. In the case of dug-amber, with each piece enclosed in a thick opaque crust, it was impossible to judge of the quality, and the purchase of a parcel of such material was a risky speculation. This objection to purchasing rough amber was overcome by removing the opaque external crust by the method already described. The dug-stone had then no disadvantages compared with sea-amber, and even possessed a certain advantage over the latter, seeing that after the treatment to which it was subjected for the removal of the crust, every piece could not fail to be absolutely sound and free from cracks.

Another new and important departure, with which the same firm is to be credited, was the system of sorting the rough amber into **trade varieties** before placing it on the wholesale market. The rough amber is sorted out with regard to colour and quality, the size of the pieces, the number that go to make up a pound, and the suitability of their form to different purposes ; merchants are thus enabled to buy rough material, the whole of which is exactly suited to their requirements. Previous to this, the rough amber had been classified to a certain extent, but not in accordance with the requirements of a more extended trade, and it is unnecessary to enumerate here the names by which these original trade varieties were known. The classification introduced by Stantien and Becker has been gradually coming into general use since 1868. The brief review of it, which now follows, is taken from R. Klebs, who has made a thorough study of the amber-producing industry, and to whom we are indebted also for much of the information already given.

One of the chief aims of the amber-worker is to produce any required article with the least possible waste of rough material. Thus, for the fashioning of a long, thin cigar-holder a piece of amber, approximating as closely as possible to the form of this object, should be chosen. The form of the pieces of rough amber is thus of the utmost importance, and the preliminary classification of the rough material is based on this character. The material thus sorted is then again classified according to size and quality, and irrespective of clearness or turbidity, these features being of little importance from this point of view. Amber of an unusual colour is not recognised in the trade classification, being too rare to possess any commercial significance. The classification given below applies specially to massive amber, but shelly, osseous, and other varieties of amber are sorted for trade purposes in the same way.

1. *Tiles* (German, *Fliessen*).—Tabular pieces, about three times as long as they are broad, and with a thickness of at least 75 millimetres and a length of 25 centimetres. The most valuable are those in which the two surfaces are approximately parallel. They are known as " work-stone-tiles " (German, *Arbeitssteinfliessen*), and are sorted into five classes :

" Work-stone-tiles," No. 1, with 10 to 12 pieces to the kilogram.

,,	,,	,,	2,	,,	30	,, ,,
,,	,,	,,	3,	,,	60	,, ,,
,,	,,	,,	4,	,,	100	,, ,,
,,	,,	,,	5,	,,	170	,, ,,

Pieces which lack the regular, rectangular form of " work-stone-tiles " are known as " ordinary tiles." Of these, ten trade varieties are distinguished : No. 0 is the first variety, and consists of material of which 2 or 3 pieces are required to make up the kilogram, while in variety No. 7, 360 pieces go to the kilogram. " Tiles " are used for the manufacture of smokers requisites, such as cigar-holders, mouthpieces for the same, and so forth.

The No. 1 variety of " ordinary tiles " fetches £7 2s. per kilogram, while the price of the No. 7 variety (the smallest pieces) is 9s. per kilogram. Variety No. 0 is so rarely met with that it can scarcely be considered as an ordinary trade variety. The specially selected " work-stone-tiles " are worth more than the " ordinary tiles." The price of varieties Nos. 1 and 2 is 33⅓ per cent. higher, that of No. 3, 50 per cent., that of No. 4, 25 per cent., and that of No. 5, 10 per cent. higher.

2. *Plates* (German, *Platten*).—Pieces of amber of the same tabular form, but thinner than " tiles." Seven trade varieties are distinguished :

No. 0 "plates." With an area of from 40 to 60 square centimetres.

,,	1	,,	,,	,,	13 to 26 ,, ,,	(about 50 pieces to the
,,	2	,,	80 pieces to the kilogram.			kilogram).
,,	3	,,	170	,,	,,	
,,	3½	,,	260	,,	,,	
,,	4	,,	350	,,	,,	

"Polangen plates." Still smaller than No. 4.

" Plates " also are used mainly in the manufacture of smokers' requisites, especially for cigarette-holders, but also for ornamental articles ; for example, " manelles," " horse-corals," crosses, and bells on the tesbih (rosaries) of Mohammedans.

3. *Ground-stone* (German, *Bodenstein*).—Large, rounded pieces of amber of good colour.

1. " Fine ground-stone " . . . 10 pieces to the kilogram.
2. " Ordinary ground-stone " . 14 to 16 ,, ,,

" Fine ground-stone " costs £2, and " ordinary ground-stone " 25s. per kilogram.

This variety of amber is often carved and used for ornamental objects of minor importance, also for the mouthpieces of Turkish hookahs, which are often decorated with gold and turquoise.

" Bockelstein " is a " flohmig " variety of " ground-stone," which is worked to form the centre pieces of necklaces, destined for export to Central Africa and South America. Large " Bockelsteine " are 10 per cent. dearer than " ground-stones."

4. *Round amber* (German, *Runder Bernstein*).—Round pieces of amber are divided according to colour into clear and turbid ("bastard"). According to size fourteen trade varieties are distinguished :

" Bastard round " and " clear round ".	.	.	No. 1,	50 pieces to the kilogram.	
,,	,,	,,	. . .	,, 2,	100 ,, ,,
,,	,,	,,	. . .	,, 3,	170 ,, ,,
" Bastard ground-stone " and " clear ground-stone " .		.	320	,, ,,	
" Bastard Knibbel " and " clear Knibbel "	.	,, 1,	600	,, ,,	
,,	,,	,,	.	,, 2,	820 ,, ,,
,,	,,	,,	.	,, 3, 1600	,, ,,

" Clear round " and " bastard round," No. 1, costs 32s. per kilogram ; No. 3, 17s. to 18s. per kilogram ; and " clear Knibbel " and " bastard Knibbel," No. 3, 1s. 7d. per kilogram.

Osseous amber, especially in " Fliessen," has a particular classification. " Osseous Fliessen " is sorted according to size and colour into four lots. It is used for the manufacture of smaller mouthpieces, and its price is 25 per cent. lower than the corresponding material in " bastard amber." " Round osseous " is classified according to the size of the pieces into three sorts.

According to its structure *shelly amber* is sorted into two lots, one being described as " large, fine shelly amber " and the other as " unsorted shelly amber." The best qualities of " large, fine shelly amber " are worth 42s. per kilogram, while the price of " unsorted shelly amber " is only 3s. per kilogram. As a general rule, however, " shelly amber " is broken up, the best pieces, according to purity and suitability of size and form, being used for mouthpieces or beads, and the remainder included with the better and smaller pieces of " varnish amber." The latter is sorted according to quality and purity into ten trade varieties. " Shelly amber " is worked to only a small extent ; specimens remarkable for their enclosures are usually placed in collections for scientific study, or are used as ornaments, this being especially the case when the specimen encloses an insect, for at one time bracelets, each bead of which contained such an enclosure, were much esteemed.

There remains now to be mentioned only the class of material known in Germany as *Brack*. This includes pieces of amber of large size, but so cracked and blebby or so impure that they are useless for ordinary purposes. This kind of material is sometimes bought as a speculation, or it may furnish the amber-worker with a cheap material for the bases of large objects. It is classed into " large Brack," including the purer pieces, and " ordinary Brack."

The trade varieties of amber and the prices quoted above were current in 1883. A comparison of these prices with those of earlier times shows that rough amber has fallen in value, this being especially true of the larger and more valuable sorts, the single exception being afforded by the small, clear amber. This fall in price is conditioned partly by the increase in production, an increase with which the manufacture and export has not quite kept pace ; partly by the manufacture of imitation amber, especially by the development of the celluloid industry in North America, and, finally, partly by the use of artificially pressed amber in place of large pieces of natural occurrence. It may be mentioned here that the largest piece of amber yet found weighed 9·7 kilograms (21⅓ pounds). It is of a very beautiful bastard colour and is valued at £1500. It was found in 1860 at Cammin, in Pomerania, and is now preserved in the Museum für Naturkunde at Berlin.

IMITATION AND COUNTERFEIT AMBER.—The legitimate trade in amber suffers to a considerable extent by the substitution of cheap imitations for genuine amber. These imitations are sometimes more, sometimes less, clever, but can always be detected by the application of a few simple tests.

The difference between genuine amber and an imitation substance is usually apparent to a practised eye at a glance. This is specially the case with bastard amber, the colour of the turbid portions being so soft and pure, and passing so gradually and imperceptibly into the clear amber interlaminated with it, that the general appearance of the stone can never be reproduced in an imitation.

If, however, mere inspection is not sufficient to determine the nature of a doubtful specimen, it must be subjected to some or all of the tests described below under the different imitation substances.

The clumsiest imitation of amber is that made of yellow **glass.** This, which is usually made in imitation of clear amber, is now scarcely ever used for smokers' requisites, but is manufactured into beads, large numbers of which are sold, for the most part in China, as amber beads. This glass imitation of amber can be distinguished by the fact that it feels cold to the touch, by its greater hardness and greater specific gravity, and by the glassy conchoidal fracture which is distinctly visible at places where splinters have broken off the edge.

Celluloid, which latterly has found such an extensive and varied application, can be manufactured to resemble very closely many kinds of amber. Imitation amber of celluloid has been called " ambre antique." The use of this substance for articles, such as the mouthpieces of cigar-holders, can usually be detected by the fact that they have been moulded, not turned or ground, to the desired form. Moreover, in the imitation substance the alternating stripes of clear and cloudy material are sharply defined, this sharpness never being seen in cloudy varieties of genuine amber. When the imitation substance is rubbed a smell of camphor becomes perceptible, but very little electricity is developed. Again, with a knife, parings may be cut off celluloid, but genuine amber gives a powder when cut. The parings adhere to a hot platinum wire and take fire, flaring up with a bright flame and with the evolution of an acid smell. Amber, owing to its higher melting-point, does not adhere to a hot platinum wire, and it burns slowly, giving off a characteristic aromatic odour. Celluloid is quickly attacked by sulphuric ether, in which it dissolves, while amber may lie in this liquid for a quarter of an hour without sustaining any serious damage.

The extremely inflammable nature of celluloid cannot be too often insisted upon, and its use for smokers' requisites must be strongly deprecated. Although the inflammability of celluloid is often denied by the manufacturers, yet the fact remains, and all attempts to render it uninflammable have been as yet unsuccessful.

Amber differs from the other **resins,** which are frequently substituted for it, in possessing a higher melting-point, greater hardness, slighter solubility in alcohol, ether, and similar liquids, as well as in the presence of succinic acid as a constituent, and the evolution of a characteristic odour when rubbed or burnt.

Of these resins the most important is **copal,** which is often dug out of the earth, and is exported in large quantity from East and West Africa, South America, and Australia. In colour and general appearance it resembles some kinds of amber, and, like this substance, often encloses insects, but articles made of copal always have a dirty appearance. Owing to its lower melting-point copal, when rubbed with the hand or on cloth, becomes sticky; it is soft enough to take an impression even of the finger-nail, and when immersed in acetic ether it loses its lustre and swells up. All these characters serve to distinguish it from amber, and, if more were needed, it becomes less strongly electrified on rubbing, and, when burnt, does not give off the smell characteristic of amber. In order to produce the latter feature copal is sometimes melted up with pieces of amber, but owing to the difference in the melting-points of the two substances they do not mix, and after the mass has solidified the pieces of amber are easily recognisable lying in the copal. This substance is unsuitable for certain of the uses to which amber is put; for example, it is too brittle to allow of the worm of a screw being cut on it.

Dammar-resin mixed with powdered amber, and clear copal-colophony mixed with Venetian turpentine, furnish materials for the manufacture of imitation amber articles, such as cigar-holders. These imitations of amber show a turbid, " kumst "-coloured ground-mass, in which lie sharply defined masses of clear material, and in which insects—often ants—are embedded. The artificiality of this substance is very apparent, for clear and bastard material is rarely seen in combination in genuine amber, and, moreover, when the mass is broken it becomes obvious that the insects are made of metal. If made of copal the substance quickly swells up when placed in acetic ether, and if it consists of dammar-resin it loses its lustre when immersed in sulphuric ether for a period of from ten to fifteen minutes. Genuine amber is only affected after a much longer subjection to this treatment. Like copal dammar-resin also becomes sticky when rubbed.

Amber containing enclosures is sometimes manufactured by boring a hole into a suitable piece of genuine amber, placing some animal, such as a lizard or a tree-frog, in the hole, and filling it up with melted dammar-resin. It is not uncommon to find objects so treated mounted in gold and highly prized by the unsuspecting owner. When treated with alcohol or ether the melted resin is, of course, easily dissolved out.

The use of the imitations of amber described above is now almost superseded by that of **pressed amber** or **ambroid**, a substance produced from genuine amber, and first brought into the market at Vienna. Attempts had been made from time to time to devise some means of utilising small pieces of amber other than by converting them into varnish. This was at last accomplished, and the smaller pieces of amber are now welded together, as it were, by the application of a great pressure and an elevated temperature.

The method employed depends upon the fact that at a temperature between 170° and 200° C. amber softens. Pieces of rough amber after being freed from all impurities are placed in a flat mould of steel, the steel cover of which is then hermetically sealed. The mould is then placed in an oven, of which the temperature can be regulated to a nicety, or in a bath of glycerine or paraffin. By means of the hydraulic press, a pressure of from 8000 to 10,000 atmospheres is applied to the cover of the mould, and this pressure welds the pieces of amber, which are softened by heat, together to form a flat cake. These flat cakes of pressed amber are known in the trade as Spiller imitations. A much finer product is obtained when the amber, softened by heat, is driven under high pressure through a metal sieve, the resulting mass being by this method more thoroughly intermixed.

The pressed amber so produced can be obtained in all the varieties in which amber occurs, the " flohmig " and clear material being remarkably similar to natural amber. There is a difference apparent under the microscope however, for the rounded cavities of natural amber are flattened, elongated, and pressed out in a dendritic fashion in pressed amber. "Flohmig" pressed amber is more like "flohmig-clear," in which turbid and clear portions occur in parallel stripes. At the junction of the turbid and clear portions there is to be observed in transmitted light a yellowish-red colour, which in reflected light and against a dark background changes to blue. This appearance is very rare in natural amber, and never seen at all in bastard or in clear natural amber. Moreover, in pressed amber the clear and turbid portions are always sharply defined instead of merging imperceptibly the one into the other, as in natural amber. In clear portions of pressed amber, too, there are almost always to be seen small brownish patches and veins, and even if these are absent the material is never glassy clear, but always exhibits clouds and streaks, such as are seen when different liquids mix or when sugar dissolves in water. Many of the characters of pressed amber are, of course, identical with those of natural amber, such, for example, as hardness, high melting-point, marked tendency to develop frictional electricity, the giving out of an aromatic odour when rubbed or burnt, and the difficult solubility in the liquids mentioned above.

Between 600 and 700 hundredweights of pressed amber is now produced annually, and the material fetches from 25s. to 30s. per pound. It is used chiefly in the manufacture of cheap articles for the use of of smokers, and recently in the manufacture of beads for exportation to Africa. It can be worked in precisely the same way as natural amber.

In the foregoing account the chief consideration has been given to the true Baltic amber, known to mineralogists as succinite. Being an almost exclusively German product it has been dealt with in considerable detail ; it may also be stated here that owing to the amber-working industry being essentially a German trade, many of the trade terms quoted above have no equivalent in the English language. On the south-east coast of England a few pieces of amber are occasionally picked up. Besides the Baltic amber there are numerous other resins of a similar nature and applied to similar purposes, which may be included in the term amber used in its widest sense. They differ from amber proper in many of their characters, especially in the absence of succinic acid, and for this reason they have latterly been distinguished by mineralogists from succinite by special names. From a commercial point of view they are quite unimportant, but as they are of local interest in the places at which they occur the most important will be briefly described below.

GEDANITE.

Several other resins accompany succinite, but only one, gedanite, is suitable for ornamental purposes. It is known to amber-workers as " brittle," " friable " or "unripe " amber, and is usually transparent, or, at any rate, strongly translucent, and of a clear wine-yellow colour, rarely dirty yellow and opaque. Most of the pieces have the appearance of having been rounded and rubbed, and are dusted over with a snow-white powder, which can be wiped off. There is no succinic acid in gedanite, and hence the fumes given off when the substance is burnt are not irritating, although the smell is very like that of burning amber. The melting-point is about $140°$ C., being thus lower than that of amber, but higher than that of copal. Its hardness, $H = 1\frac{1}{2} - 2$, is less than that of amber. Its solubility is much the same, but gedanite is more easily attacked by oil of turpentine than is amber. When rubbed it acquires a strong charge of negative electricity, when it attracts to itself scraps of paper and other light bodies.

Owing to its great brittleness gedanite does not compare favourably with amber as a workable material. It is true that it can be turned on the lathe to any desired form, but, owing to its brittleness, it cannot well be bored, nor a screw cut on it. For the same reason articles made of gedanite must be handled with great care. In consequence of these disadvantages the different kinds of gedanite are not worth more than a third of the value of corresponding qualities of amber. No distinction is made in the trade between articles made of succinite and those made of gedanite, and the services of an expert are required when it is necessary to determine whether a given article is made of gedanite or of true amber.

The occurrence of gedanite is confined to the amber-diggings of the Prussian coast. The material is found in small amount with dug-amber, but never with sea-amber, probably because it is too brittle to resist the battering of the waves and the pebbles of the beach.

ROUMANIAN AMBER. (ROUMANITE.)

Roumanite, or Roumanian amber, is rarely yellow, but is usually brownish-yellow to brown. It is transparent to translucent, and scarcely ever opaque. Fluorescent specimens are sometimes met with, finer even than Sicilian amber, which is regarded as fluorescent amber *par excellence*. A characteristic feature of Roumanian amber is the presence of numerous cracks, which, however, do not seriously affect the cohesion of the material. Notwithstanding the fact that many of these stones are quite full of cracks, they can be turned, cut, polished, and otherwise worked without being broken. The substance is brittle, and has a conchoidal fracture. When rubbed it acquires a charge of negative electricity. The hardness slightly exceeds that of true amber. Succinic acid is present in variable amount (up to 3·2 per cent.), but on an average in less amount than in Prussian amber. The presence of a comparatively large amount of sulphur amounting to 1·15 per cent. in roumanite is a characteristic feature. The substance offers a still greater resistance to solvents than amber. When heated it gives off a peculiar aromatic odour, and at the same time a smell of hydrogen sulphide (like the smell of rotten eggs), which is due to the presence of sulphur. It fuses at 300° C. without swelling up, and gives off fumes, which act as an irritant on the respiratory passages.

This resin occurs enveloped in a closely-adhering, weathered crust, which is always very thin, and of a dark yellowish-grey to reddish-brown colour. It is found as nodules in carbonaceous, laminated shales, or as interrupted layers in sandstone beds in the district of Buseo. At Buscou, on the railway line from Bucharest to Braila, it is found within a radius of about a mile in the earth on common-land; while at Valeny di Muntye rounded fragments are found among the pebbles of a brook. The strata in which roumanite occurs, or in which it originated, belong to the later Tertiary period (the Congeria beds). Most of it is sent to Vienna, where, under the name of "Roumanian amber," it is manufactured into cigarette-holders and other useful or ornamental articles. To a certain extent it enters into competition with Prussian amber, but, owing to its comparative rarity and higher price, this competition is not serious.

The substance known in Roumania as "black amber" is not amber at all, but jet, a variety of coal which will be considered presently.

SILICIAN AMBER. (SIMETITE.)

Sicilian amber differs somewhat conspicuously from Baltic amber in appearance. It is usually transparent, and, as a rule, darker in colour than the latter. Reddish-yellow shading off to clear wine-red is not uncommonly seen, clear and dark brown, yellowish-white, and garnet-red colours also occur, while some specimens are of such a deep red colour that in reflected light they appear black. Among the more abundant transparent specimens are some which are only translucent or opaque. A blue or green fluorescence is a striking feature of this amber, and one which is frequently present. Another characteristic feature is the presence of a thin crust of yellowish-red or dark-red to black weathered material, which passes gradually into the fresh and unaltered paler coloured material in the interior.

Sicilian amber has the same hardness, the same fracture, and nearly the same specific gravity as Baltic amber, and like this acquires, on being rubbed, an electrical charge. When heated it fuses without previously swelling up, and gives off dense white fumes, which contain no succinic acid and do not, therefore, irritate the respiratory passages.

Among other places, Sicilian amber occurs in rounded pebbles in the neighbourhood of the mouth of the river Simeto, south of Catania, having been washed out of Tertiary strata, in which it was originally embedded, by that river, hence the name simetite. The material, together with Baltic amber, is manufactured into ornamental articles at the town of Catania. With Sicilian amber are sometimes found pieces of a black resin with a brilliant fracture, which is less hard than the transparent simetite; when heated it gives off a different smell, and is probably an entirely different substance.

BURMESE AMBER. (BURMITE.)

In Burma, also, is found an amber-like resin which is worked for ornamental purposes. The colouring is fairly uniform, and is sometimes of a clear, pale yellow, the colour of light sherry. In darker specimens this colour passes into a reddish shade, and this, in its turn, into a dirty brown colour; this is the most usual tint, and material of this colour has the appearance of colophony or solid petroleum. A few pieces are clear and almost colourless, or of a pale straw-yellow, but the majority are somewhat turbid, and exhibit a marked bluish or greenish fluorescence, which greatly diminishes their value.

Burmite is slightly harder than succinite; it is brittle but easy to work, and is often penetrated by cracks which are filled with calcite. It is difficult to obtain large pieces free from cracks, and this again makes the material less valuable.

The district in which Burmese amber is found is in the north of the country (*see* Maps, Figs. 54 and 55). The long-famous mines are situated not far distant from the jadeite mines, in a hill three miles south-west of Maingkwan in the basin of the Hukong, the upper course of the Chindwin river, in latitude 26° 15′ N. and longitude 96° 30′ E. of Greenwich. The pieces are distributed sporadically throughout a bluish-grey clay belonging to the Lower Miocene division of the Tertiary formation. They are smooth and flat, and the largest are of the size of a man's head. The material is highly esteemed both by the inhabitants and by the Chinese, and is worked to represent animals and idols and also for ear ornaments. The yield was always scanty, and, on this ground alone, it seems scarcely probable that Burmese amber should reach the European markets. On the other hand, a large quantity of Baltic amber is exported through India to Burma, where it is sold as Indian amber more cheaply than the native product. It has been stated that the Burmese mines are not now being worked, and if this is the case, the material which still comes into the native markets must be part of an old stock. A large piece of amber, 11½ inches long, carved to represent a duck, was taken from King Theebaw's palace at Mandalay, and is now preserved with the Burmese regalia in the Indian Section of the Victoria and Albert Museum at South Kensington. One of the presents given by the King of Ava to Colonel Symes, when on an embassy to that country in 1795, was a box containing " amber in large pieces, uncommonly pure."

Amber-like resins are obtained in other countries also, one which appears to be very rich in this respect being southern Mexico. The material is brought to the coast by natives from some unknown locality in the interior, and is shipped as " Mexican amber." It is of a rich golden-yellow colour, strongly fluorescent, and so abundant as to be used by the natives as fuel. Details of the occurrence are quite unknown, and an examination of the material itself is needed before it can be decided whether it consists of succinite or of another similar resin.

JET.

Jet ("black amber"; German, *Gagat, Agstein;* French, *jais*) is a variety of fossil coal often worked for mourning ornaments and other articles. Material to be suitable for this purpose must combine a number of characters, which are not, as a rule, associated in ordinary coal. Thus it must be compact, dense, and homogeneous, characters which are expressed by a perfect conchoidal fracture. It must contain no foreign matter, and specially no iron-pyrites, a common impurity in coals. Moreover, the original internal woody structure must not be retained, the preservation of the external form of the original stem or branch from which the coal was formed does not, however, make the material unsuitable for the purpose, but only emphasises its derivation from extinct trees.

Material which is to be used for ornamental purposes must be free from patchiness, and must be of a deep, pure black colour. Jet of a fine velvety black is most esteemed, that showing a brownish shade being considered inferior. The lustre, which is of a somewhat pronounced greasy kind, should not be too feeble and should be considerably enhanced by polishing; material with only a dull, glimmering lustre is quite worthless. Finally, the substance must be tough enough to be cut with a knife, worked on the lathe, filed or ground in the usual manner, and sufficiently hard to stand a reasonable amount of wear without being damaged. The hardness of genuine jet varies between 3 and 4.

Like all varieties of coal jet is perfectly opaque; the specific gravity is 1·35; the specimens which are said to float on water are, therefore, probably porous. When heated before the blowpipe the material easily ignites, being strongly impregnated with bituminous matter. It burns for a time with a very smoky and sooty flame, giving off an unpleasant odour, and leaving behind a shining, porous, coke-like residue. Like all substances of organic origin jet is a very poor conductor of heat, and hence feels warm to the touch, this feature being especially noticeable in comparison with black minerals or with glass, and serving to distinguish jet from these substances.

A coal having the properties enumerated above is described as a pitch-coal. From ordinary pitch-coal of frequent occurrence, which is usually very brittle, jet is distinguished by its greater toughness and firmness, and by the fact that it can be worked for ornamental articles. Coals which possess the characters detailed above in a more or less perfect degree are found at various localities, and at these places the principal jet-working establishments are situated.

The principal and, indeed, the only jet-cutting works in England are situated at **Whitby,** on the Yorkshire coast. There are but few establishments of the kind on the continent, and compared with the importance of the Whitby industry they are quite insignificant. Whitby has, therefore, the same importance in England, as the centre of the jet-cutting industry, as Oberstein has in Germany as the centre of the agate-cutting industry.

The hard jet of Whitby appears to have been used in Britain in pre-Roman days; it is alluded to by Caedmon, and mentioned in 1350 in the records of St. Hilda's Abbey. It was formerly extensively mined in the cliffs of the Yorkshire coast, near Whitby, and elsewhere; in Eskdale, Danby Dale, and in several of the dales that intersect the East Yorkshire moorlands. The hard jet occurs in the black shales of the *Ammonites serpentinus* zone of the Upper Lias, frequently in the form of flattened masses or layers, which in rare cases have been found to reach a length of 6 feet. So rich are these black shales in jet that they are often referred to as the jet-rock. The richest deposit is at Robin Hood's Bay, four miles south-east of Whitby.

A considerable amount of material is annually obtained, the yield in the year 1880, for example, being 6720 pounds. Two qualities of jet are worked at Whitby, a harder and better quality ranging in price from 4s. to 21s. a pound, obtained for the most part in the neighbourhood, and a softer, poorer, and cheaper quality, which is imported in large quantities from Aragon in Spain. The value of the jet-goods turned out by the Whitby works yearly amounts to about £100,000; in the year 1855 it amounted to £20,000. Between 1200 and 1500 persons are engaged in England at the present time in this industry.

At Whitby is also worked another substance which is of English origin and similar to jet, namely *cannel-coal.* This is of a greyish or brownish-black colour, less brilliant, less susceptible of a good polish, and more brittle than jet. It is found in large masses in the Coal Measures at Newcastle, and other places in England and in Scotland. Owing to its occurrence in large masses, this substance can be used as a veneer for surfaces of considerable size, this being impossible in the case of true jet.

Jet occurs and is worked in certain parts of the Continent. The rough material obtained in **Spain** is exported to England, the jet-working industry which once flourished there having now almost died out. The jet localities are situated in Aragon, Galicia, and Asturia, and what is done in the way of working the material is performed at several places in Asturia.

In **France** the material occurs in the greensand of the Cretaceous formation in the department Aude, province of Languedoc, where also an ancient jet-working industry once flourished. As in Yorkshire the jet occurs as thin plates, which rarely reach a weight of 15 pounds. The chief localities are Monjardin, near Chalabre, on Mount Commo-Escuro, and Bugarach, on Mount Cerbeiron, where mining was carried on, though not systematically. An additional supply of rough material was obtained from Spain, and it has been stated that some of the Spanish jet is superior in quality to the French.

The jet-working industry flourished in France in the eighteenth century, and in the year 1786 there were 1200 people employed in it, principally in the communities of

Sainte-Colombe, Dourban, Segure, Payrat, Bastide, and others.   At that time about 1000 hundredweights of jet of local and foreign origin was annually worked up.   The finished articles were exported in large quantities to Spain, 180,000 francs' worth of goods being received by that country annually, also to Italy, Germany, Turkey, and other parts of the Orient.   With the fall of jet ornaments into disfavour came the decay of the industry, and in 1821 the net gain from the rough material and finished articles amounted only to 35,000 francs, while at the present day the industry is practically dead.

In **Würtemberg** jet is found, under the same conditions as at Whitby, in the *Posidonia* beds of the Upper Lias; for example, at Schömberg, Balingen, Boll, and many other places in the Swabian Alps.   Abundance of jet is here easily obtained, and efforts have been made to establish an industry to rival that of Whitby.   These attempts have received liberal Government support, but in spite of this, the workshops established at Gmünd, Balingen, and other places have been obliged to close, being unable to compete successfully with Whitby.

In **North America** material equal in quality to Whitby jet is found in the southern part of Colorado, in Wet Mountain Valley, and especially in El Paso County.   It is very little used, however, for the manufacture of ornaments, and this is also the case with the fine material found at Pictou, in Pictou County, Nova Scotia.   On the other hand, a black coal with a somewhat metallic lustre, known as anthracite, is sometimes used in the same way as is jet, especially that from Pennsylvania.   In America the use of jet for mourning ornaments is largely replaced by " black onyx," that is to say, by an artificially coloured onyx or agate, which is of a deeper black, brighter, harder, and more durable than jet, and which can be obtained from Oberstein very cheaply.   At present there has been no attempt to establish a jet industry in the United States, in spite of the occurrence of abundance of rough material of suitable quality.

Jet furnishes a material for the manufacture of mourning ornaments of all kinds, including brooches, bracelets, necklaces, cross-shaped pendants, &c.   It is also used for such articles as rosaries, snuff-boxes, ink-stands, candlesticks, and stick-handles.   These objects are usually first roughly fashioned with a knife or file, and then finished on the lathe or grinding lap.   They may be more or less artistically carved; and, finally, are polished, this operation being performed with the palm of the hand.

Jet articles, even of the best quality, are worth but little, the only exceptions being in the case of objects of special artistic design and workmanship.   Notwithstanding the cheapness of the genuine material, it is not uncommon to see mourning ornaments and other articles, purporting to be of jet, made in reality of some other black substance.   A material which frequently replaces the use of jet is glass, either natural (obsidian) or artificial.   It is never difficult, however, to decide whether one is dealing with jet or glass, for the latter is harder, heavier, and much more brilliant than the former, and, moreover, feels cold to the touch, while jet feels warm.   Other substances used in place of jet are black onyx, black tourmaline, and black garnet (melanite), each of which differs from jet in the same ways as glass.   An artificial product of much the same chemical composition as jet is vulcanite.   These substances are very similar in appearance, and both feel warm to the touch; but vulcanite, even when lightly rubbed with a cloth, becomes strongly electrified, and readily attracts bits of paper, while jet does not.   Vulcanite in a plastic condition can be pressed into a mould, and articles made of it are therefore often decorated in bas-relief.   The blunt contours of such figures, however, make it at once clear to the expert that he is dealing not with figures carved in jet, which are remarkable for sharpness of outline, but with the impression of a mould.

THIRD PART

THE DETERMINATION AND DISTINGUISHING OF PRECIOUS STONES

THE DETERMINATION AND DISTINGUISHING
OF PRECIOUS STONES

UNDER the description of each precious stone mentioned in this work there will be found a method of distinguishing it from precious stones of similar appearance or from imitations. By the application of the tests there laid down it is possible to determine whether a supposed diamond, for example, is really such, or whether it is another precious stone of similar appearance, or merely a glass imitation.

There are many cases in which the colour of a precious stone—always the first and most important feature to strike the eye—offers no obvious clue to its identity. We may, for example, meet with a red stone which might equally well be ruby, spinel, garnet, topaz, tourmaline, fluor-spar, or even red glass.

An experienced jeweller or mineralogist would seldom be in doubt in such a case, for he would be guided by external characters such as transparency, lustre, shade of colour, &c., all observable with the naked eye or with a lens. And in the case of a rough, uncut stone the identification is made more easily by the evidence afforded by the crystalline form of the stone, the character of its fractured surface, and the presence or absence of cleavage, evidence which in skilful hands usually points to a correct conclusion.

Glass can frequently be distinguished from genuine precious stones by the fact that it is warmer to the touch, and that, when breathed upon, it receives a film of moisture more easily, and retains it for a longer period, than do precious stones.

In cases where the more obvious characters of a stone are insufficient to establish its identity one has to fall back on its less conspicuous features, the observation of which may require the aid of some specially designed instruments and appliances.

These instruments and appliances to be of any practical value must be simple and substantial, as cheap as possible, and such that determinations made with their aid can be as well performed by the working jeweller as by a trained mineralogist. The methods adopted for the determination should be such as can be employed without injury to the stone, especially when the latter is cut, the avoidance of small surface injuries in uncut stones being of less importance. If the method of determination can be applied to mounted stones so much the better, but in many cases this is impossible, and gems for which large sums are asked should always be purchased unmounted.

The features which best fulfil the requirements outlined above are the optical characters and the specific gravity. The observation of these features is therefore a matter of considerable importance, and has been dealt with in detail in Part I. of this work. It is proposed here to recapitulate briefly the main facts of the subject, and for details to refer the reader to Part I., or to the special description of each precious stone given in Part II.

The **specific gravity** is best and most conveniently determined by the aid of certain heavy liquids, especially methylene iodide, which can be diluted to any extent with benzene until the stone under observation remains suspended in the liquid. The specific gravity of the liquid, and therefore of the stone, can then be determined by Westphal's balance (Fig. 7), or more conveniently by means of indicators, such as fragments of various minerals, of known specific gravity. In the case of stones which sink in pure methylene iodide, the specific gravity of which at ordinary temperature is 3·3, a heavier liquid, obtained by saturating methylene iodide with iodine and iodoform, may be used; or their specific gravity may be determined by the use of the pycnometer, the hydrostatic balance, or Westphal's balance (Fig. 5). Stones with a specific gravity exceeding 3·6 will sink even in this heavier liquid, and for them a still heavier liquid, molten silver-thallium nitrate, must be used. For ordinary purposes an exact determination of the specific gravity is unnecessary, and the four standard liquids recommended above will be found quite sufficient. These being so frequently referred to may be repeated here :

No. 1. Methylene iodide saturated with iodine and iodoform. (Sp. gr. = 3·6.)
„ 2. Pure methylene iodide. (Sp. gr. = 3·3.)
„ 3. Methylene iodide diluted with benzene. (Sp. gr. = 3·0.)
„ 4. Methylene iodide further diluted with benzene. (Sp. gr. = 2·65.)

The data furnished by the use of these liquids enable us to classify precious stones into five groups, a classification which, taken in conjunction with the other characters of a stone, is a valuable aid in determining its identity.

The specific gravity method of determination is just as applicable to rough as to cut stones; the former when under determination should be free from all adhering foreign matter, and the latter should of course be unmounted. The merest fragment of a stone, with a density less than 3·3 or 3·6, will float in one or other of the heavy liquids, and thus indicate approximately what its specific gravity is; but for the exact determination of heavier stones, unless silver-thallium nitrate (sp. gr. = 4·8) is used, a rather larger fragment will be needed.

In establishing the identity of a stone next in importance to the specific gravity come the **optical characters.** The chief of these for our present purpose is the character of the refraction, whether double or single. This may be determined sometimes by direct observation, for, with cut stones, each facet forms with the one next the eye a prism, and the image of a small flame given by this prism will be single in the case of singly refracting stones (Fig. 26b) and double with doubly refracting stones (Fig. 26a). In the case of feebly doubly refracting stones the two images overlap so much that they appear to be one, so that it is not possible by these means to distinguish such stones from stones which are singly refracting. Moreover, observations of this nature can only be made on stones which are perfectly transparent and bounded by smooth crystal-faces or plane polished facets; irregular broken fragments, or stones cut *en cabochon*, will not give sharp images of the flame.

For these reasons, therefore, it is necessary in some cases to resort to the **polariscope** (Fig. 27) for aid in the determination of the nature of the refraction of a stone. The stone is placed on the object-carrier of the instrument with the planes of polarisation crossed, that is to say, in the dark field. If during the course of a complete revolution of 360° the stone changes from light to dark four times it is indisputably doubly refracting. If, on the other hand, the stone remains during a complete rotation of the carrier uniformly dark, like the rest of the field of view, it may be singly refracting. This, however, is not sufficient proof of the fact, for doubly refracting crystals when viewed in the direction of an optic axis present the same appearances. The stone must, therefore, be placed upon the object-carrier in another position, and if on rotation it appears alternately light and dark it must be doubly refracting, while if, on the contrary, it remains dark, we may conclude that it is singly refracting. Absolutely conclusive evidence as to the refraction of a stone is only obtained, however, when the appearance of an alternately light and dark field shows that it is doubly refracting. Cut stones should be examined in two positions, but the two facets on which they successively lie should not be parallel. It should always be borne in mind that the rays of light which pass into a strongly refracting stone may undergo total internal reflection and not emerge from the stone in the desired direction. When this is the case the stone will remain dark during the rotation of the carrier, even though it be strongly doubly refracting. In order to avoid this ambiguity, the gem may be attached to the carrier with wax in such a position that the largest facet, for example, the table in brilliants or the base in rosettes, shall be directed towards the observer and parallel to the object-carrier. Another device which serves the same purpose is to immerse the stone in methylene iodide or monobromonaphthalene or other strongly refracting liquid contained in a vessel with a plane transparent bottom. This device makes it possible to examine in the polariscope not only gems with plane facets but also stones cut *en cabochon* and quite irregular fragments. For the examination of a stone in the polariscope perfect transparency is by no means a *sine quâ non* ; a considerable degree of transparency is quite sufficient to admit of the different degrees of illumination at different stages of the rotation being distinctly observed.

The existence of anomalous double refraction in certain stones must be borne in mind, but is not likely to lead to error. In all these observations with the polariscope it is necessary to cut off all light incident upon the stone from the sides by placing upon it an opaque tube of paper or by using the hand in a similar manner.

An absolute proof of the double refraction of a stone is obtained when it is shown to be dichroic. When any transparent and not too faintly coloured stone is placed in front of the **dichroscope**, Fig. 28, and held towards the light, two coloured images of the aperture of the instrument are seen. If these images be differently coloured the stone is indisputably dichroic, and consequently doubly refracting. If, on the other hand, both images appear of the same colour and remain so during a complete rotation of the stone relative to the dichroscope, then the stone may be characterised by the absence of dichroism, and it may be singly refracting. It is possible, however, in such a case that the stone is really dichroic, though so feebly so that the difference in colour of the images is not noticeable. Moreover, the light may have passed through the stone along an optic axis, which also would account for the images being of the same colour. It is necessary, therefore, to make a second observation with the stone in a different position ; and if now the two images differ in colour the stone is indisputably dichroic and doubly refracting. If, however, they are still of the same colour the stone may be devoid of dichroism and singly refracting ; but the evidence cannot be regarded as conclusive. In making observations with the dichroscope it is necessary to avoid the possibility of internal total reflection by placing the stone with its largest facet over the opening of the instrument.

The dichroism of some precious stones is strong enough to be apparent to the naked eye. The use of the dichroscope in such cases is superfluous, except to confirm one's direct observation or to determine the precise shades of colour shown by the two images. These shades are in a measure characteristic of each stone, and are therefore given a place in the tables which follow. As a means of establishing the identity of a stone, the observation of dichroism has the advantage of being applicable to stones which are mounted *à jour* as well as to cut and uncut unmounted stones.

In the possession of every person having to do with precious stones there should be a dichroscope, a polariscope of the kind described and figured above, and a series of heavy liquids with indicators, or, in place of these, a Westphal's balance. No special skill or theoretical knowledge is required in the use of these instruments, and their judicious employment constitutes a safeguard against the many cases of loss which arise through errors of determination.

It is desirable or necessary at times to take into consideration other characters of the stone, the determination of which may possibly result in a slight injury. Thus it is permissible to determine the hardness of a single cut stone, but tests for the fusibility or the behaviour towards acids can only be applied when a parcel of rough stones is at the disposal of the experimenter. In buying a parcel of gems it often pays the merchant to sacrifice one stone in order that an exhaustive determination, including perhaps a chemical analysis, may be made of it. In the case of rough stones it is often possible to detach a small splinter for detailed examination, and a slight scratch, or spot marked by acids, is of no consequence, since it will be removed when the stone is cut.

Minute differences in **hardness** cannot be relied upon for the purpose of discriminating stones, since differences of the same degree often exist between different faces of a crystal, or in different directions on the same face. The determination of the hardness of a stone is best performed by drawing a projecting edge or corner across the smooth surfaces of the different minerals which constitute the scale of hardness. We have already seen, in the earlier part of this work, that for the determination of the hardness of all ordinary precious stones pieces of only three minerals are necessary, namely, felspar (H = 6), quartz (H = 7), and topaz (H = 8), the pieces of which should have smooth faces and preferably should be artificially polished. No. 5 of the scale may be replaced by a piece of glass, lower members, as also Nos. 9 and 10, are not often needed. In the case of cut stones the edge with which the scratch is made should be chosen at the girdle, which will be enclosed and hidden within the setting. But even with this precaution the operation must be performed with great care, for the pressure may cause the stone to splinter at the edge, especially when it possesses a perfect cleavage.

It is necessary sometimes to reverse the operation and to attempt to scratch the stone itself. The instrument employed by preference in such cases is a pencil of specially hardened steel, provided with a sharp point, which will scratch quartz slightly and glass easily. It is easy in this way to distinguish glass imitations from genuine precious stones, such as the diamond, ruby, sapphire, &c., which are not scratched by the steel point, and the damage caused to the worthless imitation is not of very great consequence. Extra care is only necessary when dealing with the softer precious stones, such as chrysolite, &c.; but it is advisable in all cases to choose a spot on which to apply the test, which will be hidden when the stone is mounted. A scratch on almost any part of a transparent stone detracts considerably from its beauty, but a scratch on the back of an opaque stone is of little consequence. A rough stone may be scratched anywhere without fear, and the hardness of different faces and in different directions of the same face determined.

Instead of the steel pencil a file of hard steel is frequently employed; this readily marks the softer stones, and emits at the same time a lower note than with harder stones. Cut stones should be tested with the file only on the girdle, and even then with the greatest care. Owing to all these difficulties only a limited use can be made of the character of hardness. Recently, however, a very good method, depending upon hardness, has been discovered by which glass imitations can be distinguished from genuine precious stones. An aluminium pencil drawn across a clean, dry surface of glass leaves a metallic silvery streak, but on the surface of a precious stone the aluminium leaves no mark.

The behaviour of a precious stone towards acids is another character of which only a limited use can be made. It is useful, for example, when carbonates, such as malachite, are in question, for a drop of acid placed on a rough stone, or on the back of a polished one, causes a brisk effervescence, and does no particular damage. The fusibility can only be determined upon fragments of rough stones, and, for our present purpose, is of little importance.

The magnetic and **electrical characters** of a stone are often useful in determining its identity. The phenomenon of pyroelectricity exhibited by tourmaline to such a marked degree often serves to identify this stone, and the existence of this property can be demonstrated without the slightest risk of injury to the stone. The supposed tourmaline is warmed in an air-bath or on a sheet of paper held over a flame ; and while cooling it is dusted over with a mixture of red-lead and sulphur shaken through a sieve. Mutual friction causes the red-lead to become positively, and the sulphur negatively electrified, and if the crystal during cooling has acquired statical charges of pyroelectricity, the red-lead will be attracted to the negative pole of the crystal and the sulphur to the positive pole.

Quite recently it has been discovered that the **Röntgen (X) rays** afford another means whereby certain precious stones may be distinguished from their imitations. The most important case is that of the diamond. This is perfectly transparent to the rays, but glass, topaz, rock-crystal, &c., are opaque, and will appear in a photograph taken by X rays with sharp outlines. Corundum (ruby and sapphire, &c.) is semi-transparent, while spinel and blue tourmaline, like glass, are opaque, so that here again the former may be distinguished from the latter. Amber and other similar resins, as well as jet, are transparent to a marked degree, and can, therefore, be distinguished from glass imitations, which are opaque to the rays. The whole method is as yet in its infancy, but will no doubt become important as a ready means of establishing the identity of a diamond, whether mounted or unmounted, without injury to the stone. Moreover, a large parcel of diamonds may be photographed, and the presence of any stone not diamond readily detected.

In what follows, the characters described above will be utilised to distinguish precious stones of similar appearance one from another. The first step is to classify all stones into three main groups :

(*a*) Transparent stones.
(*b*) Translucent and opaque stones.
(*c*) Stones with special optical effects.

A classification based upon transparency or translucency is difficult to follow in practice, for large specimens of ordinarily transparent stones are sometimes only translucent, and one specimen of the same mineral species may be perfectly transparent while another is perfectly opaque. In the scheme which follows, therefore, stones which may be either transparent or translucent will be placed in both classes. The stones included in classes *a* and *b* are arranged in sub-groups according to colour, and those in the third class according to the nature of the optical phenomenon to which their characteristic appearance is due.

A. TRANSPARENT STONES.

These are classified into fourteen groups according to colour :

1. Colourless
2. Greenish-blue (sea-green)
3. Pale blue
4. Blue
5. Violet
6. Lilac and rose
7. Red

8. Reddish-brown
9. Smoke-grey and clove-brown
10. Reddish-yellow
11. Yellowish-brown
12. Yellow
13. Yellowish-green
14. Green

In the tables which follow, each markedly transparent precious stone will be found in one of these fourteen groups, and in cases where the same stone shows two or more colours, or a shade more or less in harmony with the colours characteristic of several groups, it will be included in each, so that there will be no difficulty in finding it.　The stones included in each colour-group are further classified according to specific gravity, as determined by the four standard heavy liquids.　They are placed in the tables in order of decreasing specific gravity, and in three other columns are tabulated the values of the specific gravity and hardness, and the character of the refraction of each ; while in a fourth column the presence or absence of dichroism, its strength, and, in some cases, the colours of the dichroscope images are indicated.

It should be possible from these tables to establish the identity of almost any precious stone one may happen to meet with.　The remarks appended to each table are intended as a help, mainly in cases where the specific gravity places several stones in one or other of the five sub-divisions.

The names of the more important precious stones, and those most commonly met with, are printed in the tables in heavier type, and special attention is paid to such in the remarks.

1. COLOURLESS STONES.

Division.	Name.	Specific gravity.	Hardness.	Refraction.
I. (Sp. gr. over 3·6)	Zircon Corundum Spinel	4·6—4·7 3·9—4·1 3·60—3·63	$7\frac{1}{2}$ 9 8	Double Double Single
II. (Sp. gr. = 3·3—3·6)	**Topaz** **Diamond**	3·50—3·56 3·50—3·52	8 10	Double Single
III. (Sp. gr. = 3·0—3·3)	Tourmaline	3·022	$7\frac{1}{4}$	Double
IV. (Sp. gr. = 2·65—3·0)	Phenakite Beryl	2·95 2·68—2·75	$7\frac{3}{4}$ $7\frac{3}{4}$	Double Double
V. (Sp. gr. under 2·65)	**Rock-crystal** Opal	2·65 2·0	7 6	Double Single
	Glass	Variable	5	Single

The stones included in Division I. sink in liquid No. 1, and this high specific gravity is a feature which marks them off from all the others.　They are distinguished *inter se* by differences in their optical characters, zircon and corundum (" white sapphire ") being doubly refracting and spinel singly refracting. Colourless spinel is very rare and is scarcely ever met with cut as a gem. Zircon and corundum are commoner, and not infrequently are substituted for the diamond; the latter, however, is distinguished by the fact that it floats in liquid No. 1. Zircons and "white sapphires" of some size may be distinguished by an exact determination of their specific gravities; and another difference between the two stones lies in the hardness, for the latter readily scratches a smooth surface of topaz while zircon does not. Moreover, colourless zircon has a brilliant adamantine lustre, while the lustre of corundum is of the vitreous type. Finally, cut colourless zircons are nearly always small, at the most no larger than a pea, and there is often a perceptible reddish tinge about them, left after heating the yellowish-red hyacinth, which is never exhibited by colourless corundum.

Division II. embraces topaz and diamond; the specific gravity is practically identical, but topaz is doubly refracting and diamond singly refracting; and there is also an enormous difference in hardness, diamond scratching topaz very easily and deeply. Moreover, the strong adamantine lustre of diamond is very characteristic, and the magnificent play of prismatic colours, shown by cut stones, still more so, so that it is scarcely possible to mistake diamond for topaz or for strass. The last-named is readily scratched by a steel point or marked by an aluminium pencil, and also feels warmer to the touch than does diamond.

Colourless tourmaline is the only stone embraced by the third division, and is of very rare occurrence as a gem.

Phenakite and beryl, the two stones which constitute division IV., can be distinguished only by an exact determination of their specific gravity: if performed with the aid of methylene iodide the smallest fragments of material will suffice. Characteristic of phenakite is its specially strong lustre, by which it is distinguished from colourless beryl, which is rarely cut as a gem.

Rock-crystal falls into division V. and cannot, therefore, be mistaken for other and heavier stones.

From many of the stones in this group glass can be distinguished by the fact that it is singly refracting. So, indeed, also is opal, but the transparent, glassy variety of opal known as hyalite is practically never used as a gem. It may be recognised by the fact that it is capable of scratching glass, though it itself is easily scratched by a steel point. The feeble anomalous double refraction always shown by water-clear opal also serves to distinguish it from glass.

2. GREENISH-BLUE (SEA-GREEN) STONES.

Division.	Name.	Specific gravity.	Hardness	Refraction.	Dichroism.
I.	"Oriental Aquamarine"	3·9—4·1	9	Double	Not very strong
II.	Topaz	3·50—3·56	8	Double	{ Distinct (colourless and greenish-blue)
	Diamond	3·50—3·52	10	Single	Absent
III.	Fluor-spar	3·1—3·2	4	Single	Absent
	Euclase	3·05—3·1	7½	Double	Observable
IV.	Aquamarine	2·68—2·75	7¾	Double	{ Distinct (bluish and yellowish)
	Glass	Variable	5	Single	Absent

The two members of this group most likely to be confused are aquamarine and topaz, since they are often identical in colour. There need be no confusion, however, since they differ very widely in specific gravity, the latter falling into Division II. and the former into Division IV. Euclase, a rare mineral of much the same colour, is distinguished by the same feature, namely, the specific gravity.

The distinctions between corundum (" oriental aquamarine "), diamond, and topaz have been dealt with in the first table.

The specific gravity, low degree of hardness, and single refraction unite to distinguish fluor-spar from other stones of the same group; it is always rather darker in colour than are aquamarine, topaz, or euclase. Glass is singly refracting and much softer than any of the stones included in this group except fluor-spar.

3. PALE BLUE STONES.

Division.	Name.	Specific gravity.	Hardness.	Refraction.	Dichroism.
I.	**Sapphire**	3·9—4·1	9	Double	Feeble
II.	**Topaz**	3·50—3·56	8	Double	Observable (colourless and bluish)
	Diamond	3·50—3·52	10	Single	Absent
III.	Tourmaline	3·1	7¼	Double	Distinct
IV.	**Aquamarine**	2·68—2·75	7¾	Double	Feeble
	Glass	Variable	5	Single	Absent

The remarks given under the first and second tables will enable the stones brought together here to be distinguished; those of more frequent occurrence are readily recognised by the specific gravity alone.

4. BLUE STONES.

Division.	Name.	Specific gravity.	Hardness.	Refraction.	Dichroism.
I.	**Sapphire**	3·9—4·1	9	Double	Distinct (dark blue and pale greenish-blue)
	Spinel	3·7	8	Single	Absent
II.	Kyanite	3·60	5—7	Double	Observable
	Diamond	3·50—3·52	10	Single	Absent
III.	Fluor-spar	3·1—3·2	4	Single	Absent
	Tourmaline	3·1	7¼	Double	Strong (pale blue to dark blue)
V.	Cordierite	2·60—2·66	7¼	Double	Strong (pale blue, dark blue, and yellowish-grey)
	Haüynite	2·4	5½	Single	Absent
	Glass	Variable	5	Single	Absent

By far the most important of these is sapphire; the only other blue stones of consequence as gems are spinel and tourmaline. The former is distinguished by the fact that it is singly refracting, while sapphire and tourmaline differ widely in density. The colour of tourmaline is usually very dark and inclined somewhat to green, while that of sapphire is usually a pure blue; moreover, tourmaline is very much more strongly dichroic than is sapphire.

Kyanite almost invariably exhibits a series of fine straight cracks running in one direction; it is not perfectly transparent as a rule, and, compared with sapphire, has little lustre. It remains suspended in or sinks slowly to the bottom of liquid No. 1, in which sapphire readily sinks. Diamond of a deep blue colour is extremely rare; it is recognisable by its specific gravity, hardness, and single refraction.

Cordierite is not a gem ordinarily met with in the trade; it is distinguished by its low specific gravity and strong dichroism, with the characteristic shades of colour.

Glass and haüynite are both singly refracting and not dichroic, but the specific gravity of the former is considerably higher than that of the latter.

5. VIOLET STONES.

Division.	Name.	Specific gravity.	Hardness.	Refraction.	Dichroism.
I.	**Almandine**	4·1—4·2	7¼	Single	Absent
	"Oriental Amethyst"	3·9—4·1	9	Double	Distinct
	Spinel	3·60—3·63	8	Single	Absent
III.	Axinite	3·29—3·3	6¾	Double	Strong (violet, brown, green)
	Apatite	3·2	5	Double	Feeble
	Fluor-spar	3·1—3·2	4	Single	Absent
V.	**Amethyst**	2·65	7	Double	Feeble
	Glass	Vari ble	5	Single	Absent

Apatite and fluor-spar are scarcely ever used as gems: their specific gravity and lack of hardness distinguish them from other violet stones, and the difference in optical refraction distinguishes the one from the other. Axinite is included in the same division; it is characterised by specially strong dichroism, and its colour is not pure violet, but markedly brownish in shade.

Of the stones included in this group, the one met with most commonly is the true amethyst, which constitutes Division V., and is readily distinguished from the members of the first division by its low specific gravity and feeble dichroism.

Almandine is distinguished from the other stones included in Division I. by its colour, which is better described as red inclined to blue than as pure violet. It differs both from "oriental amethyst" and from true amethyst in that it is singly refracting, and from the former in its hardness. "Oriental amethyst," amethyst, and spinel are, perhaps, most conveniently distinguished by the presence or absence of dichroism, which is strongest in the first, feeble in the second, and absent from the last. "Oriental amethyst" and spinel are capable of scratching quartz, but amethyst, of course, is not, and spinel differs from the other two in being singly refracting. The only well-marked difference between spinel and almandine is one of density; the former remains suspended or slowly sinks in liquid No. 1, while the latter quickly sinks. The slight difference in hardness may enable one to discriminate between these stones, but in certain cases it is difficult to distinguish one from another.

Glass differs in hardness from all the violet stones ordinarily used, being readily scratched by a steel point; it is distinguished both from "oriental amethyst" and from true amethyst by its single refraction.

6. LILAC AND ROSE-COLOURED STONES.

Division.	Name.	Specific gravity.	Hardness.	Refraction.	Dichroism.
I.	**Ruby**	3·9—4·1	9	Double	Feeble
	Spinel ("Balas-ruby")	3·60—3·63	8	Single	Absent
II.	**Topaz**	2·50—3·56	8	Double	Strong (red and yellow)
	Diamond	3·50—3·52	10	Single	Absent
III.	Fluor-spar	3·61	4	Single	Absent
	Tourmaline	3·02	$7\frac{1}{4}$	Double	Distinct (pale red and dark red)
V.	Rose-quartz	2·65	7	Double	Very feeble
	Glass	Variable	5	Single	Absent

Of the stones of this group, the first three are commonly met with, tourmaline is less common, and the others make only rare appearances in the precious stone market. "Balas ruby" exhibits not infrequently a milky cloudiness. Topaz and tourmaline both show differently coloured images in the dichroscope. Topaz is distinguishable also from ruby and spinel by the difference in specific gravity, and the two last named differ from each other both in the character of their optical refraction and in the fact that dichroism is feeble in the one and absent in the other. Glass is easily recognisable by its lack of hardness and by its single refraction.

7. RED STONES.

Division.	Name.	Specific gravity.	Hardness.	Refraction.	Dichroism.
I.	**Almandine**	4·1—4·2	$7\frac{1}{4}$	Single	Absent
	Ruby	3·9—4·1	9	Double	Distinct (pale red and dark red)
	Pyrope				
	("Cape ruby")	3·86	$7\frac{1}{4}$	Single	Absent
	(Bohemian garnet)	3·7—3·8	$7\frac{1}{4}$	Single	Absent
	Hessonite	3·65	$7\frac{1}{4}$	Single	Absent
	Spinel	3·60—3·63	8	Single	Absent
II.	**Topaz**	3·50—3·56	8	Double	Strong (red and yellow)
	Diamond	3·50—3·52	10	Single	Absent
III.	Fluor spar	3·1	4	Single	Absent
	Tourmaline	3·08	$7\frac{1}{4}$	Double	Strong (rose and dark red)
V.	Fire-opal	2·2	6	Single	Absent
	Glass	Variable	5	Single	Absent

Of these, red fluor-spar, red diamond, and fine opal are all rare.

Division I. embraces the garnets—almandine, hessonite or cinnamon-stone, "Cape ruby," and Bohemian garnet; the last two do not differ essentially and both belong to the sub-species pyrope. These, together with spinel, are distinguished from ruby by their single refraction, and from each other by differences in colour. Almandine is purplish-red (inclined to blue), "Cape-ruby" and Bohemian garnet are both dark blood-red with a yellowish tinge, and hessonite is pale yellowish-red. The different varieties of garnet are characterised also by not unimportant differences in density. The variety of spinel known as "ruby-spinel" is very similar in colour to ruby, but is easily distinguishable by the means indicated already. On the other hand it is often difficult to distinguish spinel from garnet, especially in the case of the reddish-yellow spinel (rubicelle) and hessonite, which agree in the character of their refraction, in the absence of dichroism, in colour, and very closely in specific gravity; it is sometimes possible, though difficult, to distinguish cut stones by their difference in hardness. There is less difficulty in distinguishing rough stones, since spinel always occurs in octahedra, while hessonite scarcely ever takes this form, and spinel is infusible and hessonite easily fusible before the blowpipe. The fusibility of hessonite and almandine distinguishes them also from the infusible pyrope.

Red tourmaline and topaz in colour are very similar to each other, and also sometimes to ruby, but each of the three differs very definitely from the others in specific gravity, and a further distinction is afforded by the fact that one of the two images shown by topaz in the dichroscope is distinctly yellow, this not being the case with red tourmaline or ruby.

In the second division diamond is distinguished from topaz by its single refraction and lack of dichroism, and these same features serve to distinguish fluor-spar from tourmaline, both of which are included in Division III.

The specific gravity of fire-opal is always much lower than that of any other stone in this group. It is singly refracting, always rather cloudy, and sometimes shows a play of colours like that of precious opal. Glass is distinguished from the stones of this group in the usual way; the most convenient distinction between glass and opal is that of the difference in specific gravity, the density of glass being scarcely ever less than 2·6.

8.   REDDISH-BROWN STONES.

Division.	Name.	Specific gravity.	Hardness.	Refraction.	Dichroism.
I.	Zircon	4·6—4·7	$7\frac{1}{2}$	Double	Very feeble
	Almandine	4·1—4·2	$7\frac{1}{4}$	Single	Absent
	Staurolite	3·73	$7\frac{1}{2}$	Double	Feeble
	Hessonite	3·65	$7\frac{1}{4}$	Single	Absent
II.	**Topaz**	3·50—3·56	8	Double	Strong (yellow and red)
III.	Tourmaline	3·1	$7\frac{1}{4}$	Double	Strong (pale brown and dark brown)
V.	Citrine	2·65	7	Double	Feeble
	Fire-opal	2·2	6	Single	Absent
	Amber	1·08	2—3	Single	Absent
	Glass	Variable	5	Single	Absent

Staurolite is rarely met with, it is seldom perfectly transparent and always very darkly coloured. Tourmaline and almandine of a reddish-brown colour are rarely cut as gems.

Zircon, hessonite, topaz, and citrine can be readily distinguished by the help of the character tabulated above and by the remarks appended to previous tables.

Amber differs from all the other members of this group in feeling warm to the touch, and in the fact that it acquires a strong charge of electricity on being rubbed.

Glass and opal are distinguished in the way indicated under the preceding table.

9. SMOKE-GREY AND CLOVE-BROWN STONES.

Division.	Name.	Specific gravity.	Hardness.	Refraction.	Dichroism.
II.	Diamond	3·50—3·52	10	Single	Absent
	Epidote	3·47—3·50	$6\frac{1}{4}$	Double	Strong (green, yellow, brown)
	Idocrase	3·4	$6\frac{1}{2}$	Double	Distinct
III.	Axinite	3·29—3·30	$6\frac{3}{4}$	Double	Strong (violet, brown, green)
	Andalusite	3·17—·3·19	$7\frac{1}{2}$	Double	Strong (yellow and red)
V.	Smoky-quartz	2·65	7	Double	Feeble
	Glass	Variable	5	Single	Absent

Of the stones of this group, practically the only one which is widely distributed is smoky-quartz. Andalusite and epidote are rare, and axinite still more so, but brown diamond is less uncommon. The characters given in the table are sufficient to distinguish any one of these stones from any other, and from glass without much difficulty.

It is sometimes a little difficult, however, to discriminate between epidote and idocrase, and one has to rely on the fact that the former is more strongly dichroic than the latter.

Andalusite is distinguishable from axinite by the greenish tint of its colour, and by the fact that the dichroscope images are differently coloured.

10. REDDISH-YELLOW STONES.

Division.	Name.	Specific gravity.	Hardness.	Refraction.	Dichroism.
I.	Hyacinth	4·6—4·7	$7\frac{1}{2}$	Double	Very feeble
	" Oriental hyacinth "	3·9—4·1	9	Double	Distinct
	Pyrope	3·7—3·8	$7\frac{1}{4}$	Single	Absent
	Hessonite	3·65	$7\frac{1}{4}$	Single	Absent
	Spinel (Rubicelle)	3·60—3·63	8	Single	Absent
II.	Topaz	3·50—3·56	8	Double	Strong (red and yellow)
V.	Fire-opal	2·2	6	Single	Absent
	Glass	Variable	5	Single	Absent

The characters which distinguish pyrope, hessonite, and spinel one from another have been given already under Table 7. Hessonite and hyacinth are sometimes very similar in colour, and are often mistaken the one for the other, in spite of the fact that hyacinth has a much stronger adamantine lustre than has hessonite ; there is also a fundamental difference between them in the character of their refraction, hessonite being singly and hyacinth doubly refracting. " Oriental hyacinth " is distinctly dichroic, but the hyacinth variety of zircon is only very feebly so.

The density of topaz places it in a division apart from all the other stones of this group, and it is further characterised by strong dichroism.

11. YELLOWISH-BROWN STONES.

Division.	Name.	Specific gravity.	Hardness.	Refraction.	Dichroism.
II.	Topaz	3·50—3·56	8	Double	Distinct (yellow and brownish-red)
	Diamond	3·50—3·52	10	Single	Absent
	Epidote	3·47—3·50	6½	Double	Strong (green, yellow, brown)
	Idocrase	3·35—3·45	6½	Double	Distinct (green and yellow)
	Sphene	3·35—3·45	5½	Double	Distinct
III.	Axinite	3·29—3·30	6¾	Double	Strong (violet, brown, green)
V.	Citrine	2·65	7	Double	Feeble
	Fire-opal	2·2	6	Single	Absent
	Amber	1·08	2—3	Single	Absent
	Glass	Variable	5	Single	Absent

Of the stones of this group, topaz, amber, and the yellow variety of quartz known as citrine are of common occurrence. Amber feels warm to the touch, is very light and soft, and when rubbed becomes strongly electrified, attracting to itself any light objects, so that it can scarcely be mistaken for any of the stones with which it is here grouped. The means whereby amber may be distinguished from resinous and glass imitations have been fully described under amber and need not be repeated.

Topaz and quartz may be distinguished by the difference in specific gravity alone, but the greater hardness and stronger dichroism of the former furnishes additional aid in this direction.

Sphene is less hard than is idocrase or epidote; these may be distinguished from each other by an accurate determination of their specific gravity. Epidote is characterised also by very strong dichroism.

12. YELLOW STONES.

Division.	Name.	Specific gravity.	Hardness.	Refraction.	Dichroism.
I.	Hyacinth	4·6—4·7	7½	Do ble	Very feeble
	" Oriental topaz "	3·9—4·1	9	Double	Feeble
	Chrysoberyl	3·68—3·78	8½	Double	Feeble
II.	Topaz	3·50—3·56	8	Double	Distinct (pale yellow and dark yellow)
	Diamond	3·50—3·52	10	Single	Absent
	Chrysolite	3·33—3·37	6½	Double	Feeble (green and yellowish-green)
III.	Fluor-spar	3·1	4	Single	Absent
IV.	Beryl	2·67—2·76	7¾	Double	Feeble
V.	Citrine	2·65	7	Double	Feeble
	Fire-opal	2·2	6	Single	Absent
	Amber	1·08	2—3	Single	Absent
	Glass	Variable	5	Single	Absent

The commonest stones of a yellow colour are citrine, topaz, and the yellow variety of corundum known as "oriental topaz," they are easily distinguished from each other by the differences in their specific gravity.   Several other stones of this group are not altogether uncommon in the precious stone market.

The colour of hyacinth is never pure yellow, but is always definitely inclined to red, while the yellow of chrysoberyl and chrysolite has always a greenish tinge.   In stones of not too small size, "oriental topaz" may be recognised by an exact determination of the specific gravity.   Hyacinth is not capable of scratching topaz, but this latter is scratched both by "oriental topaz" and by chrysoberyl : and "oriental topaz" will scratch chrysoberyl, a specimen of which may be conveniently kept for this test.

Topaz, diamond, and chrysolite are distinguished from each other by differences in hardness and in their behaviour in the polariscope, while the specific gravity is in each case sufficient to distinguish them from the stones included in other divisions.   Topaz and citrine differ both in specific gravity and in hardness ; the former sinks in pure methylene iodide, and, moreover, is capable of scratching quartz.

Although beryl and citrine fall into different divisions of this group, there is very little difference in their specific gravities, and in some cases there may be uncertainty as to the identity of these stones.   Beryl, nevertheless, is always rather heavier than citrine (quartz), and sinks in liquid No. 4, in which citrine remains suspended.   It is also characterised by its capability of scratching quartz, which, of course, is not possessed by citrine.   Beryl is both lighter and harder than chrysolite.

Amber, fire-opal, and glass are distinguished in the manner already explained in the remarks appended to preceding tables.

13.   YELLOWISH-GREEN STONES.

Division.	Name.	Specific gravity.	Hardness.	Refraction.	Dichroism.
I.	Zircon	4·6—4·7	$7\frac{1}{2}$	Double	Very feeble
	"Oriental chrysolite" }	3·9—4·1	9	Double	Distinct
	Demantoid	3·83	7	Single	Absent
	Chrysoberyl	3·63—3·78	$8\frac{1}{2}$	Double	Feeble (yellowish and greenish)
II.	**Topaz**	3·50—3·56	8	Double	Distinct
	Epidote	3·47—3·5	$6\frac{1}{2}$	Double	Strong (green, yellow, brown)
	Idocrase	3·35—3·45	$6\frac{1}{2}$	Double	Distinct (green and yellow)
	Sphene	3·35—3·45	$5\frac{1}{2}$	Double	Distinct
	Chrysolite	3·33—3·37	$6\frac{3}{4}$	Double	Feeble (green and yellowish-green)
III.	Hiddenite	3·17—2·20	$6\frac{3}{4}$	Double	Feeble (pale green and dark green)
	Andalusite	3·17—3·19	$7\frac{1}{2}$	Double	Strong (yellow, green, red)
	Tourmaline	3·1	$7\frac{1}{4}$	Double	Strong (yellow and green)
IV.	**Beryl**	2·67—2·76	$7\frac{3}{4}$	Double	Distinct
V.	**Moldavite**	2·36	$5\frac{1}{2}$	Single	Absent
	Glass	Variable	5	Single	Absent

Of the stones included in the first division of this group, chrysoberyl is most frequently met with : it is characterised by great hardness, being capable of scratching topaz, and often by the possession of a milky sheen. "Oriental chrysolite" is also capable of scratching topaz, but is much more strongly dichroic than is chrysoberyl. Zircon of this colour is rare; it has a strong adamantine lustre, and is far denser than any other stone in the division, so that an exact determination of the specific gravity of this stone is sufficient to establish its identity. Demantoid is the only singly refracting stone included in this division.

In Division II. topaz is characterised by its hardness, and is the only stone in this division which is capable of scratching quartz. The characters of many of these minerals have been considered already; chrysolite is harder than sphene, and much less dichroic than epidote or idocrase. Not infrequently the specific gravity of chrysolite is exactly the same as that of methylene iodide, the stone will therefore sink slowly as the liquid is warmed or rise to the surface as it is cooled, the contact of the vessel with the warm hand being sufficient to cause the stone to rise. "Oriental chrysolite" is much heavier, much harder, and much more strongly dichroic than is true chrysolite.

Division III. includes hiddenite, andalusite, and tourmaline; all three are dichroic, but andalusite is most strongly so, and is characterised further by one of the dichroscope images being red. Andalusite and tourmaline scratch quartz, but hiddenite does not.

The variety of beryl known as "aquamarine-chrysolite" is recognised by its low specific gravity and feeble dichroism.

Moldavite is singly refracting, and is distinguished from artificial yellowish-green glasses only by its lower specific gravity.

14. GREEN STONES.

Division.	Name.	Specific gravity.	Hardness.	Refraction.	Dichroism.
I.	Zircon	4·6—4·7	$7\frac{1}{2}$	Double	Very feeble
	"Oriental emerald"	3·9—4·1	9	Double	Distinct (green and brown)
	Demantoid	3·83	7	Single	Absent
	Alexandrite	3·68—3·78	$8\frac{1}{2}$	Double	Strong (green, yellow, red)
II.	Diamond	3·50—3·52	10	Single	Absent
	Epidote	3·47—3·50	$6\frac{1}{2}$	Double	Strong (green, yellow, brown)
	Idocrase	3·35—3·45	$6\frac{1}{2}$	Double	Distinct (green and yellow)
	Sphene	3·35—3·45	$5\frac{1}{2}$	Double	Distinct (yellow, green, reddish-brown)
	Chrysolite	3·30—3·37	$6\frac{3}{4}$	Double	Feeble (green and yellowish-green)
III.	Diopside	3·2—3·3	6	Double	Feeble
	Dioptase	3·29	5	Double	Feeble
	Apatite	3·2	5	Double	Feeble
	Hiddenite	3·17—3·20	$6\frac{1}{2}$	Double	Distinct (pale green and dark green)
	Andalusite	3·17—3·19	$7\frac{1}{2}$	Double	Strong (yellow, green, red)
	Tourmaline	3·1	$7\frac{1}{4}$	Double	Strong (yellow and bluish-green)
	Fluor-spar	3·1	4	Single	Absent
IV.	**Emerald**	2·67	$7\frac{3}{4}$	Double	Distinct (green and bluish-green)
V.	**Moldavite**	2·36	$5\frac{1}{2}$	Single	Absent
	Glass	Variable	5	Single	Absent

In the first division the green variety of corundum, known as "oriental emerald," is very rare. Zircon and demantoid are uncommon, and dark green chrysoberyl (alexandrite) more frequent. The latter is characterised by its strong dichroism and the possession of red as a second colour. Zircon, although doubly refracting, is practically not dichroic at all, while demantoid differs from all three in being singly refracting.

The only singly refracting stone in the second division is diamond. For the distinguishing features of epidote, idocrase, sphene, and chrysolite the reader is referred to the remarks appended to previous tables. Epidote, always of a rather dark green colour, differs from the other stones in that one of the images it gives in the dichroscope is of a dark brown colour. Under certain circumstances, however, it may be difficult to decide whether a stone is idocrase or epidote.

Diopside and chrysolite are sometimes almost precisely the same, both in colour and specific gravity, and in such cases one has to rely on the difference in hardness. Chrysolite is capable of scratching a crystal of diopside, but a cut diopside of course is not; a crystal of diopside may conveniently be kept at hand for this purpose. Dioptase is always very deeply coloured, and never perfectly clear. Fluor-spar is singly refracting. Apatite is usually of a dark green colour tinged with blue. Andalusite and tourmaline are the most strongly dichroic stones of this division, the former shows a characteristic red colour in the dichroscope, and the latter a bluish-green. Both scratch quartz, while the distinctly but less strongly dichroic hiddenite does not.

The true emerald falls into Division IV.; stones of similar colour, such as " oriental emerald," hiddenite, and alexandrite, cannot therefore be mistaken for it. The specific gravity of much fissured emeralds is lower than that of quartz, and such stones would therefore fall naturally into division V.; but this is never the case with faultless stones. The true emerald, however, is sufficiently characterised by its wonderful green colour; and small variations in the specific gravity are of no consequence for purposes of determination.

Moldavite, like artificial glass, is singly refracting, and is sometimes difficult to distinguish from glass, although it is, as a rule, harder and less heavy.

Alexandrite and green andalusite are very similar both in the colours of cut stones and in those of the dichroscope images; but besides the difference in the specific gravity, which is considerable, there is a difference in hardness, alexandrite being capable of scratching topaz while andalusite is not. Zircon is always recognisable by its strong adamantine lustre and high specific gravity, which latter character distinguishes the stone from diamond. Alexandrite, tourmaline, emerald, and moldavite, the green stones most commonly met with, are sufficiently distinguished by their specific gravity alone, each falling into a different division. The means whereby they may be distinguished from artificial glass have been given already in the remarks appended to previous tables.

B. TRANSLUCENT AND OPAQUE STONES.

In the distinguishing of translucent and opaque stones, we are obliged to dispense with the aid afforded in the case of transparent stones by the character of the refraction, and to rely in a large measure upon the specific gravity. In dealing with opaque stones, however, more use can be made of the character of hardness, since there is less fear of damaging such stones by scratches. The translucent and opaque stones here considered are classified, according to their colour and lustre, into eight main groups, each of which embraces five divisions, into which fall stones of different specific gravities. The main groups are :

1. White, faintly coloured, and grey stones.
2. Blue stones.
3. Green stones.
4. Black stones.

5. Yellow and brown stones.
6. Rose-red, red, and lilac stones.
7. Stones exhibiting more than one colour.
8. Stones with metallic lustre.

1. WHITE, FAINTLY COLOURED, AND GREY STONES.

Division.	Name.	Specific gravity.	Hardness.
III.	Jadeite	3·33	$6\frac{1}{2}$—7
IV.	Nephrite	3·0	$5\frac{1}{2}$—6
V.	Chalcedony Opal	2·6 1·9—2·2	$6\frac{1}{2}$ 6
	Glass	Variable	5

The stones of this group are distinguished by their specific gravity.

Jadeite has almost exactly the specific gravity of liquid No. 2, and nephrite that of liquid No. 3, so that they remain suspended in these respective liquids, or in some cases slowly sink. Cut jadeite and nephrite can usually be distinguished by the difference in specific gravity, and sometimes also by a difference in hardness. Rough specimens may be distinguished by the difference in fusibility, for fine splinters of jadeite fuse even in an ordinary candle-flame without the application of the blowpipe.

Chalcedony is both harder and heavier than opal, and glass is invariably softer than chalcedony and heavier than opal.

2. BLUE STONES.

Division.	Name.	Specific gravity.	Hardness.
I.	Chessylite	3·8	$3\frac{3}{4}$
III.	Lazulite Bone-turquoise	3·1 3·0—3·5	$5\frac{1}{2}$ 5
IV.	Turquoise	2·6—2·8	6
V.	Agate (artificially coloured) Lapis-lazuli	2·6 2·4	$6\frac{1}{2}$ $5\frac{1}{2}$
	Glass	Variable	5

Chessylite is distinguished from all other blue stones by its very dark colour, high specific gravity, low hardness, and by the fact that it effervesces when brought into contact with a drop of hydrochloric acid.

Lazulite and turquoise, and specially turquoise, are always lighter in colour than chessylite, and turquoise is never very brilliant. The difference in specific gravity is sufficient to distinguish these two stones one from another, and the characters which distinguish them from bone-turquoise have been detailed under the special description of that substance.

Lapis-lazuli is always of a dark blue colour, which has sometimes a more or less pronounced tinge of green. It frequently exhibits yellow, metallic specks of iron-pyrites and encloses patches and veins of calcite. Its hardness and specific gravity readily distinguish it both from the artificially coloured blue agate ("false lapis-lazuli") and from all the other stones of this group.

Turquoise and lapis-lazuli are sometimes imitated in glass; the means whereby such imitations may be recognised will be found under the respective descriptions of these stones in Part II. of this work.

3. GREEN STONES.

Division.	Name.	Specific gravity.	Hardness.
I.	Ceylonite	3·8	$7\frac{1}{2}$
	Malachite	3·7—3·8	$3\frac{1}{2}$
II.	Chloromelanite	3·4	$6\frac{1}{2}$—7
III.	**Jadeite**	3·33	$6\frac{1}{2}$—7
IV.	**Nephrite**	3·0	$5\frac{3}{4}$
	Prehnite	3·28—3·0	$6\frac{1}{2}$
	Turquoise	2·6—2·8	6
V.	**Prase**	2·65	7
	Chrysoprase	2·65	7
	Plasma (and heliotrope)	2·6	$6\frac{1}{2}$
	Jasper	2·6	$6\frac{1}{2}$
	Amazon-stone	2·55	6
	Opal	1·9—2·2	6
	Glass	Variable	5

In Division I. malachite is recognisable by the fact that it effervesces when brought into contact with hydrochloric acid; it is also characterised by a peculiarity of structure, being built up of curved concentric layers which are alternately lighter and darker. Ceylonite is of a dark blackish-green colour, almost black in fact; it is harder and heavier than any other stone in this group.

Nephrite and jadeite have figured already in Table 1. Chloromelanite is nothing more than a ferruginous, and therefore rather darkly coloured and heavy variety of jadeite, possessing otherwise all the features characteristic of that stone.

Prehnite is intermediate in density between nephrite and turquoise, and floats in liquid No. 3. It is capable of scratching felspar, but nephrite and turquoise are not. Nephrite and prehnite are both characterised by a fibrous structure; the green of prehnite has usually a pronounced tinge of yellow, while that of nephrite inclines more to grey. Turquoise, on the other hand, is never fibrous in structure.

The first four of the stones which constitute Division V. of this group are all varieties of quartz. Prase is of a dark leek-green colour, and chrysoprase of a pale apple-green; plasma and jasper are of a pure green colour, which is always dark in shade. The colour of true chrysoprase is so characteristic that it is difficult to mistake any other stone for it, except artificially coloured apple-green chalcedony, which, to all intents and purposes, is chrysoprase. Prase, plasma, and green jasper, though scarcely distinguishable when cut and polished, are easily recognised in the rough, or when thin sections are examined under the microscope. All three have about the same value as ornamental stones, and their distinguishing features are not therefore of very great practical importance. Prase is slightly harder and denser than is plasma or jasper. It is usual to apply the term jasper to stones which are perfectly opaque, and plasma to those which are slightly translucent. Plasma with red spots is known as heliotrope.

Amazon-stone is of a bluish-green colour, which is never very dark in shade. In hardness it is a whole degree lower than the varieties of quartz and is easily scratched by any of these. It possesses a distinct cleavage, the existence of which is very evident in rough stones, and is indicated in cut specimens by the presence of numbers of straight cracks.

Green opal (prase-opal) of the colour of prase is not very common, and is characterised by its very low specific gravity.

Glass is distinguishable from all the members of this group by its low degree of hardness.

4. BLACK STONES.

Division.	Name.	Specific gravity.	Hardness.
I.	**Hæmatite**	4·7	$5\frac{1}{2}$
	Ceylonite	3·8	$7\frac{1}{2}$
II.	Diamond	3·50—3·52	10
V.	**Obsidian**	2·5—2·6	$5\frac{1}{2}$
	Jet	1·35	3—4
	Glass	Variable	5

In Division I. hæmatite is possessed of a metallic lustre and a red streak, while ceylonite is sufficiently distinguished by its great hardness and high specific gravity.

Black diamond is characterised by still greater hardness and by the brilliant lustre possessed both by natural crystals and by cut stones.

Obsidian and black glass are of very minor importance; they cannot be distinguished from each other by mere inspection, but require a microscopical examination of thin sections.

Jet feels warm to the touch and can be cut with a knife; the characters in which it differs from vulcanite have been given already under the description of jet.

5. YELLOW AND BROWN STONES.

Division.	Name.	Specific gravity.	Hardness.
I.	Iron-pyrites	5·0	$6\frac{1}{2}$
V.	**Carnelian** (and sard)	2·6	$6\frac{1}{2}$
	Natrolite	2·2—2·3	$5\frac{1}{4}$
	Fire-opal	1·9—2·2	6
	Amber	1·08	2—3
	Glass	Variable	5

Iron-pyrites differs from all the other minerals of this group in the possession of a high specific gravity and a metallic lustre.

Carnelian varies in colour from yellow to yellowish-brown and reddish-brown; when of a pronounced chestnut-brown colour it is known as sard. It frequently exhibits alternating bands of different colours. Opal, though similarly coloured, is never banded, and is less hard and less dense than is carnelian. Natrolite is of an isabel-yellow colour inclining to brown; it has but little lustre and is always fibrous in structure. Amber feels warm to the touch, acquires a strong charge of electricity when rubbed, and can be cut with a knife; it is thus easily distinguished from glass.

Glass is always heavier than natrolite or fire-opal, and is readily scratched by carnelian.

6. ROSE-RED, RED, AND LILAC STONES.

Division.	Name.	Specific gravity.	Hardness.
II.	Rhodonite	3·55	5½
IV.	Lepidolite	2·8—2·9	2—3
V.	Rose-quartz	2·65	7
	Jasper	2·65	7
	Carnelian	2·6	6½
	Fire-opal	1·9—2·2	6
	Glass	Variable	5

Rhodonite is rose-red, but of a darker shade than rose-quartz, which is less dense, harder, and more lustrous than rhodonite.

Lepidolite is of a lilac shade and may be scratched even with the finger-nail, and still more readily with a knife.

Jasper is perfectly opaque and ranges in colour from dark red to brownish-red. Carnelian is translucent and usually of a darker or lighter shade of yellowish-red. Fire-opal is sometimes very similar in colour to carnelian, but is lighter and less hard.

Glass is softer than any member of this group with the exception of lepidolite.

7. STONES EXHIBITING MORE THAN ONE COLOUR.

Division.	Name.	Specific gravity.	Hardness.
I.	Malachite	3·7—3·8	3½
V.	Riband-jasper	2·6	6½
	Agate	2·6	6½
	Heliotrope	2·6	6½
	Amber	1·08	2—3
	Glass	Variable	5

The curved concentric layers of which malachite is built up appear, in section, as bands coloured alternately light and dark green, the latter, indeed, so dark as to be sometimes almost black. It is distinguished from the other minerals of this group by its high specific gravity and small degree of hardness ; and can be identified unmistakably by the fact that it effervesces when touched with hydrochloric acid.   The green solution of the mineral thus obtained imparts a blue colour to the flame of a spirit lamp.

Riband-jasper is opaque, and is characterised by a banded structure, the bands being straight, and coloured alternately green and brownish-red.   Agate is always more or less translucent, the bands of which it is built up show great variety of arrangement and marked contrast in colour.

Heliotrope is the name given to dark green chalcedony (plasma) spotted with red.   Amber sometimes exhibits spots and clouds of a brown or yellow colour ; it is distinguished from the other stones by the characters enumerated in preceding tables.

Glass of variegated colours is sometimes met with, but is not likely to be mistaken for any of the stones included in this group.

8.   STONES WITH METALLIC LUSTRE.

This group comprises iron-pyrites and hæmatite (including iserine) : these minerals are respectively yellow and black in colour, and have been dealt with in Tables 5 and 4.

C.   STONES WITH SPECIAL OPTICAL EFFECTS.

These optical effects are so characteristic of the stones which exhibit them that the consideration of other distinguishing features is almost superfluous.

1. **Stones exhibiting a chatoyant star.**—Star-sapphire, star-ruby and "star-topaz."   These all belong to the species corundum and are generally referred to as star-stones or asterias.   Having a hardness of 9 they are capable of scratching topaz. The specific gravity of each is about 4.   They are respectively blue, red, and yellow in colour, while in other features they are identical.

2. **Stones exhibiting a chatoyant band.**—Girasol-sapphire, girasol-ruby, "girasol-topaz," adamantine-spar ; cymophane, cat's-eye, tiger-eye, and hawk's-eye ; moon-stone and chatoyant obsidian.

The four first mentioned stones, like those under the preceding division, belong to the mineral species corundum, and possess all the features characteristic of that mineral, including great hardness.   Adamantine-spar is characterised by its small degree of trans-lucency and its hair-brown colour, features which distinguish it from the other three stones, which are only slightly clouded.   The remaining stones are all softer and specifically lighter than corundum.

Cymophane or oriental cat's-eye is very similar in appearance to ordinary quartz-cat's-eye, but differs from it both in hardness and specific gravity, the former being $8\frac{1}{2}$ and 7, and the latter 3·7 and 2·65 in the two minerals respectively.   The greater lustre and transparency of cymophane are usually sufficient to distinguish it from quartz.

Moon-stone resembles cymophane in the possession of a band of chatoyant light, but differs from it in being colourless and almost transparent, much lighter (sp. gr. = 2·6) and softer (H = 6).

Tiger-eye and hawk's-eye, like quartz-cat's-eye, have the characters of quartz, namely, a specific gravity of 2·65 and a hardness of 7.   Both possess a markedly fibrous structure, but tiger-eye is of a bright golden yellow colour and hawk's-eye dark blue. Cat's-eye never exhibits the magnificent golden lustre characteristic of tiger-eye, the light reflected from it being more milky and opalescent in character.

The sheen of chatoyant obsidian is always less brilliant than that of the other stones of this group, from which it is also distinguished by a lower density (sp. gr. $= 2\cdot5 - 2\cdot6$), and a lower degree of hardness ($H = 5\frac{1}{2}$).

3. **Stones with a metallic sheen.**—Under this head come hypersthene, bronzite, bastite, and diallage, all belonging to the pyroxene group of minerals. Their hardness is rather less than 6, and their specific gravity varies between $3\cdot3$ and $3\cdot4$. The metallic reflection of hypersthene is copper-red in colour, that of bronzite is bronze-yellow, green, or brown, while that of bastite and diallage ranges from green to brown. Hypersthene is readily identified, but it is less easy and of but little practical importance to distinguish between the remaining stones. Their distinguishing features have been pointed out in the special description of these stones.

4. **Stones flecked with metallic shining points.**—Avanturine-quartz and sun-stone (avanturine-felspar) are flecked with specks exhibiting a red metallic sheen. They are distinguishable by their difference in hardness, the hardness of avanturine-quartz being 7 and that of sun-stone 6. The artificial avanturine-glass is a close imitation of these stones, the metallic reflection being given by the enclosure of small octahedra of metallic copper. These crystals, with their regular triangular faces, are distinctly observable when the glass is examined with a lens, and their presence serves to distinguish the imitation from the genuine stone. Avanturine-quartz possessing all the essential characters of the ordinary red variety, but of a green or blue colour, occurs as a rarity.

5. **Stones exhibiting a play of variegated colours.**—Precious opal, iridescent quartz, labradorite. Opal is usually white, less commonly yellow or red, rarely black; the play of variegated colours extends over areas of varying size. It has a hardness of 6 and a specific gravity which ranges from $1\cdot9$ to $2\cdot2$, features which serve to distinguish it from water-clear, iridescent quartz, with a hardness of 7 and a specific gravity of $2\cdot65$.

Labradorite is dark grey; the play of colour is confined to one surface, and frequently to straight bands running across this surface. It has a hardness of 6 and a specific gravity of $2\cdot7$. In rough pieces of labradorite a perfect cleavage in one direction is always noticeable. It would be impossible to mistake labradorite for opal, or *vice versâ*, but it is possible to confuse precious opal and iridescent quartz, although the characters mentioned above are sufficient to distinguish one from another. Labradorescent felspar is very similar to labradorite, but its play of colour is less fine.

APPENDIX

PEARLS AND CORAL

PEARLS.

THE NATURE AND FORMATION OF PEARLS.—We now come to the consideration of pearls, those objects which, on account of their beauty and costliness, may well rank next to the most splendid jewels. They are products of the life and activity of certain molluscs, the insignificant-looking inhabitants of warm seas and of the rivers and streams of many temperate regions; they thus differ in an important way from precious stones, but like the latter have been used from time immemorial for every kind of decorative purpose. In form they are spherical, ovoid, or pear-shaped, or sometimes quite irregular; they are never transparent, but at most only translucent; they vary considerably in size, and, as a rule, are colourless. Their beauty depends largely upon the peculiar lustre of their surface. Though to appreciate this beauty it is by no means necessary to be familiar with the minute structure and mode of origin of pearls, yet a knowledge of these is an aid to the comprehension of their characteristic features. Before beginning the consideration of these features, therefore, we will describe the mode of origin of the pearl.

A substance which stands in the closest relationship to pearls, and which, moreover, has derived from them its name *mother-of-pearl*, is also much used in decorative work. It has a lustre exactly similar to that of pearls, and frequently exhibits a more or less marked and beautiful play of colours. It forms the inner coating of the shells of many bivalve molluscs, and a mass of similar nature occurs also in the shell of certain univalve molluscs. As the formation of pearls by univalve molluscs is a rare phenomenon, and without significance from the point of view of pearl-fishers, we will confine our attention to the pearl-forming bivalves.

The thin outermost layer, the periostracum or so-called "epidermis," of the shell of any bivalve mollusc, consists of a horny material known as conchiolin. Beneath this comes the shell proper, consisting of two layers which differ in structure. The outer or prismatic layer is formed of minute prisms of calcium carbonate separated by thin layers of conchiolin; the inner forms the internal part of the shell, and is built up of alternate layers of calcium carbonate and conchiolin arranged parallel to the surface. The laminated internal layer is what is known as nacre or mother-of-pearl, and varies in thickness in different molluscs. The laminæ of which it is built up consist of that variety of calcium carbonate which is known to mineralogists as aragonite. The calcareous prisms, which form the middle layer of the shell, consist, on the other hand, of calcite, another modification of calcium carbonate which is softer and lighter than aragonite.

The laminæ, of which the pearly or nacreous layer of the shell is built up, are small compared with the size of the shell, and overlap one another something like the tiles on the roof of a house. This finely laminated structure is the cause of the peculiar lustre which characterises mother-of-pearl, a lustre which is exhibited by all substances having the same structure, that is to say, all substances which are built up of transparent, overlapping laminæ.

The laminæ which form the nacre or mother-of-pearl layer of the shell do not lie in planes parallel to the surface of the shell, but are always more or less bent and curved. Their edges intersect the inner surface of the shell, and produce a very fine striation, which can sometimes be seen with a simple lens, but which usually requires the help of the compound microscope. These striæ or furrows take zig-zag or quite irregular courses, sometimes indeed forming small closed rings; the distance between them averages $\frac{1}{3000}$ inch, but varies within small limits. It is to these striæ that the play of colours so characteristic of mother-of-pearl is due, while the peculiar lustre of the substance is due to its laminated structure, as has been already explained. This play of colours, then, is quite independent of the presence of pigment of any kind. It is due to purely physical causes, the fine striæ so acting upon the rays of ordinary daylight or candle-light as to split them up into their coloured constituents; these reach the eye singly, and so produce the sensation of colour. That this is the true explanation is proved by the fact that the same play of colours is to be seen upon sealing-wax after it has been pressed upon the natural surface of the mother-of-pearl layer, or, better still, upon a section of this layer cut obliquely to the surface, and has thus received an impression of the striæ.

The body of a bivalve mollusc is produced into two lateral flaps, the so-called mantle, which lie in immediate contact with the inner surface of the shell. The different parts of the shell of the mollusc are secreted by the mantle, the cells of which have the power of separating out the calcium carbonate dissolved in the water in which the animal lives. Different areas of the mantle have somewhat different functions in this respect: thus the edge of the mantle secretes the periostracum, the so-called "epidermis"; the outer surface lays down the calcareous laminæ, of which the nacre is composed; while the prismatic layer, between the periostracum and the mother-of-pearl or nacre, is secreted by a narrow zone round the margin of the mantle.

This, then, is the process of shell-formation which goes on under ordinary conditions, and so long as these obtain no pearls are produced. The formation of a pearl is the response of the mollusc to the stimulus afforded by a local irritation of the mantle. Being unable to eject the cause of the irritation, the mollusc obtains relief by enveloping it with a deposit of mother-of-pearl substance. The rounded aggregation of mother-of-pearl substance so produced is then known as a pearl.

That the formation of pearls is an abnormal occurrence in the life of a mollusc is shown by the fact that among the pearl-forming molluscs, only about one in thirty or forty is found to contain pearls. Moreover, the observations of pearl-fishers all point in the same direction; for they state that there is little prospect of finding pearls in a well-formed, normal shell, and that the shells most likely to contain pearls are those which are irregular and distorted in shape, and which bear evidence of having been attacked by some boring parasite. That the formation of pearls is caused by some disturbance of the normal conditions of life may, therefore, be regarded as a well-established fact; but the exact causes which bring about this secretion are not in all cases satisfactorily explained.

The nucleus of many pearls is a tiny grain of sand, and in such cases it is obvious that this foreign body set up an irritation of the mantle and caused an abnormal secretion of calcium carbonate, in the same way as a particle of dust in the eye causes a copious flow of water in that organ, or as the presence of a trichina (flesh-worm) in a muscle-fibre causes the latter to secrete a calcareous cyst around the intruder. Many observers, following

Möbius, the author of a valuable work entitled "The True Pearls," consider the formation of pearls to be due most frequently to the irritation caused by the presence of a grain of sand. Of fifty-nine pearls examined by Möbius himself, and obtained both from sea and from fresh-water molluscs, a certain number were found to possess a nucleus consisting of a crystalline, granular, calcareous substance, but that of the majority was brown in colour and of organic origin, being possibly the remains of small intestinal worms, while in no single case was the nucleus found to consist of a grain of sand. Not only is the formation of pearls actually caused by the presence of parasites in the mollusc, but also by the attacks of water-mites, small fishes, boring sponges, and worms, which penetrate the shell from without, by the growth of algæ, and even by the eggs of the mollusc itself.

Pearls formed by different parts of the mantle often differ in shape and appearance. For example, a pearl formed in the inside of the soft part of the mantle, perhaps in response to the irritation caused by a parasite of some kind, will be more or less perfectly spherical, and will lie free in the mantle. If, however, the pearly matter is laid down around the orifice made by some boring external parasite, the result will be a wart-like protuberance. In the former case is produced a pearl proper, which can be used as an ornament without preliminary treatment ; in the latter, the so-called *button* or *wart-pearls*, which are very irregular in shape, but, notwithstanding this, are detached from the shell and utilised in various ways as *fantasy-pearls*. In exceptional cases these button-pearls are hollow and contain a beautiful spherical pearl, lying freely in the cavity.

The number of pearls found in a single mollusc varies according to circumstances. If the mantle is irritated at one spot only, then only one pearl is formed ; if at several spots, then several pearls will be formed ; and in exceptional cases a single shell may contain a large number. Among remarkable cases of this kind is that of a pearl-oyster from the Indian Ocean, which contained eighty-seven pearls of good quality ; while in another from Ceylon sixty-seven of various sizes were found. As a general rule, the greater the number of pearls found in a single shell the smaller they are.

The first person to point out that the structure of pearls is identical with that of the molluscan shell was the French naturalist Réaumur (1683–1757). This structure can be best made out by examining under the microscope a thin section of a pearl taken right through its centre. It will be seen to be built up of concentric coats like an onion. Each of these concentric layers consists of overlapping laminæ, exactly like those which build up the mother-of-pearl layer in the shell of the mollusc. This concentric structure points to the fact that the secretion of calcium carbonate was not continuous, but that there were longer or shorter intervals of rest, coinciding, perhaps, with certain seasons of the year, when growth also was arrested. Thus, each concentric layer corresponds to a period of growth, and each interruption between neighbouring layers to an interval of rest. These concentric coats are sometimes visible to the naked eye, but, as a rule, they are of microscopic thickness. When a pearl is raised to a red-heat, the concentric coats peel off and the laminæ separate from each other. Much the same thing happens in the case of pearls which have been worn for a long time strung together to form a necklace. At the spot where each pearl is perforated, the concentric layers begin to fall off as scales, so that the orifice through which the string passes becomes gradually wider and wider. The substance of pearls has a hardness of between 3 and 4 on Mohs' scale, so that being comparatively soft they cannot be expected to resist a large amount of hard wear.

The majority of pearls, and certainly all the most beautiful, are formed in this way, and with the exception of the nucleus consist wholly of layers of nacre. Not infrequently, however, pearls are met with in which the nucleus is surrounded by a dark layer corresponding with the periostracum or "epidermis" of the shell; enclosing this is a layer showing a columnar or fibrous structure, which agrees in every respect with the prismatic layer of the shell, while the external layer only consists of finely laminated mother-of-pearl or nacre. Such a pearl, therefore, has the structure of a molluscan shell, but the layers are arranged in the reverse order, periostracum inside and the nacreous layer outside. In some cases the nacreous layer is entirely absent, and the outermost layer is constituted by the prismatic layer or by the periostracum, the pearl then being of a dark brown or black colour, without lustre and, therefore, without value. Not infrequently also there are several nacreous layers, separated from each other by prismatic or conchiolin layers.

One can easily see the connection between these differences in structure and the processes by which the pearls were formed. If a pearl arises wholly in that region of the mantle which secretes mother-of-pearl, as is generally the case, it will consist wholly of mother-of-pearl substance, the nature of which has been already described. A pearl formed in this way does not, however, remain indefinitely in the same position. From one cause or another it may be brought into contact with that part of the mantle which secretes the prismatic layer, when it will become invested with a layer of this description, or with the extreme edge of the mantle, when a layer corresponding to the periostracum of the shell will be deposited upon it. If the object, which eventually becomes the nucleus of the pearl, lie first in contact with the edge of the mantle, a layer of material corresponding to the periostracum will first envelop it. If the pearl during its growth moves slowly inwards it becomes coated successively with a prismatic and a nacreous layer, while if it move back again these same layers will be again laid down, but in the reverse order. In this way is attained great variety in the arrangement of the concentric coats of pearls, this variety depending solely upon the way in which the pearl moved about within the shell of the mollusc.

The pearl is not only identical with mother-of-pearl in structure, but also in **chemical composition, hardness, and specific gravity.** Like that substance it consists of that modification of calcium carbonate which is known as aragonite. Besides calcium carbonate, there are always present small amounts of other inorganic substances, and, in addition, up to 12 per cent. of conchiolin, the organic material of which the periostracum of the molluscan shell is formed. Layers of conchiolin alternate with layers of calcium carbonate, binding the whole together. The specific gravity of fresh, white, brilliant sea-pearls varies between 2·650 and 2·686. The hardness is nearly 4; it varies somewhat in different pearls, but is always rather less than that of mother-of-pearl. Owing to the admixture of conchiolin, both the hardness and specific gravity of pearls are rather less than those of aragonite.

As a result of their composition pearls dissolve readily in acid, even in acetic acid, with the evolution of carbon dioxide, which causes a brisk effervescence. Upon this is founded the story that Cleopatra, the Egyptian queen, at a banquet dissolved a priceless pearl in vinegar and drank the solution. In ordinary table-vinegar, however, the acetic acid is too weak to dissolve completely a pearl even of small size in the time that a banquet might be supposed to last, so that the story is probably apocryphal. Moreover, pearls do not dissolve completely in acid, the calcium carbonate only is extracted and the conchiolin left behind as a scaly, soft, somewhat swollen, but still shining mother-of-pearl-like mass, which has the form, size, and colour of the original pearl, but upon which the acid has no further effect.

Pearls are also affected by perspiration from the skin, and if they are much handled or worn for long in contact with the skin they gradually lose their lustre and much of their beauty. Old, much-worn pearls never possess the freshness and beauty of newly-fished maiden-pearls so-called. Not only do pearls, which are strung together, suffer from the rubbing of one upon another, and from the friction of the string, but the organic constituent, like all other animal substances, in course of time completely decays. An example of the way pearls are affected by the lapse of time is furnished by the state of those found in the grave of a daughter of Stilicho, a Roman statesman and general. The pearls had lain in the grave from the year 400 to the year 1544, a period of eleven centuries, and when brought to the light of day they fell instantly to dust. In this respect, therefore, pearls are in no way comparable with the unchanging precious stone, which preserves its brilliancy and lustre through age after age.

The perishable nature of pearls, as much a result of their softness as of their organic origin, is the more regrettable, seeing that when once the outer surface of a pearl is damaged its beauty is irrevocably lost. A precious stone which has sustained serious injury can be restored to all its original beauty by re-cutting and re-polishing, but not so with a pearl. These beautiful objects, therefore, need to be handled with great care, in order to preserve the surface layer in its natural condition for as long as possible. This outer layer when discoloured or damaged may, indeed, be sometimes peeled off, a perfect but smaller pearl being thus obtained. This operation, however, requires all the care of a skilled artist, and even then is seldom completely successful.

The **surface** of a pearl, like that of mother-of-pearl, is not perfectly smooth, but shows many irregularities, elevations, and depressions, delicate ridges and grooves, and so forth ; these are all of microscopic dimensions, and, as we shall see further on, have an important bearing on the appearance of the pearl.

The surface especially exhibits a peculiar **lustre**, which is not very brilliant, but beautiful and delicate, and which defies verbal description ; it is known to jewellers as the " orient " of a pearl, corresponding to the " fire " of a diamond. This pearly lustre is due, as in mother-of-pearl itself, to the laminated structure already described. The not wholly transparent, but very translucent layers lying near the surface allow some light to pass through, which is reflected outwards again from the deeper layers. This light reaches the eye in company with that reflected directly from the surface, and the two together produce the impression which we have learnt to call a pearly lustre. The thinner the calcareous laminæ are the more beautiful is the lustre, and the more valuable the pearl.

Individual pearls differ considerably in lustre ; those formed by marine molluscs are far superior to the pearls formed by the fresh-water pearl-mussel, which, compared with the former, appear dull and lifeless. Pearls which are unusually brilliant are also somewhat harder than more ordinary specimens, this character depending upon the more or less intimate association of the single layers.

As we have already seen, the lustre and brilliancy of pearls disappear with the lapse of time, and these objects, therefore, undergo a constant depreciation in value. Many efforts have been made to discover a means whereby the freshness and beauty of maiden-pearls may be restored, but in vain. Any attempt to remove the outer coat of a pearl, whose lustre has been dimmed by time, is doomed to disappointment, for the underlying coat is usually as dull and lustreless as the eye of a dead fish. It is, therefore, very rarely that a dull pearl can be improved by peeling off its outer coat. Attempts to restore their pristine beauty to old and faded pearls have been made by immersing them in the sea for long periods, by allowing them to be picked up and eaten by hens and doves, and in other foolish and irrational ways, but unsuccessfully. The inimitable freshness and delicacy of maiden-pearls once lost is lost for ever.

No pearl is completely transparent, but among different specimens there are many degrees of **translucency**. Different qualities of pearls are described, as in the case also of diamonds, as being of different " waters," and the difference between them depends upon the amount of light transmitted in each particular case.

The **colour** of a pearl has an important bearing on its value. Those used for ornaments are mostly white, yellowish-white, or bluish-white, more rarely reddish- or blackish-grey. A pearl with a perfect pearly lustre, or, as the jewellers say, a "ripe" pearl, has the colour of the mother-of-pearl layer in the shell in which it was formed, but each pearl has sometimes a certain individuality as regards colour. The pearls formed by the true pearl-oyster (*Meleagrina margaritifera*) are white, and it is these and the silvery, milk-white pearls which are most valuable. The smaller the dimensions of the microscopic irregularities in the surface of a pearl the finer will be its appearance, for on these rugosities of the surface depends the scattering of the incident light and the gleaming white colour of the pearl. The fine colour of "ripe" pearls is due partly also to the fact that they consist wholly of colourless nacre. Should there be present a nucleus consisting of the substance of the prismatic layer, it shines through the strongly translucent nacreous layer, making the pearl appear dull and dark, and this is specially the case when the outer pearly layer is thin. Such grey or brown translucent pearls are described as being "unripe." Of a large and beautiful Indian pearl it was said that it rolled about on a sheet of white paper "like a globule of quicksilver," surpassing even the metal itself in lustre and whiteness. Silvery, white, transparent pearls of this description are classed as pearls of the finest water : they always possess a thick outer nacreous layer. Many true pearls, however, are slightly tinged with yellow, or are even of a pronounced yellow colour, this being more frequently the case with those from the Persian Gulf than with the pearls from Ceylon. These yellowish pearls are much esteemed in Asia, India, China, &c., being considered to be less perishable than the white. It has been said that white pearls, if enclosed in the shells of decaying molluscs, acquire a yellow tinge, but according to special research in this direction this is not, or at any rate not always, the case. Pearls with a faint bluish tinge are also frequently met with, this colour being the same, as is also the case with yellowish pearls, as that of the mother-of-pearl layer of the shell in which such pearls were formed.

Beautiful black pearls are sometimes found in molluscs from the South Seas and the Gulf of Mexico, having been formed probably near the border of the mantle. They are the hardest of all pearls, and when of a beautiful and uniform colour and perfect form are worth almost as much as pearls of the purest white. Such black pearls are used in Europe in mourning jewellery. Intermediate between white and black are lead-coloured pearls, which are not uncommon. Reddish-brown pearls containing some iron come from Mexico. The hammer-oyster (*Malleus*) from the Gambia Islands yields pearls with a bronze-like sheen. Not infrequently greyish-brown pearls, in which the nacreous layer is absent, are found in the fresh-water pearl-mussel, *Unio margaritifer*. In the fan-mussel, *Pinna nobilis*, are found light and dark brown pearls, in some of which also the nacreous layer is absent, the outer coat consisting of a substance corresponding with the prismatic layer of the shell. The same shell, however, may contain garnet-red pearls, in which the pearly nacreous layer is present, and which are highly prized both by the Hindoos and by the Jews. Pale, rose-red pearls with delicate white wavy lines, like the most beautiful pink velvet, come from the Bahamas. Pale blue pearls are often met with in the edible mussel (*Mytilus edulis*) ; greenish-white and pale rose-red pearls in *Spondylus gædaropus* ; violet in the ark-shell (*Arca noæ*) ; purple in *Anomia cepa* ; and lead-coloured in *Placuna placenta*.

Dull white pearls, in which the beautiful lustre of typical pearls is absent, are probably produced by all molluscs, the shells of which have a white inner surface. Such pearls are found in the pilgrim-shell (*Pecten jacobæus*), the giant-clam (*Tridacna gigas*), in many varieties of the common fresh-water mussels (*Unio* and *Anodonta*), in the razor-shell (*Solen*), and many others. Dull white pearls have indeed been found in the edible oyster (*Ostrea edulis*) in spite of the absence of mother-of-pearl in its shell; and it has been related by a dredger how he found a pearl in an oyster he was eating, and obtained for it 22 thaler (66s.). The occurrence of pearls in certain univalves has been mentioned already. The large West Indian *Strombus gigas*, and the East Indian *Turbinella scolymus* yield very beautiful rose-red pearls, in which, it is true, there is no mother-of-pearl layer, and which, therefore, are not true pearls. They are distinguished from the latter by the fact that their colour, like that of the univalve shells in which they originate, gradually fades away with the lapse of time.

Some pearls exhibit the colour-effects characteristic of mother-of-pearl, but only to a slight degree and of very pale bluish, greyish, and reddish shades. In such pearls the outer layer of mother-of-pearl substance is not continuous over the whole surface, but is confined to small, irregularly bounded areas separated by areas from which it is absent. In addition to the elevations and depressions usually present on the surface of a pearl, there are also delicate and irregularly curved grooves, which pursue either approximately parallel courses or form closed curves of irregular form. The colours exhibited by such pearls are due, as in mother-of-pearl, to physical causes connected with the structural peculiarities of the surface.

The value of a pearl, and the uses to which it is applied, are considerably influenced by its **form**, and in this respect there is a certain amount of variety among pearls. They may be perfectly spherical, more or less ovoid, or pear-shaped. Pear-shaped and elongated ovoidal pearls are referred to as *pear-pearls*, and the former more especially as *bell-pearls*, while spherical pearls are known as *pearl-drops* or *pearl-eyes*. Ovoidal pearls sometimes contain two nuclei, each of which is invested by its own series of concentric coats, only the outermost coats of the pearl enclosing both nuclei. We have evidently, in such cases, two originally separate pearls which have become enclosed in a common coat, and thus formed a single ovoidal pearl.

Pearls of very irregular form are known as *baroque pearls* ("barrok pearls"), and are specially abundant in the fresh-water pearl-mussel. These also are used as ornaments and for other purposes, but are less valuable than those of a more regular form. As extreme examples of irregularity in form may be mentioned two pearls described by the Parisian jeweller Caire, one of which resembles in form the head of a dog, and the other the Order of the Holy Ghost.

The **size** of pearls is very variable. One of the largest known is the property of the Shah of Persia; it is pear-shaped in form, and is 35 millimetres long and 25 thick. A still larger one formed part of the famous collection of pearls and precious stones of Henry Philip Hope, and is figured and described by B. Hertz in his catalogue (London, 1839) of this collection. This pearl weighs 3 ounces (troy), or about 454 carats; it is irregularly pear-shaped in form, measuring 2 inches in length, $4\frac{1}{2}$ inches in circumference at its thicker end, and $3\frac{1}{4}$ inches at its narrower end, and is thus almost as large as a hen's egg. For a length of about $1\frac{1}{2}$ inches this gigantic pearl is of a fine bright "orient," the remainder being of a fine bronze tint or dark green shaded with copper colour. A portion of the shell on which it grew is left adhering to it, but this of so fine an "orient" and so well polished that it is scarcely recognisable as such. In the Austrian Emperor's crown there is a large pearl weighing 300 carats, but it is of medium quality only. A pearl brought to the Spanish Court from Panama in the sixteenth century is said to have been as large as a pigeon's egg.

Admittedly the most beautiful of large pearls is one in the museum of Zosima in Moscow, which has received the name " La Pellegrina." It was found in India, is perfectly spherical, of a pure white colour, and almost transparent, and weighs 28 carats. More remarkable than any hitherto mentioned is the " Great Southern Cross," which consists of nine large pearls of a pure white colour and beautiful lustre, naturally joined together to form a cross over an inch in length. This beautiful natural object was found in 1886 in a pearl-oyster off the north-west coast of Western Australia. It was shown at the Colonial Exhibition in London in 1902, and is valued at £10,000.

The pearls of ordinary, every-day occurrence are very much smaller than the exceptionally large specimens described above. Special terms are used for pearls of particular sizes: thus those with the dimensions of a walnut or thereabouts are termed *paragon-pearls*, and those of the size of a cherry as *cherry-pearls*. *Piece-pearls* are smaller, but not too small to be handled and dealt with separately, each exceeding a carat in weight. *Seed-*, *shot-* or *ounce-pearls* are dealt with in the trade only in parcels ; and the same is the case with *sand-* or *dust-pearls*, the largest of which are no bigger and sometimes even less than a grain of millet. Fine Indian pearls are usually from one and a half to three times the size of a pea.

THE APPLICATION OF PEARLS.—This differs in nowise from that of precious stones, which is mainly for purposes of ornament. Pearls have been valued as ornaments from the earliest ages, and the favour in which they were held, especially by the Romans, is evidenced by the writings of ancient authors.

Although pearls and diamonds are alike in the purpose to which they are devoted, they differ very widely in that while the latter need to be cut and polished to fit them for use as ornaments the former need no preliminary treatment, for they leave the hand of nature with a beauty which cannot be enhanced by artificial means. Polishing adds no lustre to the surface of a dull pearl, nor can the form of an irregular pearl be made more regular by grinding without destroying its lustre.

Pearls are set in various ways, but never *à jour*, because of their imperfect transparency. Fine pearls of large size are often mounted with a border of small diamonds or coloured stones, and, on the other hand, a border of small pearls makes an effective setting for a fine diamond or other precious stone. More frequently, however, pearls are bored and worn in strings and ropes. The perforation is easily performed, the substance of pearls being comparatively so soft, but as we have seen already the pearly laminæ are very apt to scale off around the perforation. To attain the most beautiful effect possible in a string of pearls, only such as associate well in form, size, and colour should be strung together. It is by no means necessary that each pearl should be the exact counterpart of all the others, for small differences between individual pearls do not affect the beauty of the *tout ensemble* of a pearl necklace. For this reason, it is usual when choosing one pearl from a string to cover over the adjacent pearls in order that their proximity may not prejudice or enhance the effect of the pearl to be chosen.

The fantastic forms often taken by baroque pearls make it possible to convert them with slight additions and alterations into caricatures of men and things. A large collection of objects of this kind, among which are some pearls of unusual size, is preserved in the " Green Vaults " in Dresden. One represents the figure of a Court dwarf, whose body is formed of a suitably-shaped baroque pearl, the size of a hen's egg. A proof that popular taste for articles in the rococo style has not yet died out is furnished by the fact that baroque pearls are sometimes even now treated in the same way.

Not only button-pearls, which are found attached to the shell, but also the equally irregular fantasy-pearls are sometimes of sufficient beauty to admit of their employment in jewellery. They usually have a hemispherical form when detached from the shell, and are frequently cemented together by their flat surfaces in pairs and worn in ear-rings, or strung together for necklaces. The absence of lustre on the flat surface of such pearls serves to distinguish them from pearls which, though formed free from the shell, are yet hemispherical in form.

The **value** of fine pearls is quite comparable with that of the costliest gems, and enormous sums have been paid for large pearls of singular beauty. As in the case of precious stones the value of pearls varies with their size, form, and the general beauty of their appearance. The more perfect the form of a pearl the more valuable does it become, and other things being equal a pearl of irregular form is considerably less valuable than one which is perfectly regular. Arranged in order of the degree of esteem in which they are held, we have: first, perfectly spherical pearls; secondly, those of an equally symmetrical pear-shaped form ; and, finally, ovoidal or egg-shaped pearls. A pearl of the " first water " must possess, beside a symmetrical form, a smooth surface and a perfect " orient," the latter, a feature which depends upon the existence of a finely laminated structure ; it must be free from blemishes and fractures, very translucent, and possessed of a fine white colour and a perfect pearly lustre. Provided the lustre is good, fine black pearls, as also those showing deep shades of red, yellow, and other colours, are as costly as those of the purest white. Pearls of even the most beautiful colour and perfect form if they are lacking in " orient " are comparatively valueless.

The prices of exceptionally large or beautiful pearls are subject to no fixed rules, and depend rather upon the eagerness of the purchaser and the length of his purse. The prices of pearls of ordinary size, such as are bought and sold every day, are governed by the laws of supply and demand, and the same general principles apply as in the case of precious stones. The relation between the size and the price of a pearl is in moderately close agreement with Tavernier's rule; that is to say, the price of a pearl varies with the square of the weight, the usual unit of weight being the " pearl-grain " ($=\frac{1}{4}$ carat). A pearl of unit weight will be worth from 2s. to 10s., according to the quality ; the appended table of prices applies to pearls of a quality such that a pearl of unit weight is worth 6s.

Weight in "pearl-grains."					Value in shillings.
$\frac{1}{3}$	.	.	.	.	$\frac{1}{3} \times \frac{1}{3} \times 6 = \frac{2}{3}$
$\frac{1}{2}$	.	.	.	.	$\frac{1}{2} \times \frac{1}{2} \times 6 = 1\frac{1}{2}$
$\frac{2}{3}$	.	.	.	.	$\frac{2}{3} \times \frac{2}{3} \times 6 = 2\frac{2}{3}$
1	.	.	.	.	$1 \times 1 \times 6 = 6$
2	.	.	.	.	$2 \times 2 \times 6 = 24$
3	.	.	.	.	$3 \times 3 \times 6 = 54$
4	(= 1 carat)	.	.	.	$4 \times 4 \times 6 = 96$

To convey an idea of the relative sizes of pearls of different weights, it may be mentioned that one weighing 3 carats has approximately the size of a pea.

The price of a string of carefully matched pearls is more in proportion than that of a single pearl would be, for the reason that there is often considerable difficulty in finding a sufficient number of pearls of appropriate size, quality, form, and colour. Considerable time is often needed to make up the required number, during which those already procured and paid for lie unused and depreciating in value. Möbius states that in his time (the end of the 'fifties) a string of from seventy to eighty three-carat pearls was worth from 4000 to 6000 thaler, that is to say, about 70 thaler (old German thaler = 3s.) for each pearl. This was approximately twice the price current at that time for a single pearl of the same size and quality.

In conclusion, it will be interesting to note the value of some of the more famous pearls. That in the Hope Collection was valued at £12,500. At the valuation made in 1793 of the French Crown jewels, a round maiden-pearl weighing $27\frac{5}{16}$ carats, and of magnificent lustre, was valued at 200,000 francs (£8000); while 300,000 francs was the value attached to two well-formed, pear-shaped pearls of very beautiful " water" which together weighed $57\frac{11}{16}$ carats; and 60,000 francs to four differently shaped pearls the combined weight of which was $164\frac{6}{16}$ carats. Another pear-shaped pearl weighing $36\frac{10}{16}$ carats, and therefore much larger than the one first mentioned, was valued at 12,000 francs only, on account of its being flat upon one side. At the International Fishery Exhibition held at Berlin in 1880, the Berlin jewellers exhibited a string of yellowish Indian pearls worth £4000, a string of white pearls from Panama which cost £5000, and a string of black pearls from the Pacific Ocean which was valued at £6000. The value attached to the " Great Southern Cross " pearl found off the north-west coast of Western Australia in 1886 is £10,000.

PEARL-FISHING.—Though as we have seen there are many molluscs which occasionally produce pearls, there are only two species which produce these objects in numbers sufficient to make the systematic collection of them a profitable industry. By far the more important of the two is the marine pearl-oyster *Meleagrina* (= *Avicula*) *margaritifera*, which inhabits the warm seas of many tropical regions. The other and less important mollusc is the fresh-water pearl-mussel *Unio* (= *Margaritana* = *Alasmodon*) *margaritifer*, and a few near relations inhabiting the streams of northern regions, or, at any rate, of extra-tropical countries. By far the largest number of pearls, and those also of the finest quality, are, and have always been, obtained from the pearl-oyster, which in addition supplies the bulk of the mother-of-pearl for industrial purposes. Only a small number of pearls, generally of poor quality, are obtained from the pearl-mussel.

The Pearl-Oyster.—According to the opinion of the majority of conchologists the salt-water pearl-forming molluscs from almost all regions of the world belong to one and the same species. Individuals from different regions do, indeed, show differences in the size and thickness of the shell, in the roughness of the exterior, the colouring of the inner surface, &c., but these differences are not of specific importance. A small, thin-shelled form from which no mother-of-pearl is profitably obtained is sometimes distinguished from the large thick-shelled *Meleagrina margaritifera* by the name *Avicula margaritifera*, but there are many transitional forms between the two, so that they cannot be sharply separated.

The pearl-oyster, like the common oyster, lives with large numbers of its fellows, forming the so-called oyster-banks. These lie usually from 3 to 5 fathoms, occasionally from 6 to 10 fathoms below the surface of the water, rarely any deeper. The bank has a foundation of calcareous material; very often, indeed, it is built on a coral-reef. The molluscs are not free but attach themselves firmly to any suitable object by means of the beard, or byssus, a bundle of tough horny threads, which reach the exterior through a hole in the hinge line of the shell. The pearl-oyster banks are inhabited also by corals, univalve molluscs, and many other marine animals; the temperature of the water in which the whole community lives is never much below 25° C. (= 77° F.).

The pearl-oysters are brought up from the sea-bottom by divers, who descend into the depths, in some cases, with no artificial appliances to aid them in their difficult and dangerous calling. The provision of the best and newest divers' apparatus makes their descent into not too great depths comparatively easy and safe, and renders a longer stay under water possible. Pearl-fishing is not a profitable industry in every place where the pearl-oyster is to be found, and, moreover, the individuals from which numbers of pearls are obtained do not always supply in addition mother-of-pearl of good quality. The best mother-of-pearl known, which is worth from £80 to £150 per metric ton (= 1000 kilograms), is obtained from the pearl-oysters found in the sea surrounding the Sulu Islands, between the Philippines and the north point of Borneo; comparatively few pearls are yielded by the pearl-oysters found in this region. On the other hand, the pearl-oysters inhabiting the Gulf of Manar, off the island of Ceylon, which yield a very large number of the finest pearls in the world, have shells so thin that they are useless in the mother-of-pearl industry.

These pearl fisheries in the Gulf of Manar, on the north-west coast of **Ceylon**, also known as the fisheries of Aripo, after an old fort, are more important than any other The fishery is not confined entirely to this gulf, a small number of pearls being obtained also from the Coromandel Coast, lying opposite on the mainland of India. The sea in this district is well sheltered by the islands and by sand-banks lying to the north, so that the pearl-oysters are undisturbed by buffeting waves. The most important banks lie between 8° 30' and 9° north latitude, at a distance of three miles from the coast; the largest are two miles long and two-fifths of a mile wide. The banks extend for a length of ninety miles along the coast, and the furthest out are about twelve miles from the shore. The best shells lie at a depths of from 18 to 40 feet beneath the surface of the water. Pearl-oysters have been brought up by divers from this spot for time immemorial, and much the same methods as were adopted in the time of the Romans, and even earlier, are still employed. Since the earliest times these pearl-fisheries have been under Government control, first native, then Portuguese, then Dutch, and finally English. The head-quarters of the divers in these districts is Kondachchi, and during the fishing-season, six weeks in March and April, when the sea is quietest, the place is inhabited by from 15,000 to 20,000 people. These, drawn from all parts of India, include divers, fishers, shark-charmers, merchants, and so on, who help to people this strip of shore, which at other seasons of the year is entirely unoccupied.

As many as 300 boats, each carrying ten divers, row about the fishing-ground, the boundaries of which are accurately marked out under the superintendence of the auth orities Each boat can collect in one day on an average 20,000 oysters, and a yield of two or three shillings' worth of pearls per thousand oysters is sufficient to make the thing pay, while a haul which yields half as much again is considered good. The oysters are seldom opened directly they are landed, but are safely secured and allowed to rot, the atmosphere being pervaded during the process with an indescribably evil smell. The pearls are collected as they fall out of the decaying masses, and are there and then sorted according to size by means of sieves provided with meshes of various sizes. Many are bored and sold on the spot. The number of pearls suitable for ornaments is but a small proportion of the total yield. Those which are not adapted for ornaments are used in India and elsewhere in the East for medicinal purposes, serving, for example, as a costly substitute for the ordinary shell-lime used in the preparation of betel, a luxury which only the richest can afford. The pearl-oyster of Ceylon is small, no larger than the palm of the hand, and the shell itself is so thin that although the inner surface is extraordinarily lustrous and beautiful it has no value as a source of mother-of-pearl.

Pearl-fishing is not restricted in this district to the Gulf of Manar, but is carried on also in Trincomalee, on the east coast of Ceylon, and at other places, all, compared with the Gulf of Manar, unimportant. No part of a district is fished oftener than once in six or seven years, this arrangement giving the fishing-grounds ample time to become repopulated, by the development of young individuals. If an oyster-bank is allowed to remain much longer than seven years many dead individuals are found in it, so that this period appears to be the normal limit of their life.

The pearl-oyster banks in the **Persian Gulf** are also of great importance, and have been known and fished since ancient times. They are situated chiefly on the Arabian side, and are wholly controlled by the Arabs, who do not brook any interference with their rights. The methods they employ are exactly the same as those adopted in Ceylon. Very productive banks are situated in the neighbourhood of the Bahrein islands on the Arabian coast (about 26° North latitude), and further south, skirting the Pirate coast for a length of seventy geographical miles. At both places the banks lie at a depth of about 40 feet. Oysters, from which beautiful pearls are obtained, are found also at a greater depth off the Persian coast opposite, one such place being situated between the islands Kharak and Gorgo, north-west of Bushire. The pearl-oysters of the Persian Gulf are double the size of those of Ceylon, thicker and also smoother on the outside. The pearls found in this region have a yellowish colour as compared with the pure white of Indian pearls, but in other respects they are in no way inferior.

Pearl-oyster banks appear to be scattered over the whole of the **Red Sea**, except the most southerly part, and at several places flourishing fisheries have been established. One such is situated near the island of Dahalak, not far from Massaua, and another near the Farsan islands, which lie just opposite, off the Arabian coast. Pearls of an inferior quality are obtained also from the sea near Jedda, west of Mecca, in Arabia. The pearl-oysters of the Red Sea yield a large quantity of excellent mother-of-pearl.

Every part of the **Indian Ocean** yields pearl-oysters, but the fisheries have not the importance of those already mentioned. Pearls of small size and inferior quality are obtained from the Gulf of Cutch, on the north coast of the Kathiawar peninsula, and from the sea near Kurrachee, on the western mouth of the Indus; while pearls of rather better quality come from the coasts of the Mergui Archipelago, off the coast of Lower Burma. The Sulu Islands have been mentioned above; from them and from the neighbouring Tawi-Tawi Islands pearls equal to Indian are obtained. The pearl-oysters of this region are, however, more valuable as a source of mother-of-pearl. The shells are very large, averaging ¾ pound in weight, and the largest weigh as much as 2 pounds. They are remarkable not only for their size but also for the purity and lustre of the mother-of-pearl layer. In reference to the fact that many of these shells reach the market through Macassar, they are known as Macassar shells. Pearls are also obtained from oysters found off the coasts of New Guinea and the neighbouring island groups, especially the Aru Islands.

The pearl and mother-of-pearl fisheries of the north-west coast of **Western Australia** have constituted for the last quarter of a century one of the most valuable assets of the Colony. The value of the pearls exported in 1896 was estimated at £20,000 and that of pearl-shell at £30,000. Two definite varieties of the pearl-oyster are found off these coasts: the larger inhabits the north-west coast as far as Exmouth Gulf, in latitude 22° S.; while the smaller is confined to Sharks Bay, in latitude 25° S. The largest and finest pearls are yielded by the larger variety; those obtained from the smaller shells are perfect in form and lustre, but are of a bright golden- or straw-yellow colour, and find a readier sale in India and China than in Europe. The smaller variety is, or was formerly,

the more abundant, and the oysters are obtained by dredging; while those belonging to the larger variety are collected by diving.

According to Möbius, the whole of the **Pacific Ocean** is a pearl-sea, for the inhabitants of most of the islands north and south of the equator have been observed by sea-farers to wear pearls and mother-of-pearl as ornaments, and to use fish-hooks fashioned out of mother-of-pearl.

South of the equator the pearl-oyster is known to occur in the vicinity of the Solomon, Society, and Marquesas Islands; also in the Low Archipelago, south of which lie the small Gambier Islands, particularly important as a locality for pearl-oysters. North of the equator these molluscs are found near the Marianne Islands and the Marshall Islands; while from the Sandwich Islands are obtained pearls of small size and inferior quality. Those from the latter islands are, however, not of marine but of fresh-water origin; the mussels in which they were formed inhabiting among other streams the Pearl river, distant fourteen miles from Honolulu in the island of Oahu.

On the west coast of Central America and of Mexico there are extensive oyster-banks, which were fished by the original inhabitants even before the discovery of the new world, especially near Tototepec, in the Mexican State of Oajaca. The Spaniards reaped a rich harvest from both the Gulf of California and the Gulf of Panama, and even now both pearls and mother-of-pearl are obtained from the same parts. In the Gulf of Panama, the neighbourhood of the Archipelago del Rey and of Taboga is rich in pearl-oysters; these are the Pearl Islands (Islas de las Perlas) of the Spanish conquerors, but the banks in the neighbourhood are now almost exhausted. On the coast of Costa Rica the Gulf of Nicoya is mentioned.

The islands of Cubagua and Margarita in the **Caribbean Sea** on the east side of America were once noted for their pearl-fisheries. The pearls found there surpass all others of American origin in size and beauty; but these so-called "occidental" pearls are never quite equal to the "oriental" or Indian. Though often larger they are usually less perfect in form and more lead-coloured in tint. The oyster-banks surrounding these islands are now completely exhausted; and New Cadiz in Cubagua, the town founded by Diego Columbus, the son of the discoverer, in 1509, once a busy centre of the pearl-fishing industry, is also deserted. From the Colombian coast between Rio Hacha and Maracaibo beautiful pearls are now obtained in greater numbers than from the west coast. These pearls originate, however, in another mollusc, namely, *Avicula squamulosa*, the shell of which, though beautifully lustrous, is too thin to be of any value as a source of mother-of-pearl. In the West Indian Sea the island of St. Thomas is mentioned as a locality for the pearl-oyster, but it is of little importance.

Möbius estimates that about 20,000,000 marine pearl-oysters are fished annually, and of these about 4,000,000 contain pearls. Even on the supposition that from every thousand pearl-oysters only one fine pearl is obtained, the yearly yield of costly specimens would be 20,000, and when the table of prices quoted above is referred to it becomes obvious that such an average yield allows of a considerable margin for profit. Moreover, the enormous quantities of shells supplied yearly to the trade, as a source of mother-of-pearl, are worth at least as much as the total annual yield of pearls. In spite of the millions of pearl-oysters taken every year from the sea, there is apparently no exhaustion of the banks. Although the fishing of certain spots has had to be abandoned because of the exhaustion of the banks, yet this has made no appreciable difference in the total yield.

Many attempts have been made in the Dutch East Indian seas to establish pearl-oyster banks, just as the development of edible-oyster banks is commenced by planting a number

of animals at a suitable spot. Although the colony of animals thus planted may thrive and grow, it does not follow that pearls will be formed in the individual oysters, for to provoke the attack of a boring parasite, or the entrance of a grain of sand, or other cause of irritation is at present somewhat beyond the control of man.

The Pearl-Mussel.—Compared with the marine pearl-oyster, the fresh-water pearl-mussel as a pearl producer is of very subordinate importance. The variety inhabiting rivers resembles the common fresh-water mussel very closely, especially in the corrosion of the shell near the hinge line, but is somewhat larger. It is estimated that one pearl will be found in every hundred mussels, and that one per cent. of the pearls so found are of good quality. River-pearls of even the best quality often possess a lead-coloured tinge, or are lustreless and of a greyish-brown colour in consequence of the absence of the nacreous layer ; they are never equal to pearls of marine origin. As in the case of the marine pearl-oyster it is useless to expect to find pearls in well-grown, regularly formed shells, and even in distorted and abnormal shells they are often sought for in vain.

The pearl-mussel inhabits rivers in all parts of the world, but in contrast to the pearl-oyster is more abundant in temperate than in warm countries; its favourite habitat is in streams and small rivers of clear, fresh water.

It is absent from the southern countries of Europe and from the Alpine regions, but is abundant in **Germany** in the many water-courses which drain the Bohemian and Bavarian forests and the Fichtelgebirge, Erzgebirge, and Riesengebirge. Specially remarkable for the number of their pearl-producing inhabitants are the Ilz and the Regen in Lower Bavaria ; the Oelschnitz above Berneck, and the Perlenbach in the upper reaches of the Main ; the Elster and its tributaries in Saxon Voigtland especially in the neighbourhood of the town of Oelsnitz ; the Queiss and the Juppel in Silesia ; the Moldau above Frauenberg and its tributary the Wattawa in Bohemia. For centuries the pearl-mussels found in these rivers have been turned to the best possible account by the Government of the different countries, especially by that of Saxony. The yield has always been small and is gradually diminishing both in quantity and quality. In the year 1893 the total yield was 55 pearls ; in 1894 only 13 ; and in 1895 there were 68 found, of which 21 were bright, 22 half bright, and 25 dull and useless.

The famous collection in the " Green Vaults " at Dresden shows, however, that at one time beautiful pearls were obtained from the mussels inhabiting the rivers of Voigtland. A necklace preserved in this collection has been valued at 3000 thalers ; it consists of 177 pearls obtained from the river Elster. For a necklace of pearls from Voigtland, said to have been the property of a duchess of Sachsen-Zeitz, 40,000 thalers (£6000) was once offered.

Pearls have also been obtained from true pearl-mussels inhabiting certain streams in the north of Germany ; for example, the Wipperau, Gerdau, and Barnbeck on the Lüneburg heath, between Celle and Uelzen. From *Unio crassus*, another species of mussel, have been derived the pearls occasionally found in the Tapps-Aa near Christiansfeld on the northern border of Schleswig, in the neighbourhood of the Rheinsberg, and in the lake near Lindow in the province of Brandenburg.

The pearls of **Britain** were mentioned by Tacitus and by Pliny, and a breast-plate studded with British pearls was dedicated by Julius Cæsar to Venus Genetrix. We find a reference to Scotch pearls as early as 1355, in a statute of the goldsmiths of Paris, and in the reign of Charles II. the Scotch pearl trade was sufficiently important to attract the attention of parliament. After languishing for years, the pearl-fishing industry of Scotland was revived in 1860 by a German named Moritz Unger, who visited the country and bought

up all the pearls he could find in the hands of the peasantry, thus stimulating the search for more. It is estimated that the produce of the season's fishing in 1865 was worth at least £12,000. This yield, however, was not maintained; the rivers were over-fished, and the industry was discouraged as it tended to interfere with the salmon fishing, and in some cases to cause damage to the banks of the stream. At the present time a pearl is found now and again by an occasional fisherman.

The Scotch rivers which have yielded pearls are the Spey, the Tay, and the South Esk, and to a lesser extent the Doon, the Dee, the Don, the Ythan, the Teith, and the Forth.

In North Wales the Conway was at one time celebrated for its pearls; and it is related that a Conway pearl, which it is believed now occupies a place in the British crown, was presented to the queen of Charles II. by her chamberlain, Sir Richard Wynn. In Ireland the rivers of the counties of Donegal, Tyrone, and Wexford have yielded pearls. It is said that, in England, Sir John Hawkins, the circumnavigator had a patent for pearl-fishing in the Irt in Cumberland.

The pearl-mussel is abundant also in Sweden and Norway from Schonen and Christiansand as far north as Lapland; and in the north of Russia, from the sources of the Don and the Volga to the White Sea. The rivers of these regions yield a certain number of pearls, among which are many of good quality.

A pearl-mussel which differs in no essential particular from the European *Unio margaritifer* is found in **North America**, and is specially abundant in the New England States, but yields very few pearls. On the other hand, there are to be met with in the river-system of the Mississippi a number of species of the genus *Unio* from which many pearls are obtained. The first European discoverers of this region in the sixteenth century found immense numbers of pearls, the largest being the size of a nut. The occurrence of the fresh-water pearl-mussel in the Sandwich Islands has been noted already.

The pearl industry of eastern Asia and particularly of **China** is of peculiar interest. Pearls are greatly esteemed by the Chinese as ornaments, and for that reason have been eagerly sought for centuries. Pearl-producing mussels are said to inhabit some of the rivers of Manchuria and East Siberia, but to what genera and species they belong is not exactly known. The pearl-mussel which inhabits the water-courses near Canton and Hu-che-fu further to the south is *Cristaria plicata*. This mollusc possesses a peculiar interest in that for centuries it has been experimented upon by the Chinese in their attempts to induce it to form pearls in response to an artificial stimulation. Thousands of Chinamen make this a regular occupation, but their efforts are never quite successful. The manner in which they proceed is to insert into the carefully opened mussel without injury to the animal a small hemispherical object, or a thin image of Buddha in tin, which they place between the mantle and the shell. These objects when invested with a layer of nacre acquire a pearly appearance. After remaining inside the shell for a period ranging from 10 months to three years, their nacreous coat is from $\frac{1}{10}$ to $\frac{1}{5}$ millimetre thick, and the objects can be utilised for purposes of ornament. They are removed from the shell, to which they have become firmly attached, and mounted in a suitable manner.

Other similar attempts have been made to induce by artificial means the formation of pearls, such as, for example, by the introduction of a grain of sand or of small spheres of mother-of-pearl, but never very successfully. It is related that a method was known to Linnæus, which he had described in writing, but the details have never become known.

Not only have attempts been made to induce the formation of natural pearls by artificial means, but no efforts have been spared to produce less costly substitutes which shall resemble genuine pearls as closely as may be. With the consideration of this subject we shall close.

IMITATION PEARLS.—One of the best imitations of a pearl is furnished by a polished sphere of mother-of-pearl, but it differs very essentially from a natural pearl inasmuch as the coats of which it is built up are not concentric. In 1680, or possibly even earlier, in 1656, Jacquin, the Parisian rosary maker, discovered a means whereby imitation pearls, which reproduce the beautiful pearly lustre of genuine pearls, may be manufactured. His method is still employed, and on it depends a flourishing industry. The first step is the production of hollow, thin-walled glass beads of "girasol," a colourless, easily fusible glass manufactured for the express purpose. In form these may be spherical, ovoid, or pear-shaped, or they may be made to resemble baroque pearls. The inner surface of these beads is then coated with a silvery-white material obtained from the scales of certain fishes, for example, from the white-fish and the bleak (*Alburnus lucidus*). This material lies just beneath the scales, from which it is separated by shaking the same with water. The microscope reveals the fact that it consists of numberless minute, irregular, rhombic plates. One pound of the substance is yielded by seven pounds of fish scales, to provide which from 18,000 to 20,000 fishes are necessary. This silvery-white material, when mixed with a solution of isinglass, yields a thin, glutinous pulp, known in the trade as "essence d'Orient." It is introduced into the hollow glass balls, over the inner surface of which it is spread uniformly. When dry, this inner coating of silvery-white material gives the beads very much the appearance of real pearls, so that even an expert may be deceived at the first glance. In order to make them more substantial, the beads are filled with wax. When carefully made, they resemble very closely good Indian pearls, and are much worn in place of these, although on account of the care necessary in their manufacture they are not at all cheap. Less carefully made beads are employed in the cheaper articles of jewellery, and though they imitate real pearls less perfectly, yet the beautiful pearly lustre is always reproduced with more or less fidelity.

Very beautiful artificial pearls, with a lustre like satin, are sometimes made from the incisors of the Dugong, a whale-like, aquatic mammal belonging to the group which includes sea-cows. It inhabits the sea in the neighbourhood of the Dahalak Island, which lies in the Red Sea off Massaua, already mentioned as a pearl-fishing station. Imitation pearls of this kind are not, however, frequently met with.

In recent times the so-called opaline glass has been used for the manufacture of imitation pearls, the resemblance to real pearls being given by a careful treatment with hydrofluoric acid.

Black pearls can be imitated very successfully in hæmatite, as we have already seen under this mineral. A polished ball of hæmatite often resembles a black pearl very closely, especially when not too highly polished. It is readily distinguished, however, by its high specific gravity and by the fact that it feels cold to the touch. Red spheres cut from the shell of the great West Indian univalve mollusc, *Strombus gigas*, have a lustre which is somewhat pearl-like, but they are more easily passed off as coral than as pearls.

CORAL.

Red or precious coral is a material almost as important for decorative and ornamental purposes as are precious stones and pearls. It constitutes the substance of the calcareous axial skeleton of the coral polyp, *Corallium rubrum*, a lowly organised animal belonging to the class Anthozoa. The Mediterranean Sea is the principal habitat of this coral; and not only is it fished almost exclusively by Italians, but the working of the rough material is for the most part in the hands of the same people, so that we are dealing with what is practically an Italian industry.

Corallium rubrum has been sometimes known by other names, such as *Corallium nobile*, *Isis nobilis*, &c., but these terms are no longer employed. This coral is not a solitary organism, but forms branching colonies. Each colony is supported by an axial rod of red calcareous material, the so-called coral, and this is invested by a layer of soft, living material known to zoologists as the cœnosarc. At intervals along the branches are situated the individual polyps, embedded, as it were, in the cœnosarc, with the material of which their bodies are continuous. The cells of the cœnosarc have the power of separating out calcium carbonate from the sea water, in which the coral lives, and this forms the substance of the axial skeleton of the colony. From this the cœnosarc can be peeled off "like the bark of a willow twig in spring," leaving it a clean, red, branching rod, the coral of commerce. It is proposed, first, to consider the structure and life-history of a colony of *Corallium rubrum* in some detail, and afterwards the methods employed for collecting coral and its manufacture into ornamental articles.

The Coral Skeleton.—The calcareous skeleton of a colony of *Corallium rubrum* is red—more rarely white or black—in colour and arborescent in form. It is attached firmly by a disc-shaped foot to any suitable object in the sea, such as a rock or a large stone. Colonies have also been found fixed to cannon-balls, bottles, shells, or other corals, and in one case a human skull. The disc-like foot affords a firm base for the whole structure. It cannot penetrate the rock or other object to which it is attached as do the rootlets of a plant, but any depressions, furrows, or cracks in the surface are filled up with calcareous material, which thus cements, as it were, the two together. The main axis and its lateral branches are not straight, but curved, like the trunk and branches of a tree or shrub. The former seldom exceeds a foot in length and an inch in diameter. From a single disc-like foot there may grow out, not one, but several independent colonies, which are then usually of small size. This is especially the case with the corals off the coast of Provence.

The growth of a coral-colony is not affected by gravity in the same way as is that of plants; it always grows in a direction perpendicular to the surface to which it is attached, regardless of its inclination to the vertical. If attached to some object on the bottom of the sea, its direction of growth will be vertical and upwards; if to the vertical face of a rock it will grow out in a horizontal direction; and when attached to the roof of a rock cavity, it grows vertically downwards, this being, indeed, a direction commonly taken.

The main axis of the skeleton tapers gradually to a blunt point, so gradually, indeed, that a short length of it may appear perfectly cylindrical. It may commence to branch quite near the foot or at a distance of some centimetres away from it. The primary branches may bear secondary branches, and these, again, produce lateral offshoots, and so on. These lateral branches, like the main stem, do not follow a straight course, but twist and turn about in a more or less irregular manner; they also terminate in irregularly blunted points. The branching follows no fixed law, but the branches spread as much as possible, and never arise two at the same level on the parent stem.

The angle at which the branches are inclined to the parent stems varies considerably. It is often between 40° and 50°, but may be obtuse, so that the branch is directed backwards, or, on the other hand, so acute that it runs close beside the parent stem and may actually fuse with it, only to separate again after a short distance. Such an intergrowth may take place also between branches of the same colony which have originated at different spots, or between branches of neighbouring but independent colonies.

Though two branches never originate at exactly the same level on the parent stem, the difference in level between them amounts, as a rule, to only a few millimetres, though it may be several centimetres. The branches are usually crowded together at the attached end of the parent stem, and are less numerous towards the free end. At the point where a branch originates, the parent stem is somewhat flattened, so that its outline in section is oval instead of circular. The same flattening may be observed also in other parts of the stem where there is no obvious reason for it. This is often considered to be due to some temporary disturbance in the economy of the colony.

The general form of a coral-stock, the system according to which it branches, and the actual direction of the branches vary according to the locality in which it grows, its depth beneath the surface of the sea, and so on. Colonies which have grown under similar conditions usually resemble each other in form, so that we may reasonably conclude that the form taken by a colony depends upon the conditions under which it grows. So characteristic is the form of coral-stocks from different districts of the Mediterranean that an expert confronted with a collection of corals from the same locality has no difficulty in naming at once the place from whence they came. Corals from the Algerian and Tunisian coasts, from Sicily, and especially from Sciacca, from Spain, and from Provence, show remarkable differences in this respect, differences which are important, inasmuch as they render the coral more or less suitable for certain purposes, and therefore more or less valuable.

The furrowed surface of precious coral is another of its characteristic features. These fine furrows run in a direction either parallel to the length of the stem and its branches or in more or less of a spiral. When a furrow reaches the point of origin of a branch, it either passes to one side of it or divides into furrows, which enclose the spot, and which frequently unite again after a short course. These furrows are always more numerous towards the base of the stem, since a certain proportion disappear in their course from the base towards the apex. The distance between adjacent furrows is always small, never less than $\frac{1}{4}$ millimetre and never more than $\frac{1}{2}$ millimetre.

Another equally striking feature of a piece of natural coral is the presence of small, circular, shallow depressions, measuring at most two millimetres in diameter. They may lie so close that their edges touch, or, on the other hand, they may be separated by a space of a centimetre. They mark the spots where grew the individual polyps of the colony of which the piece of coral was the axis.

These characteristic ridges and depressions are an invariable feature of the natural surface of precious coral. When absent, it is quite certain that they have been removed by polishing or by some other artificial process. Besides the depressions in the surface of coral, there may be noticed also small pits no larger than a pin-prick. These, which are frequently absent in perfectly fresh coral, have been caused by marine organisms, such as boring-sponges and boring-worms. Dead coral is frequently so much bored and eaten away by such creatures that it is useless for commercial purposes.

A piece of coral which has not been attacked by boring animals appears perfectly homogeneous, compact, and solid, and free from internal cavities. It may occasionally enclose foreign bodies of various kinds, which have by some chance found themselves in such a position as to become enveloped by the calcareous material secreted by the colony, of which the piece of coral formed the axis. The fractured surface of a piece of fresh coral is uneven and splintery. In many cases, when a transversely fractured surface is examined with the naked eye or with a lens, it becomes obvious that the piece of coral is built up of thin, concentric layers, the whole mass, indeed, consisting, as it were, of a number of hollow tubes fitting closely one inside another. This structure is more clearly demonstrated by the examination under the compound microscope of thin, transverse sections; and one can make out, in addition, the fact that each concentric layer is built up of numberless fine fibres, the general direction of which is radial, that is to say, they run outwards from the common centre towards the periphery. These fibres have an extraordinary power of double refraction, in which respect, and in others to be mentioned presently, they agree with the mineral species calcite, to which they are probably to be referred. The concentric structure of a piece of coral is also demonstrated very clearly when the latter is ignited, the concentric layers separating from each other and peeling off.

The observation of thin, transverse sections of coral under the microscope also discloses the fact that the concentric layers are coloured alternately bright red and white. In very thin sections the red colour is scarcely apparent at all, only appearing in slices of a certain thickness. The precise shade of colour of coral is different in different specimens. That from which the cœnosarc has been newly stripped ranges in colour from pure white to the brilliant tint of red lead. Pure white coral is rare, and the absence of colour is said to be the consequence of a diseased condition of the organism. Yellow coral is also rare. In Italy, the home of the coral industry, special terms are used to distinguish coral of different shades of colour. Thus, first comes pure white (*bianco*), next a fresh, pale, flesh-red (*pelle d'angelo*), then pale rose (*rosa pallido*), then bright rose (*rosa vivo*), which is followed by "second colour" (*secondo coloro*), red (*rosso*), dark red (*rosso scuro*), and finally the darkest red of all (*carbonetto* or *arciscuro*). It is unusual to find different colours or different shades of colour in the same piece of coral, this being usually of one uniform tint in all its branches.

Red coral when reduced to a fine powder is pale reddish in colour, the shade being more pronounced the deeper the colour of the piece, and *vice versâ*.

The death and decay of the living portion of a coral-colony is accompanied by a change in colour of the skeleton, that is to say, of the red substance commonly known as coral. Coral which has lain at the bottom of the sea in muddy water for any considerable length of time is almost sure to assume a more or less dark brown or black colour, and is said by the Italians to be "burnt" (*bruciato*). Some of the black coral of commerce is, however, something quite different, and will be considered later. In exceptional cases dead coral, instead of becoming black, turns white or yellow. An immense amount of dead coral is rendered useless for industrial and ornamental purposes by the attacks of boring worms, sponges, and other marine organisms.

The black or brown colour of a piece of dead coral does not always extend over the whole of its surface nor penetrate to its innermost layers. There may be black patches and spots here and there, or it may be red outside, but quite black inside, or the reverse may be the case, while occasionally pieces of coral are met with in which the core and the external layer are black and the intermediate layer is red, so that a cross-section shows a central black spot, surrounded by an inner red and an outer black ring. It is sometimes stated that the original colour can be restored to coral which has turned black by allowing it to lie in water and then exposing it to the sun. The investigations which have been made do not, however, tend to support this statement.

The substance of coral consists for the most part of calcium carbonate, which, as we have already seen, is probably present in the form of calcite. This is impregnated with a small amount of organic material, the presence of which accounts for the fact that coral becomes black when heated. The specific gravity of coral is very near that of calcite, much nearer than to that of aragonite, the other crystallised modification of calcium carbonate. The specific gravity of pure calcite is 2·72, while that of precious coral, irrespective of colour, lies between 2·6 and 2·7; the values found by Canestrini, for example, are 2·671 and 2·68. Precious coral is rather harder than calcite, being placed between 3 and 4 on Mohs' scale, but rather nearer to 4, so that we may write $H = 3\frac{3}{4}$. This greater degree of hardness is no doubt due to the admixture of foreign material. Coral is soft enough to be easily worked with a knife or file, or turned on the lathe, but it is not hard enough to admit of a very brilliant polish, the material depending for its beauty more upon its fine colour.

The chemical composition of coral is given by the following analysis by Tischer:

	Red coral.	Black coral.
Water (H_2O)	0·550	0·600
Carbon dioxide (CO_2) . . .	42·235	41·300
Lime (CaO)	48·825	48·625
Magnesia (MgO)	3·240	3·224
Ferric oxide (Fe_2O_3)	1·720	0·800
Sulphuric anhydride (SO_3) . . .	0·755	0·824
Organic matter	1·350	3·070
Deficiency, &c.	1·325	1·557
	100·000	100·000

From this can be calculated the following constituents:

	Red coral.	Black coral.
Calcium carbonate ($CaCO_3$) . . .	86·974	85·801
Magnesium carbonate ($MgCO_3$) . .	6·804	6·770
Calcium sulphate ($CaSO_4$) . . .	1·271	1·400
Ferric oxide (Fe_2O_3)	1·720	0·800
Organic matter	1·350	3·070
Water (H_2O)	0·550	0·600
Phosphates, Silica, &c., and deficiency .	1·331	1·559
	100·000	100·000

Earlier analyses give similar, though somewhat varying, results; it is clear therefore that, broadly speaking, coral consists of calcium carbonate, with a small amount of magnesium carbonate. The proportions in which these two constituents are present are not always the same; young coral contains only about 1 per cent. of magnesium carbonate, while older material may contain as much as 38 per cent. The greater the amount of magnesium present, the harder does coral become, and the extent to which its hardness exceeds that of calcite probably depends upon the proportion of magnesium carbonate present in it. We learn also from the analyses quoted above that the differences in the composition of red and of black coral are quite unessential, the chief being that black coral contains a larger proportion of organic material, which also probably differs in character from that present in red coral.

The colour of black coral is no doubt due to the presence of this organic material; it has, indeed, been referred to the presence of manganese dioxide, but this substance was not detected in the above analysis. It has also been suggested that the action of hydrogen sulphide upon red coral might result in a change of colour from red to black; but this requires proof. Attempts to refer the red colour of fresh coral to the presence of inorganic constituents have been equally unsuccessful. It has been said to be due to the presence of iron oxide, of which Tischer found 1·720 and other chemists up to 4·75 per cent. This, however, is not very probable, seeing that the red colour is destroyed on ignition, just as is the black. The red colouring matter is thus presumably organic in nature, as is the pigment of variously coloured univalve and bivalve shells, and the change of red coral first to black and then to dirty yellow mark different stages in an oxidation process.

The Living Coral.—As already stated, the branching axis of a living coral-stock is invested by the cœnosarc, a layer of soft, red, living material, with a surface like velvet, and a thickness of less than a line. In the substance of the cœnosarc are embedded calcareous spicules, which can be felt by pressing a portion of it between the thumb and finger. They are so small, however, that their form can only be made out by examining thin sections under the microscope. The cœnosarc extends beyond the ends of the branches of the skeleton, and the blunt prolongations of the branches so formed are soft and flexible, and can be cut through with a sharp knife, while the portion containing the calcareous skeletal rod cannot be so treated. The foot of young colonies is invested with cœnosarc, just as are the trunk and the branches, but in old stocks, not only the foot, but the lower part of the stem and the lower branches, are bare and often much corroded and eaten away. When a branch of living coral is allowed to dry in the air, the cœnosarc assumes the appearance of dry, rough skin of a brick-red or red-lead colour, raised here and there into wart-like protuberances. From the central orifice of each protuberance radiate eight short grooves, which divide it into eight wedge-shaped portions, and give the surface of the wart the appearance of an eight-rayed star.

These little protuberances mark the spots occupied by the individual polyps of which the colony is built up. Nothing more than what has been described can be seen of the polyps when a branch of living coral is examined in water which has been disturbed or is in motion, and the branch appears of a uniform red colour. When the water has come to rest, however, one may watch the wart-like prominences gradually expand and open, and the white cylindrical bodies of the polyps, each with eight white, pinnate tentacles come into view. The position of each polyp is marked in the skeleton by a shallow depression, to which attention has been directed already. The polyps respond readily to external irritation; no matter how slightly the water be disturbed, or how softly one of their number be touched, each one is instantly retracted into its pocket-like depression in the cœnosarc, and then reassumes its original aspect of a wart-like protuberance. It may be some hours before the polyps may again be seen in a fully expanded condition, and even then they expand, only to retract again on the slightest disturbance of the surrounding medium. When fully expanded, the coral-polyps look like white starry blossoms on a coral-red background, and, indeed, formerly the coral-stock was regarded as a plant of which the polyps were the flowers. The true nature of the organism was first recognised, in the year 1723, by the French physician and naturalist, Peyssonel, whose perspicacity was not, however, recognised by his brother zoologists.

Into the single body-cavity of each polyp project eight radial partitions, the outer ends of which are united to the body-wall and the inner to the stomodæum or gullet. This opens to the exterior in the mouth, which is surrounded by eight tentacles. The nourishing of the polyp is performed by the ingestion of food into the body-cavity. The rapid and continual motion of the tentacles induces currents in the sea-water which sweep small organisms of all kinds into the mouth, down the gullet, and into the body-cavity of the polyp, where they are digested, forming a white milk-like fluid. This circulates in a complicated system of tubes throughout the cœnosarc, so that all members of the colony share equally, and none is better nourished than another. The fine grooves so characteristic of the surface of natural coral are due to this tube-system ; they correspond to vessels lying in the innermost part of the cœnosarc, close to the skeleton.

Reproduction takes place in two ways, by budding and by the development of eggs. By budding, new individuals in an already established colony are formed, while from each egg arises a perfectly independent daughter-colony. The polyps of a single stock are usually all male or all female, but occasionally polyps of both sexes may be found in the same colony, though, as a rule, on different branches. The fertilising cells of the male are discharged into the water and eventually find their way into the body-cavity of a female. The eggs of the latter after fertilisation develop into small oval larvæ scarcely visible to the naked eye, which escape through the mouth of the parent, and for a time swim about freely in the sea. After a time each larva attaches itself to some suitable object on the sea-bottom and gradually assumes the form of an adult polyp, which by repeated budding forms a colony.

The first stage in the development of the polyp after the fixation of the larva, is the appearance of a tiny knob or swelling upon the free end. This becomes gradually larger and larger, and, as calcareous spicules appear in its substance, assumes a red colour, finally becoming transformed into a perfect adult polyp. This transformation is accompanied by the deposition of calcium carbonate in a circular area of the surface of attachment of the polyp, and the formation of the foot, the organ of attachment of a coral-colony. During the development of the colony, calcium carbonate is continuously separated out from the sea-water and deposited in the living cœnosarc in the form of spicules. These become embedded in a dense calcareous, cement-like substance and thus the hard, axial skeleton of the colony is formed. This continually increases in diameter by the deposition of fresh layers of calcium carbonate, and gives off new lateral branches with the development of new polyps. The foot also, which is invested by cœnosarc, increases in circumference proportionately with the increase in size of the colony. The laminated structure of the coral skeleton points to the fact that the deposition of calcareous matter does not proceed uninterruptedly, but that periods of activity alternate with periods of rest, these periods being possibly seasons of the year. Thus each layer of calcareous material corresponds to a period of activity, and the interruption between successive layers to a period of quiescence. The coral-skeleton grows also in length as well as in diameter, for fresh material is always being deposited at the free ends of the stem and branches.

At an early stage in its development the coral-colony consists only of one polyp, but, by a process of budding, multiplication of the polyps and the formation of a larger or smaller colony is effected. The new polyps arise at various spots in the cœnosarc as knob-like swellings, which gradually unfold, take on the characters of the adult, and perform their share of the work of nutrition and reproduction. The process suggests the formation and unfolding of a leaf-bud, and for that reason is known to zoologists as budding or gemmation.

Not only from a scientific but also from an economic standpoint it is important to know the period required by a coral-colony to attain its full growth, but there is much diversity of opinion upon the point. The length of the period probably varies with the conditions under which the colony lives, and more especially with its depth beneath the surface.

In the opinion of many experts a coral-stock reaches its full size only after a period of thirty years, but others maintain that well-grown stocks are to be found on coral-banks which have been fished very much more recently than thirty years previously. Thus at the beginning of the nineteenth century the fishing of the coral-banks on the north coast of Africa was abandoned, in consequence of war, for a period of four years, and when fishing was again commenced, an unparalleled amount of coral was harvested, many pieces of exceptional size being found.

Near Vico Equense, in the neighbourhood of Sorrento, in the Bay of Naples, is a coral-bank situated six miles away from the coast, on which the coral, at a depth of 60 feet, is supposed to require eight years to attain its maximum size, while at greater depths it requires still longer. The coral-bank which stretches along the coast of Sicily, from the Point of Faro to a point many miles south of the town of Messina is divided into ten sections, one of which is fished every year. A period of ten years therefore elapses between consecutive fishings of each of these sections, and this is sufficient to admit of the coral-stocks reaching their full size. A rich yield of coral was obtained from a bank still further south, near San Stefano, which had not been fished before within the memory of man. Notwithstanding the fact that the bank had remained undisturbed for centuries, the coral-stocks were no larger, and only about one-third thicker than those known to have developed within a period of ten years. It may be concluded from this that a coral-stock continues for a time to grow in thickness after it has ceased to grow in length.

The fact that the development of a coral-colony is greatly influenced by the conditions under which it lives is pointed to by the observations of coral-fishers in the Bay of Naples, among other regions. The coral formed off the west side of the bay is more beautiful and regularly formed than that obtained off the opposite coast near Sorrento ; a difference which is attributed to the fact that the coast near Sorrento on the east consists of limestone rocks, while the islands of Nisida, Procida, Ischia, &c. on the west are constituted of volcanic tuffs, which are obviously more suitable for corals than the former.

The discovery of the conditions most favourable for the development of fine coral has been, and is, the subject of much experimental study, but no very valuable result has yet been obtained.

The polyp, *Corallium rubrum*, which forms precious coral is placed by systematists in the sub-kingdom Cœlenterata, class Anthozoa, sub-class Alcyonaria or Octactinia, all the members of which have eight tentacles and eight mesenteries.

Precious coral, as we have seen, forms the skeleton of small isolated colonies, but the reef-building corals, by whose industry are reared the coral-reefs and islands (atolls) of the Pacific Ocean and other warm seas, do not differ essentially in their organisation from precious corals, the chief difference being the possession of six mesenteries and of six, or some multiple of six, tentacles. Those corals which agree with the reef-building corals in this respect, constitute the other division of the class Anthozoa, namely the sub-class Zoantharia or Hexactinia. This sub-class comprises three orders of corals : the Madreporaria or stony corals, which includes most of the reef-building corals ; the Actiniaria or sea-anemones, which are destitute of a calcareous skeleton and remarkable for their gorgeous colouring ; and the Antipatharia, including the " black corals," the skeleton of which is horny.

The Distribution of Coral : Coral-banks.—Our attention must now be turned to the distribution of coral in nature, to the manner in which it is collected, and to some aspects of the coral-industry.

The colonies formed by *Corallium rubrum* are always found growing in groups, which are known as coral-fields or coral-banks. The shrub-like branches of living coral form with sea-weeds submarine forests, which afford food and shelter to numberless marine creatures. The coral-stocks are situated usually in clefts, crevices, or cavities in the rocks near the shore. They prefer a steep face of rock directed towards the south and never settle upon rocks which face north, though they are sometimes found in situations the aspect of which is east or west. The true precious coral, *Corallium rubrum*, is probably confined to the Mediterranean Sea and its inlets, and the red coral found in other localities belongs, as we shall see in passing later on, to another species.

The depth at which precious coral is found varies between wide limits ; very little grows at a depth less than 3, or greater than 300 metres. At depths such as the latter the development of the colonies is slow ; they never reach a large size and are always pale in colour. The depth most favourable for the growth and development of coral-colonies lies between 30 and 50 metres, but varies according to locality, the best grown corals in the Straits of Messina, for example, being met with at depths of from 120 to 200 metres.

A coral-field once discovered is not a perennial source of coral, but after being fished for a longer or shorter period becomes exhausted. Banks, which at one time were productive, are now exhausted through over-fishing or other causes ; while in some cases they have to be abandoned, usually on account of the roughness of the sea at that particular spot, and the consequent danger to the fishing-boats. The deficit in the total yield occasioned by the exhaustion of previously fruitful fields is usually compensated for by the discovery of fresh banks. The coral-producing portions of the Mediterranean Sea will be separately treated in some detail below ; they include the coasts of east Algeria and of Tunis, the west coasts of Sardinia and Corsica, part of the south and of the west coasts of Sicily, and the Straits of Messina from whence coral-banks stretch along the whole of the west coast of Italy, the coast of Provence, the whole of the Mediterranean coast of Spain, and finally in the neighbourhood of the Balearic Islands. The yield of coral from the Italian coasts of the Adriatic Sea is very meagre, and very little more is found off the Dalmatian coasts opposite ; the stocks are so small and scarce as hardly to repay the searcher for his trouble. A small amount of coral is found further east in the sea round Corfu and Cyprus, and at various isolated spots off the coast of Asia Minor, but these localities are all unimportant.

The largest amount of coral is obtained from the coasts of **Algeria** and **Tunis,** the yearly yield from these regions being 10,000 kilograms. The coral-banks stretch eastward from Cape Ferro (Cap de Fer), a little west of Bona, to Cape Bon, and from thence southwards as far as the neighbourhood of Sfax in the Lesser Syrtis. An exhaustive search for coral has been made on the west coast of Algeria, west of Cape Ferro, but hitherto in vain. At places where coral-banks exist they are fished for a distance of six or eight miles out to sea and at depths varying from 90 to 900 feet, but the coral obtained at great depths is paler and less brilliant. In the districts mentioned above the coral-fishing industry has been carried on for centuries ; it has flourished and declined, and at times has almost reached a standstill, owing to the passive neglect or active opposition of the barbarian rulers of the coast. The headquarters of the fishing-fleet was formerly the island of Tabarca, which lies near the coast and very near the prolongation of the boundary line between Algeria and Tunis. The island is still of importance, as also is the island of Galita, lying a little further from the coast. The most important centre, however, at present is La Calle, a town on the neighbouring Algerian coast, where the industry is promoted in every possible way by the French Government. Other places in the same neighbourhood where abundance of coral is to be found are Vieille-Calle or Bastion de France, a fort erected in former times for the protection of coral-fishers; Cape Rosa, a little to the west of La Calle; further to the west the Gulf of Bona, on which Bona itself stands; then Calle-Traversa and other places; and still further in the same direction Cap de Garde and Cap de Fer, which constitute as far as is known at present the western boundary of the coral-producing stretch of these seas. Towards the east we have the coast of Biserta; the neighbourhood a little to the south of Cape Bon, near Kelibia; and further to the south Mansuria (Sidi Mansur), which lies to the north of Sfax, opposite the Kerkenna Islands in the Gulf of Cabes. The coral found on the African coast is distinguishable from that found at other localities, and specially from Sicilian coral, by certain peculiarities of form. The main stem of the stock is almost perfectly straight, rising up like a column, and bearing upon its sides perfectly straight branches, some pieces being like a hand with outstretched fingers. Pieces of coral of this form are much sought after as they can be worked with very little waste of material.

A large amount of coral has been obtained in recent years from the coasts of **Sicily**. Also from the neighbourhood of the small islands Linosa and Pantellaria south of Sicily, and from the sea between Malta and Cape Passaro, the southernmost point of Sicily. The most important localities for coral lie, however, near Sciacca, a little to the west of Girgenti, where coral-fishing has been carried on since the middle of the 'seventies; on the west coast near Trapani; and in the vicinity of the neighbouring Ægadean Islands. Near Sciacca there are three banks of different sizes, and at an average depth of about 200 metres (148–221 metres), which are remarkable in that all the coral-stocks are dead, and the skeleton in many cases has turned black. The process is of course progressive, and since the discovery of the banks in 1875 a gradual increase in the proportion of black and red coral has been observed, the deterioration being noticeable as early as the beginning of the 'eighties. Moreover, the old stocks not being replaced by young ones, the bank is quickly becoming exhausted. It is the presence of a thick layer of mud which has proved fatal to the corals of these banks, for the organism never thrives unless living in clear, still water. This covering of mud may be traced to the violent submarine volcanic eruptions which took place at intervals during a period of three months in the year 1831 in the sea between Pantellaria and Sciacca. To these eruptions was due the appearance of the small crater-island known as Graham (= Ferdinandea = Julia Island), now washed away by the sea, and the mud which killed the corals was probably laid down over great stretches of the sea-floor in the form of fine volcanic ash. The coral-stocks of Sciacca are so gnarled and distorted that, from the commercial point of view, they are of little use, since to work them into any of the objects ordinarily made of coral involves a considerable loss of material.

There are coral-banks in the neighbourhood of Ustica and the Lipari Islands (Lipari, Vulcano, Stromboli, Basiluzzo, &c.), small islands lying to the north of Sicily. They are moderately productive, but the stormy seas render the work of fishing difficult or impossible. They are less important than the coral-banks in the Straits of Messina, which extend from Cape Faro, the north-east point of Sicily, for a distance of six miles, that is to say, to a point three miles beyond Messina and opposite Chiesa della Grotta, while a few corals are found even as far south as San Stefano.

The opposite **Calabrian coast**, especially the neighbourhood of Scilla and Palmi, is famous for the beauty of colour of its corals, which on this account are specially valuable. This coast is productive as far north as Cape Vaticano and Tropea and the Gulf of San Eufemia, and in the south in the neighbourhood of Altafiumana, on Cape dell' Armi, and of

Melito.   On the east coast coral is found off Capes Spartivento and Bruzzano, and off Cape Rizzuto, in the Gulf of Squillace, and on the eastern margin of the Gulf of Taranto, off Cape Santa Maria di Leuca.

In the **Bay of Naples** several coral-banks are known.   They are situated five or six miles from the shore, in the neighbourhood of Capri ; near the small island of Nisida, between Pausilipp and Pozzuoli ; in the neighbourhood of Cape Miseno ; around the island of Ischia ; between Naples and Vico Equense ; and near Castellamare, on the promontory of Sorrento.

The coral-fisheries next in importance to the North African have been for a long time those in the waters surrounding the islands of **Sardinia** and **Corsica**, the fishing being confined almost exclusively to the west coasts.   The islands of San Pietro, to the south of Sardinia, with the localities of Carloforte, San Antioco, and del Toro must also be mentioned ; while further north there is the Cape San Marco, and the stretch of coast between Bosa and Alghero.   A still richer locality for coral is the Strait of Bonifaccio, between Sardinia and Corsica.   It is obtained on the Sardinian side from Asinaro, Castelsardo, Longosardo, and from the Maddalena and Caprera islands.   On the Corsican side it is found in the neighbourhood of Bonifaccio ; and also along the whole of the western coast of the island as far north as Cape Corso, for example, in the Gulf of Propriano (Gulf of Valinco), and at many other places.   A little coral is found on the east of Sardinia, near Cape Corallo, but none on the east of Corsica.

A certain amount of coral is obtained from the islands of Elba and Giglio, and from that part of the **Tuscan coast**, between Monte Argentaro in the south and San Stefano in the north, opposite to which these islands lie.   Here also lies Alsidonia (= Ansidonia), famous even in Pliny's time for its coral-fishery.   Another coral-producing part of the Tuscan coast stretches from Cecina, near Livorno, in the south to La Spezia in the north. Monte Nero, near Livorno, was at one time a somewhat important centre, but fishing appears now to be discontinued.

There are coral-banks, though only of minor importance, on certain parts of the **French coast** ; in the neighbourhood of the Hyères, near Toulon, on the coast of Provence ; on both sides of the bay of St. Tropez, south of Cannes ; and a little further to the west, near Point Riche, east of the mouth of the Rhone.   French coral is usually short and thick ; the foot of a stock frequently bears a large number of branches, so that the whole stock looks like a tuft of hair.

The whole of the **Spanish coast**, from the French border to Gibraltar, forming parts of the provinces of Catalonia, Valencia, Murcia, and Granada, is fringed with coral-banks, the yield of which is not without importance.   The localities deserving of special mention lie quite in the north ; they are situated near Cabo de Creus, in the Gulf of Rosas, and in the vicinity of Cape Bagur, of Cape Sebastiano, and of Palamos, all near latitude 42° N.   The seas surrounding the Balearic Islands are also productive of coral.   The broad foot in Spanish coral-stocks supports several branching stems growing up like columns, and the coral itself is often of a particularly dark-red colour.

Pliny and other ancient writers, and also certain modern authors, state that precious coral is to be found in other districts ; for example, in the Red Sea and Indian Ocean.   It does not seem to exist in those parts of the ocean at the present time, however, although reef-building corals abound.   It is possible that true precious coral (*Corallium rubrum*) is really confined to the Mediterranean, but other seas are inhabited by many corals, which, though belonging to different species, are yet very like *Corallium rubrum*, and yield coral which is used for similar purposes.   One of these, *Corallium lubrani*, is said to furnish the coral which for some time past has been obtained near the Cape Verde Islands, especially from the neighbourhood of the Island of São Thiago, and also of that from the Canary Islands. The coral-banks near São Thiago lie at a depth of from 90 to 190 metres, and at a distance of from 400 to 1000 metres from the coast. The fishery was established in the 'sixties by Antonio Lubrano, an Italian, from whom this coral takes its name. Coral of a red colour, though lighter in shade than precious coral, is found near the Sandwich Islands. It ranges in colour from pale rose-red to white, and the branches are sharply pointed, which is not the case with Mediterranean coral. The material, which at the present time figures in the trade as Japanese coral, is probably formed by a polyp of the same species, known, however, as *Corallium secundum* (Dana). This differs in no essential particular from *Corallium johnstoni* (Gray), found in the neighbourhood of Madeira. These and all corals other than those living in the Mediterranean are so unimportant from a commercial point of view that they need no further consideration.

Coral-Fishing: Application of Coral: Trade.—The methods adopted for coral-fishing are unique, and cannot be compared with those employed in any other kind of fishing. Coral-fishers pursue their occupation during the six summer months, and only in exceptional cases all the year round. The arrival of the autumn storms makes their trade too dangerous, and the ships then return to harbour laden with their spoils, to return to the fishing-grounds in the following spring.

These ships are constructed specially for coral-fishing, and are all of the same pattern in external form and in general equipment, but of various sizes. They are very solidly built, seaworthy, fast-sailing boats of 6 to 16 tons burden, and adapted for rowing when necessity arises. The largest measure from 13 to 14 metres in length, $3\frac{1}{2}$ in breadth, and $1\frac{1}{2}$ in depth. They are manned by a crew of from 6 to 12 men, who work hard for 18 hours a day for miserable wages. There are just a few quite small boats engaged in coral-fishing, with a capacity of less than six tons, and manned by no more than two or three men.

The fisheries lie, as a rule, from four to six nautical miles out to sea. Nearer the shore, where the depth of water is at most only ten metres, a small amount of coral is collected by divers; but the largest and most beautiful coral grows only at the depths which are inaccessible to the diver, even when provided with every possible appliance.

For fishing up the coral from these greater depths, there is in use all over the Mediterranean a special instrument, which in all probability has been employed by coral-fishers for centuries, and is known to the Italians by the term "ingegno," and to the Provençals as "engin."

Its construction and method of use are almost identical in all parts of the Mediterranean. Two massive beams of oak, with tapering ends and a length of from $2\frac{1}{2}$ to less than 1 metre, according to the size of the boat to which the instrument belongs, are bound firmly together crosswise. In order to make this sink in the water, a heavy weight is attached to the centre of the cross, or its four arms are let into a centre-piece of iron. The end of each of the four arms is grooved, and in each groove is fastened a strong line 6 to 8 metres in length, while a fifth, still longer line, is attached to the central point of the cross. These lines carry the catching apparatus proper, which consists of a number of very coarse, four-cornered nets, made of loosely twisted hemp-string of the thickness of a finger, and with a mesh several centimetres in size. Each net is gathered together at one end and bound firmly, so as to form a tassel-like bundle of string open at one end, similar to the mops used for swabbing a ship's deck. The length of the nets vary with the size of the "ingegno" to which they are attached, but may measure 2 metres, or even more. Each line has a certain number of nets attached to it at suitable distances apart, and there are about thirty or forty on each " ingegno." Between these hang similar but finer and closer-meshed bags, which are commonly made of old sardine-nets.

The " ingegno " is attached to the ship by a long rope, and in fishing is thrown overboard. In its descent through the water the netting-bags open, and when the sea-floor is reached, branches of coral, as well as other animals, plants and stones, become entangled in their meshes, and as the boat moves on are torn up. Such objects are caught chiefly in the wide-meshed nets, the largest of them remain there, but the smaller fall through and are again caught by the small-meshed nets.

The manipulation of the heavier " ingegni " is very arduous, especially when performed for long hours under the burning summer sun of the north African coast, or indeed of any other part of the Mediterranean. In the largest boats the " ingegno " weighs upwards of two hundredweight, and is worked by a winch, but in the small boats these instruments are of less size and are hauled about with no mechanical aid. The nets often get so inextricably entangled with objects on the sea-floor, that a special appliance is provided for hauling up the " ingegno " in such cases. The use of this even does not always loosen the instrument from its moorings, so that the fishermen are often obliged to abandon the whole of the gear, including the rope which attaches it to the boat, a proceeding which involves a loss of 200 francs, if the " ingegno " is a large one.

Each time the " ingegno " is thrown out it is allowed to remain a certain length of time in the water, the fishermen meanwhile controlling its movement along the sea-bottom in various ways designed to increase the catch. When hauled up, whatever has been caught in the nets is picked out, and the latter, which naturally are apt to be much torn, are repaired in readiness for another cast. The mending of the nets involves the expenditure of a large amount of time, and in one working-day, under average conditions, the " ingegno " cannot be let down and hauled up more than from seven to fourteen times. Small boats return to harbour after their day's work, but the larger ones remain at sea, cruising about from bank to bank, and only returning to harbour for provisions or repairs.

Coral-fishing is an almost exclusively Italian industry ; nearly all the boats are built in Italy and winter in Italian harbours. The fishing fleet in the 'eighties included 260 large, and 200 small boats. The most important harbour is Torre del Greco, near Naples, from which 300 boats set out every year. Next comes Santa Margherita, east of Genoa, on the Riviera di Levante, which harbours 49 boats, though 50 years ago there were 200 boats at anchor there ; then Alghero and Carloforte in Sardinia with 19 each, then Trapani with 8, Livorno with 6, and Messina with 3. The number of boats which respectively make their headquarters at these different harbours varies from time to time, but not to any great extent.

The most productive coral-banks lie within French domains on the North African coast, and no efforts have been spared by the French Government to establish a coral-fishing industry in Algeria. So favourably are Italian coral-fishers treated, that a considerable number have now settled in La Calle and other African harbours, from whence about 100 boats, all flying the French flag, though manned by Italians, annually set forth. The only coral not obtained by Italian fishers is the comparatively small amount found off the coast of Provence, which is fished by Frenchmen, and that collected by Spaniards, off the coast of Spain, who employ about 60 small boats.

The amount of coral collected yearly by one boat varies greatly according to circumstances and to the locality at which the fishing has been carried on. The coral obtained from different localities differs not only in amount, but also in quality, and not infrequently the one compensates for the other. A boat working on the African coast collects on an average 150 kilograms per annum of coral worth 75 francs per kilogram; on the Sardinian coast a total weight of 190 kilograms, worth 50 francs per kilogram, will be collected annually, while in the neighbourhood of Corsica is obtained 210 kilograms of material valued at 45 francs per kilogram.

The total weight of coral collected during the last decade by boats flying the Italian flag amounted to not less than 56,000 kilograms, the value of which was 4,200,000 francs. In Algeria the yearly yield amounted to 10,000 kilograms, and was valued at 750,000 francs. The 60 Spanish ships brought in about 800,000 francs' worth, or 12,000 kilograms. The amount collected by vessels other than Italian boats is estimated to have been 22,000 kilograms, of the value of 1,550,000 francs; so that, all sources considered, there must have been collected every year of that particular decade on an average 78,000 kilograms of coral, of a total value of 5,750,000 francs (£230,000).

Rough coral before being placed on the market is sorted, and there is often a considerable difference between the mean market values of different classes of material. There is also a very striking discrepancy between the price of rough coral and that of finished ornaments.

The different qualities of rough coral are as follow:

i. *Dead or rotten coral.* This includes the disc-like foot of the coral-stock and also the lower portion of the main stem. The foot is often found still clinging to the stone to which the colony was attached; both it and the base of the stem are so encrusted with animal and vegetable growths of various kinds that it is impossible to judge of the quality of the coral beneath. The purchase of this class of material, at a price of from 5 to 20 francs per kilogram, is of the nature of a speculative transaction, for the removal of the unpromising outer crust sometimes discloses very valuable pieces of coral. The broad foot-plates are not infrequently worked into small dishes and bowls.

ii. *Black Coral.*—This includes coral in which the change of colour is not merely superficial but extends through the whole or a large part of its substance. When of good quality it is worth from 12 to 15 francs per kilogram, and is used for mourning ornaments. Black coral of quite another kind will be mentioned further on.

iii. *Ordinary Red Quality.*—In this class is placed red coral of all kinds, irrespective of shape and of form and size; it embraces alike whole stocks and broken fragments. The price ranges from 45 to 70 francs per kilogram, according to the character of the coral.

iv. *Selected Pieces.*—Pieces of coral of exceptional size or beauty constitute this class. They are sold by weight either separately or in parcels, and fetch as much as, or even more than, 500 francs per kilogram.

Among the factors which determine the price of a single piece of coral, colour is the most important. According to the caprice of fashion, now this shade of colour, now that, commands the highest price. Thus, at one time bright-red coral found most favour, but in Europe at the present time the coral most admired is that of a fresh rose colour, known to the Italians as *pelle d'angelo*. This same shade stands first also in the regard of the inhabitants of East India and China, the countries which receive the largest exports of coral. Well-formed pieces of coral of this colour, even though of small size, will fetch 100 francs or more. The taste for coral of this shade is not universal however, the Arabs, for example, still prefer the bright-red shade.

The **working of coral** is for the most part also an Italian industry, very little being worked in Spain or in France. It is very common, however, for coral which has been worked in Italy to be sent to France and Spain to be set, sometimes with diamonds and other precious stones, in articles of jewellery. There are about sixty workshops in Italy for the manufacture of coral ornaments and articles of various kinds. These are exported to all parts of the world and their manufacture gives employment to about 6000 workpeople. The centre of the coral-working, as of the coral-fishing industry, is Torre del Greco in which are forty workshops giving employment to 3200 persons, of whom 2800 are women. Other places at which the industry is established are Genoa, Naples, Livorno, Trapani, and, to a less extent, Rome.

A staple branch of the coral-working industry is the piercing of pieces of coral of all shapes and sizes; these being threaded on a string and used as necklaces, bracelets, rosaries, &c. When the coral is fashioned into beads, they are either spherical or ovoid in form—in the latter case being known as " olives "—large, medium, or small in size, with or without facets. Faceted beads are less in demand now than at the beginning of the nineteenth century. " Arabian beads " are from $1\frac{1}{2}$ to 2 centimetres in length; they are bored through lengthways but not otherwise worked, and are threaded to form long strings, which are much worn in the East as girdles. Beads of this description are often bored crosswise.

Italians are very skilful in the art of carving coral, and in the production of beautiful cameos and articles carved to represent trees, animals, figures, and other objects; they know how to utilise the small irregularities of the surface of coral to produce the greatest possible effect. Objects carved in this way are either set separately in brooches, pins, &c., or in rows in bracelets and necklaces. The ends of small branches are often polished and mounted in their natural form in ear-rings and pins, or utilised for the decoration of various other objects.

Pieces of coral of sufficient size supply material for the handles of sticks and umbrellas, and even of still larger objects. Pieces of fine quality are often carved, which add considerably to their intrinsic value. The carved coral handle of a sunshade belonging to the Queen of Italy is worth £360.

The value of worked coral depends not only upon the quality of the coral, but also upon the artistic worth of the workmanship. The price of a necklace or bracelet varies with the number and size of the beads and with their uniformity of colour. At the International Fisheries Exhibition held at Berlin in 1880, a coral necklace was exhibited which was valued at £6000.

The **uses** to which coral is put in Europe are by no means insignificant, but the material is employed to a much greater extent in India and China, the two countries to which the largest proportion of the total annual yield of coral is sent.

In Europe coral ornaments are worn, for the most part, by children; they are in general use only in certain parts of Italy, Austria, and Hungary, in Poland, and specially in Russia. In Turkey, however, coral serves not only for the personal adornment of men as well as of women, but also for mural decoration and the ornamentation of pipes, weapons, harness for horses, and various other articles. It is used to a large extent over the whole of north Africa, especially in Morocco, and also in Arabia, but is not much esteemed in Egypt.

Coral ornaments are as much in favour with Persians as with Japanese or Chinese. They are very generally worn by Chinese men and women, and incredible sums are often paid for coral buttons of large size and fine quality, intended for the decoration of mandarins' caps. The demand for coral is largest, however, in India, and several million francs' worth is imported annually to Bombay, Calcutta, and Madras. It is worn in this country in the form of necklaces, bracelets, anklets, and other ornaments; it is used for rosaries and charms, and is often offered to the dead, with the belief that it has the power of preventing the occupation of the body by evil spirits.

A moderately large amount of coral is exported to America, especially to South America, and also to Australia. The taste for coral for ornamental purposes would appear to be confined to partially civilised peoples, for attempts to substitute coral for glass beads in dealing with uncivilised tribes have been wholly unsuccessful, the more glittering and much cheaper glass being in every case preferred to coral.

Genuine coral possessing a considerable intrinsic value is of course subject to imitation, and a large amount of **imitation coral** of various kinds is sold at very low prices. Some of the substances used for the manufacture of imitation coral-beads are red gypsum, which, however, can be readily distinguished from genuine coral by the fact that it does not effervesce when touched by acid, and can be scratched even with the finger-nail ; bone, burnt and coloured red ; powdered marble mixed with isinglass and coloured with cinnabar or red-lead ; and even red sealing-wax.

The black coral which has been previously mentioned represents the first stage in the decomposition of precious red coral. The **black coral**, to be now briefly described, is black by nature, and constitutes the skeleton of the coral known to zoologists as *Antipathes spiralis* (Pall.), belonging to the order Antipatharia. The shining pitch-black, branched skeleton of the stock formed by this coral may reach a length of 2 feet, and a thickness of some inches. It is formed not of a calcareous, but of a horny substance, and single pieces of this kind of black coral can therefore be moulded to form armlets and such like. The material is found in the Indian Ocean, and is known in that region by the name of " akabar." It is much esteemed there, and often constitutes the substance of which the sceptres used by native kings and princes are made, hence the term king's-coral, which is also applied to it Black coral of a similar kind is found in the Mediterranean, where it is known as " giojetto."

On the Cameroon and Gulf coasts, **blue coral** was formerly fished and worked into ornaments, which were much prized by the negroes. It is known to the natives as " akori," and to zoologists as *Allopora subviolacea*. It has disappeared from the West African markets for a long period, and therefore needs no further consideration.

INDEX